# *Prehistoric Cultural Ecology and Evolution*

## Insights from Southern Jordan

*INTERDISCIPLINARY CONTRIBUTIONS TO ARCHAEOLOGY*

Series Editor: Michael Jochim, *University of California, Santa Barbara*
Founding Editor: Roy S. Dickens, Jr., *Late of University of North Carolina, Chapel Hill*

---

*Current Volumes in This Series:*

THE ARCHAEOLOGY OF GENDER
Separating the Spheres in Urban America
Diana diZerega Wall

CHESAPEAKE PREHISTORY
Old Traditions, New Directions
Richard J. Dent, Jr.

DIVERSITY AND COMPLEXITY IN PREHISTORIC MARITIME SOCIETIES
A Gulf of Maine Perspective
Bruce J. Bourque

EARLY HUNTER–GATHERERS OF THE CALIFORNIA COAST
Jon M. Erlandson

FROM KOSTENKI TO CLOVIS
Upper Paleolithic–Paleo-Indian Adaptations
Edited by Olga Soffer and N. D. Praslov

HOUSES AND HOUSEHOLDS
A Comparative Study
Richard E. Blanton

ORIGINS OF ANATOMICALLY MODERN HUMANS
Edited by Matthew H. Nitecki and Doris V. Nitecki

PREHISTORIC CULTURAL ECOLOGY AND EVOLUTION
Insights from Southern Jordan
Donald O. Henry

PREHISTORIC EXCHANGE SYSTEMS IN NORTH AMERICA
Edited by Timothy G. Baugh and Jonathon E. Ericson

REGIONAL APPROACHES TO MORTUARY ANALYSIS ,
Edited by Lane Anderson Beck

STYLE, SOCIETY, AND PERSON
Archaeological and Ethnological Perspectives
Edited by Christopher Carr and Jill E. Neitzel

*A Chronological Listing of Volumes in this series appears at the back of this volume.*

---

# *Prehistoric Cultural Ecology and Evolution*

## Insights from Southern Jordan

DONALD O. HENRY
*University of Tulsa*
*Tulsa, Oklahoma*

*PLENUM PRESS • NEW YORK AND LONDON*

Library of Congress Cataloging-in-Publication Data

On file

ISBN 0-306-45048-8


A Division of Plenum Publishing Corporation
233 Spring Street, New York, N. Y. 10013

10 9 8 7 6 5 4 3 2 1



Printed in the United States of America

To my mother Katherine and my wife Nancy

# Contributors

*Nancy R. Coinman* • Department of Anthropology, Iowa State University, Ames, Iowa 50011-1050

*Aline Emery-Barbier* • Laboratoire de Paleobotanique, Musee de l'Homme, 75116 Paris, France

*Fekri A. Hassan* • Institute of Archaeology, University College London, London WCIH OPY England

*Richard G. Klein* • Department of Anthropology, Stanford University, Stanford, California 94305-2145

*Daniel E. Lieberman* • Department of Anthropology, Rutgers University, Douglass Campus, New Brunswick, New Jersey 08903-0270

*David S. Reese* • Department of Anthropology, Field Museum of Natural History, Chicago, Illinois 60605-2496

*Arlene Miller Rosen* • Archaeological Division, Ben Gurion University, Beer Sheva, Israel 84105

*John J. Shea* • Department of Anthropology, State University of New York, Stony Brook, New York 11794-4364

*Chen Shen* • Department of Anthropology, University of Toronto, Toronto, Ontario M5S 1A1

# Foreword

Modern fieldwork in archaeology stresses the necessity for long-term regional studies. Understanding local human adaptations through time can only be achieved when specific geographic areas are well known. This goal in turn cannot be achieved without a long-term investment of personal energy and commitment on the part of a researcher. The end result, and the motivation behind this dedication, is that such regional archaeological investigations provide the basic data for reconstructing cultural changes. They raise important questions concerning such issues as human movements, societal restructuring, changes in strategies of resource exploitation, and the formation of local cosmologies or religions.

This volume is a fine example of such a regional study of prehistoric cultures, in this case dealing with a specific region within the Near East. It presents in a clearly written, well-illustrated format the results of Donald Henry's original study of several specific regions in southern Jordan that uncovered prehistoric sites from the Middle Paleolithic through the Chalcolithic periods. Over the course of the last twenty years, Henry— an experienced field archaeologist— has single-handedly maintained this project, a fine example of in-depth prehistoric scholarship.

The material presented here holds significance for many aspects of prehistoric research. For example, until recently most of the evidence for the Mousterian in the southern Levant was derived from open air sites in the Negev. Now the excavations at Tor Faraj and Tor Sabiha have provided for the first time stratified and dated Mousterian sequences from a cave and rock-shelter. Despite the difficulties of excavating stratified cave sites, preservation in this type of locale is often better than in open air sites. Thus these excavations have been able to provide data from sources such as animal bones and the remains of hearths, in addition to the more commonly found lithics.

Upper Paleolithic, Epipaleolithic, and Chalcolithic sites were also extensively excavated. Indeed, the report on the Epipaleolithic sites presented here makes one of the most important recent contributions to the study of the terminal Pleistocene, a period of diversifying human adaptations in the Near East immediately preceding the emergence of agriculture. The current hypothesis, which places the origins of sedentary farming communities in the Levant, relies on a better knowledge of cultural variability of the Epipaleolithic of this region— which these new detailed reports go a long way toward providing.

The Chalcolithic period, during which full-fledged pastoral economy encompassed both the desert and the town, occupies a special place in Levantine prehistory. The emergence of pastoral societies was a process that took place from the middle of the ninth millennium through the sixth millennium B.P. (uncalibrated). Henry's reports of his excavations of Chalcolithic materials will stand among the cornerstones for the reevaluation of the significance of the Chalcolithic sites along the margins of the Syro-Arabian Desert— a region which, lamentably, has been investigated less than the Sinai and the Negev until only recently.

The combination of site reports, artifact counts, and discussions of the palynological, faunal, and geomorphic observations makes this volume of great interest not only to Near Eastern archaeologists but also to other Old World prehistorians. Those who are interested in human adaptations in desertic areas of the world will find in this book an array of insights into human cultural modification that were either influenced or enhanced by the special characteristics of this semiarid area.

Ofer Bar-Yosef
*Peabody Museum*
*Harvard University*
*Cambridge, Massachusetts*

# Preface

This book describes a research effort in which archaeological and paleoenvironmental evidence are used to reconstruct synchronic and evolutionary aspects of the cultural ecology of the prehistoric inhabitants of southern Jordan. The primary evidence collected in the study is unusual in several ways. Not only does it fill a virtual void in our knowledge of the vast reaches of southern Jordan, it also provides the most comprehensive understanding of any area of the Levant east of the Rift Valley. The detailed results of the study stem from both the large sample of sites, numbering over a hundred, and their diverse environmental settings. Recovery of evidence from archaeologically synchronous sites located in distinct, closely packed macroenvironments forms the basis for the ecologically oriented study. Moreover, the data are drawn from archaeological periods stretching from the Middle Paleolithic to the Chalcolithic. This allows the region's prehistoric cultural ecology to be examined from an evolutionary perspective encompassing a temporal sweep of some 70,000 years.

The volume contributes to the discipline in three broad areas. First, evidence recovered from ten years of excavation and survey is presented in considerable detail for sites representing Lower Paleolithic, Middle Paleolithic, Upper Paleolithic, Epipaleolithic, Neolithic, and Chalcolithic periods. This largely centers on descriptions of chipped stone artifact assemblages and associated evidence in conjunction with their site and stratigraphic contexts. Ancillary data are also described for studies that emphasize areal geology, palynology, phytoliths, cementum layering of teeth, shells, and vertebrate fauna.

Another contribution of the volume rests in the reconstructions that are presented for the prehistoric cultural ecology of the region. These are based on the integration of a wide range of archaeological and paleoenvironmental evidence. This examination of cultural ecology becomes especially important when viewed from an evolutionary perspective that extends over most of the Late Pleistocene. The definition of a time-transgressive strategy of transhumance by the research is not surprising given the region's high elevations and marked relief. But the scope and the detail of the reconstructions are unique as are the comparisons of the archaeological evidence to the modern Bedouin ecology of the region.

The discovery of such a large number of sites situated within four elevationally distinct study areas allowed for detecting prehistoric occupations related to different seasons and settlement modes. Moreover, many of the sites rest in rockshelters and caves filled with thick cultural deposits. These often contain multiple, stratified occupations associated with well-preserved economic and environmental evidence. Patterned differences in the settings of sites and associated evidence led to definition of prehistoric transhumance. This adaptive strategy persisted from Middle Paleolithic through Chalcolithic times and, in fact, was also followed by modern Bedouin groups of the region. The strategy involved seasonal adjustments in the sizes and permanency of camps in order to accommodate fluctuations in the abundance of critical resources.

These adjustments were also coordinated with seasonally scheduled moves to different elevational belts. Although transhumance was maintained in the region over most of the Late Pleistocene, the global rise in temperature accompanying the terminal Pleistocene–early Holocene prompted a marked change in the schedule by which certain elevational belts were occupied.

The volume also makes an important theoretical contribution in showing that cultural historical and processual approaches can form complementary rather than competing aspects of an archaeological research effort. Frequently it is argued that archaeological research programs can be divided between those that pursue the definition of ethnic groups and those that emphasize the understanding of adaptive responses of populations. Often this theoretical dichotomy is extended to a comparison of Old World (ethnic) versus New World (functional) paradigmatic biases. As might be expected, this perception also finds its way into the realm of specific interpretations and methodology. Especially important here are the ways we interpret variability in artifact assemblages and what we choose to examine. For example, ardent traditionalists (few of whom exist) argue that typologic variability is derived solely from ethnic differences. In contrast, fervid functionalists see all variability as stemming from differences in activity or the degree to which stone tools have been reduced.

There are compelling reasons for using both cultural historic and processual approaches for addressing questions of prehistoric cultural ecology. Given our understanding of extant cultures, it appears likely that variability in the Late Pleistocene archaeological record comes from both ethnically induced behaviors and behaviors resulting from functional responses to the environment. This complexity of behaviors presents an important task to archaeologists interested in cultural adaptation. In order to understand those behaviors that formed responses to external environmental stimuli, it is necessary that they be distinguished from those that were environmentally insensitive or neutral. One might also argue that ethnicity, as a "mechanism by which interest groups symbolize their within-group organization in opposition to and in competition with other interest groups" (Hodder 1979:452), simply provides hominids a means of controlling critical resources. This, coupled with the notion that group over individual selection characterizes cultural adaptation, makes the definition of ethnicity—and attendant specific social groups—a fundamental step in any detailed study of prehistoric cultural ecology.

# Acknowledgments

It seems nearly impossible to properly thank all of those who had a hand in the project during the years 1977–1988. The dozens of students and volunteers who assisted in the fieldwork and laboratory analysis formed the backbone of the research effort. To them, I offer a special thanks. I am also grateful to my friends, colleagues, and students who took the time to comment on the book. Tony Marks, Ofer Bar-Yosef, Dan Kaufman, Lilliane Meignen, Steve Rosen, Itzik Gilead, George Odell, Nancy Coinman, Ian Kuijt, Meredith Chesson, and Greg Walwer were especially helpful in this regard.

I also offer my appreciation to Dale Phelps for her good humor and untiring wizardry in preparing the manuscript. Chen Shen, Dorothy and Greg Walwer, Patricia Thomas, and Steven Mack furnished great assistance in what seemed to be endless computations and analysis; for this I offer my thanks. Also my gratitude is extended to Lucy Addington, Chen Shen, Rita George, and Amy Finnegan for their illustrations of artifacts and petroglyphs. Moreover, Mike Jochim, Eliot Werner, and Marian Manaois are thanked for their efforts in bringing the book to press.

The Office of Research, of the University of Tulsa, the National Science Foundation, the National Endowment for the Humanities, and the National Geographic Society made the project possible through their support. Assistance from the Department of Antiquities of Jordan and the American Center of Oriental Research in Amman also was crucial to the success of the seven field seasons of research in the Wadi Hisma.

Finally, I offer my sincere gratitude to Nancy for her continual encouragement and good cheer.

## FROM CONTRIBUTORS

I am deeply grateful to Don Henry for graciously allowing me to study specimens from Wadi Judayid and Tor Hamar and for extending the invitation to publish my results here. Thanks are also due to Ofer Bar-Yosef, Richard Meadow, Eitan Tchernov, and David Reese of the Field Museum for their help in providing data and/or comments. I also thank Aaron Stutz for his help preparing the samples. This research was funded by grants to the author from the National Science Foundation (BNS 9015973), the Wenner-Gren Foundation, the L.S.B. Leakey Foundation, and the CARE Foundation.—*D. Lieberman*

I would like to thank Stephen Weiner of the Weizmann Institute for providing me with laboratory facilities at the Structural Biology Department. Michael Friedlander and Sara Segi from the Biology Department at Ben Gurion University kindly allowed me access to the departmental microscopes.—*A. M. Rosen*

# Contents

# *Prehistoric Cultural Ecology and Evolution*

Insights from Southern Jordan

# Chapter 1

# Introduction and Overview: Evolutionary Ecology and Archaeology

The population geneticist and evolutionist, Theodosius Dobzhanksky (1972:422), observed that "culture is the most potent method of adaptation that has emerged in the evolutionary history of the living world." His remark finds dramatic support in the archaeological record as shown by the progressively accelerating tempo of cultural evolution over the last 4 million years accompanied by the expansion of hominids throughout all of the earth's diverse environments. The fundamental characteristic distinguishing the cultural method of adaptation from the biologic one is that it is behaviorally based and thus inorganic. Learned, patterned behaviors and their material consequences as opposed to physical changes have come to be depended upon by hominids for their survival. Unlike biologic adaptation, behavioral responses to external environmental forces can be acquired, transmitted, and modified within the lifetimes of individuals. This Lamarckian aspect of cultural adaptation provides its greatest advantage over biologic responses to the environment. When compared to the genetic recombinations and generational successions necessary to induce widespread organic changes within populations, the adoption and transmission of behavioral changes are extraordinarily rapid. Rather than the genetic basis that grounds physical adaptation, behavioral adaptation relies upon social interaction. But like biologic evolution, cultural evolution would appear to entail more than the behavioral outcomes of natural selection alone (Kirch 1980:131; Clark 1991:436–437; Rosenberg 1994:3–4). Some human behaviors are perpetuated as a consequence of cultural inertia or tradition, Grant's (1985:434) "cumulative cultural heritage," and the "evolutionary potential" of culture described by Sahlins and Service (1960). Such behaviors experience evolutionary trajectories that are inexplicable with reference to their interaction with the environment alone.

The uniqueness and potency of cultural adaptation furnish a fascinating subject not only for those concerned with the remarkable development of our lineage, but for anyone interested in the broader subject of evolution. Of the disciplines concerned with understanding the past, anthropological archaeology is singularly positioned for examining the issues of cultural adaptation and evolution. Following a trend established in studies within the biologic realm of evolution, archaeologists have come to adopt a paradigm of "evolutionary ecology" (Clark 1991). Pianka (1978:15) sees this as a means of synthesizing the different time-scales of adaptation, that is, between thinking in an "ecological" time-scale (now time) or an "evolutionary" time-scale (geological time). These different ways of thinking about adaptation, however, are conceptually linked within the paradigm of evolutionary ecology. As noted by Kirch (1980:103), evolutionary change cannot be understood fully outside of its ecological context and the functional integration of ecological systems is comprehensible only as the product of evolution and natural selection.

## EVOLUTIONARY ECOLOGY AND CULTURAL HISTORY

Perhaps the most important and often the most vexing task of archaeological research is understanding the temporal and spatial distributions of the material under study. Such controls seem especially important to those research programs centered on questions of evolutionary ecology, given the necessity of tracing synchronic spatial variability as well as diachronic changes in behaviors. This is certainly stressed in Kirch's (1980:132–133) outline for developing a research design for examining adaptation archaeologically. In reference to time, he suggests that the smallest possible temporal units, those equivalent to individual stratigraphic and assemblage components, should be employed in order to closely measure variation. He goes on to argue that:

> For the archaeologist concerned with adaptation, a traditional cultural-historic approach, wherein periodization is a primary aim, is not merely inadequate but inappropriate because it obscures rather than elucidates variation over time (Kirch 1980:134).

Here, before condemning the contribution of the cultural-historic approach toward an understanding of cultural adaptation and evolution, it is important to recognize what this traditional approach does provide in the realm of chronology. Whereas the stratigraphic or assemblage unit is the appropriate scale of analysis within short time-frames over confined space (often site-specific), this is too small a scale for examining widespread variability over temporal sweeps covering millennia. In some ways, this recalls Pianka's (1978) distinction between scales for examining adaptational (now time) and evolutionary (geological time) questions. Clark (1991:438), in commenting on Kirch's (1980) research design, makes a similar point in observing that "its successful application is contingent on the definition of temporal, spatial, and behavioral analytic units appropriate to given problem contexts." Clearly, comparisons at the scale of the stratigraphic/assemblage unit are appropriate and, in fact, required for discovering archaeological taxa and linking these to environmental and economic data. Such connections in data-sets, associated with fine resolution temporal controls, enable us accurately to reconstruct prehistoric behavioral systems and compare these to specific environmental settings. Ironically, these also are the initial steps of doing cultural-history, at least as practiced by many scholars, and they lead to more than merely establishing periodization.

Perhaps an even more important contribution of cultural-history as a foundation for processual inquiry rests in the framework it provides for defining specific social or ethnic groups. As will be argued, defining the social boundaries of prehistoric populations at some scale is fundamental to understanding their behavioral adaptations and evolution. Before exploring this, however, a criticism of using a cultural-historic framework as a basis for understanding cultural process needs to be addressed. In objecting to the notion that "the inductively-based cultural historical approach . . . serves as a basis for the deductively-oriented cultural processual approach" (Sharer and Ashmore 1979:535), Sabloff *et al.* (1987:205) rightly argue that while historical and processual approaches may be complementary:

> . . . . culture history as it has been developed at the level of methodology, cannot provide the factual foundation for processual studies, since that cultural history has not been "read" directly from the archaeological record. . . . the "facts" of culture history are interpretations based upon assumed culture process.

A question then emerges as to whether there might be different kinds of culture history and, more to the point, whether or not cultural historic frameworks can be developed through deductive reasoning. It seems that there are two kinds of cultural historic studies in archaeology. There are those that focus on the description of materials and their organization in time-space dimensions. And there are those that attempt to explain variability in the archaeological record. Those involved in the description and the time-space organization of "facts" (artifacts, ecofacts, etc.) are the taxonomists of archaeology and, by and large, they follow deductive reasoning in pursuit of their goals. Multiple lines of evidence encompassing artifact/attribute, chronometric, stratigraphic, and environmental data are typically compared in testing the integrity of a taxon and its fix in time and space. Therefore, in many ways, this kind of culture history is "read" rather directly from the archaeological record.

The second kind of culture history is concerned with explanation of the archaeological record. As described by Sharer and Ashmore (1979) and criticized by Sabloff *et al.* (1987), such explanations typically are inferred through inductive reasoning. Distinguishing between these two kinds or domains of culture history seems important because it is generally in the case of the latter where cultural historians and processualists find their greatest disagreement. The classic debate between Francois Bordes and Lewis Binford did not center on the taxonomy of the Mousterian (although each of them chose to measure the variability differently) or the time-space placements of the various taxa, but on how the variability between the Mousterian facies was to be explained. With the adaptive paradigm of a processualist, Binford found a functional explanation whereas Bordes favored one tied to ethnicity as would be expected of a traditional cultural historian. Obviously, the variability in the archaeological record comes from both ethnically-induced behaviors and those resulting from functional responses to the environment. This complexity of behaviors presents an important task to archaeologists interested in cultural adaptation. In order to understand those behaviors that formed responses to external environmental stimuli, it is necessary that they be distinguished from those that were environmentally insensitive. Thus it seems important here that processualists not confuse the objective with the methodology in their criticisms of cultural history. Although there is certainly a valid and important questioning of when ethnicity emerged in the archaeological record (Binford 1989; Gamble 1986; Roebroeks *et al.* 1992; Chapter 5 this volume), this raises rather than lowers its significance as a topic of inquiry.

The task of defining ethnicity at some scale requires far more than comparing artifact inventories and tracing these through time and space. The "functional" sensitivity of the artifactual attributes forming what are construed as ethnically diagnostic patterns should be examined with the intent of determining if they do in fact represent "functionally neutral" attributes (Jelinek 1976; Dunnell 1978; Henry 1977, 1992; Close 1989). The distribution of these attributes, in turn, can be compared across various site types and settings in an effort to discern if functional relationships exist. If, after such examination, patterned variability in artifact attributes are found to remain constant while cross-cutting functional domains (i.e., different functional artifact classes, intrasite activity areas, site types, and environmental settings) such patterning is most likely to represent ethnicity.

Biologists, sociobiologists, and anthropologists generally contend that "group" over "individual" selection characterizes cultural adaptation (Alexander 1974; Alexander and Borgia 1978; Grant 1985; Jochim 1981; cf. Durham 1976, however). If indeed this is true, the definition of ethnicity and attendant specific social groups appears fundamental to any detailed study of prehistoric cultural adaptation and evolution. There are several characteristics of cultural adaptation that favor group selection (Jochim 1981; Dunnell 1980; Kirch 1980; Grant 1985):

(1) Unlike organic adaptation, dependent upon the exchange of genetic material, behavioral changes offering a selective advantage are rapidly disseminated to all or most members of a group through social processes.

(2) "Culture" in the sense of the total repertoire of behavioral adaptation, or in following Jochim's (1981:19) concept of "a valid set of solutions to a variety of problems," is never fully controlled by an individual, but rests only at the level of a social group.

(3) Human groups compose functional units with respect to information sharing, decision making, and economic pursuits. The individual constituents of such units typically fill non-redundant, often specialized roles in these areas based upon their age, sex, status, and talents.

(4) The behavior of an individual is strongly governed by the norms of a group to the extent that an individual's learned behavioral repertoire is dominated by cultural over idiosyncratic behaviors.

(5) Cultural adaptation is in part a consequence of social-group selection and the relative degree of success a group might achieve in competition with neighboring groups. While not a uniquely human characteristic (Grant 1985:143–144;430), social-group selection appears to be particularly important to hominid evolution because of a dependence upon learned, behavioral adaptation and residence in groups.

One also might mention the pragmatic importance of focussing on group over individual selection in an archaeological context. In most archaeological situations we simply are incapable of assessing the behavior of individuals. Even if selection operates to some degree at the individual level in cultural adaptation, however, Jochim (1981:97–98) points out that what we see archaeologically likely reflects an aggregate of individual decision making and, in turn, differential adaptive success at the group level.

## GOALS, CONCEPTUAL LOCKS, AND RESEARCH DESIGN

In many ways, it would be unrealistic to discuss the goals of some ten years of research in southern Jordan as if they had been neatly defined and nicely arranged for study season after season. Although the overarching concern of the project was with prehistoric cultural ecology, at a specific level the objectives of research for the project were ever-changing. Sites or groups of sites came with their own problems and practical considerations. Quite apart from the strictly archaeological concerns, factors related to funding, personnel, and time available for fieldwork played a large role in shaping the objectives for a given season. In retrospect, the field seasons seem to have been linked by a series of decisions related to such basic goals as establishing better stratigraphic controls at a certain site, recovering pollen samples from deposits of a specific time-frame, or collecting a larger artifact sample from a particular layer. Each season also came with its own disappointments and pleasant surprises that prompted new research targets to emerge as old ones were abandoned. If this seems somewhat chaotic, I suppose it was, but in most respects probably not greatly different from other long-term projects. Field archaeologists appear to spend the large majority of their time grappling with various untidy tactical problems, leaving their lofty and more abstract strategic concerns to be addressed in grant proposals, major papers, and summary volumes such as this.

In keeping with the earlier discussion of behavioral adaptation and evolution, the principal paradigmatic bias that guided the research is a functionalist and ecological interpretation of culture (Steward 1955; White 1959; Durham 1976; Harris 1979; Kirch 1980; Jochim 1981; Konner 1983). This is a rather widely held notion among anthropological archaeologists. It primarily involves culture being seen as a set of learned, behavioral solutions that undergo selection on the basis of the degree by which they enhance the survival of a human population, especially through the balancing of numbers with available resources. Under this theoretical canopy, the program of research in southern Jordan set out to isolate certain prehistoric behavioral solutions to problems of human survival and to see how these solutions changed through time relative to external environmental stimuli and to internal cultural inertia.

From a practical archaeological perspective, the broader goals of the study required a setting in which a great number of sites, representing a substantial and near-continuous temporal sweep, could be related to significant environmental changes. The mountainous region flanking the Wadi Hisma on the southern edge of the Jordanian Plateau furnishes such a setting (Figure 1.1). With its elevational diversity and distinct plant communities, the region shows considerable environmental sensitivity to climatic changes. Because of this it was selected along with two other regions, the Jafr Depression and the Wadi Yutm, for reconnaissance during a two-week visit to Jordan in 1977.

The sites found in the Jafr Depression and the Wadi Yutm proved disappointing, however, as they consisted only of a few small surface scatters of heavily patinated lithic artifacts. Buried, *in situ* occupations, if present, appeared to be uncommon. But in only two days of searching one of the high basins located along the northern edge of the Wadi Hisma, several *in situ* sites were found (Henry 1979). Discovery of sites in the Judayid Basin beneath the little village of Ras en Naqb not only confirmed the high density of prehistoric occupations for the area, but the sites also suggested that a substantial time-depth (Middle Paleolithic through Epipaleolithic) could be traced with further work. Plans then were made for beginning a full-scale program of research there in 1979.

As discussed earlier, the definition of prehistoric behavioral systems (social groups, cultures) and their sequencing is for many reasons prerequisite to examining those forces that drive cultural evolution. Therefore, in order to address the broader aims of the study, a fundamental task involved constructing a cultural-historic framework for those prehistoric occupations discovered during the course of the investigation. This primarily involved classifying lithic assemblages based upon technotypologic attributes and fitting these into existing taxa or forming new ones. This also involved placing the assemblages in chronologic order given their local and regional stratigraphic contexts and their associated chronometries.

Coupled with the results of the reconnaissance of the Judayid Basin, previous work in the region hinted of a rich prehistory, but southern Jordan was virtually unexplored archaeologically (Stockton 1969). As survey and testing of sites progressed from 1979 on, the job of developing a cultural-historic framework was in part simplified by being able to compare newly discovered assemblages to those defined for existing archaeological taxa from neighboring regions, especially the Negev and Sinai. But the discovery of many assemblages that failed to fit within previously defined taxa necessitated the recognition of new taxa and the construction of a cultural-historic framework incorporating certain units unique to southern Jordan (Henry 1982a, 1986, 1988a, 1988b, 1994).

Efforts directed toward taxonomic restructuring have been especially common in the Levant since the 1970s when intensive research began in the Negev and Sinai. At that time our understanding of cultural-historic frameworks primarily came from studies in the northern and central Levant, regions covered by Mediterranean woodlands. Extension of research to the south not only enlarged the prehistoric data-base and geographic coverage, but projects in the Negev and Sinai generated evidence from the arid zone, an environmental setting poorly represented in the earlier studies. Given this background, well-conceived archaeological unit construction in southern Jordan is especially important because the region represents a geographic and environmental extension of the southern Levant to the east across the Rift Valley.

Earlier geographic growth in Levantine prehistory stimulated questions essentially related to north–south, circum-Mediterranean interaction. A north–south dichotomy emerged as a common theme in archaeological unit construction for virtually every prehistoric period. In offering an opportunity for examining east–west, inland-coastal interactions, the evidence from

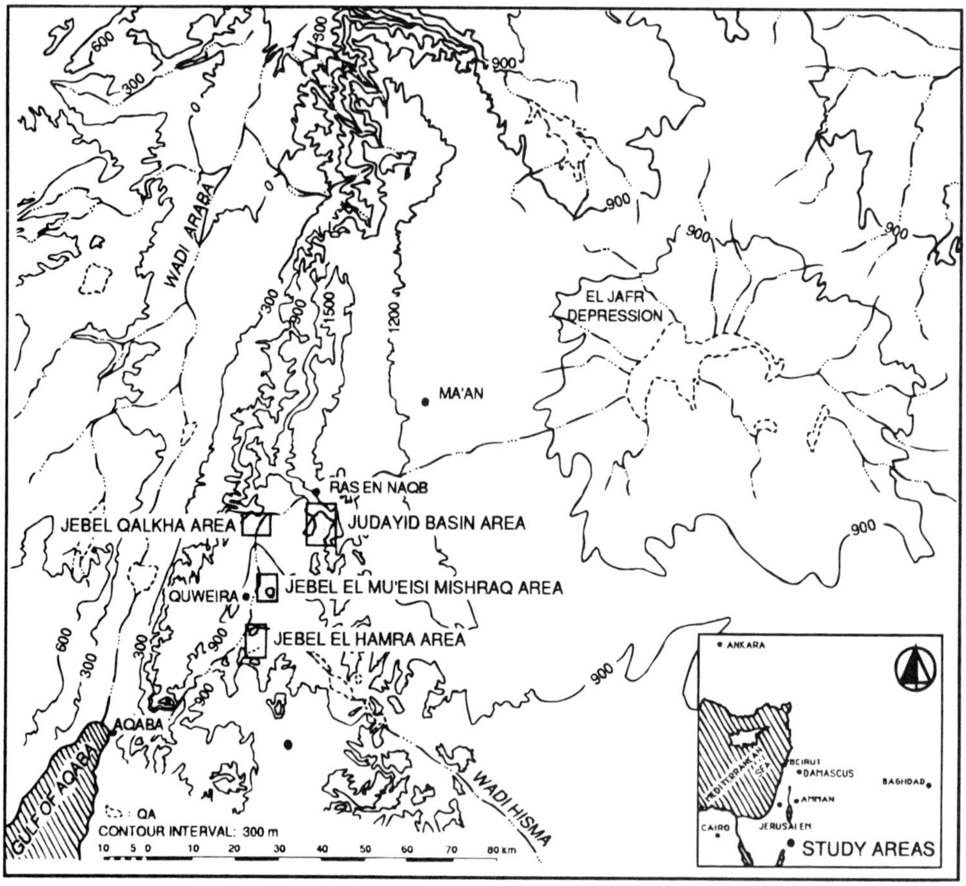


Figure 1.1. Regional map of southern Jordan showing the four study areas.

southern Jordan and elsewhere east of the Rift Valley produces a more holistic perspective of Levantine prehistory. An especially significant question to be answered relates to the degree that the Rift Valley served as a barrier to cultural interaction. By the same token, it is important to know whether or not cultural manifestations of archaeological taxa tied to the arid zone in the Negev and Sinai followed the sweep of desert across the rift northward through eastern Jordan and into Syria.

## RESEARCH DESIGN: STRATEGY AND METHODS

The main objectives of the research conducted in the initial summer field season of 1979 involved: (1) locating and assessing prehistoric sites, (2) ascertaining their chronological order, (3) reconstructing environmental settings coeval with the sites, (4) identifying prehistoric economic and settlement patterns, and (5) to the extent possible, examining correlations between environmental changes and settlement-procurement patterns. In many ways, problems related to achieving these rather fundamental aims not only dominated the work of 1979, but continued as central themes of the study through the last field season in 1988.

Early in the program of research, a striking asymmetry in site varieties tied to elevational differences was recognized. As study progressed and more sites representing a greater range of time were discovered, it was further observed that this elevationally governed asymmetric pattern was time-transgressive. It was found to represent an extraordinarily long temporal sweep stretching from the Middle Paleolithic through historic Bedouin, that is, some 60,000–70,000 years. These patterned differences in natural settings, site plans, and artifact inventories were thought to reflect elevationally structured, seasonal shifts in residence accompanied by changes in group size and camp permanence—that is, transhumance. The evaluation of this interpretation and the specific definition of transhumance for certain time-frames emerged as the central focus of the broader theoretical aims of the overall study. As an adaptive strategy within the region, transhumance bridged foraging and food-producing economies and perhaps even different human taxa. Thus the significance of understanding the evolution of transhumance was recognized early in the project. This acted to pull together the research plans for the various sites and time-frames season after season.

## Survey Strategy

A systematic pedestrian survey of the western portion of the Judayid Basin was carried out during the summer of 1979, followed by similar surveys of three other naturally defined areas (Jebel Qalkha, Jebel El Mueisi-Mishraq-Muheimi, and Jebel Hamra) during seasons in 1980, 1983, 1984, 1985, and 1988 (Figure 1.2). The study areas were selected in order to collect evidence from different elevations, drainage networks, physiographic settings, and modern plant communities.

A systematic sampling strategy focussing on contiguous survey units was chosen over strategies favoring random or stratified units primarily because of logistical considerations. The study areas that were selected in order to provide the greatest diversity of settings had to be reached on rough tracks or overland across rugged terrain. Once in the study areas, many settings could be reached only on foot. Had a random or stratified sampling strategy focussing on isolated survey units been employed, a significantly greater amount of time would have been expended in locating and gaining access to the survey units. Even when investigating contiguous survey units from tent-camps within study areas, a considerable amount of time (1–3 hrs) was spent in daily travel by foot or vehicle between camp and survey unit.

In short, a decision was reached to sacrifice some of the statistical integrity of the site data in an effort to generate a larger overall sample. As we shall see, once the sites were partitioned into archaeological periods, local settings, and site types, a very large global sample of sites was required for meaningful representation to be achieved within each of these subsets.

## Survey Methods

The study areas were examined by surveyors walking in a "picket line" formation. Depending on the visibility of the surface, members of the survey teams were spaced some 5–10 minutes apart. When artifact concentrations were encountered, a series of observations were recorded on site forms and sketch maps were drawn of the artifact distributions and the local topography. Artifact concentrations, definitive of sites, were identified on the basis of artifact densities that significantly exceeded the general "background" density of some 1–2 specimens per 100 m$^2$. Such isolated artifacts, ranging from handaxes to Roman sherds, were found in low densities within most of the study areas. Although they were not plotted or recorded, they did furnish information on the antiquity of the various geologic surfaces under investigation. Artifact concentrations associated with sites typically displayed from 10 to over 100 specimens per square meter and extended over areas greater than 100 m$^2$.

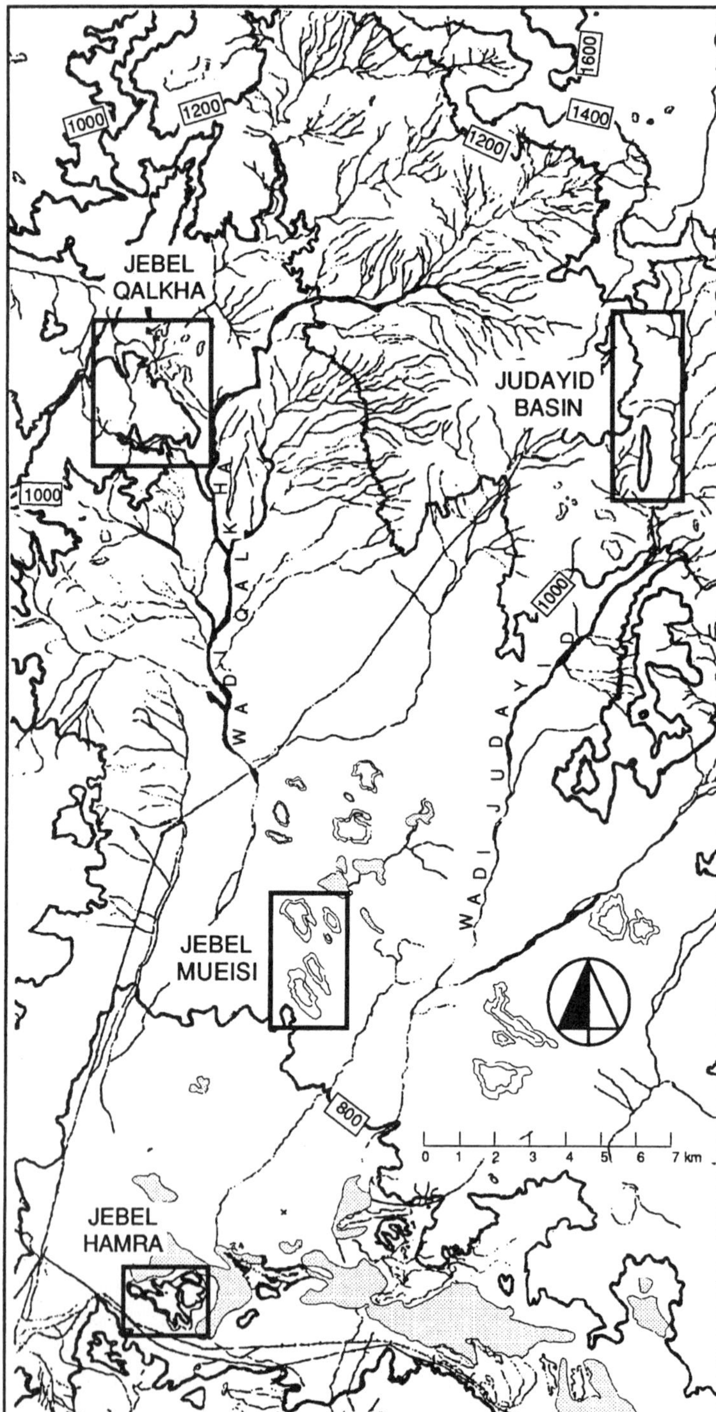


Figure 1.2. Map of the western end of the Wadi Hisma showing the locations of the four study areas relative to major topographic features.

When sites were found, an effort was made to determine if they were confined to the surface or if buried occupations might be present. If sites were determined to be limited to the surface, controlled surface collections of artifacts were made. All other sites were test excavated.

## Study Areas and Varieties of Sites

Survey of the four study areas, encompassing 32 km$^2$, resulted in the discovery of 109 sites (Table 1.1). All but four were single component sites that contained occupations representative of archaeological periods ranging from the Lower Paleolithic through the Chalcolithic. Maps and systematic surface collections were made at 35 sites with each archaeological period being represented. Test excavations were undertaken at 32 sites with all but the Lower Paleolithic being included. Excavations were conducted at six sites containing Middle Paleolithic, Upper Paleolithic, Epipaleolithic, and Chalcolithic occupations.

*Judayid Basin.* The Judayid Basin study area was the largest and furnished the greatest elevational and environmental diversity (Figure 1.3). Elevations surveyed above 1,400 m, confined to the limestone hills on the lip of the plateau, display lag deposits of limestone and flint as well as stony Terra rossa. Plant cover mainly consists of low shrubs and relic trees reflecting a deforested Mediterranean woodland. Much of the area is dry-farmed by villagers from Ras en Naqb and by mobile Bedouin tent-groups. They cultivate fields of wheat and barley. Water is obtained from springs and shallow wells located near the rim of the plateau. Isolated artifacts were discovered in abundance and they represented most time-frames, although Middle Paleolithic specimens seemed especially common. Artifact concentrations indicative of sites, however, were next to nonexistent. A single, small surface scatter of Levallois material, located on a bench beneath Ras en Naqb, was the only site to be recorded above 1,400 m. Given the abundance of isolated specimens, the near absence of sites in hills on the plateau appears more related to the area having experienced severe erosion and surface stripping than to its settlement history.

Within the study area, those elevations surveyed between 900–1400 m fell within the basin proper and its western flanks. Here sandstone bedrock is covered by silts and sands supporting a steppe vegetation. Small fields restricted to silt deposits adjacent to wadis on the floor of the basin are occasionally planted in wheat and barley. Surface water is limited to pools located in wadi beds that are charged by winter runoff from rain and melting snow in the nearby uplands.

**Table 1.1. The Elevation, Phytogeographic Zone, Area, and Number of Sites Discovered Within Each of the Study Areas**

| Study area | Elevation (masl) | Phytogeographic[a] zone | Area (km$^2$) | No. of Sites (density No./km$^2$) | |
|---|---|---|---|---|---|
| Judayid Basin | 1,500–1,600 | M | 3.0 | 1 | (0.3) |
| | 1,300–1,500 | M/I | 4.0 | 5 | (1.3) |
| | 1,200–1,300 | I | 2.1 | 6 | (2.9) |
| | 1,200–1,200 | I | 4.3 | 15 | (3.5) |
| | 1,000–1,100 | I | 2.6 | 21 | (8.1) |
| | 900–1,000 | I | 1.5 | 2 | (1.3) |
| Jebel Qalkha | 960–1,020 | I | 6.5 | 23 | (3.5) |
| Jebel Mueisi | 830–870 | S | 5.0 | 10 | (2.0) |
| Jebel Hamra | 790–810 | S | 3.0 | 4 | (1.3) |

[a]M—Mediterranean, I—Irano-Turanian, S—Saharo-Arabian

Figure 1.3. Photograph of Judayid Basin looking south from near Ras en Naqb (elevation ca. 1,600 masl).

The sandstone jebels creating the western edge of the basin provide windbreaks and rockshelters. Sites, occurring in high densities, were discovered eroding from terraces, alluvial fans, and shelter deposits. Although sites of Middle Paleolithic through Chalcolithic periods were recorded, Epipaleolithic and Chalcolithic occurrences were especially numerous.

*Jebel Qalkha.* The environment and bedrock geology of the Jebel Qalkha study area, encompassing 6 $km^2$, closely resembles the lower elevations (i.e., 900–1,100m) of the Judayid Basin. Where Middle and Upper Paleolithic deposits are relatively rare in the Judayid Basin, such deposits are widespread over the plain of Wadi Qalkha and within the sandstone canyons that dissect the jebel. Even older depositional units representing Last Interglacial sediments (yielding Late Acheulean artifacts) were found just north of Jebel Qalkha in a deep headcut of the Wadi Ahemir. The presence of numerous rockshelters and overhangs in the study area contributed to a high density of sites that often display thick cultural deposits. Late Lower Paleolithic through Chalcolithic periods were represented by finds in the area, but Upper Paleolithic and Epipaleolithic sites were the most common.

*Jebel Mueisi/Mishraq/Muheimi.* This study area includes three clusters of sandstone inselbergs that rise from the northern edge of the valley of the Wadi Hisma. The bases of the inselbergs, falling between 800–900 m, are surrounded by drift sand and often are situated near the dry beds of Pleistocene lakes. Dwarf desert shrubs sparsely cover the landscape. Even seasonal surface water, available in potholes, is rare because of the small catchments of the jebels which are limited to the meager local rainfall. During abnormally wet years, the nearby mud pans are charged by run-off from the uplands. Rockshelters and overhangs appear as the most common site types with only the Epipaleolithic and Chacolithic periods represented.

*Jebel Hamra.* This, the lowest of the study areas (700–800 m), includes another jebel cluster positioned near the southern edge of the valley. Although Jebel Hamra is formed of sandstone, granite mountains define the flank of the valley less than half a kilometer to the south. In all other directions the jebel overlooks dry lake beds that are separated from its base by drift sand covered in desert shrubs. Without exception, the recorded sites were associated with rockshelters or overhangs and temporally tied to Epipaleolithic or Chalcolithic periods.

### Excavation Methods

Essentially three plans were followed for the excavation of those sites thought to contain *in situ* artifacts. These included: (1) excavation of isolated 1 $m^2$ test units, (2) deep excavation of large blocks of contiguous 1 $m^2$ units, and (3) shallow excavations of large blocks of contiguous 1 $m^2$ units. Two to six test units were excavated at each site depending on the surface area covered by artifacts. The placement of test units was intended to provide information on the subsurface extent and stratigraphy of a site.

Based on the results of the test excavations, one of three options for additional work at a site was selected. For those sites failing to yield *in situ* artifacts, no additional work was undertaken. For those sites that yielded deeply buried *in situ* material, plans were made for conducting larger block excavations. None of these exceeded 16 $m^2$, however. Finally, for those sites from which *in situ* artifacts were recovered from shallow deposits, large block excavations (20–30 $m^2$) were undertaken.

Uniform recovery procedures were followed for the excavation plans. All excavated matrix was sieved through 3 mm mesh screens for the recovery of minute artifacts and organic remains. Moreover, sediment samples from selected features were wet sieved through 1 mm mesh and floated in an effort to recover paleobotanic specimens. Other sediment samples were collected for geologic, pollen, and phytolith studies.

## HISTORY OF RESEARCH IN SOUTHERN JORDAN

Prior to the systematic survey of the Judayid Basin in 1979, prehistoric investigations in southern Jordan had been limited mainly to nonsystematic, reconnaissance level surveys (Rhotert 1938; Kirkbride and Harding 1947; Zeuner *et al.* 1957; Kirkbride 1959; Field 1960; Huckreide and Weisemann 1968; Copeland and Hours 1971; Price and Garrard 1975; Henry 1979). Test excavations had been conducted at the Epipaleolithic rockshelter of Wadi Madamagh (Kirkbride 1958) and at the small PPNB tell of Ain Abu Nekheileh (Kirkbride 1960; 1978), but the only intensive and well reported research to be conducted for the region (and for that matter Jordan east of the Rift as a whole) was the work at the Natufian and Neolithic site of Beidha (Kirkbride 1966; Mortenson 1970). Beyond the results of several seasons of excavation that focused on the Neolithic village at Beidha, southern Jordan had contributed little to our understanding of the prehistory of the Levant.

During the early 1980s, this situation changed greatly. Systematic surveys were launched or continued as part of large-scale research programs in the Wadi Hisma (Henry 1982a, 1982b, 1985b, 1987a, 1988a), the Petra area (Schyle and Uerpmann 1988; Gebel 1983–84; 1986a, 1986b; Gebel and Starck 1985; Gebel *et al.* 1988), and the Wadi Hasa (MacDonald 1988; MacDonald *et al.* 1982; MacDonald *et al.* 1983; Coinman *et al.* 1988; Clark *et al.* 1988a; Clark *et al.* 1988b). Each of these projects involved test excavations and full-scale excavation was resumed at Beidha with an emphasis on the earlier Natufian horizons (Byrd 1989). Beyond forming the basis for the development of a regional cultural-history stretching from Middle Paleolithic through Chalcolithic times, these projects contribute significantly to the prehistory of the Levant on a broad front of issues. In reviewing the collection of papers that recently summarized Jordanian prehistory, Marks (1988:816) underscored their significance in noting that:

> . . . . Jordan has become the centre stage for Levantine prehistoric studies and that, in the future, it must be taken into account in all considerations of Levantine prehistory.

## Middle Paleolithic

Levantine Mousterian sites thus far discovered in southern Jordan have yielded only D-Type assemblages (Henry 1982a, 1986; Lindly and Clark 1987). This is significant in that it strengthens the suggested ties between this industry and the arid zone (Henry 1982a; Marks 1983b). Based upon cultural stratigraphic and chronometric evidence from the Mediterranean woodlands of the central and northern Levant, the D-Type Industry has been viewed as the earliest of the industrial variants although its absolute age is ambiguous (Bar-Yosef and Meignen 1992; Jelinek 1992). As discussed in this study, dating of the Tor Faraj and Tor Sabiha deposits has added weight to the proposed continuity of the D-Type Industry (at least in the arid zone) from early Levantine Mousterian to transitional Upper Paleolithic times (Henry and Miller 1992; Marks 1992).

## Upper Paleolithic

Upper Paleolithic occupations, revealed by excavations at five sites in the Wadi Hisma (discussed in this study) and two sites in the Wadi Hasa along with four sites discovered in-survey of the Petra area, have extended the known ranges of both Ahmarian and Levantine Aurignacian industries east of the Rift Valley. By generating radiocarbon dates for late Ahmarian occurrences the Wadi Hasa research also contributed to a better chronological understanding of this poorly dated interval (Olszewski and Clark 1990; Clark *et al.* 1988b). A recent comparative study has revealed strong technological affinities between assemblages from southern Jordan and Ahmarian and Levantine Aurignacian assemblages recovered from the Negev and Sinai (Coinman 1990). The site of Wadi Aghar (Chapter 8, this volume) also promises to shed new light on the transitional phase between the Middle and Upper Paleolithic.

## Epipaleolithic

The best represented and, perhaps because of this, the most complex archaeological period to be addressed by recent research efforts in southern Jordan is the Epipaleolithic. Each of the major projects has resulted in the discovery of occupations attributed to this roughly 10,000 year-long interval.

The most significant results of these discoveries relate to the definition of new Epipaleolithic taxa (Qalkhan), the extension of the geographic ranges of known taxa (Kebaran, Geometric Kebaran, Mushabian, Late Natufian), the extension of the time depth of the Early Natufian, and the demonstrated contemporaneity of the Early Natufian and the Late Mushabian.

The discovery of a series of assemblages from sites in the Wadi Hisma triggered the definition of a new taxa, the Qalkhan, that appears to represent a very early Epipaleolithic industry (Henry 1982a, 1990; Jones 1983). Qalkhan assemblages have also been reported from two sites in the Petra area (Schyle and Uerpmann 1988).

Several other assemblages recovered from stratified deposits in the Wadi Hisma also failed to fit the techno-typologic definitions of existing Epipaleolithic taxa, at the level of "industry," thus prompting the creation of new taxa (Henry 1982a, 1990; Jones 1983). Unlike the Qalkhan Complex, however, these taxa have strong affinities to taxa defined for areas west of the Rift Valley and are most likely regional variants of the Kebaran and Geometric Kebaran, at the level of "complex." In denoting their affinities to complexes recognized west of the Rift, these taxa were given new industry, as opposed to complex, labels (Henry 1988a, 1989a). Although undated, Early, Middle, Late, and Final Hamran assemblages occur in stratified contexts in several rockshelter deposits in the Wadi Hisma. An Early Hamran horizon has been found to stratigraphically overlay a Qalkhan horizon at the site of Wadi Humeima, area B (Henry 1988a).

The Early Natufian has been known from southern Jordan since Kirkbride's (1966) work at Beidha. The recent excavations of the Natufian occupations at Wadi Judayid (Henry and Turnbull 1985; Sellars 1989) and Beidha (Byrd 1987, 1989), however, have produced the earliest radiocarbon dates (ca. 12,500 B.P.) for the Natufian in the Levant. These early dates, coupled with what appears to be a gradual transition from a local variant of the Geometric Kebaran (the Final Hamran), perhaps even point to the initial beginnings of the Natufian in the mountains of southern Jordan. Another Early Natufian site, Sabra 1, has been recorded in the Petra area (Gebel and Stark 1985). Late Natufian occupations, previously unknown in the region, also have been discovered in the Wadi Hisma (Wadi Humeima, J406a) and the Petra area (Sunah 1).

Another line of Epipaleolithic development, outside the Kebaran—Geometric Kebaran—Natufian succession, is represented by the Mushabian. This complex, previously known only from the Negev and Sinai, was identified at two rockshelters (Tor Hamar and Tor Fatma) in the western Wadi Hisma (Henry and Garrard 1988; Henry 1988a; Shen 1992). Both sites contained thick deposits that yielded rich artifact inventories and fauna. Techno-typological analysis of the assemblages showed strong affinities to Mushabian industries of the Negev and Sinai. These characteristics are also shared by the assemblage from Wadi Madamagh rockshelter which was originally identified by Kirkbride (1958) as Kebaran many years before the Mushabian had been discovered. Radiocarbon dates (ca. 12,700 B.P.) from the Tor Hamar deposit are also consistent with a Mushabian designation. A new industry-level taxon, the Madamaghan Industry, was proposed for these southern Jordan Mushabian assemblages.

The contemporaneity of occupations at the Early Natufian site of Wadi Judayid and the late Mushabian horizon at the nearby site of Tor Hamar is indicated by overlapping radiocarbon dates from the two deposits (Henry and Garrard 1988). Although interaction between late Mushabian mobile foraging groups and early Natufian communities has been proposed on the basis of shared artifact styles, Wadi Judayid and Tor Hamar provide the first evidence of nearby occurrences with synchronous radiocarbon dates.

## Neolithic

Within this study, the early Neolithic is known only from the small rockshelter occupation of Jebel Queisa (Henry 1982a, 1988a), but impressive Pre-pottery Neolithic B villages have been recorded at Beidha (Kirkbride 1966; Mortensen 1970), Basta (Gebel *et al.* 1988; Gebel 1986a) and Baga (Gebel *et al.* 1988) at high elevations around Petra and also from the small oasis site of Abu Nakheileh in the lowlands of Wadi Rum (Kirkbride 1978). The site of Ain Jamal, situated near a spring flowing from the escarpment ca. 2 km west of the Judayid Basin, appears to represent another nearby PPNB village (Gebel *et al.* 1988). The apparent scarcity of Pottery Neolithic occupations in southern Jordan parallels the "hiatus palestinienne" observed for the Levant as a whole and adds weight to the postulated widespread decline in population during the late Neolithic.

## Chalcolithic

In contrast to the late Neolithic, Chalcolithic occupations occur in great numbers in southern Jordan (Henry 1982a, 1990; MacDonald 1988). The architecture, chipped stone industry, and pottery of the Chalcolithic occupations in the Wadi Hisma strongly resemble those described for the Timnian of nearby northeastern Sinai (Kozloff 1974, 1981). A radiocarbon date of ca. 5,700 B.P. from one of these occupations falls near the early end of a series of dates for the Timnian. Settlement and economic evidence for the Timnian sites of the Wadi Hisma indicate that the groups subsisted on sheep/goat herding augmented with hunting and gathering. They appear to have followed a transhumant settlement cycle quite similar to that followed by the historic Bedouin groups of the area.

## ENVIRONMENTAL SETTINGS

Mediterranean woodland, steppe, and desert settings provide surprising environmental contrasts for a region that is dominated by such a dry climate. In occupying a land-bridge joining Africa with Asia and Europe, the region's biotic diversity is heightened by representatives from each of the continents. Furthermore, the environmental contrasts of the region are sharpened by the presence of quite distinct physiographic settings in close proximity. Three major landforms are represented by the rolling hills of the limestone plateau, the rugged landscape of the sandstone inselbergs that front the plateau to the south, and the granite mountains that border the Rift Valley to the west (Figure 4). Elevations fall from as high as 1,700 masl in the hills on the edge of the plateau to under 800 masl on the floor of the Wadi Hisma only a few kilometers from the foot of the escarpment. These striking edaphic and elevational differences have created a setting in which the three major biotic communities of the Levant are closely packed into parallel elevational belts often separated by less than 20 km. Combined with the spatial variability in environmental settings, the region's Mediterranean climate makes for marked seasonality in precipitation with a short winter wet season followed by a nine-month-long dry season.

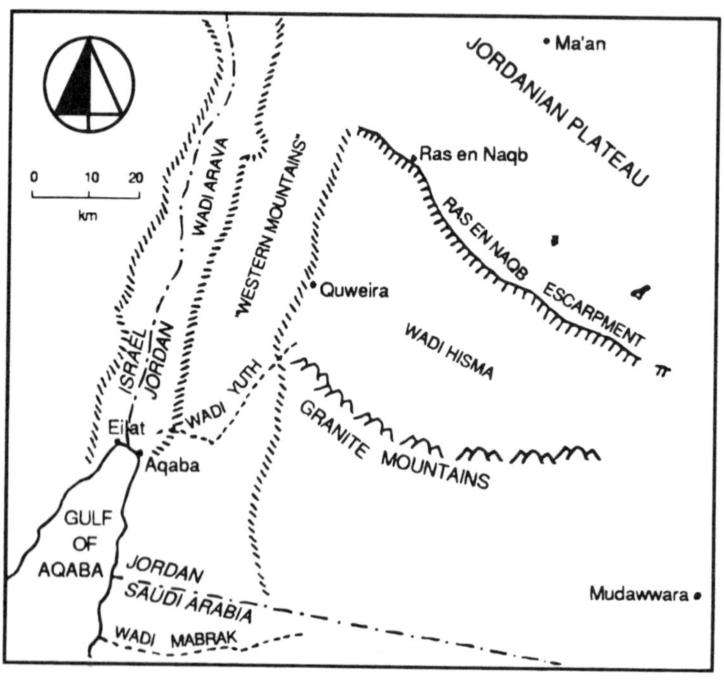


Figure 1.4. Map of southern Jordan showing the physiography of the region (after Osborn 1985).

## Physiography

Exposures along the Ras en Naqb escarpment show Cretaceous limestones and sandstones overlying strata of Paleozoic sandstone (Selley 1972; Bender 1974; Osborn 1985; Chapter 2, this volume). This suite of strata, forming much of the Jordanian Plateau, originally extended south over what is now the inselberg landscape which covers the broad Wadi Hisma and its many branch canyons (Osborn 1985). The southern limit of these strata east of the rift in Jordan and Saudi Arabia is unknown, but their remnants west of the rift have been recorded as far south as the southern tip of the Sinai Peninsula (Osborn 1985).

Following the stripping of the Cretaceous limestones, subsidence and faulting associated with the Rift led to differential erosion, exposure of Precambrian granites, and isolation of sandstone inselbergs. Osborn (1985) suggests that the Wadi Yutm, the major drainage outlet of the area, probably existed throughout the development of the landscape. The temporary blockage of Wadi Yutm, however, appears to have prompted aggradation of sand and playa sediments. These sediments form a relatively flat plain on the floor of the Wadi Hisma above which rise erosional outliers of the sandstone. Evidence of Pleistocene lakes are evident as mud pans or playas which dot the plain.

## Climate and Environment

The climate of the region is characterized by a short winter wet season followed by a long dry season, both accompanied by relatively mild temperatures. In that topography is the single most important factor governing the precipitation and temperature patterns of the region (Shehadeh 1985), there is considerable variability in climate from the high reaches of the plateau to the floor of the Wadi Hisma (Table 1.2). These climatic belts in turn govern the zonation of distinct biotic communities (Figure 5).

*Uplands.* The rolling hill country of the plateau, reaching elevations of 1,700 m near Ras en Naqb, is dominated by shrubs and low bushes of Garigue and Batha Mediterranean vegetation (El-Eisawi 1985; Zohary 1973). Remnants of former Mediterranean woodlands are seen as isolated trees or small stands of juniper (Juniperus phoenicea) and pistacia (Pistacia atlantica). Overgrazing is thought to have degraded the woodlands and induced massive erosion of the uplands as evidenced by flint regalith or hammada surfaces (El-Eisawi 1985). Harlan (1988) challenges the popular notion that the southern highlands of Jordan were recently forested, however. In reviewing the observations of several 19th century visitors to the region, he concludes that

**Table 1.2. The Different Elevational Belts Encompassed by the Study Areas Showing the Local Characteristics and Correlations with the Koppen System of Climatic Classification.**[a]

| Koppen Classification | Local characteristics |
|---|---|
| Cool temperate rainy | Mediterranean climate with average annual temperature ranging from 12–17°C, average July temperature is 22°C, average annual precipitation near 300 mm. Elevation: 1,300–1,700 masl. |
| Cool Steppe | Semi-arid climate with avergae annual temperature 18°C or less and average annual precipitation less than 300 mm. Elevation: 1,100–1,300 masl. |
| Cool desert | Arid climate with average annual temperature is 18°C or less and average annual rainfall is less than 200 mm. Elevation: 900–1,100 masl. |
| Warm desert | Arid climate with average annual temperature ranging from 19–25°C and average annual rainfall is less than 100 mm. Elevation: –900 masl. |

[a]Also, note the correspondences between these climatic classifications and the phytogeographic zones identified in Figure 1.5.

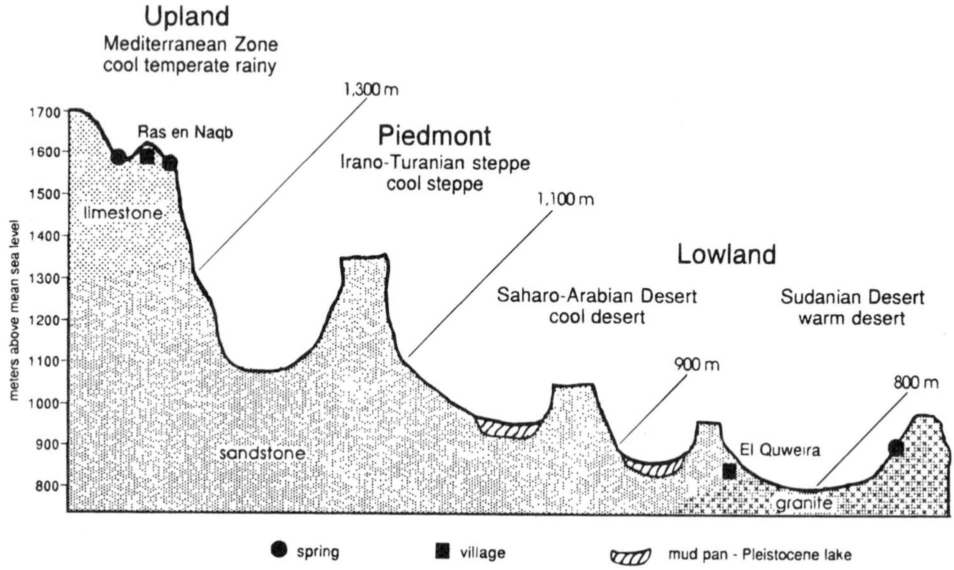


**Figure 1.5.** Schematic transect of area showing the relationships of elevation, bedrock geology, biotic zones, and climate.

most of the plateau was covered in grassland. Woodlands would have been confined to exceptionally well watered areas such as very high elevations and where ground water seeps laterally along the edges of the plateau (Harlan 1988). In a pristine setting these areas are likely to have included trees such as pine (Pinus halepensis), oak (Quercus calliprinus), laurel (Laurus nobilis), maple (Acer syriacum), cedar (Cedrus libani), and Daphne linearifolia in addition to the juniper and pistacia. Understory vegetation would have consisted of shrubs, grasses, and bulb and tuber plants common to the Mediterranean woodlands. Recent efforts of reforestation have been launched with the planting of pines and cedars below Ras en Naqb and establishment of extensive fruit tree orchards north of the village. Much of the uplands presently are cultivated in wheat.

Remnants of *Terra rossa* soil confirm mean annual precipitation levels of over 300 mm. With mean annual temperatures for the uplands ranging from 12–17°C, the area falls within the "Cool Temperate Rainy" climate of the Koppen Classification (Shehadeh 1985) and the "Arid Mediterranean (Cool Variety)" bioclimate of El-Eisawi (1985). Mean minimum temperatures for January range from 0–3°C with the precipitation often coming as snow. July temperatures average only about 22°C; thus transpiroevaporation rates are relatively low, leaving sufficient soil moisture for sustaining green grasses and shrubs long after they have died back at lower elevations. The clayey soils of the uplands further enhance moisture retention.

*Piedmont.* A series of high basins formed between the steep escarpment of the plateau and sandstone outliers create a piedmont at elevations between 1,000–1,300m (Figure 6). Alluvial fans drape the steep flanks of the basins and coalesce to form broad, near level floors up to 3 km across. The large Judayid Basin, typical of those fronting the plateau, shows clear differences between its escarpment and outlier sides. Beneath the escarpment, very steep boulder strewn and sparsely vegetated slopes fall to the floor of the basin. The alluvial fans fronting

**Figure 1.6.** Photograph of piedmont from the floor of the Judayid Basin (elevation 1,100 masl) looking north toward the escarpment and Ma'an Plateau.

the sandstone outliers on the western edge of the basin are more gently sloped, better vegetated, and often composed of sand. Beyond the alluvial fans an older depositional unit composed of silt is exposed on the basin's floor. Because of its water retention, the silt supports a denser plant cover and is also favored for cultivation of wheat and barley by contemporary Bedouin groups.

Steppe vegetation, characterized by small shrubs and bushes, dominates the piedmont. Sage (*Artemesia herbae-albae*), saltwood (*Haloxylion articulata*), bean caper (*Zygophyllum dumasum*), white broom (*Retama raetam*), and *Anabasis articulata* are the most common species. Better watered areas along wadis and shaded settings beneath cliffs support denser vegetation occasionally accompanied by pistacia (*Pistacia atlantica*), the principal tree of the steppe zone. It should be noted, however, that such trees represent remnants or relics of earlier distributions during moister intervals, for nowhere in the area have seedlings or young trees been found (Zohary 1962).

With mean minimum temperatures in January ranging from 3–6°C and rainfall measuring some 200–300 mm annually, the piedmont falls within the "Cool Steppe" of the Koppen Classification and "Arid Mediterranean (Warm Variety)" bioclimate of El-Eisawi (1985).

*Lowlands.* The lowlands essentially occupy the floor of the broad Wadi Hisma which parallels the edge of the plateau (Figure 1.7). In falling between 800 and 1,000 masl, this area is low only in a relative sense. The floor of the valley is some 10 km wide at its western end and narrows to the east before joining a maze of narrow sandstone canyons and inselbergs. The southern edge of the valley is defined by granite mountains. These are cut by the deeply incised Wadi Yutm which drains the area to the Red Sea some 40 km to the south.

Figure 1.7. Photograph of the floor of the Wadi Hisma (elevation 800 masl) showing the landscape of inselbergs, drift sand, and playas.

The near level floor of the valley contains numerous mud flats, the remnants of Pleistocene lakes, and is occasionally punctuated by sheer walled jebels. These spectacular sandstone inselbergs rise 200–300m above the valley floor and often attain elevations of over 1,000 m. Surface sediments within the Hisma form a mosaic of talus scree from the inselbergs and granite mountains, silty-sand of the lake beds, and extensive fields of sand dunes. The sparse plant cover of the valley is associated with desert vegetation of the Saharo-Arabian and Sudanian phytogeographic zones (El-Eisawi 1985). The Saharo-Arabian vegetation is composed mainly of shrubs such as sage (*Artemesia herbae-alba*), saxual shrub (*Haloxylon persicum*), *Calligonum comosum*, and *Archillea fragrantissima*. This vegetation dominates the valley with the densest stands occupying areas of drift sand and the shore-lines of mud pans. Sudanian species, characterized by acacia (*Acacia sp.*) and thorn (*Zizyphus spina-christi*) trees, are more common to the lower, drier settings that are covered by granite scree along the southern edge of the valley.

The daily temperatures of the lowlands average some 3°C higher than the piedmont and over 6°C higher than the uplands. With minimum January temperatures averaging 5–8°C and annual rainfall measuring less than 50 mm, the lowlands fall mainly within the "Cool Desert" classification of Koppen and the "Saharan Mediterranean (Warm Variety)" bioclimate of El-Eisawi (1985). The lower (near 800 masl) southern edge of the valley, with higher average annual temperatures of 20–22°C, would be included within the "Warm Desert Climate" of the Koppen Classification. As with the uplands, the meager rainfall of the lowlands is restricted to the winter; many years go without measurable precipitation. Water sources are limited to springs along the valley's southern margin and to the lake beds which are charged for a few weeks by runoff from the uplands during abnormally wet years.

## SUMMARY

In the Levant, prehistoric studies have traditionally focused on the rather narrow geographic area of the central and northern regions. Research undertaken over a half century ago in coastal Lebanon (Ewing 1947; Garrod and Kirkbride 1961), the Carmel (Garrod and Bate 1937), and the Judean Hills (Neuville 1934, 1951) formed the basis for assumed pan-Levantine cultural-historic sequences, environmental successions, and even the underpinnings of certain contemporary processual models. Over the last twenty years, however, ongoing investigations in the central and northern Levant have been complemented by intensive and long-term studies of previously ignored areas in the south. The areal oriented projects conducted in the Negev (Marks 1976, 1977a, 1983; Goring-Morris 1987; Gilead and Bar-Yosef 1987) and Sinai (Bar-Yosef and Phillips 1977; Phillips 1987a, 1987b, 1988) have furnished a body of evidence that has prompted reevaluation (and often abandonment) of long-held cultural-historic and environmental schemes. This relatively recent research in the southern Levant has revealed a much richer and more complex cultural-historic mosaic than was previously entertained. And, perhaps of even greater significance, the area-wide approaches of the investigations in the south have provided a better understanding of the variability within site types for discrete time-frames. This in turn has brought much greater precision to our understanding of settlement-procurement patterns and prehistoric cultural ecology for the region as a whole.

While the new, widespread evidence from the southern Levant is correcting many of the misconceptions introduced through previous geographic bias, one common trend of the past has continued in these studies. This trend relates to examining cultural-historical and adaptive relationships within the region from a north to south, circum-Mediterranean perspective. Such a perspective has inevitably come about as a result of the near absence of data from the eastern periphery of the southern Levant. For a combination of reasons, perhaps best described as historical accident, only a handful of prehistoric studies were conducted in Jordan east of the Rift Valley prior to the 1980s; and these were confined to terminal Pleistocene and Holocene deposits.

In the context of Levantine prehistory, the results of the research in the Wadi Hisma is of importance at two scales. At a regional scale, the study provides the first detailed and near continuous cultural-historic framework from an area east of the Rift Valley. This allows for comparisons with similar frameworks developed for areas west of the rift. Such comparisons, in turn, provide an alternative to the traditional north to south, circum-Mediterrean perspective in the reconstruction of relationships among prehistoric populations of the Levant. Consideration can now be given to what degree populations interacted between inland areas and the coast. The study also provides initial clues as to the importance of the Rift Valley as a barrier to prehistoric cultural interaction.

By extension these data also enable us to test propositions concerning the taxonomic (and presumably behavioral) differences between the northern/central and the southern Levant. Within the Levant, synchronous industrial differences have been recognized for most prehistoric periods and these differences are linked to north–south, woodland–steppe/desert distinctions. These "parallel phyla" have been attributed to either functional requirements tied to environmental adaptation or to geographic proximity expressed in cultural territoriality. The general correspondence of the Mediterranean woodlands with the northern/central Levant and the steppe-desert zone with the southern Levant, however, has made it difficult to distinguish between adaptation and regional proximity as forces responsible for the observed taxonomic differences. In that much of the evidence from southern Jordan comes from a Mediterranean woodland situated in the south, these data break the logical conundrum posed by the redundancies between archaeological taxa, environmental zones, and geographic regions west of the rift. Without exploring the details at this point, the results of the study suggest that environmental similarities were more important than geographic proximity in determining archaeological (and presumably behavioral) affinities. In this light, research into the prehistoric relationships of woodland and steppe-desert adapted peoples of the Levant are likely to prove more worthwhile than concerns with north–south (or, for that matter, east-west) territorial interactions.

The second scale at which the research in the Wadi Hisma makes a significant contribution relates to local prehistoric ecology. The dramatic relief of the area acts to compress all of the Levant's major biotic zones into tightly spaced elevational belts separated by less than 20 km. Extensive erosion of Pleistocene deposits within the area has resulted in the exposure of a high density of archaeological sites stretching back to the Middle Paleolithic. The area's combined characteristics of striking biotic diversity and abundance of prehistoric sites provide an ideal laboratory for examining cultural ecology from synchronic and diachronic perspectives.

Study areas, selected from different elevational belts and representing each of the modern biotic zones, were surveyed for sites. Results of the surveys coupled with evidence obtained through excavation of selected sites revealed patterns in the natural and cultural data that were linked to differences in elevation. These patterns are thought to reflect the presence of transhumance, an adaptive strategy that involved a group scheduling adjustments in its size and mobility along with determining when (season) and where (elevational belt) it would reside.

Beyond the prehistoric evidence, an unanticipated source of information on transhumant patterns came from ethnoarchaeological observations of contemporary Bedouin encampments. In studying both occupied and abandoned camps, striking parallels were seen between Bedouin and prehistoric (especially Chalcolithic) settlement patterns.

Given the high density of sites spread over a considerable time-depth, it was not only possible to reconstruct specific prehistoric patterns of transhumance, but also to trace changes in these patterns over much of the late Pleistocene. A reconstruction of past environments is a fundamental step toward understanding such changes in human adaptation, for even if environmental forces were not linked to behavioral changes they have to be known to be eliminated as causal factors. Again, certain natural characteristics of the area made this a reasonable task. The area's elevationally governed biotic zones provide a sensitive record of past climatic changes. These zones retreated or expanded up- or down-slope in concert with fluctuations in precipitation and temperature patterns. Therefore, when paleoenvironmental evidence pointing to a certain biotic community was recovered from a dated deposit at a specific elevation, an area-wide reconstruction of the prevailing climate and environmental settings could be made with considerable certainty.

The presence of numerous rockshelters is another natural feature of the area that aided in environmental reconstruction, as well as in refining the chronologies of cultural and natural successions. Shelter deposits, with their thick stratigraphic sequences and tendencies for the preservation of organics, are rare in the southern Levant. Such features are common to the Wadi Hisma, however, because of the sandstone geology and weathering forces of the area (Osborn and Duford 1981:9). Rockshelters, in fact, formed the most common site type of those study areas selected from the lowlands and lower piedmont. Many of these revealed stratified, multicomponent cultural successions accompanied by well preserved pollen, phytoliths, shell, and bones of vertebrates.

Differences in site types, settings, and artifact inventories have been linked to transhumance in other areas of the Levant (S. Binford 1968; Vita-Finzi and Higgs 1970; Scott 1977; Goring-Morris 1987; Kaufman 1992; Lieberman 1993). Such suggestions have been based upon evidence from single or small groups of sites thought to represent individual segments of an annual round. The site types, settings, and artifact inventories for other segments have been hypothecated based upon assumptions concerning the availability of resources and creature comfort. The data pointing to prehistoric transhumance in the Wadi Hisma are unique in that they are derived from sites representing multiple segments of an annual cycle within a compact geographic area. While an entire annual-round within a transhumant cycle probably has not been captured by the study due to the absence of evidence from above 1,400 m and below 700 m, representative warm and cold-season segments clearly have been identified. Beyond the comprehensive and detailed reconstruction of the settlement-procurement strategies associated with prehistoric transhumance, the study offers an unusual opportunity to examine specific changes in these strategies over the last 70,000 years.

*Chapter* 2

# *Late Quaternary Geology and Geomorphology of the Area in the Vicinity of Ras en Naqb*

FEKRI A. HASSAN

## GEOLOGICAL SETTINGS

Southern Jordan lies at the edge of the stable shelf which borders the Arabo-Nubian massif. To the west of the Rift Valley, the cratonic shelf is unstable and transitional between the geosynclinal conditions of the stable area of the shelf and the Arabo-Nubian shield.

During the geological past and since the later part of the Early Cambrian, the Tethys Sea transgressed southward reaching as far as Saudi Arabia during the late Ordovician and Silurian times. The rocks of the Arabo-Nubian Massif (Picard 1943; Said 1962) consist of igneous and metamorphic rocks which crop out in the southwestern part of Jordan, plunging in a NE direction from the eastern side of the Gulf of Aqaba. This basement complex consists of aplite granites, granodiorite, quartzporphyry, alaskite, and quartz diorite. Metamorphic rocks include gneisses, paragneiesses, and a series of epimetamorphic slate-graywackes. The igneous rocks have been subjected to precambrian faulting. These faults strike in a N 40°–60° E and can thus be distinguished from the later faults which strike in N 15° E and 30°–160° directions (Bender 1974:19).

The Basement Complex is overlain uncomformably by a sequence of classic sedimentary rocks (Figure 2.1). The rocks are mostly sandstones that belong for the most part to the Paleozoic and the Cretaceous. Sands were deposited during the periods of sea transgressions of Early-Middle Cambrian age, Early Ordovician, and Late Ordovician-Silurian times. During the Cretaceous, clastic sediments were formed in Early Cretaceous times. Upper Cretaceous limestones and sandy limestone were formed in later Cretaceous times. The Paleozoic strata show a gentle regional dip from the uplifted eastern side of the Wadi Arabia rift, in a ENE direction. They are separated from the overlying Cretaceous beds by an unconformity.

The Paleozoic sandstone crops out in buttes and, as in Wadi Hisma, inselbergs. Farther to the east beyond Madwara the sandstone forms a continuous plateau.

The Lower Cretaceous sandstones (180 m) show a decrease in thickness toward the east. The Upper Cretaceous sediments consist predominantly of calcareous Cenomanian and Turonian facies at Ras en Naqb, but change laterally into a sandy facies toward the east and the southeast.

The Ras en Naqb escarpment consists of Cenomanian limestones near the top (150 m) underlain by Lower Cretaceous sandstone (an upper unit of dark sandstone "*Bunte Sandstein*" and

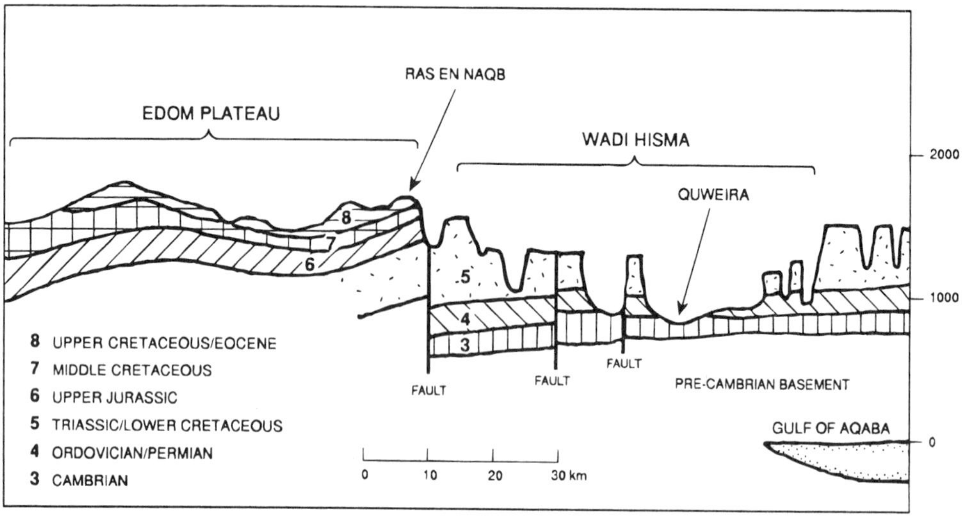


Figure 2.1. Geological cross-section of southern Jordan showing the major formations and landforms.

a lower unit of White sandstone). The Lower Cretaceous sandstones (150 m) are underlain by Ordovician sandstones.

## GEOMORPHOLOGICAL SETTING

The geomorphological setting of the Ras en Naqb area is the product of the tectonic events which led to the formation of the Aqaba-Dead Sea Rift. Geomorphologically the rift zone is represented by the Wadi Araba-Dead Sea depression which runs for about 360 km in a NE direction from the Gulf of Aqaba to Lake Tiberias. The depression is about 15 km in width and rises in elevation to about 250 m above sea level 80 km NNE of Aqaba. The depression is bordered from the East by an uplifted block (East Araba Highlands). In southern Jordan, the highlands reach a maximum elevation of 1,800 m above sea level at Jebel ash Sharah, west of Ras en Naqb. The slope toward the rift depression in the west is unlike the eastern part which extends over most of central Jordan. In the area of Ma'an-Ras en Naqb, the plateau is flat with a hamada surface dissected by a few shallow wadis and includes depressions occupied by mud pans. To the East, a central depression (about 80 km in radius) contains an extensive playa with endoeric drainage. This is the Jafr Depression. At Ras en Naqb the plateau is bounded by an escarpment (a fault escarpment) leading to a rapid drop in elevation of 600 m over a distance of 1 km. The depression south of the Ras en Naqb escarpment extends south to near Quweira and forms the western end of the Wadi Hisma. The area from Quweira halfway to Madwara is occupied by mud pans over a distance of 45 km. South of Quweira the area is dominated by the hills of Jebel Ram and Jebel Sahm which consists of buttes and inselbergs of Paleozoic sandstone overlying granites of the basement complex.

## QUATERNARY GEOLOGIC HISTORY: A BACKGROUND

Following the regression of the Tethys Sea during the Late Eocene times, the initial formation of the Aqaba–Dead Sea–Jordan Valley rift took place. "Syntectonic" conglomerates of Oligocene-Miocene age accumulated in the rift depression. The occurrence of such gravels to a thickness of more than 270 m in the highland of the eastern rim of the rift suggests that the highlands may not have been elevated until the rejuvenation of the tectonism in Pliocene-Pleistocene time. Bender (1974), however, suggests that the conglomerates were probably deposited in structural lows under marine conditions or in lacustral environments within the structurally elevated zone.

Bender (1974:26) also suggests that a marine connection existed between the Red Sea–Gulf of Aqaba–Wadi Araba across the Dead Sea-Jordan rift and the Beisan–Jesreel depression to the Mediterranean. Sediments of marine origin include the Oligocene (?)-Pliocene evaporites of the Sodom formation and the evaporite series of el Lisan Peninsula. Oolitic limestones containing foraminfera on Miocene/Pliocene age indicate a littoral, marine environment were located south of the watersheds of Jebel er Risha, at Jebel er Khurieji. The sediments (more than 40 m thick) occur 200 m above sea level.

It seems that it was during the initial phase of rifting that the Aqaba–Ras en Naqb–Medwara triangle, now represented by the southern desert of Jordan, was affected by faulting. Faults are N 15° E and S 50° E with small throws, but are traceable over many kilometers. One of the fault lineaments, running for over 70 km from Quweira to Medwara, facilitated erosion along the lines of structural weakness. Rapid downcutting and denudation, presumably under wetter climatic conditions than at present, created the spectacular canyons and steep cliffs well known for travelers in the Wadi Rum and to the Bedouins who gave the area the name of Hisma (canyonland, in Arabic). The vertical erosion has been substantial, perhaps as much as 1100 m, in judging from the total removal of the Eocene strata coupled with the differences in elevation of the plateau (Beihiri 1973:140–144).

The Ras en Naqb escarpment was created by slope retreat from the Oliogocene to the Late Pleistocene. Outliers and inselbergs attest to the rapidity of the process. Drainage today is toward the Gulf of Aqaba through Wadi Yutm. The erosional basin is now blocked from the west by the highlands of the eastern rim of the rift as represented by Jebel Baqir in the south and Jebel Aheimer in the north. The block might have been of lower elevation during the earlier phase of the rifting, rising higher during the later episodes, as indicated by the occurrence of Miocene/Pliocene marine sediments 200 m above sea level at Jebel Khurieji.

The uplift of the Eastern rift highlands leads to a gentle regional dip of the Eastern limestone plateau from the highlands toward the east and from Ras en Naqb to the NE. Erosion by water action during wet periods and wind action during the dry periods created a number of depressions, including the Jafr depression.

The Quaternary environmental and palaeoclimatic history of Jordan are at present poorly understood. The rift depression was occupied by a number of lakes during the Quaternary period at various times. The lakes range from fresh to brackish. The rift area represented a subsiding taphrogenetic depression which received a considerable amount of sediments reaching in parts more than 2000 m. Bender (1974:95ff) recognizes an Upper Pliocene-Basal Pleistocene unit, the Shagur Formation which consists of conglomerates intercalated with sandstone (100 m) deposited under fluvio-lacustrine conditions (*Melanopsis* and date palm fossils were found in this unit). The unit was subjected to faulting and is overlain uncomformably by the Gohr el Katar series, which consists of fluvio-lacustrine conglomerates, sandstone, marl, and marly clay. The loamy matrix of the conglomerates resemble in color the Mediterranean red soil. The Gohr el Katar series is in turn overlain uncomformably by the Abu Habil unit which consists of conglomeratic limestone. The limestone is in part pisolitic. Pebble tools (Oldowan?) are supposedly reported from this unit. The Abu Habil deposits are overlain with Middle Pleistocene basalts. The basalts also occur at Zarqa Marin where they overlie fluviatile gravel (at Jurf and Darawish). The gravels, according to Bender, contain Abbevillian/Lower Acheulian artifacts. Near the top, the artifact assemblage suggests to Bender "Mousterian of Acheulian tradition." The Jurf ed

Darawish gravels, also exposed near Shubuk an Uneize, are capped by caliche. The caliche layer is overlain with loose gravel which contains Upper Paleolithic artifacts.

In the depression, lacustrine deposits attributed to the Upper Pleistocene are referred to as the Upper Lisan Beds. The beds include a lower unit of grey/greenish marl (at Damiya) which include Mousterian artifacts and an upper unit of "varved" clay, marl, chalk, gypsum, silt, and sulfur concretions suggesting saline water conditions. Deposits containing Middle Paleolithic artifacts are also noted at Qala, at Wadi al Hasa, and the Jafr Depression.

The Terminal Pleistocene-Holocene witnessed the shrinkage of the rift lakes to their modern relics, probably a result of both rapid subsidence and the onset of dry climatic conditions (Schuldenrein and Goldberg 1981; Koucky and Smith 1986).

## LOCAL GEOLOGY

The geomorphological units in the Ras en Naqb area include: the Ma'an-Ras en Naqb Plateau, Ras en Naqb escarpment, alluvial fans, outliers and inselbergs, wadi terraces, modern wadi, and aeolian sand (Figure 2.2).

### The Ma'an-Ras en Naqb (Edom) Plateau

This plateau represents a part of the Central Desert Area of East Jordan in the geomorphological classification of Bender (1974:9). It is bounded in the west by the Ash Sherah highlands of the Eastern rim of the rift valley and slopes eastward to the Jafr Depression. From the north it is bounded by Wadi el Hasa and from the south by the Ras en Naqb escarpment.

The plateau reached maximum heights just north of Ras en Naqb with an elevation of 1670 masl at Shudayid and Naqb Ishtar. The plateau slopes both to the east and the north to elevations slightly more than 1000 m. The surface is a hamada desert, i.e., a stony desert covered with classic detritus. At Ras en Naqb, the stones are mostly made of flint and are variable in size from a few centimeters in diameter. The gravel is angular and poorly sorted and is embedded in a sandy matrix of brown-reddish yellow 7.5 YR 5.5/5 color. The surface deposits also include carbonate nodules. The surficial deposits are only tens of centimeters thick and seem to represent residual lag deposits of a Mediterranean red soil. In places, the stones have been removed to clear plots for cultivation.

### The Ras en Naqb Escarpment

At Ras En Naqb the Ma'an plateau drops precipitously toward the Quweira-Disa depression, the western end of the Wadi Hisma. The escarpment is controlled by S 33° E–N 35° W trending faults. Faults also run in a 30° W–N30°E direction and guide the drainage toward the Quweira depression. The escarpment shows three pediments. The top pediment (I) is at 1600 masl and its relicts are represented by Tilal Ras en Naqb el Quibliya and its dissected surface represents the surface of the Ma'an plateau described above. The middle pediment (II) of Naqb Ishtar is at 1400 masl and also is represented by Jebel el Jill. The Naqb Ishtar Pediment (II) represents a rock surface cover with a thin (0–40 cm) mantle of a stony lag deposit. The stones range in size from about five to 25 cm. They are platy in shape, angular, and are made of chert, siliceous limestones, and dolomites. The matrix is loamy sand of yellowish brown-brown color (10YR 5.5/6) and seems to represent a residuum of a Mediterranean or Steppe Yellow soil. The low, Sumei's Pediment (III) is covered by a surficial mantle about 30 cm in thickness. The surface is stony with a loamy sand matrix of brownish yellow-yellow color (10YR 6.5/6). The stones consist of siliceous quartzone sandstone and occasionally chert. This reflects the lithography of the bedrock that consists of bedrock which consists of sandstone. The surficial deposits seem to represent the residue of a steppe yellow soil.

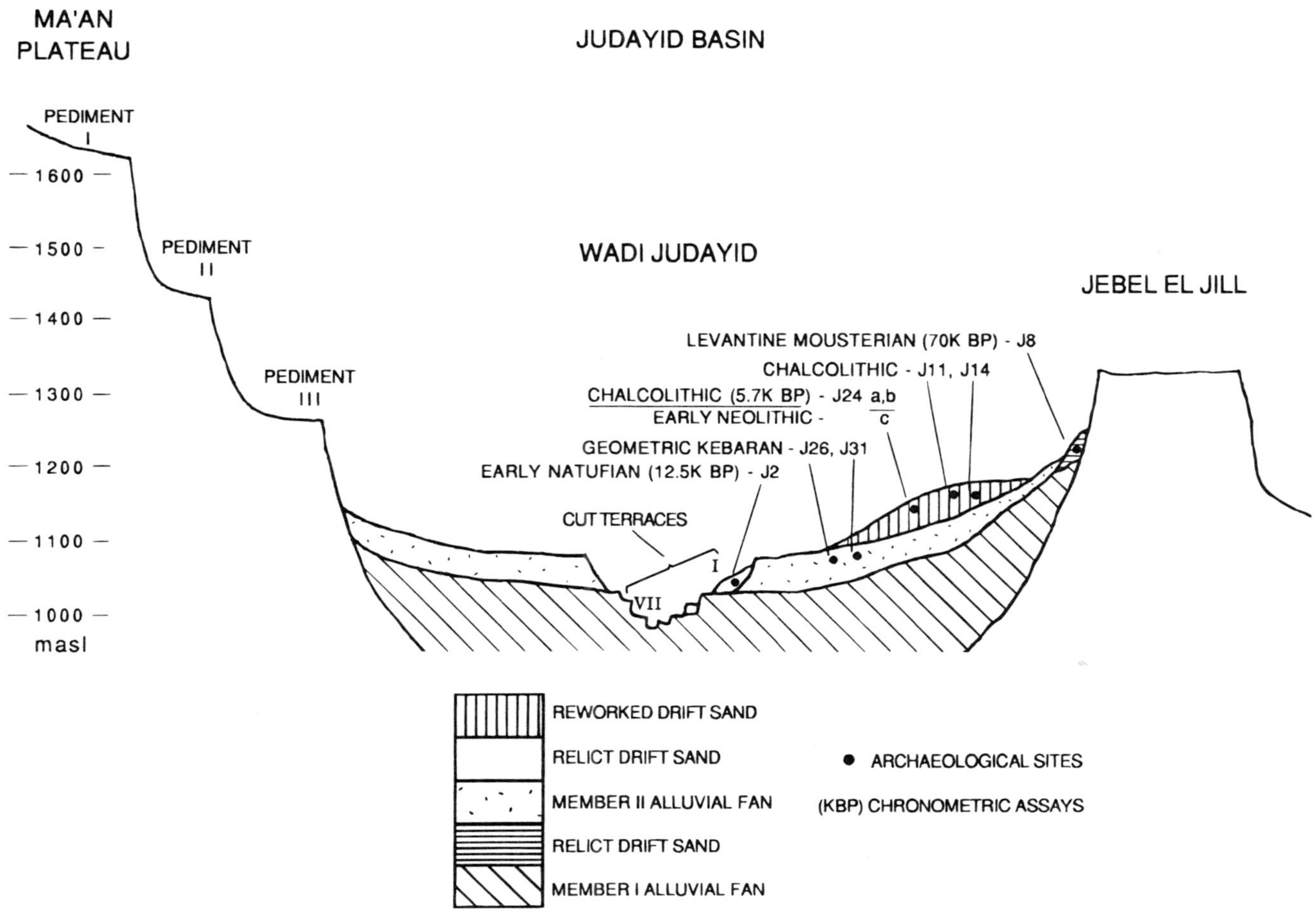


Figure 2.2. Cross-section of the Judayid Basin showing the landforms, Upper Pleistocene depositional units, and positions of archaeological sites.

The slopes of the Ras en Naqb escarpment are therefore composite showing, when the surface of the older pediments are preserved, a discontinuous cliff face of vertically connected J-shaped slopes. The formation of the escarpment and the retreat of the slope seems thus to have progressed in cycles each terminated by a period of relative stability and relaxation of down-cutting with lateral planation by stream action. The present surface of the basin bounded by the Ras en Naqb escarpment seems to represent a stage in the development of a fourth pediment.

## Alluvial Fans

The Ras en Naqb escarpment overlooks a basin representing the headwater of Wadi Judayid. The basin is filled with alluvial fan deposits. The fans coalesce to form a bajada dissected by more recent gulleys and are eroded toward the center of the basin by stream action.

The fans show the typical profile of alluvial fans in semiarid regions. The apex of the fan is steep (gradient 36%) and covered with large stones and grades upward in talus cones (gradient 50%). The middle segment is less steep (10%, i.e., 10 meters per 100 m of horizontal distance). The toe of the fans had been modified by stream downcutting and its morphology is not clear. It should be also noted that near gullies the surfaces of the fans show a reduction of elevation. The fans consist of massive sand laminated in parts. The sand is medium grained and contains numerous angular flat pebbles and cobbles as well as spheridal clasts of sandstones. Pockets of large stone blocks are also present. The fans seem to consist of two units. A lower unit is composed of hard compact sand with carbonate nodules. The color of the sediments ranges from light brown 7.5 YR 6.5/4 to strong brown-reddish yellow 7.5 YR 5.5/6. The upper units consist of friable sand, reddish yellow to light brown or pink in color (7.5 YR 7/5 YR 6/6).

## Outliers and Inselbergs

These features which represent remnants on the Ma'an Plateau include conical and elongate hills rising in elevation to 1350 at Jebel el Jill and 1200+ in other hills.

## Wadi Terraces

Six cut-terraces stand above the wadi floor at 10 m (Terrace I), 4.5 m (Terrace II), 1.7 m (Terrace III), 0.5 m (Terrace IV), 1.7 m (Terrace VI), 1.2 m (Terrace VII). The fill terrace (Terrace V) rises at 3.5 m above the floor of Wadi.

The highest terrace is the most widespread in the wadi basin. The other terraces are restricted to a band of 200 m or less along the course of modern wadi. The gradient of the Highest terrace is slightly less than that of the modern wadi. It measures 3% compared with 3.8% for the wadi bed at present. The gradient for the lower terraces approximates that of the modern wadi.

The fill-terrace consists of friable sand, in places laminated with a relatively high content of rubble. The color of the sediments is reddish yellow (7.5 YR 7/8, 7.5 YR 6/6) and is apparently inherited from the sediments of the alluvial fan.

## The Modern Wadi

The modern wadi is incised into the deposits of the older alluvial fan fill and the young alluvium of the fill-terrace. The channel is sinuous and quite narrow. It ranges in width from 7 m to 17 m. Where the channel widens it often consists of a point bar about 20 cm above the incised channel, which can be as narrow as 4.6 m. On the concave side of the point bar chutes cut across the bar. The walls of the wadi are destroyed by undercutting and mass-wasting. The floor of the wadi channel is covered with sand, whereas the point bars are strewn with flat angular cobbles.

### Aeolian Sand

Wind-blown sand consists mostly of drift sand that consists generally of a thin cover with a maximum thickness of 30 cm. The sand is medium grained and moderately sorted. It has accumulated in the wadi floor, on the top of terraces, and in places drapes over the middle and the top segments of the fans. Hummock fields have also formed where sand has built up around bushes. The sand, for the most part, has been reworked from the top sands of alluvial fans and from sands of an earlier episode of mechanical weathering. These sources have produced an abundant supply of fresh sand. A slightly moist and warm, yet still desertic, climate seems to have succeeded the episode of weathering. That episode could have been characterized by a dry, cool climate. The Ordovician and Lower Cretaceous sandstones provide an ample supply of sand. The strata are jointed and the sandstone is semi-consolidate and crumbles easily.

## SEQUENCE OF GEOLOGIC EVENTS

Following the formation of the Wadi Judayid depression, alluviation by tributaries draining the adjoining catchment basin led to the formation of a valley fill of coalesced alluvial fans. The Levantine Mousterian site of Tor Sabiha (J8), dated to ca. 70,000 B.P. and found in an overlying dune unit, may represent the period following the major episode of alluviation that led to the formation of the lower member of the alluvial fan. The formation of the depression might thus have taken place in the early part of the Late Pleistocene or earlier. The early fan deposits of hard, alluvial sand are cemented by carbonates and include zones rich in carbonate nodules. The deposits seem to indicate a subarid environment, moist and warm with rainfall perhaps in the magnitude of 300 mm. The development of a reddish yellow color is suggestive of the warm, moist conditions that were sufficient to mobilize iron oxides. Seasonal dryness seems also to be indicated by the precipitation of carbonates, which were presumably introduced by solution during the rainy season. The vegetation was perhaps represented by steppe with wooded areas in places, especially at high elevation.

Downcutting and accumulation of dunes from freshly weathered sandstone under dry, cool conditions ensued. A relict of these dune sands is preserved in an embayment in the Jebel el Jill. This deposit contains the Levantine Mousterian occupation of Tor Sabiha (J8) in association with faunal and pollen evidence pointing to steppic conditions. Amino acid racemization dates of ostrich eggshell place the occupation at ca. 70,000 B.P. (Henry and Miller 1993).

Alluviation resumed under conditions of more open vegetation, warm and somewhat moist conditions; rainfall was perhaps in the range of 250 mm. In the lower parts of the depression, finer sediments accumulated and thin crusts of calcium carbonates were developed. The lack of carbonate concretions and the presence of thin crusts suggest a drier climate than that which prevailed during the alluviation of the lower unit of the fan deposits. The climate was probably similar to that under which Yellow steppe soils are formed. Frequent cut-and-fill structure, rapid lateral and vertical changes in facies, and frequent occurrence of large stone blocks and boulders suggest occasional torrential rainfall. Middle Hamran (Geometric Kebaran Complex) sites (J26 and J31) are located *in situ* in the top part of the upper unit of the fan thus indicating that alluviation lasted until ca. 14,000–13,000 B.P.

The geologic history of the area since that time seems to have been predominantly erosional, with the exceptional episode of wadi alluviation. A series of cut-terraces were established with the lowest visible terrace at 0.5 m above the modern wadi floor. The erosional processes seem to have taken place under dry climatic conditions, a thin vegetation cover, and occasional rains feeding streams of high velocity. Following the formation of the highest terrace (Terrace I), an early Natufian settlement, the site of Wadi Judayid (J2), was established. This provides a limiting date for the termination of the episode of erosion. Radiocarbon dates fix the occupation to ca. 12,500 B.P. The erosional period then would appear to have lasted from as early as 14,000 B.P. to as late as 12,500 B.P.

Cooler, drier conditions followed, leading to intensive mechanical weathering and aeolian activity. This interval is likely to have corresponded to the late Natufian–early Neolithic timeframes (i.e., 11,000–9,000 B.P.). Dune sands were formed covering the artifact horizons at the site of Wadi Judayid and the early Neolithic occupation at Jebel Queisa (J24) is also found in drift sand. A subsequent episode of warmer and somewhat more moist conditions led to the liberation of iron oxides. Reddish sands were reworked into drift sand on top of earlier deposits. The wadi aggradation may date to that period. By correlation with other areas in the Near East and North Africa it is possible that this moist phase dates back to ca. 9,000–7,000 B.P., correlated with the Pre-pottery Neolithic B moist phase.

Drier and cooler conditions appear to have characterized later time periods with most recent intensification of aridity beginning ca. 1890 A.D. (Burdon 1959:Fig. 7) and this seems to have been culminating in the last few years according to observations by local informants. Springs near the Natufian site of Wadi Judayid which were flowing in the recent past (about 60 years ago) are now dried up, for example.

## GEOARCHEOLOGICAL IMPLICATIONS

The choice of the settlement location with the respect to the geomorphological and other geologic aspects of an area is the subject matter of the geokistics (Hassan 1979). In the Ras en Naqb area, the geological factors which could have influenced settlements include: (1) availability of water (hydrogeology), (2) natural shelters (not very critical) and protected slopes, (3) well-drained dry, flat surfaces, (4) proximity to geomorphological units where natural exploitable resources are located (e.g., areas visited by game animals, chert, arable land, etc.).

Water in the Ras en Naqb area today is available from seasonal rainfall and springs. Depending upon elevation, mean rainfall ranges from 150 to 300 mm annually (Beheiri 1973; El-Eisawi 1985). Surface water mostly is confined to wadi beds and pools where it occurs as run-off from the uplands during the winter, wet-season. Aquifers in the area include the Paleozoic-Cretaceous sandstones (Bender 1974:182–183). Small amounts of water may also be obtained from springs at the base of the alluvial fans.

Rockshelters develop in the area as a result of the erosion of the bedrock by fluvitile processes. Karsitic phenomena do not seem to be pronounced in the Cenomanian-Turonian limestones, in part as a result of the predominance of fluvitile processes and the relatively low rainfall during most of the Pleistocene.

Since the dominant wind directions during the warm season are from the west and the southwest, east- and northeast-facing slopes would have been preferred for camp locations. During the winter, with predominantly northern and northwestern winds, southern and western exposures would have provided maximum wind protection and available sunlight.

With their well-drained surfaces and commanding views, the higher slopes of the alluvial fans provide good settings for habitations. However, over the dry season ground moisture would not have been a factor influencing site placement. During alluviation of the fans, the central part of the basin would have contained surface water and higher ground moisture thus creating an excellent habitat for vegetation and game animals. Accordingly, settlements may have been established at a distance to avoid alarming game animals.

The surface of the lower member of the alluvial fans is the most suitable for agricultural activities. It provides a loamy sand soil with a moderate content of carbonates and it lacks a pronounced content of stones. The residual soils on the pediments are also suitable for cultivation, but stones, which are quite abundant, must be cleared.

Beyond water and food stuffs, chert was probably the most important natural resource used on a daily basis by the prehistoric inhabitants of the area. Chert, occurring in the Cenomanian, Turonian, and Campanian-Maastrichtian limestone formations of the Ma'an Plateau, is confined mainly to elevations above 1,400 masl. Chert derived from *in situ* sources also appears in the lag deposits on the pediments of the escarpment and in stream cobbles found in the beds of wadis.

## Geomorphology and Site Preservation

Sites older than the Middle Paleolithic are not likely to be located in the basin since the excavation of the depression seems to date to the Upper Pleistocene. Older sites on the Ma'an Plateau or the pediments would have been subjected to deflation, although in places sites could be preserved under the protective armor of desert lag.

Middle Paleolithic sites are likely to have been removed from the central part of the basin on account of the terminal Pleistocene and Holocene erosional processes. Upper Paleolithic settlements could have been established from the time when the upper part of Member I of the alluvial fans was deposited to the time when the lower part of Member II was formed. Sites dating to the time of Member I would have been subjected to erosion in the central part of the basin during the erosional interval separating the two episodes of basin fill. Sites may be found deflating from that surface. The absence of Upper Paleolithic sites, most likely coincident with the deposition of the lower part of the Member II fill, may be explained by the removal of this fill over most of the central basin coupled with its deep burial within the upper part of the fan deposits.

Early Epipaleolithic sites, located in the upper part of Member II of the fan deposits, also would have been removed from the central part of the basin as a result of the Holocene erosional episodes. Late Epipalaeolithic (e.g., Natufian) sites established following the leveling of the fan deposits to about 10 m above the modern wadi floor will be preserved on the surface of this high terrace where it has not been subjected to downcutting in later Holocene erosional episodes. In places, as in parts of the Early Natufian site of Wadi Judayid (J2), drift sand may have accumulated and buried the surface of the terrace under thick sand deposits.

Early to mid-Holocene archaeological sites are likely to be preserved unless they were established on the narrow strip now occupied by the wadi where downcutting and later planation have taken place. Higher on the fans, sites of this age may well be found buried within reworked drift sand. Early Neolithic and Chalcolithic occupations at the site of Jebel Queisa (J24), found within a deposit of over 3m of drift sand, represent such a setting.

# Chapter 3

# Cultural-Historic Framework

A brief overview of the cultural-historic framework by which the results of the investigation were organized is presented at this point for two reasons. First, in following a hierarchical or taxonomic structure that departs somewhat from traditionally held schemes for the Levant, some discussion of unit building seems warranted. But perhaps more importantly from a practical perspective, a concise description of the cultural-historic relationships of the assemblages and the nature of the data-sets upon which these relationships were determined seems especially important to understanding a framework spanning six major prehistoric periods and much of the Upper Pleistocene.

## TAXONOMIC STRUCTURE AND CULTURAL-HISTORIC INTEGRATION

With the extension of research into the arid zone of the Sinai and the Negev during the 1970s, nearly all of the prehistoric periods of the Levant underwent major revisions of their cultural-historic frameworks. The discovery of new taxa (e.g., Ahmarian and Mushabian complexes, Harifian Industry) and lateral facies of taxa known previously only from the Mediterranean woodland zone (e.g., Levantine Aurignacian, Geometric Kebaran) prompted the abandonment of traditionally held unilineal schemes (Garrod 1954; Garrod and Bate 1937; Neuville 1934, 1951; Perrot 1968). The restructured frameworks reflected a dichotomy in taxa for each major prehistoric period stretching from the Middle Paleolithic through the Chalcolithic. The distributional boundaries of these "parallel phyla" essentially were found to correspond to the boundaries of the wooded and the arid zones of the Levant. The taxonomic revisions also prompted the adoption of a new paradigm tied to explaining material culture variability. In this, synchronic functionalist interpretations of patterned variation in artifact assemblages emerged as alternatives to evolutionary and ethnic interpretations.

The revision of traditionally held taxonomic schemes was accompanied by some rather fundamental theoretical changes in the ways Levantine prehistorians explained variability in the archaeological record. Traditionally, differences between assemblages primarily were explained as resulting from "change through time." This diachronic or evolutionary interpretation of material culture variability placed a great emphasis upon establishing archaeological sequences as mirrored in field and analytic methodologies. Deeply stratified caves and tells with their long prehistoric records were sought out for intensive excavation. Because of the physiography of the region these sites were situated for the most part in the wooded Mediterranean zone. Coupled with the field searches for deposits containing long prehistoric sequences, analytic programs concentrated upon discovering *fossile directeurs* and other time-sensitive attributes.

The spread of research into the arid zone was accompanied by theoretical and methodological shifts from traditional diachronically oriented studies to ones giving greater attention to the roles of synchronous functional forces in shaping material culture inventories. Areal research programs, stressing the recovery of artifact samples from numerous sites, emerged with an objective of identifying variability in assemblage/site types. The inter-and intra-site comparative studies that were common to these programs contained a strong functionalist bias as part of their operating paradigm. Differences in prehistoric activities within and between sites and regions were viewed as potential sources of artifact assemblage variability. At this time, there appears to have been a decided methodological shift from "vertically oriented" to "horizontally oriented" archaeology. Concerns with populations, sampling, ecology, and systemic relationships became common to most research programs. In many ways this change was facilitated if not prompted by adoption of some of the tenets of New Archaeology and by the availability of radiocarbon dating.

The widespread use of radiocarbon dating not only enabled areally oriented researchers to establish temporal controls in areas lacking deeply stratified sites, but it also precipitated the discovery of parallel phyla. With the recognition of synchronous taxa, Levantine prehistorians have come to adopt explanations based either on functional or ethnic differences. In many cases these are not clear-cut and most scholars entertain alternative explanations situationally. Certainly, the linkage between personality and theoretical stance that epitomized the Bordes–Binford debate on precisely this issue has not developed in the Levant. But neither have many solutions to the problem of precisely identifying factors responsible for material culture variability. Behavioral differences associated with differences in time, ethnicity, and activity will not be easily or accurately correlated with patterned variability in material culture until some fundamental changes are made in the ways we go about cultural-historic integration. Perhaps the two greatest obstacles encountered in meshing material and behavioral cultures rest in: (a) not distinguishing between the different fields of data that go into the construction of classificatory units that in turn are integrated for building cultural-histories and (b) not recognizing different scales of classification in the formation of these units.

## FIELDS OF CLASSIFICATION

Cultural historians as well as processualists organize their evidence drawn from three large fields of data into material culture units, temporal-spatial units, and ultimately socioeconomic units. The synthesis and integration of these units is prerequisite to constructing cultural histories and ultimately identifying cultural processes, yet the initial use of independent classificatory systems for each of these fields of data is needed to avoid circular reasoning and terminological ambiguities. Within Levantine prehistory an informal classification has been developed for material culture and temporal-spatial phenomena, yet a corresponding classification of evidence from the related socioeconomic field is limited to the terminal Pleistocene (Henry 1989a). Such terms as "industry," "complex," "culture," "population," "groups," "people," "economy," and "times" are often used interchangeably, with no real attempt to distinguish between material culture, temporal-spatial, and socioeconomic fields.

The classificatory scheme presented here follows an approach used in an earlier study confined to the late Epipaleolithic (Henry 1989a). The scheme is essentially based upon a classificatory hierarchy developed by Clarke (1968), but with modification. Clarke's approach to organizing archaeological phenomena into hierarchical scales of socioeconomic units (i.e., culture, culture group, and technocomplex) implicitly incorporated environmental, economic, demographic, temporal, and spatial variables (Clarke 1968). While he described in great detail the procedures by which patterned variation in artifacts might be used to define each of the scales, he was much less specific in how patterned variation in the other lines of evidence (i.e., environment, economy, demography) would be used.

I have suggested two modifications to Clarke's approach. First, discrete classificatory schemes should be used for the material culture and the prehistoric socioeconomic data (Table 3.1). This insures that the definitions of different hierarchical scales of material culture are based only upon similarities and differences in artifact inventories and not upon *a priori* assumptions about the correlations that existed between prehistoric behavior and its material residue. Second, specific environmental, economic, demographic, spatial, and temporal evidence should be used in conjunction with the already developed material culture classification for the definition of the socioeconomic scales (i.e., culture, culture group, and technocomplex). By explicitly incorporating these lines of evidence into the definition of prehistoric social scales, a realistic correspondence is maintained between the detail of the scale and the resolution of our socioeconomic inferences. Archaeologists are often prone to ignore such ancillary data and uncritically use the term "culture" for almost all material culture scales regardless of their time and space dimensions (Clarke 1979:164).

## SCALES OF CLASSIFICATION

Lithic assemblages, common to all prehistoric sites in the Levant, represent by far the largest artifact category of those periods spanning the Lower Paleolithic through the Chalcolithic. For this reason, chipped stone artifacts form the basis of material culture classification and archaeological unit construction for the prehistoric Levant. In this study, as elsewhere (Henry 1986, 1988a, 1988b, 1989a), a three-scale taxonomy is used to organize the assemblages and identify their varying degrees of affinity (Figure 3.1). The scales of *complex*, *industry*, and *phase* or *facies* furnish a hierarchy that defines increasing affinities between lithic assemblages as determined by the increasing number and specificity of attributes and criteria used in the identification

**Table 3.1. Outline of the Sequence Used in Classifying the Material Culture, Temporal-Spatial, and Social Phenomena that were Associated with the Archaeological Sites under Study**

| | Sequence of classification | | |
|---|---|---|---|
| | Step 1 | Step 2 | Step 3 |
| Field | Classification of material culture based on affinities of artifact assemblages | Ordering of material culture units into time—space scales | Classification of social field based on material culture and ancillary data |
| Evidence drawn from | Material culture data field (e.g., lithic artifacts) | Stratigraphic, chronometric, archaeostratigraphic, and biostratigraphic units | Environmental, economic, demographic, material culture units |
| Scales of classification | Into scales of increasing specificity: Complex industry phase/facies | Into decreasing time-space scales | Into scales of techno-complex, culture group, culture |

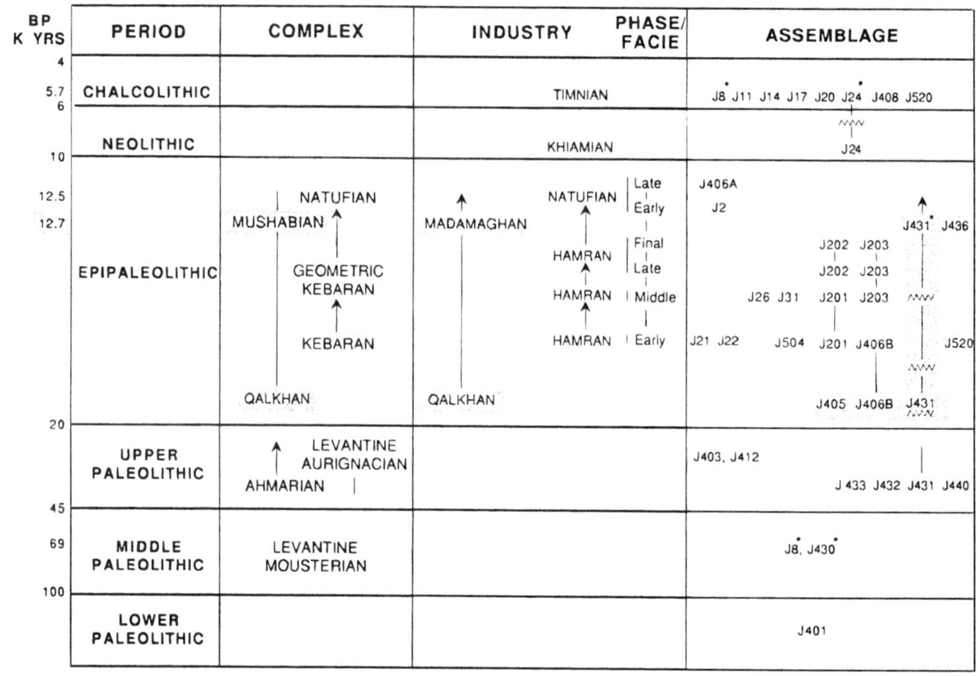


**Figure 3.1.** Hierarchical archaeological unit classification of the major assemblages investigated within the study. Note that for the Epipaleolithic "parallel phyla" are indicated by the Hamran-Natufian and Qalkhan-Madamaghan successions.

of taxa within these scales. Although not formally considered in the classification, the temporal and spatial scopes of the various scales are inversely related to the degree of affinity displayed by taxa belonging to those scales (Figure 3.1). That is, increasing specificity of taxa is accompanied by a reduction in the temporal and spatial scope of those assemblages forming the taxa.

Although an artifact *assemblage* is not a scale of classification, it is the primary building block of the system of classification and the only unit common to the material culture, temporal-spatial, and socioeconomic fields. In order for an assemblage to serve as a reliable means of correlating these fields it should consist of some part of the material culture residue of a spatially limited and temporally brief occupation by a group representing a discrete culture. The largest and most loosely bound scale in the classification is that of *complex*. Lithic artifact assemblages forming taxa at the level of complex have a high level of general technological affinity and a low level of typological affinity as determined by their sharing of a polythetic set of attributes related to blank production and tool manufacture. An *industry* represents an intermediate scale in the classificatory hierarchy. Assemblages classified to this scale have a high level of affinity in sharing specific technological attributes in conjunction with typological attributes at the class and type levels. The lowest and most internally homogeneous scale is the *phase/ facies*. Assemblages grouped into taxa at this scale share several specific technological, typological, and stylistic attributes.

The material culture classification as employed here not only allows for defining the different degrees of affinity of assemblages through their inclusion into taxa at different scales, but it also enables realistic correlations to be made between material culture and socioeconomic units of classification (Figure 3.1).

## CHARACTERISTICS OF TAXA

Descriptions of diagnostic technological and typological lithic attributes are presented for the lowest-order taxon of each of the major prehistoric periods. Ancillary information related to chronological placement, artifact samples, site dimensions and distributions, and associated evidence is also presented.

### Lower Paleolithic–Late Acheulean

*Site*: Wadi Qalkha (J401), surface assemblage.

*Technotypologic Characteristics*: Cordiform, amygdaloid, and lanceolate handaxes with regular edges fashioned from large flakes dominate the assemblage. Presence of Levallois flakes.

*Chronometry*: None.

*Stratigraphy*: Artifacts are eroding from near the top of a 30 m section of alluvial silts and sands (Qalkha Formation) that overlie a congeliofractate bed of indeterminate thickness.

*Area of Site*: Indeterminate.

*Environmental/Economic Evidence*: None.

*Artifact Sample Sizes*: 66 (27 tools).

*Site Setting*: Site is in the open located at the edge of the piedmont in a deeply incised wadi that forms an extension badland topography.

*Features and Other Artifacts*: None.

*Taxa and Assemblages with Closest Affinities*: Late Acheulean—Fajaje, Kissufim, Rephaim-Baqa, Ain el-Assad.

*Socioeconomic Scale*: Undefined.

### Middle Paleolithic–Levantine Mousterian–"D" Type Late

*Sites*: Tor Faraj C (J430) and Tor Sabiha C are both rockshelters.

*Technotypologic Characteristics*: Levallois points (40–67%) and tools on blades dominate the assemblages. Levallois flake production is very low. Inverse retouch is common (ca. 50%) for both assemblages.

*Chronometry*: Amino acid racemization of ostrich eggshell furnished dates of 69,500 ± 5000 B.P. for both sites. Uranium series dates for Tor Faraj (28,900 ± 3,800 B.P. and 62,400 ± 14,000 B.P.) and Tor Sabiha (31,500 ± 5,000 B.P.) are problematic.

*Stratigraphy*: Tor Faraj contains a single component occupation, whereas Tor Sabiha's Mousterian horizon (Layer C) underlies a thin Chalcolithic layer.

*Areas of Sites*: Tor Faraj—200/250 m; Tor Sabiha—100/150 m.

*Environmental/Economic Evidence*: Pollen, fauna, phytolith, sediment analyses were conducted.

*Artifact Sample Sizes*: Tor Faraj—3,564; Tor Sabiha—6,663.

*Site Settings*: Tor Faraj (900 masl) is located at the lower edge of piedmont, whereas Tor Sabiha (1,300 masl) is situated at the upper edge of piedmont.

*Features and Other Artifacts*: Unlined hearths and hammerstones were found at Tor Faraj.

*Taxa and Assemblages with Closest Affinities*: General technological similarities to Levantine Mousterian "D" Type (Rosh Ein Mor, Nahal Aqev); typological similarities to "B" Type assemblages (Kebara IX–XII).

*Socioeconomic Scale*: Technocomplex.

### Upper Paleolithic–Levantine Aurignacian and Ahmarian

*Site*: Levantine Aurignacian—Tor Fawaz (J403) and Jebel Humeima (J412); Ahmarian—Tor Hamar (J431) and Tor Aeid (J432); other Upper Paleolithic assemblages—J433 and J440.

*Technotypologic Characteristics*: Levantine Aurignacian assemblages are dominated by endscrapers and burins on thick blades. Ahmarian assemblages also primarily consist of endscrapers and burins, but El Wad points and a substantial bladelet component distinguish these assemblages from those of the Levantine Aurignacian.

*Chronometry*: None.

*Stratigraphy*: Lower piedmont sites rest in a pink silt underlying early Epipaleolithic age red sands. The Ahmarian horizon at J431 underlies a Qalkhan layer.

*Area of Sites*: Range 156–1,054 $m^2$. Mean 615 $m^2$.

*Environmental/Economic Evidence*: Pollen, fauna, phytolith, and sediment analyses were undertaken.

*Artifact Sample Sizes*: Range 3,983–8,850. Mean 5,507.

*Site Setting*: Sites appear on terraces of rockshelters in the mid- and lower piedmont.

*Features and Other Artifacts*: None.

*Taxa and Assemblages with Closest Affinities*: Levantine Aurignacian—Sde Divshon (Ferring 1976) Arkov (Marks and Ferring 1976); Ahmarian—Lagaman sites (Gilead 1983, 1991) and Abu Noshra sites (Phillips 1988).

*Socioeconomic Scale*: Technocomplexes.

## Epipaleolithic–Qalkhan

*Sites*: J405, Wadi Humeima (J406b), J407, Tor Hamar layer (J431).

*Technotypologic Characteristics*: Although the technology is devoted to the production of relatively long, narrow bladelets, many tools occur on blades. A unique triangular point is diagnostic of the complex as is a consistent use of the microburin technique.

*Chronometry*: None.

*Stratigraphy*: Qalkhan assemblages appear in red sands overlying pinkish silts containing Upper Paleolithic horizons in the Jebel Qalkha area. At J406b Qalkhan artifacts occur within a 50 cm thick deposit underlying a 60 cm thick Kebaran (Early Hamran) horizon. At J431 Qalkhan artifacts are stratified between Upper Paleolithic (Ahmarian) and Mushabian (Madamaghan) horizons.

*Areas of Sites*: Range 50–200 $m^2$. Mean 120 $m^2$.

*Environmental/Economic Evidence*: Pollen and phytoliths.

*Artifact Sample Sizes*: Range 436–5,390. Mean 2,622.

*Site Setting*: Sites occur in both open and sheltered settings of the piedmont.

*Features and Other Artifacts*: None.

*Taxa and Assemblages with Closest Affinities*: Sabra 3 and Adh Daman near Petra (Schyle and Uerpmann 1988), Uwaynid 14, Uwaynid 18, and Jilat 6 of the Azraq Basin of northeastern Jordan (Garrard *et al.* 1985, 1986, 1987; Garrard and Gebel 1988; Byrd 1990; Garrard and Byrd 1992), the Jabrud III/layers 4–7 rockshelter of western Syria (Rust 1950) and sites in the El Kowm Oasis in northeastern Syria (Cauvin *et al.* 1979; Cauvin 1981; Cauvin and Coqueugniot 1990).

*Socioeconomic Scale*: Technocomplex.

## Epipaleolithic–Kebaran–Early Hamran

*Sites*: J21, J22, Jebel Hamra (J201), Wadi Humeima (J406b), Jebel Mishraq (J504), and Jebel Muheimi (J520).

*Technotypologic Characteristics*: The production of relatively broad bladelets used in the manufacture of backed non-geometric microliths (especially straight backed forms truncated opposite snaps) defines the industry. The narrow micropoints that typify other Kebaran industries (Bar-Yosef 1981) are absent or rare.

*Chronometry*: Radiocarbon date on hearth charcoal from J504 of 11,985 ± 110 B.P. is obviously too young for a non-geometric assemblage.

*Stratigraphy*: Early Hamran horizons stratigraphically underlie Middle Hamran horizons at J26, J201, and J504. At J406b an Early Hamran horizon was found to overlie a Qalkhan horizon.

*Areas of Sites*: Range 260–600 $m^2$. Mean 430 $m^2$.

*Environmental/Economic Evidence*: Pollen.

*Artifact Sample Sizes*: Range 357–3,601. Mean 1,963.

*Site Setting*: Sites are found in both open and sheltered settings in the lowlands and piedmont.

*Features and Other Artifacts*: Hearths, bedrock and mobile mortars, possible petroglyphs.

*Taxa and Assemblages with Closest Affinities*: Bar-Yosef's (1981) "Cluster D" (e.g., Ein Gev I–II, Hadera V, Fazael III,4) dominated by obliquely truncated backed bladelets shows the greatest similarities to the Early Hamran.

*Socioeconomic Scale*: Culture Group.

## Epipaleolithic–Geometric Kebaran–Middle Hamran

*Sites*: J26, J31, Jebel Hamra (J201), El Quweira (J203), J437 (not investigated), Jebel Mueisi (J504) surface only.

*Technotypologic Characteristics*: Assemblages are typified by the production of relatively narrow bladelets from which trapeze/rectangles were manufactured without the use of the microburin technique.

*Chronometry*: None.

*Stratigraphy*: Middle Hamaran assemblages overlie Early Hamran horizons at J504 and J201, and underlie a Late Hamran horizon at J203.

*Area of Site*: Range 320–650 $m^2$. Mean 485 $m^2$.

*Environmental/Economic Evidence*: Pollen.

*Artifact Sample Sizes*: Range 445–2,300. Mean 1,226.

*Site Setting*: Sites appear in both open and sheltered settings throughout lowlands and piedmont.

*Features and Other Artifacts*: Hearths, basalt pestle, ornamental shells.

*Taxa and Assemblages with Closest Affinities*: Middle Hamran assemblages resemble those included in Bar-Yosef's (1981) Group II (such as Mushabi XVIII, Lagama North VIII, Nahal Lavan VI), but bladelet and trapeze/rectangle dimensions of the Middle Hamran are narrower, falling between Group I and II.

*Socioeconomic Scale*: Culture Group.

## Epipaleolithic–Geometric Kebaran–Late and Final Hamran

*Sites*: Qa Salab (J202) and El Quweira (J203).

*Technotypologic Characteristics*: Late and Final Hamran assemblages differ from those of the Middle Hamran in the production of much shorter bladelets accompanied by the intentional use of the microburin technique and the appearance of lunates. The Final Hamran is distinguished from the Late Hamran by the replacement of trapeze/rectangles by lunates, a marked increase in the microburin technique, and the presence of some Helwan retouch.

*Chronometry*: Accelerator radiocarbon date of 5,810 ± 220 B.P. (V-1905,AA-1023) on charcoal from hearth at J202 is obviously too young.

*Stratigraphy*: Late and Final Hamran assemblages are stratified at J202 and J203, overlying a Middle Hamaran assemblage at J203.

*Area of Site*: 390 $m^2$.

*Environmental/Economic Evidence*: Pollen and fauna.

*Artifact Sample Sizes*: Range 3,354–8,213 Mean 4,750.
*Site Setting*: Sites appear in sheltered settings in the lowlands.
*Features and Other Artifacts*: Stone-lined hearths, ornamental shells.
*Similar Assemblages*: None known.
*Socioeconomic Scale*: Culture Group.

## Epipaleolithic–Natufian–Early and Late Natufian

*Sites*: Early—Wadi Judayid (J2); Late—Wadi Humeima (J406a).

*Technotypologic Characteristics*: The J2 assemblage is characterized by the production of relatively wide bladelets that were subsequently sectioned by the microburin technique and then fashioned into predominantly Helwan lunates. The J406a assemblage contains significantly shorter bladelets that were sectioned by the microburin technique and then formed into minute normally backed lunates.

*Chronometry*: Radiocarbon assays of charcoal from layer C of J2 yielded dates of 12,461 ± 481 B.P. (SMU-803), 12,435 ± 646 B.P. (SMU-805), and 13,181 ± 816 B.P. (SMU-806).
*Areas of Sites*: J2-530 m² (undetermined part of site covered by drift sand) J406a-400 m².
*Environmental/Economic Evidence*: Pollen, fauna, phytoliths.
*Artifact Sample Sizes*: J2—35,148; J406a—2,042.
*Site Settings*: Sites appear in open settings in the piedmont.
*Features and Other Artifacts*: Bedrock mortar, ornamental shells.
*Taxa and Assemblages With Closest Affinities*: Early—Beidha, Late—Rosh Zin.
*Socioeconomic Scale*: Culture.

## Epipaleolithic–Mushabian–Madamaghan.

*Sites*: Tor Hamar (J431), Jebel Fatma (J436).

*Technotypologic Characteristics*: Technologically, the assemblages are characterized by relatively long, wide bladelets and high frequencies of microburins. The typological profile includes moderate percentages of points (La Mouillah, microgravette, and arched backed varieties) and low percentages of geometric microliths consisting of trapeze rectangles, normal and Helwan lunates.

*Chronometry*: Two radiocarbon assays on charcoal yielded dates of 12,683 ± 320 B.P. (SMU-1399) and 12,320 ± 95 (ETH-806) at J431.
*Stratigraphy*: Mushabian horizon overlies a Qalkhan horizon at J431.
*Areas of Sites*: Range—214–300 m². Mean—257 m².
*Environmental/Economic Evidence*: Pollen, fauna, phytoliths.
*Artifact Sample Sizes*: Range—2,451–6,333. Mean—4,260.
*Site Setting*: Both appear as sheltered sites in the piedmont.
*Features and Other Artifacts*: Mobile mortars, bone points, ochre, ornamental shells.

*Taxa and Assemblages With Closest Affinities*: Wadi Madamagh (Kirkbride 1958); Wadi el-Jilat 8 (Site E2 of Waechter *et al.* 1938) (Garrard *et al.* 1985b, 1986, 1988) near the Azraq Basin of eastern Jordan; Sinai Mushabian sites—Mushabi V, Mushabi XIV, Nahal Hadera I (Phillips and Mintz 1977); Negev Mushabian (Ramonian) sites—G9, G14, K5, K6, K7 (Marks and Simmons 1977; Goring-Morris 1987).
*Socioeconomic Scale*: Culture Group.

## Neolithic–Early Neolithic (Khiamian?)

*Site*: Jebel Queisa (J24).

*Technotypologic Characteristics*: The assemblage displays long, relatively thin bladelets struck from opposed platform, naviform-like cores. Diagnostic tools occur as Helwan points and a Netiv Hagdud truncation.

*Chronometry*: None.

*Stratigraphy*: Neolithic horizon at J24 underlies 1.5–2 m of Chalcolithic deposit.

*Area of Site*: Undetermined, but likely under 100 $m^2$.

*Environmental/Economic Evidence*: None.

*Artifact Sample Size*: 3,379.

*Site Setting*: Occupation rests on the terrace of a rockshelter in the piedmont.

*Features and Other Artifacts*: None.

*Taxa and Assemblages With Closest Affinities*: Mureybet (Upper Layer II, Lower III); Nahal Oren; Nahal Lavan.

*Socioeconomic Scale*: Technocomplex.

## Chalcolithic–Timnian

*Sites*: J11, Jebel el Jill (J14), J17, J20, Jebel Queisa (J24), J408, J521.

*Technotypologic Characteristics*: Technologically, the assemblages are characterized by the production of microlithic flakes and short blades struck from small, globular (often bipolar) cores. Large amounts of angular debris, thick platforms, high platform to inverse surface angles, and pronounced bulbs of percussion indicate an emphasis upon hard-hammer percussion. Diagnostic tools include tabular scrapers, minute lunates, transverse arrowheads, and small bifacial points.

*Chronometry*: A radiocarbon assay on charcoal from J24 yielded a date of 5,720 ± 149 B.P. (SMU-804).

*Stratigraphy*: A Timnian horizon overlies an Early Neolithic (Khiamian?) horizon at J24.

*Areas of Sites*: Range—115–8,000 $m^2$ Mean—2,317 $m^2$.

*Environmental/Economic Evidence*: Pollen and fauna.

*Artifact Sample Sizes*: Range—264–4,601. Mean—1,665.

*Site Setting*: Both sheltered and open sites appear throughout the lowlands and piedmont.

*Features and Other Artifacts*: Stone walled corrals, stone lined pithouses, querns, handstones, worked bone, pottery (thick, buff-tan plainware with sand, shale, chert tempers).

*Taxa and Assemblages With Closest Affinities*: Timnian Industry of northeastern Sinai (Kozloff 1974, 1981).

*Socioeconomic Scale*: Culture Group.

# SUMMARY

Excavations at thirty prehistoric sites yielded 37 assemblages representing all of the major prehistoric periods stretching from the Lower Paleolithic through Chalcolithic. These assemblages were classified as belonging to eleven complexes and eight industries. Phase designations were possible only for those assemblages of the Hamran and Natufian industries.

Socioeconomic correlations were made for all but the Late Acheulean assemblages. The Levantine Mousterian, Ahmarian, Levantine Aurignacian, Qalkhan, and Khiamian assemblages were assigned to the scale of technocomplex. Hamran (early—final), Madamaghan, and Timnian assemblages were correlated with culture groups; and the Natufian assemblages were linked to the scale of culture.

*Chapter* 4

# *The Lower Paleolithic Site of Wadi Qalkha*

## SETTING

Although isolated handaxes were found in the uplands overlooking the Wadi Judayid basin and within the deeply incised wadis north of Jebel Qalkha, the site of Wadi Qalkha (J401) represents the only discovery of concentrated Lower Paleolithic artifacts. The site is actually located in a deeply incised wadi near the extreme headwaters of the Wadi Aheimir and close to the point of an impending drainage capture of the Wadi Qalkha (Figure 4.1). During survey a light scatter of artifacts was found in the wadi bed and on the steep slopes of the east bank distributed over an area of ca. 2,300 $m^2$. Given the steepness of the wadi banks and the low density of artifacts, this area of artifact coverage seems likely to be greatly exaggerated over the original occupation area, however.

Artifacts were found eroding from near the top of a ca. 30 m section of alluvial silts and sands forming the Qalkha formation which overlies a congeliofractate bed (calcareous conglomerate with frost shattered flint inclusions) of unknown thickness. Hassan (Henry *et al.* 1983) places this lowest alluvial fill in the last interglacial. Although most of the artifacts were collected from the surface, their fresh edges and light patination suggest recent exposure and limited transport from their primary context. A few specimens that were partially buried exhibit patination only on that portion of the artifact that was exposed.

## ARTIFACT ASSEMBLAGE

A systematic collection of the area resulted in the recovery of only 59 artifacts, represented by 27 tools, 21 pieces of debitage, and 12 pieces of debris. Small artifact samples present a common problem to Lower Paleolithic research and this seems especially true of the southern Levant (Goren 1981; Goren-Inbar 1990c). All of the specimens display light patina over, at least, a portion of their surfaces, but only one specimen shows heavy abrasion. In following the designations of the *Munsell Soil Color Chart*, the patina ranges from light gray (N 7/0) to pale yellow (2.5Y 7/4).

The paucity of small flakes and chips in the assemblage is probably an expression of the steepness of the slopes beneath the occupation level and the high velocity of sheet wash during times of rainfall. Once exposed the small, light artifacts would have been washed rapidly downslope to the wadi bed and removed from the site.

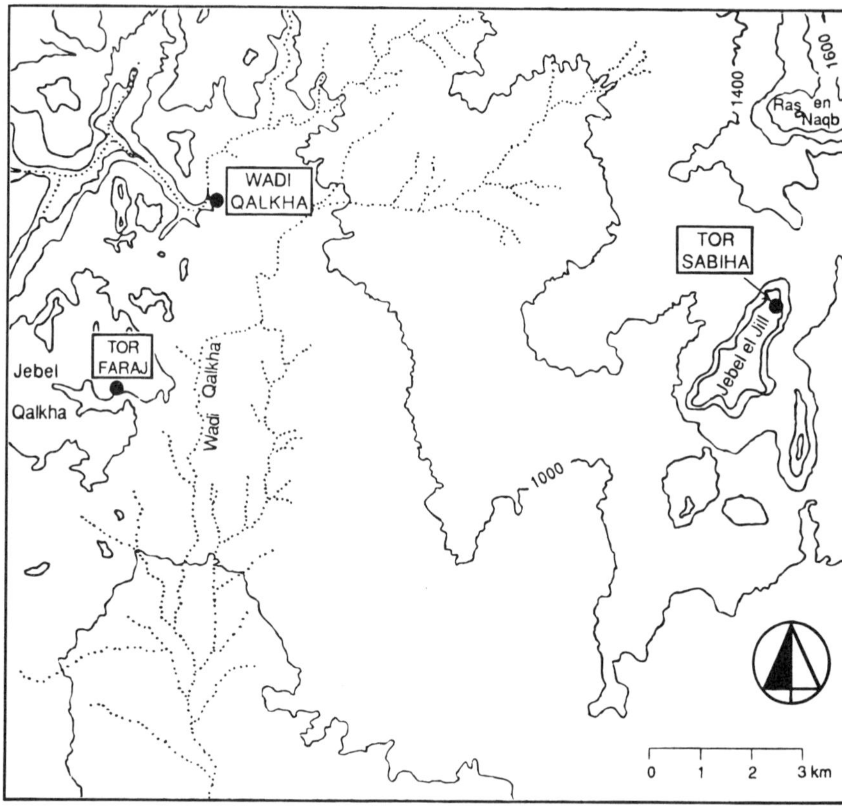


Figure 4.1. Map of the Jebel Qalkha and Judayid Basin study areas showing the locations of the Late Acheulean site of Wadi Qalkha (J401) and the Levantine Mousterian sites of Tor Faraj (J430) and Tor Sabiha (J8).

## Tool Assemblage

The tools are dominated by bifaces, followed by lesser numbers of retouched pieces, sidescrapers, truncations, and notches (Table 4.1).

The *bifaces* fall into two distinct size ranges with the five specimens of the large group (+100 mm in length) representing cardiform and amygdaloid forms (Figures 4.2 and 4.3). The small group includes ovate, amygdaloid, broken, and irregular forms (Figure 4.3). Four of the six large specimens exhibit large, unfacetted butts, indicative of their having been manufactured from large flakes. Most of the specimens have rather uniform, nonsinuous edges formed by fine marginal retouch. Interestingly, only two of the bifaces display cortex and these are from the small group. All but one of the specimens are made on chert and it occurs on a fine red basalt not found in the area.

The *sidescrapers* are roughly balanced between straight, convex, and dejete forms with all but one bearing obverse retouch. The retouch is characteristically scaliform (Figure 4.4). With the exception of one specimen fashioned from a primary element, all are made on flakes exhibiting large, unfacetted platforms; only two facetted platforms were recorded.

The *retouched pieces* were dominated by obversely retouched flakes. All of the specimens displayed facetted platforms. The *truncations* and a single *notch* also appeared on flakes, but without facetted butts.

**Table 4.1. Tool Assemblage from the Late Acheulean Site of Wadi Qalkha**

| | | | |
|---|---|---|---|
| Bifaces | 10 | 37.0% | |
| Cardiform | | | 4 |
| Amygdaloid | | | 2 |
| Ovate | | | 2 |
| Broken | | | 1 |
| Irregular | | | 1 |
| Scrapers | 7 | 20.0% | |
| Simple straight | | | 1 |
| Simple convex | | | 2 |
| Déjeté | | | 3 |
| Transverse | | | 1 |
| Truncations | 3 | 11.5% | |
| Truncation on flake, obverse | | | 2 |
| Truncation on flake, inverse | | | 1 |
| Retouched Pieces | 6 | 23.0% | |
| Unilateral retouched flake, obverse | | | 4 |
| Unilateral retouched flake, inverse | | | 2 |
| Notches | 1 | 3.8% | |
| Single notch on flake | | | 1 |
| | | TOTAL | 27 |

## Debitage and Debris

Flakes, as anticipated, occur in relatively high frequencies, but blades represent an unexpected high proportion of the debitage given their absence from the toolkits. While this could be an expression of the small artifact sample, it may well represent preferential selection of blanks for tool manufacture. Two Levallois flakes are included in the assemblage. Half of the flakes bear facetted butts, but none occurs on the blade sample. The debris consists of large chunks.

## SUMMARY

The small, but reasonably diagnostic artifact assemblage from Wadi Qalkha has importance beyond merely providing a cultural-stratigraphic marker for the thick Qalkhan Formation. The configuration and geographic location of the assemblage have bearing upon both taxonomic and cultural-ecologic issues.

The finely fashioned bifaces, dominated by cardiform and amygdaloid forms, in conjunction with evidence of a Levallois technology point to a Late Acheulean affiliation for the assemblage (Gilead 1970; Ronen 1979; Muhesen 1981; Rollefson 1981, 1982). Metrical and qualitative attributes of the bifaces (Table 4.2) further reinforce this placement.

A more detailed taxonomic placement of the assemblage is problematic. At this level the relatively high proportion of sidescrapers (especially *déjeté* scrapers) brings up the question of affinity with the Acheulo-Yabrudian facies (industry) of the Mugharan Tradition (Jelinek 1982), a complex bridging the Lower and Middle Paleolithic (Bar-Yosef 1989). At this point it should be noted, however, that the Wadi Qalkha *déjeté* scrapers are not as finely or intensively retouched as those illustrated from Tabun and Jabrud.

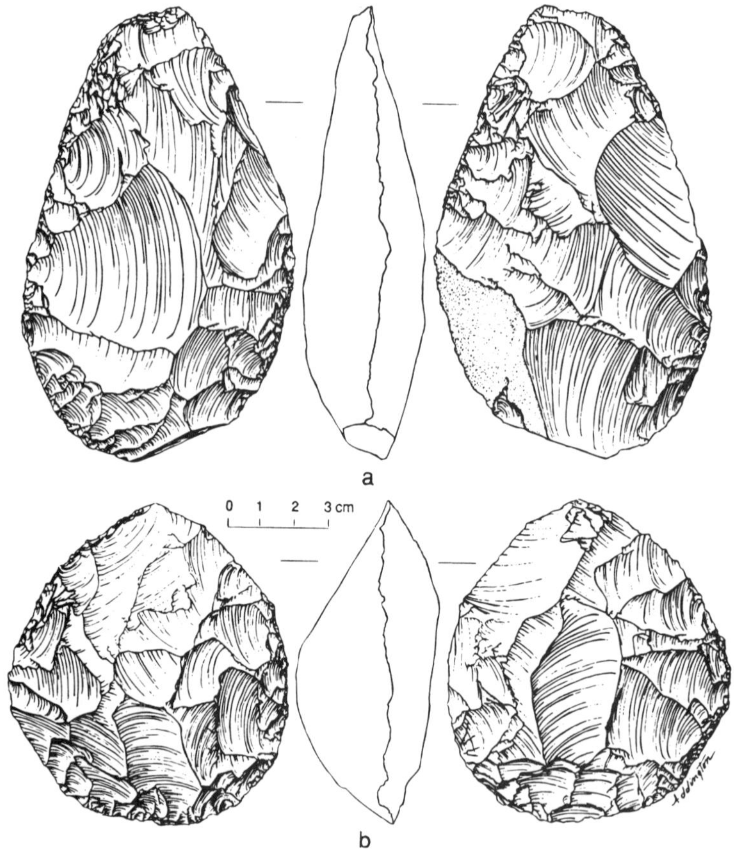


**Figure 4.2.** Illustrations of handaxes from Wadi Qalkha (J401): (a) amygdaloid, (b) cardiform.

Jelinek's (1982) study of the Yabrudian and Late Acheulean beds at Tabun showed that *déjeté* scrapers do not occur in significantly higher proportions in Yabrudian assemblages, although scrapers as a whole do. He, in fact, found scrapers and bifaces to vary inversely throughout the lower part of the long Mugharan sequence and argued that the variability in these tool classes was correlated with fluctuations in paleo-temperatures. Although unable to identify the specific functional links between tool classes and resources, Jelinek (1972:1373) nevertheless interpreted the variability in assemblages as a "response to shifting resources."

Also, in considering the question of relationships between these complexes, Copeland and Hours (1981) cautioned against overstressing the importance of the "biface-*déjeté* scraper" dichotomy and suggested a Late Acheulean *ensemble flou* composed of several sub-assemblages of which the Yabrudian is a component. While they entertained the notion of functionally driven assemblage variability, they also pointed to possible ethnically (*de traditions culturelles*) induced differences in space and time. In examining the geographic distributions of their sub-assemblages, they found that the biface dominated assemblages containing a significant Levallois element (i.e., the Samoukien) were confined for the most part to the northern inland portions of the Levant. With its bifaces and Levallois element, the Wadi Qalkha assemblage extends this distribution substantially to the south.

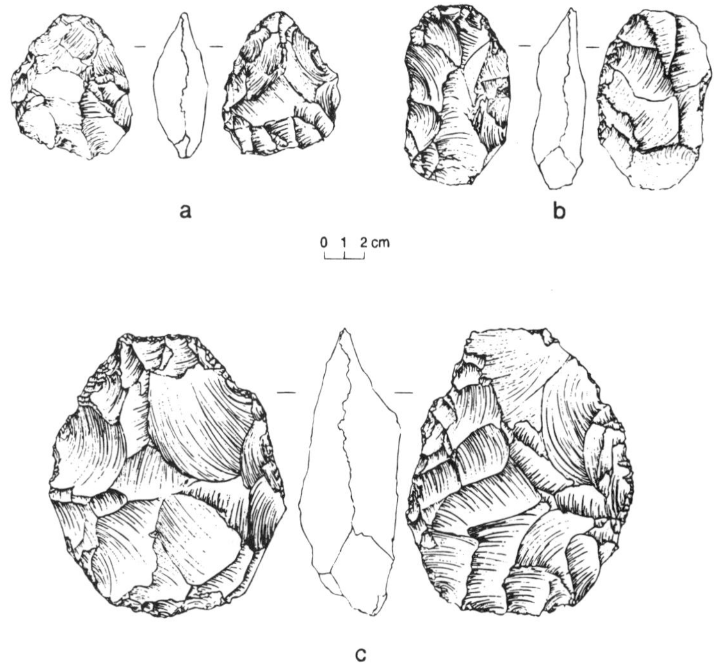


**Figure 4.3.** Illustrations of handaxes from Wadi Qalkha (J401): (a) small cardiform, (b) small ovate, and (c) large cardiform.

Interestingly, the assemblage from Fajaje (another high elevation site in southern Jordan) displays a technotypologic profile resembling that of the Wadi Qalkha assemblage (Rollefson 1981), whereas the Late Acheulean assemblage from Ain el Assad (located at the Azraq oasis in northern Jordan) lacks a Levallois element and contains a substantial cleaver component (Rollefson 1982, 1983).

Presently, the Late Acheulean information-points appear on a very large time-space grid and are so few and variable that only the most general of patterns can be seen. While the assemblage from the site of Wadi Qalkha is limited in its information content by the three problems common to most Lower Paleolithic sites (i.e., small artifact sample, poor chronology, and derived context; Henry 1986; Goren-Inbar 1990c), it may nevertheless still contribute to our understanding of late Lower Paleolithic assemblage variability. Wadi Qalkha and nearby Fajaje share the greatest number of technotypologic attributes with distant assemblages from the northern reaches of the region, mainly associated with steppe and light woodlands. These two high elevation southern Levantine assemblages differ, however, from the assemblage recovered from Ain el Assad which is in closer proximity yet tied to a lower elevation desert oasis setting.

Therefore, even during the Lower Paleolithic from which our evidence is scanty and imprecise, the assemblages recovered from high elevations in southern Jordan hint of connections with those from environmentally similar settings further to the north. These similarities are most likely to have resulted from adaptive convergence experienced by *Homo erectus* foragers. A distributional pattern in which Late Acheulean find-spots suggest a preference for shrub-steppe settings bordering small, intermontane plains and plateaus (Gilead 1970, 1975) is con-

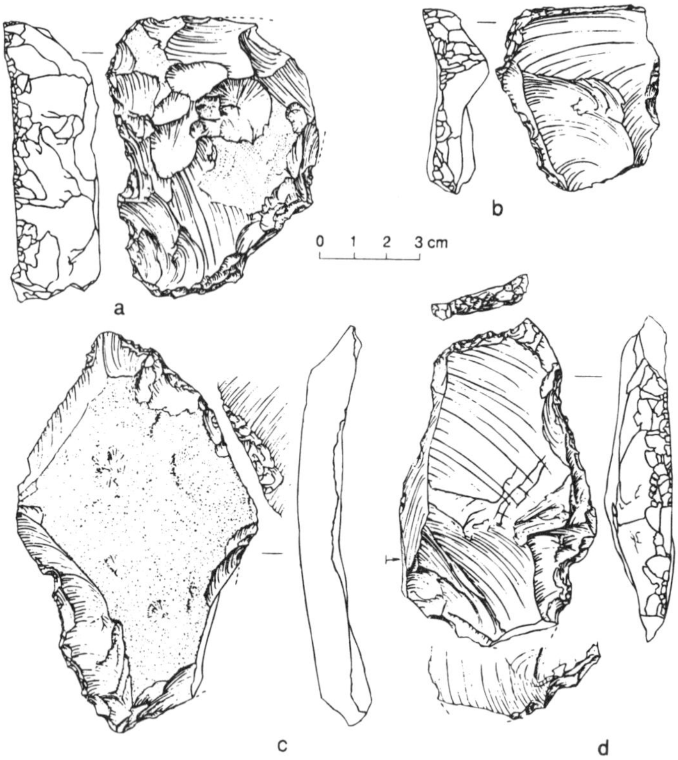


**Figure 4.4.** Illustrations of artifacts from Wadi Qalkha (J401): (a,b,d) sidescrapers, (c) heavily retouched primary flake.

**Table 4.2. Debitage from the Late Acheulean Site of Wadi Qalkha**

| | | |
|---|---|---|
| Cores (Levallois) | 1 | 4.8% |
| Blades | 8 | 38.1 |
| Flakes | 6 | 28.5 |
| Levallois Flakes | 2 | 9.5 |
| Primary Elements | 4 | 19.0 |
| Total | 21 | |

sistent with the location of Wadi Qalkha. The markedly different environmental composition of an oasis setting, such as at Ain el Assad, may well have involved different foraging activities and selected for a somewhat different set of implements. This would seem to be especially true if, as suggested by Bar-Yosef (1989a), it was only during the late Middle Paleolithic that prehistoric foragers were able to exploit extremely arid steppe settings.

# Chapter 5

# The Middle Paleolithic Sites

The Middle Paleolithic is represented by five Levantine Mousterian sites found in the Judayid Basin and the Jebel Qalkha study areas. Only two of the sites, Tor Sabiha and Tor Faraj, yielded *in situ* materials and these became the focus of research for the period (Figure 4.1). The other sites might best be described as find-spots where low densities of diagnostic artifacts (e.g., Levallois points and cores) were recorded on the surface.

Several lines of evidence indicate that the occupations of Tor Faraj and Tor Sabiha were archaeologically synchronous, yet other data point to substantial differences in the nature of their prehistoric occupations, especially in relation to settlement and procurement behaviors. Patterned differences in the configurations and contexts of the artifact assemblages parallel other differences in the natural settings of the sites. This asymmetry between the sites in both cultural and natural evidence prompted a view that the occupations reflect different segments of an annual cycle of Middle Paleolithic transhumance. Beyond having regional paleoecologic significance, such a notion also is pertinent to questions of hominid cognitive development and the emergence of cultural behaviors as we know them.

This chapter essentially describes the evidence from the deposits of Tor Sabiha and Tor Faraj, along with discussions of recovery and analytic procedures. Chapter 6 (Shea) follows in this vein, although it is devoted to a wear-pattern study of artifacts from Tor Faraj. The evidence presented in these chapters is then brought together in the predominantly interpretative Chapter 7 which examines the issues surrounding Middle Paleolithic human ecology and cognition.

## SITE SETTINGS AND RESOURCES

The sharp contrasts in the natural settings of the two sites, especially those related to elevation and exposure, and the resources available in their catchments initially inspired the idea of differences in prehistoric site usage and occupational behaviors.

### Tor Faraj

Tor Faraj (ca. 900 masl) is a large rockshelter located on the north wall of a deeply incised, narrow, sandstone canyon containing a major stream—the Wadi Aghar (Figures 5.1 and 5.2). The terrace fronting the shelter is relatively narrow and drops off steeply to the wadi bed some 12 m below. With its southern exposure, much of the shelter receives sunlight throughout the day. At present, the site rests at the lower edge of the steppe zone near to where it gives way to desert. Modern sources of surface water within a five kilometer radius include, of course, the Wadi Aghar, numerous pot-holes worn in bedrock channels throughout the Jebel Qalkha area, and the Wadi Qalkha. Although all of these sources are seasonal, the largest pot-holes have been

**Figure 5.1.** Photograph of the mouth of the Wadi Aghar.

**Figure 5.2.** Photograph of Tor Faraj from the bed of the Wadi Aghar.

observed to hold water as late as April during the driest of years. Chert sources within the site's catchment are limited to occasional stream cobbles within the Wadi Qalkha (located about 2 km from the site) and to small nodules from a limestone remnant located about 5 km distant.

## Tor Sabiha

Tor Sabiha consists of a small, shallow cave and adjacent terrace set within a deep sandstone embayment high (1,300 masl) on the western wall of the Judayid Basin (Figure 5.3). The narrow embayment opens to the east looking across the basin to the horizon formed by the 500 m high wall of the Ma'an Plateau. Although the center of the basin's floor is only two kilometers away it rests some 300 m below the cave. With its eastern exposure the site is limited to a few hours of sunlight in the morning.

Relative to modern biotic communities, Tor Sabiha rests at the boundary of woodland and steppe zones. The immediate catchment of the site (5 km radius) is dominated by steppe vegetation which covers the basin and its flanks. Although less common, Mediterranean vegetation would have fallen within the site's catchment on the nearby flat top of Jebel el Jill and in the uplands on the plateau. Sources of surface water include small natural depressions in the bedrock above the cave on the top of Jebel el Jill, the wadis on the floor of the basin, and springs along the edge of the plateau. The rock-basins and the wadis contain water only during and immediately after the winter wet-season, whereas the springs flow year round. The western wall of the basin, formed by sandstone outliers such as Jebel el Jill, is devoid of chert sources. But vast quantities and varieties of chert are available in the stream cobbles of the wadi beds on the nearby floor of the basin as well as in seams that outcrop along the high escarpment of the plateau located some 2–3 km distant.

**Figure 5.3.** Photograph of Tor Sabiha.

## STRATIGRAPHIES AND PALEOENVIRONMENTS

Test excavations followed by a larger block excavation were conducted at Tor Sabiha during field seasons in 1979 and 1980, while excavation at Tor Faraj was restricted to a stratigraphic trench dug in the spring of 1984.

### Tor Faraj

The construction of a stone storehouse or "gorfu" in the western end of the shelter and the leveling of the floor in the eastern end by Bedouin herdsmen resulted in the disturbance of some of the site's deposit. Ironically, it was the artifacts in the backdirt piles from these Bedouin "excavations" that initially brought attention to the site (Figure 5.4). Outside of the high density of specimens eroding from the backdirt piles, only a few artifacts were found on the steep talus in front of the shelter. Nondiagnostic flint artifacts and bone were also found cemented to the backwall of the shelter some 75–100 cm above the present floor. This would suggest that prior to the leveling activities, the deposit sloped up from the floor to this point on the backwall. In construction of the gorfu, Bedouin removed sediments within the confines of the structure, exposing the backwall of the shelter and its bedrock floor some 3 m below the surface. As in the eastern end of the shelter, chert and bone fragments are cemented to the backwall and a Levallois point was recovered from sand some 3 m below the surface on the bedrock floor.

#### Stratigraphy

In addition to making surface collections from the backdirt piles, all of the sediment in the piles was screened through 3 mm mesh sieves for artifact collection. In order to gain an understanding of the stratigraphy of the deposit and to recover artifacts *in situ*, a 1 × 4m trench was excavated in the eastern end of the shelter, running roughly parallel to the backwall. The excavation confirmed a +150 cm thick deposit containing three major strata (Table 5.1; Figure 5.5).

#### Depositional History and Paleoenvironment

Amino acid racemization dates of ostrich eggshell fragments suggest that the earliest occupation of the shelter, as defined thus far for Layer C, apparently began sometime ca. 69,000 B.P. (Henry and Miller 1993). Although artifacts were recovered throughout the deposit, the highest densities accompany ash lenses and burned horizons in the middle and upper parts of the layer. The ash lenses, never more than 10 cm thick, appear to represent hearths that were neither rock-lined nor dug-in. The best defined of these, found in Units 2, 3, and 4 at 40–50 cm below datum, is roughly circular in form and over two meters in diameter. It is accompanied by a few burned stones and is underlain by fire-reddened sediments. Another much thinner lens occurs near the top of the layer and closely parallels the stratigraphic profile of the lower hearth. This indicates quite similar topographies for the floors of the shelter during these two occupations and also implies relatively rapid deposition over this interval.

Prior to the deposition of Layer B, Layer C was truncated by erosion, perhaps induced by the leveling activities of the Bedouin herdsmen. The Layer B–C contact slopes off to the west and deviates sharply from the near level bedded ash lenses of Layer C. This erosion could have occurred naturally, but the removal of the protective wedge of sediment (the top of Layer C) that sloped up to shelter's backwall would certainly have accelerated erosion. This would explain the presence of a few Levantine Mousterian artifacts that could have been incorporated into the Layer B fill from remnants of Layer C that still adhered to the wall of the shelter after the leveling activities. The same is true for Layer A, which contains lenses of recent dung also accompanied by Levantine Mousterian artifacts.

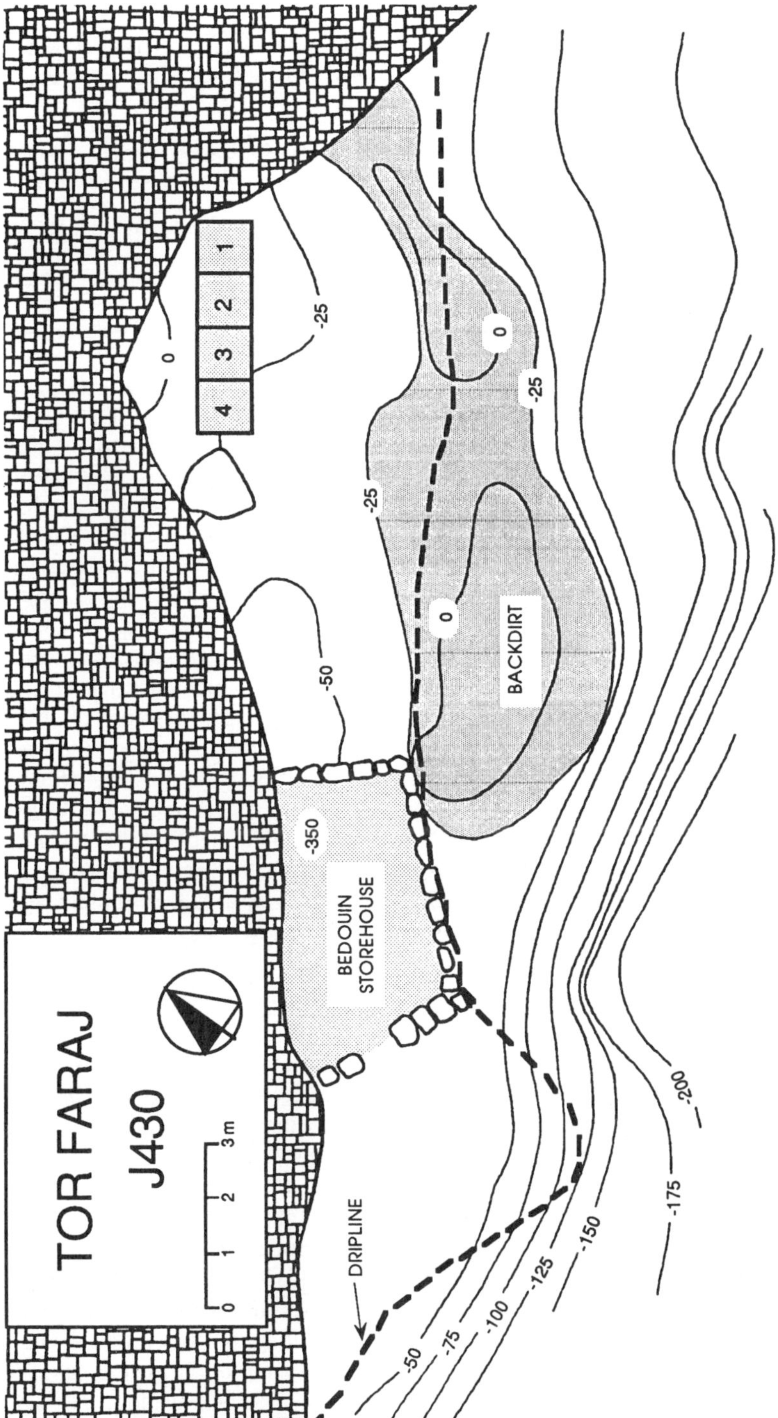


Figure 5.4. Contour map and excavation plan of Tor Faraj.

Table 5.1. Description of the Stratigraphic Layers of Tor Faraj (Munsell Readings on Dry Sediments)

| Layer | Description |
|---|---|
| A | Loose, light red (2.5 YR 6/6) sandy silt with lenses of sand and organics, most likely decomposed dung; 2–35 cm thick; contains sparse Levantine Mousterian artifacts; clear boundary with Layer B. |
| B | Friable red (2.5 YR 5/6) sandy silt; 3–20 cm thick; contains sparse Levantine Mousterian artifacts; sharp contact with Layer C. |
| C | Coherent, red (2.5 YR 5/6) sandy silt; +135 cm thick; contains abundant Levantine Mousterian artifacts associated with ash and burned lenses; AAR dates on ostrich eggshell 69,000 ± 6000 B.P.; Uranium series date on ostrich eggshell of 28,900 ± 3,900 B.P. (obviously too young); deteriorated sandstone increases with depth. |

The backdirt piles represent Layer C sediments that were removed during construction of the storehouse and leveling of the shelter's floor. A uranium series date on ostrich eggshell recovered from the backdirt yielded a date of 62,400 ± 14,000 B.P. (C. McKinney, pers. comm.). Although at variance with the obviously too young 31,000 B.P. U-series date of eggshell recovered *in situ* in Layer C, this date is in agreement with the ca. 69,000 B.P. AAR date for the layer.

Paleoenvironmental evidence recovered from the deposit of Tor Faraj indicates a generally arid climate, but cooler and more moist than that of today. Although pollen was not found in the single pilot sample, a rich phytolith assemblage dominated by Festucoid C3 (cool environment) grasses was identified (Rosen, Chapter 19). The rarity of phytoliths of flowering plant parts is also of seasonal significance, suggesting that the occupation occurred outside of the flowering season which peaks in the spring and early summer. The presence of phytoliths of canes, reeds, perhaps palm, and rushes points to nearby standing water and the use of such plants for matting or bedding as well as food-stuff (e.g., edible tubers). Although poorly preserved and sparse, the faunal remains include gazelle and equid teeth and ostrich eggshell fragments. Also, a poorly preserved segment of a longbone was identified as either that of *Homo* or red deer based upon the specimen's size and wall thickness (E. Trinkaus, pers. comm.). The remains of gazelle and equid confirm a generally open, relatively arid environment, as does the ostrich (*Struthio camelus*). With this reconstruction, it is still possible that red deer could have been found in woodlands on the plateau north of the site.

## Tor Sabiha

Headward erosion of a small gully beneath the cave contributed to the exposure of the occupation layer and resulted in the dispersal of artifacts along the banks of the channel (Figure 5.6). This led to the discovery of the site in 1979 and, in that first season, eight test units were excavated up the slope from the artifact concentration near the edge of the gully to the mouth of the cave. These test excavations confirmed that the occupational horizon had been entirely eroded in the area of the gully and remained as only a thin (3–5 cm thick) layer overlying bedrock near the entrance to the cave. However, the intermediate area displayed a relatively deep deposit (1.3 m) with a 20–30 cm thick cultural horizon. During the following season in 1980, 13 more units were excavated with 11 of these placed in the intermediate area forming a 3 × 4 m block. The other two units were placed along a line running east on the broad, gently sloping terrace in an effort to determine the eastward extent of the occupation. While material was recovered from the Unit 21, the one nearest the block excavation, Unit 20 failed to yield any artifacts.

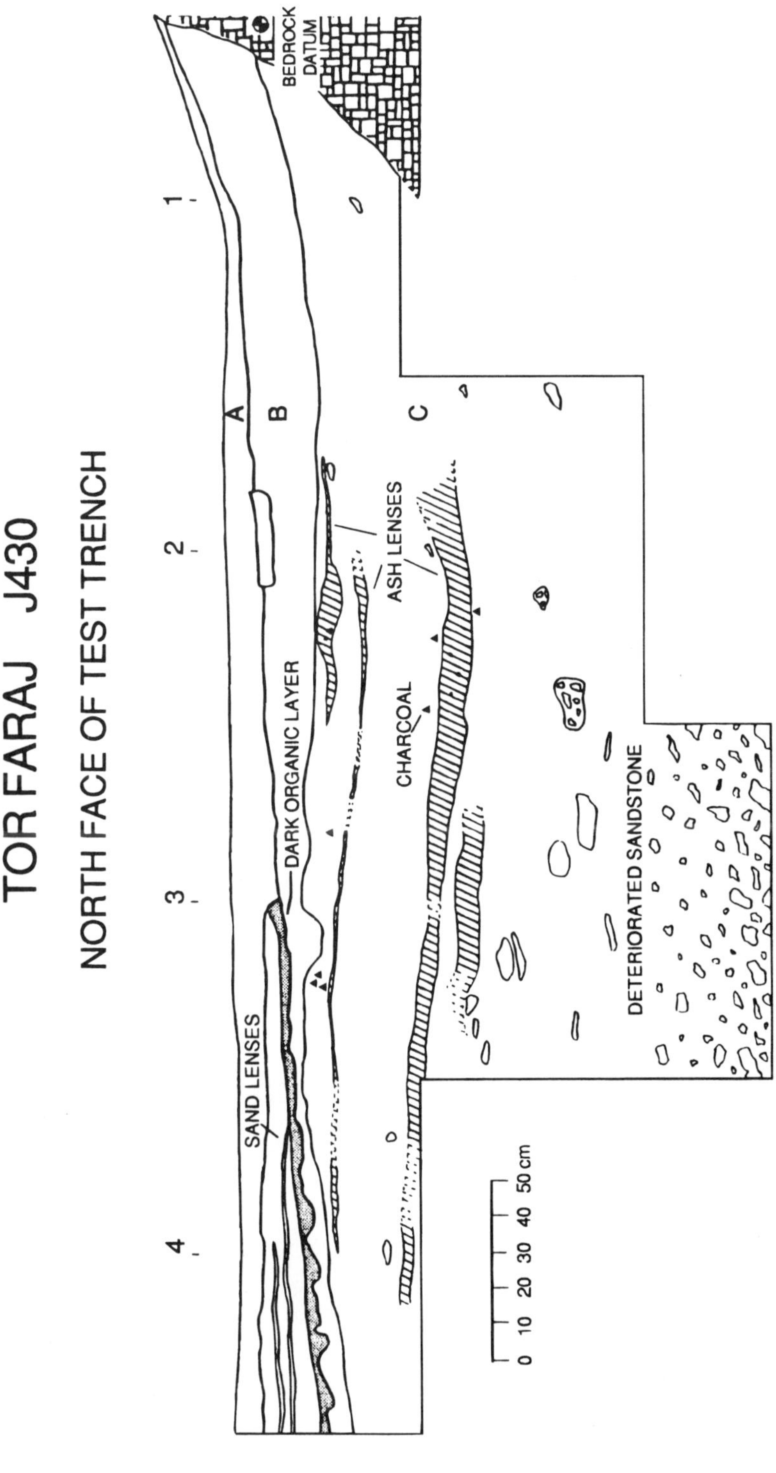


Figure 5.5. Stratigraphic profile of Tor Faraj.

Compass orientations and inclinations of large artifacts recovered from the cultural horizon indicate that the paleosurface was more gently sloped than that beneath the cave today. Most of the artifacts displayed inclinations from 0–5° oriented to the southwest in contrast to the modern slope of about 15° which falls away to the southeast.

## Stratigraphy

The deposit of Tor Sabiha is composed of four major stratigraphic layers (A through D) overlying sandstone bedrock (Figure 5.7). Layers A, B, and C contain artifacts, whereas Layer D is culturally sterile. Although all of the strata consist of weakly weathered sand from the

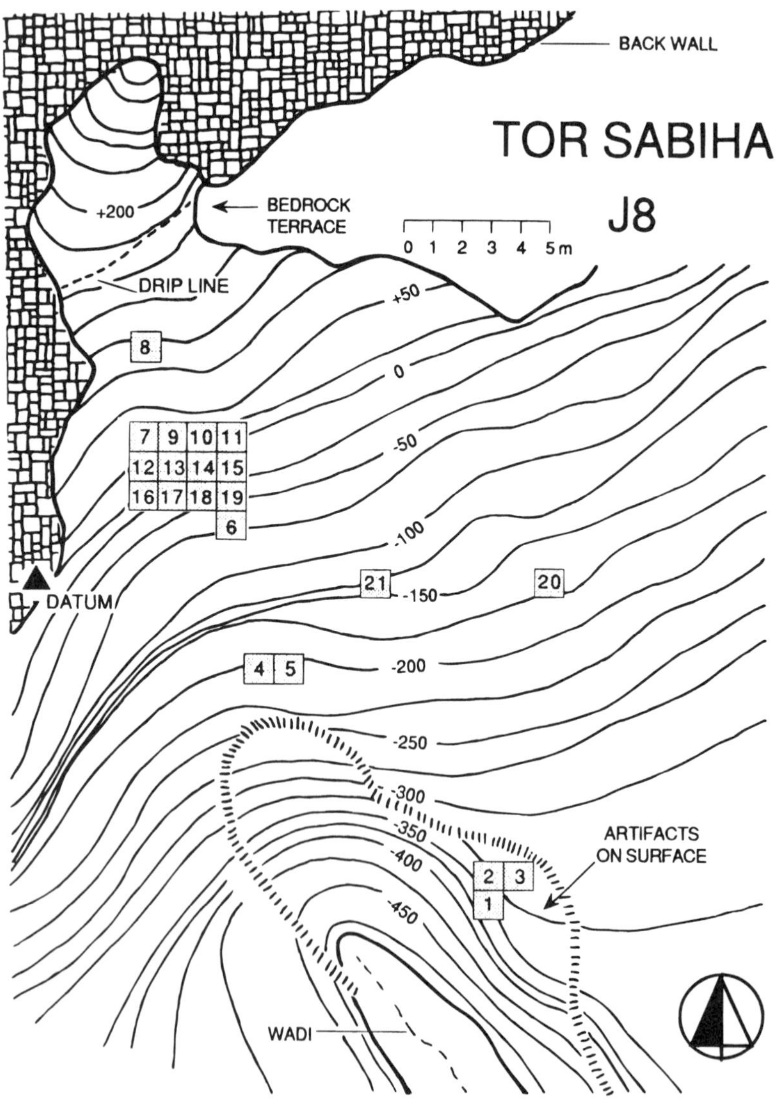


Figure 5.6. Contour and excavation plan of Tor Sabiha.

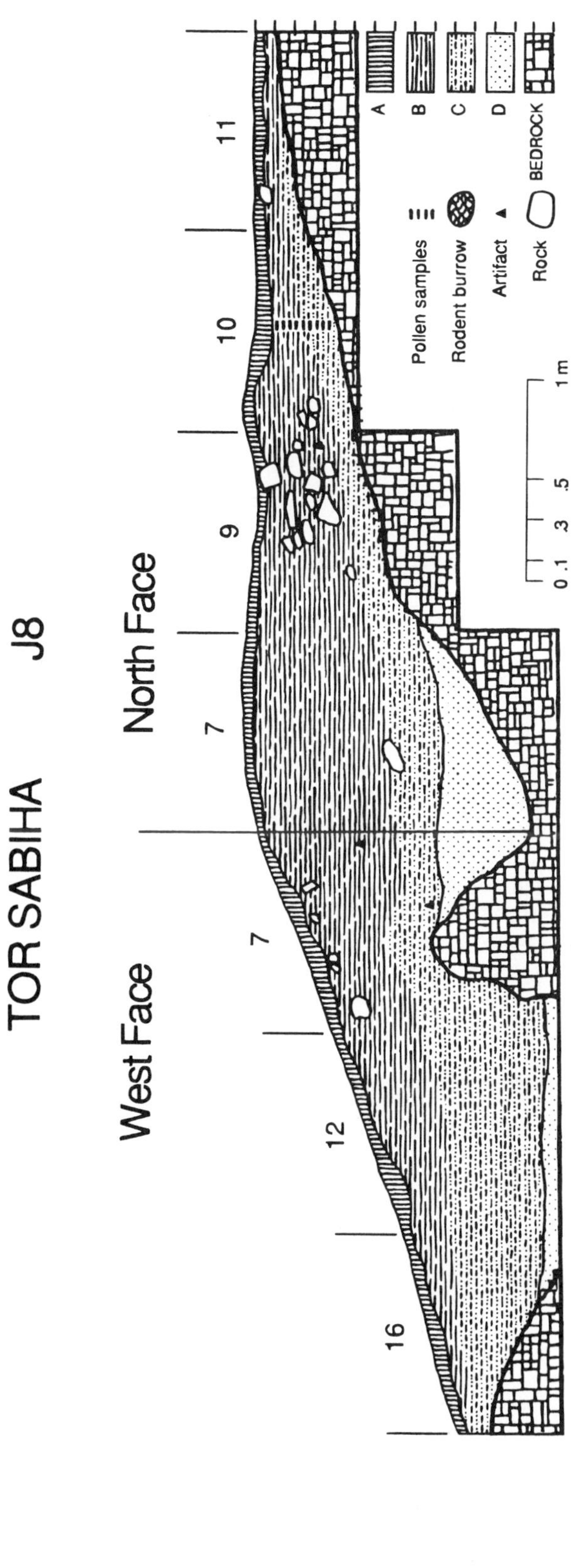


Figure 5.7. Stratigraphic profile of the deposit of Tor Sabiha.

bedrock that forms the cliff faces surrounding the site, the individual layers can be differentiated according to differences in color, hue, and compaction (Table 5.2).

## Depositional History and Paleoenvironments

The depositional succession at Tor Sabiha indicates that the occupation began following a cold-dry episode represented by the weakly weathered sand of Layer D. The angularity and light oxidation of the sand argues for less moisture and possibly lower temperatures than today. The site was apparently first occupied during an episode of slight climatic amelioration dated to ca. 69,000 B.P. by amino acid racemization of ostrich eggshell fragments (Henry and Miller 1993). Increased temperature and precipitation are indicated by the greater weathering of the Layer C sediments.

Pollen (Emery-Barbier 1988, Chapter 16) and faunal evidence from Layer C are consistent with a generally arid environment, but more moist and cooler than the area today. In more specific terms, the basin and hilly uplands are likely to have supported grassland with pockets of woodland in protected, well watered areas and along major streams. The presence of alder and elm pollen suggests that the nearby wadis on the basin's floor were lined with stands of trees and flowed for much of the year. The pine pollen denotes some woodlands, most likely in the uplands of the plateau and perhaps on Jebel el Jill. Pollen fluctuations within Layer C also may denote a rise in temperature as reflected in the rise of the thermophilous chenopodiaceae type *Noaea* and the attendant rise of grasses.

Faunal remains include gazelle, bos, equid, and ostrich eggshell fragments. As environmental indicators these are consistent with the geoarchaeological and palynological interpretations for Layer C. Gazelle, bos, and equid confirm a generally open, relatively arid environment, but with available water.

Ostrich (*Struthio camelus*) are also arid adapted and require dunes or drift sand for egg-laying. The presence of ostrich remains also are helpful in establishing paleotemperatures. Ostriches are not cold tolerant and enjoy a modern range of 15–35°C (Brown 1982). Given that the modern annual mean minimum temperatures for the piedmont and uplands range from 15–12 °C, the presence of ostriches in the area suggest that temperatures are unlikely to have been depressed much beyond 3 °C during the deposition of Layer C. In this regard, it is noteworthy that when the paleotemperatures associated with a high resolution chronostratigraphy developed by Martinson *et al.* (1987) are reviewed, the only interval with a temperature depression of 3 °C or less between 59,980 and 72,640 B.P. falls at 68,990 B.P. This interval, with a date nearly identical to the 69,500 B.P. AAR date for Layer C, denotes a slight rise (ca. 1 °C) in tem-

**Table 5.2. Description of the Stratigraphic Layers of Tor Sabiha (*Munsell* Readings on Dry Sediments)**

| Layer | Description |
|---|---|
| A | Loose, reddish brown sand with decomposed dung; 3–5 cm thick; contains Chalcolithic and Levantine Mousterian artifacts along with modern Bedouin ceramics; grades to Layer B. |
| B | Loose, light reddish brown (5YR 6/4) sand; 0–75 cm thick; contains Chalcolithic and Levantine Mousterian artifacts; $C^{14}$ date on charcoal 4,040 ± 175 B.P. (Beta-1264); sharp contact with Layer C. |
| C | Friable, pinkish gray (5 YR 6/4) sand; 0–30 cm thick; contains Levantine Mousterian artifacts; AAR dates on ostrich eggshell 69,000 ± 6000 B.P.; Uranium series date on ostrich eggshell of 31,500 ± 500 B.P. (obviously too young); sharp contact with Layer D. |
| D | Coherent, white (10 YR 8/2) sand; 0–50 cm thick; culturally sterile; rests on bedrock. |

perature over the preceding two millennia and this rise would have accommodated the modern minimum temperature demanded by *Struthio camelus*.

Although Levantine Mousterian artifacts were recovered in low numbers from Layers A and B, they appear to have been derived at some time in the past from Layer C sediments which were truncated upslope, above the block excavation and in the vicinity of the cave mouth. The presence of Chalcolithic artifacts in association with Levantine Mousterian specimens confirms the disturbance. In all likelihood, Layer C sediments were all but removed from the cave during the Late Holocene by Chalcolithic shepherds, a practice common to Bedouin pastoralists of today. The Layer C sediments and Levantine Mousterian artifacts would have been dispersed downslope and incorporated into an aggrading Layer B deposit along with Chalcolithic material. The ca. 4,000 B.P. $C^{14}$ date and a pollen spectrum resembling other Chalcolithic deposits with its warm, hyper-arid configuration (Emery-Barbier 1988, Chapter 16) are also consistent with an essentially Late Holocene age for Layer B.

## ARTIFACTS

The behavioral residue recovered from the deposits of Tor Sabiha and Tor Faraj is principally in the form of chipped stone artifacts. The specimens recovered from the relatively thin (20–30 cm thick) occupation of Layer C at Tor Sabiha are treated as a single assemblage, whereas four assemblages are defined at Tor Faraj. These include the artifacts recovered from the limited excavation of the lower part of Layer C (70–150 cm), the two occupation horizons that accompany the hearths in the middle (30–70 cm) and upper (10–30 cm) parts of Layer C, and the material collected from the Bedouin backdirt piles. While this latter assemblage was not recovered *in situ*, its primary provenance is rather tightly delimited to the top of Layer C near the backwall of the shelter.

### Tor Faraj

Although the artifact inventory of Tor Faraj is partitioned into four assemblages, the size of the sample from Lower Layer C is too small for any quantitative analyses to be meaningful (Table 5.3). From a qualitative perspective, however, the presence of Levallois points and point cores in this assemblage confirms a Levantine Mousterian occupation for the deepest excavated portions of Layer C.

In general, the assemblages suitable for quantitative study from the middle, upper, and top of Layer C share several technotypologic characteristics. The relatively high frequencies of primary elements denote a significant amount of initial processing activities. These involved the initial shaping of cores followed by a full reduction stream encompassing blank production, tool manufacture, tool maintenance and rejuvenation. The central focus of the technology was on the production of Levallois points from both unidirectionally and bidirectionally prepared cores. This emphasis finds typologic expression in the toolkits which are dominated by Levallois points accompanied by moderate proportions of tools on blades. A noteworthy, apparently stylistic, characteristic of the assemblages is the common use of inverse retouch used in fashioning tools across all of the major classes.

#### Technology

*Raw Material.* Within the area raw material suitable for knapping is restricted mainly to cherts occurring in seams and nodules found in the Late Cretaceous limestones that form the plateau and flank the Rift Valley. The foothills and lowlands fronting the plateau lack cherts as they are composed of sandstones and granites. Derived or gravel sources also occur in the stream

Table 5.3. Artifact Inventories of the Assemblages Recovered from Tor Sabiha and Tor Faraj

| | Tor Sabiha | | Tor Faraj | | | | | | | |
|---|---|---|---|---|---|---|---|---|---|---|
| | C | | Lower C | | Mid C | | Up C | | Top C | |
| | N | % | N | % | N | % | N | % | N | % |
| *A. Tools* | 129 | 1.9 | 15 | 10 | 54 | 4 | 53 | 7 | 203 | 17 |
| 1. Levallois points retouched | 19 | 14.7 | 1 | 6.7 | 8 | 14.8 | 7 | 13.2 | 40 | 19.0 |
| 2. Levallois points unretouched | 34 | 26.4 | 4 | 26.7 | 27 | 50.0 | 29 | 54.7 | 62 | 29.5 |
| 3. Side scrapers | 4 | 3.1 | 1 | 6.7 | 3 | 5.6 | 1 | 1.9 | 8 | 3.8 |
| 4. End scrapers | — | — | — | — | — | — | — | — | 6 | 2.9 |
| 5. Burins | 1 | .8 | 1 | 6.7 | — | — | 1 | 1.9 | 14 | 6.7 |
| 6. Perforators | 1 | .8 | — | — | — | — | — | — | 1 | 0.5 |
| 7. Truncations | 3 | 2.3 | — | — | 1 | 1.9 | — | — | 3 | 1.4 |
| 8. Notches | 10 | 7.8 | 1 | 6.7 | 6 | 11.1 | 2 | 3.8 | 8 | 3.8 |
| 9. Denticulates | 13 | 10.1 | — | — | — | — | 2 | 3.8 | 5 | 2.4 |
| 10. Retouched pieces | 40 | 31.0 | 7 | 46.7 | 9 | 16.7 | 11 | 20.8 | 56 | 26.7 |
| 11. Other | 4 | 3 | — | — | — | — | — | — | 7 | 3.3 |
| *B. Debitage* | 1,748 | 26.2 | 90 | 57.3 | 515 | 35.9 | 321 | 40.0 | 828 | 70.9 |
| 1. Cores | 10 | .6 | 4 | 4.4 | 12 | 2.3 | 2 | 0.6 | 41 | 4.9 |
| 2. Blades | 652 | 37.3 | 20 | 22.2 | 66 | 12.8 | 15 | 4.7 | 162 | 19.4 |
| 3. Flakes | 843 | 48.2 | 48 | 53.3 | 277 | 53.8 | 246 | 76.6 | 435 | 52.2 |
| 4. Levallois Blades | 11 | .6 | — | — | 7 | 1.4 | 9 | 2.8 | 22 | 2.6 |
| 5. Levallois Flakes | 48 | 2.7 | 3 | 3.3 | 11 | 2.1 | 6 | 1.9 | 17 | 2.0 |
| 6. Core Trim Elements | 32 | 1.8 | 1 | 1.1 | 10 | 1.9 | 2 | 0.6 | 15 | 1.8 |
| 7. Primary Elements | 152 | 8.7 | 14 | 15.6 | 132 | 25.6 | 41 | 12.8 | 136 | 16.3 |
| 8. Burin spalls | — | — | — | — | — | — | — | — | 5 | 0.6 |
| *C. Debris* | 4,786 | 71.8 | 52 | 33.1 | 867 | 60.4 | 429 | 53.4 | 125 | 10.7 |
| 1. Chips | 4,041 | 84.4 | 38 | 73.1 | 833 | 96.1 | 387 | 90.2 | 89 | 71.2 |
| 2. Chunks | 745 | 15.6 | 14 | 26.9 | 34 | 3.9 | 42 | 9.8 | 36 | 28.8 |
| *Total artifacts* | 6,663 | | 157 | | 1,436 | | 803 | | 1,168 | |
| *D. Indices* | | | | | | | | | | |
| I L | 4.4 | | | | 9.3 | | 12.0 | | 13.8 | |
| I F | 37.5 | | | | — | | — | | 46.5 | |
| I LAME | 37.1 | | | | 20.2 | | 5.4 | | 26.4 | |
| *E. Ratios* | | | | | | | | | | |
| Tools: Debitage | 1:13.6 | | | | 1:9.5 | | 1:6.1 | | 1:4 | |
| Cores: Tools | 1:12.9 | | | | 1:4.5 | | 1:26.5 | | 1:5 | |
| Cores: Debitage | 1:173.8 | | | | 1:42.9 | | 1:159.5 | | 1:19 | |
| Debitage: Chips | 1:2.3 | | | | 1:1.6 | | 1:1.2 | | 1:0.1 | |
| Prim. El.: Debitage | 1:115 | | | | 1:3.9 | | 1:7.8 | | 1:6.1 | |

cobbles of wadi beds that drain the plateau, but these are rarely found more than 2–3 km from their origins. The closest of these sources to Tor Faraj is 16 km distant and most are found more than 20 km away. Chert nodules imbedded in the limestone of a small hill located near Jebel Humeima about 4 km walking distance from the site provide an exception to this general pattern of chert distribution. Although the chert from this nearby source is of high quality, it is poorly represented in the Tor Faraj assemblages. This may be because the nodules are relatively small (50 × 80 mm maximum dimension) and difficult to extract.

An attribute analysis of the cherts in the Tor Faraj assemblages identified seven varieties that were utilized regularly. Five of these came from the plateau with others coming from the Rift Valley and the hill near Jebel Humeima. The plateau cherts consistently dominate the assemblages in accounting for 51–84% of the varieties. All of the varieties are excellent for knapping and all but the Jebel Humeima source occur in thick seams or large nodules.

In discussion of raw materials, it is noteworthy that a small number of artifacts display clear evidence (i.e., changes in luster and color) of their having been annealed. These are confined to those elements occurring near the end of the reduction stream and are most common among Levallois points with 3% showing evidence of annealing. Given that these specimens cross-cut chert varieties and display little or no secondary retouch, it seems unlikely that they were intentionally heat-treated to improve their knapping quality. A more likely explanation rests in the points having been heated accidentally. This could have taken place if they had been used in skewering meat or if they were hafted in mastic, a procedure that typically requires heating of the mastic (Shea 1991:194).

*Indices.* As traditionally defined (Bordes 1961), the Levallois Indices (IL) of the Tor Faraj assemblages are low. Levallois blades and flakes constitute only a small part of the assemblages, while Levallois points form by far the greatest part of the Levallois products (Table 5.3). This raises a rather fundamental question of whether unretouched Levallois points should be classified as tools or debitage. The issue has importance beyond classificatory semantics, for in assemblages dominated by points their placement strongly influences various quantitative measures (e.g., tool:debitage ratio; tool:core ratio) that may be helpful in defining prehistoric activities (Munday 1979; Henry 1989b; Kuhn 1990). In Bordesian systematics, points are included with other Levallois products in the type list but excluded (along with other unretouched and inversely retouched specimens) in the computation of restricted indices. In order to facilitate standardized comparisons with other Mousterian assemblages, those conventions have been followed here in the computation of traditional indices. In all other computations, however, unretouched Levallois points are included within the tool category.

The logic for categorizing unretouched Levallois points as tools stems from two observations. First, unlike the other Levallois products, Levallois points could have served predetermined and rather standardized functions without any additional modification through secondary retouch. While one might argue that the same holds true for other unretouched tools (e.g., naturally backed knives), these tool forms do not exhibit characteristics that clearly trace a complicated set of steps of core preparation designed to control the form of the delivered element. Additional support for including Levallois points in the tool category comes from wear-pattern studies in which points are shown to have been used to a proportionately far greater extent than any other class of debitage (Shea 1990, Chapter 6).

Blade Indices (ILame) show considerable variability with the assemblages from the Top and Middle portions of Layer C yielding relatively high values, while the Upper C assemblage furnishes a low value. Faceting Indices (IF) average a moderate 46.5 and on further breakdown show points and blades to have received the greatest attention in platform preparation followed by flakes and primary elements (Table 5.4).

Table 5.4. Platform Faceting by Debitage Classes from Tor Faraj

| | Prim Elem | | Blade | | Flake | | Point | |
|---|---|---|---|---|---|---|---|---|
| | N | % | N | % | N | % | N | % |
| Single | 22 | 44.0 | 24 | 27.3 | 85 | 45.2 | 8 | 9.3 |
| Dihedral | 7 | 14.0 | 11 | 12.5 | 42 | 22.3 | 9 | 10.5 |
| Multiple | 4 | 8.0 | 28 | 31.8 | 35 | 18.6 | 69 | 80.2 |
| Crushed | 11 | 22.0 | 24 | 27.3 | 20 | 10.6 | — | — |
| Cortex | 6 | 12.0 | 1 | 1.1 | 6 | 3.2 | — | — |
| | 50 | 100% | 88 | 100% | 188 | 99.9% | 86 | 100% |

*Reduction Stages, Cores, and Debitage Attributes.* Given the significant distances to the most utilized chert sources, the emphasis upon initial processing activities is somewhat surprising. For an occupation some 16–20 km from raw material, even the modest numbers of cores are unanticipated and the relatively high frequencies of primary elements are even more unexpected (Table 5.3). For example, in ranging from over 12–25%, the relative frequencies of primary elements overlap or exceed those frequencies reported for other Levantine Mousterian assemblages from sites on or very close to chert sources (Crew 1976; Munday 1976; Meignen and Bar-Yosef 1989).

While a considerable amount of initial processing was undertaken at Tor Faraj, it is evident that much of the raw material had been trimmed outside the excavation area and most likely near the distant chert sources. Most of the primary elements display cortical coverage over one-third to two-thirds of their obverse surfaces, but none were found to be totally covered by cortex. Beyond trimming the chert near its sources for weight reduction and improved portability, the occupants of the site also apparently economized on raw material use through extending the productive lives of cores.

A comparison of the dimensions of core facets with those of primary elements, flakes, blades, and points shows that the great majority of cores were well exhausted at the time of their abandonment (Table 5.5, Figure 5.8). The relative dimensions of the debitage also underscores the fact that most of the Levallois points were produced early in core reduction accompanied in part by the delivery of elongated primary elements and blades. Levallois flakes and blades also were most likely produced at this time as part of the point production, but the small number of complete specimens prohibits any meaningful metric comparisons. Some flakes undoubtedly were generated in core preparation, but the relative dimensions of flakes and core facets indicate most flakes were produced following point production.

The point cores that were recovered were apparently near exhaustion as their facet dimensions fall near the bottom of the distribution for point dimensions (Figure 5.9). Levallois point core facets and overall dimensions nevertheless are by far the largest of any of the core varieties. This again suggests that subsequent to the production of points, cores continued to be reduced, but mainly for the production of flakes.

In classifying the cores, five types were recognized based upon platform configurations, facet patterns, and overall forms (Table 5.6, Figure 5.10). The Levallois point cores, with their diagnostic converging scar patterns opposite well developed platforms on domed faces, have the largest facets of the class. Dimensionally these are followed by the opposed platform and 90-degree cores which show scar patterns oriented to blade production. Although their average facet dimensions fall within the flake range, their narrow, elongated faces indicate blade production early in their lives. The other three core types (on flakes, split cobbles, and amorphous) are smaller and signal a transition from the production of points and blades to flakes. The dom-

Table 5.5. Dimensional Data for Complete Specimens of Debitage from Top C, Tor Faraj

| | Prim Elem ($N$ = 59) | | | Points ($N$ = 86) | | | Blades ($N$ = 98) | | | Flakes ($N$ = 201) | | |
|---|---|---|---|---|---|---|---|---|---|---|---|---|
| | *r* | *x* | *SD* | *r* | *x* | *SD* | *r* | *x* | *SD* | *r* | *x* | *SD* |
| Max Length | 21–105 | 57.4 | 16.9 | 22–100 | 60.5 | 14.7 | 23–89 | 52.7 | 15.5 | 19–103 | 38.2 | 14.3 |
| Max Width | 11–64 | 33.1 | 9.8 | 20–58 | 33.2 | 8.7 | 11–58 | 24.5 | 8.8 | 12–75 | 31.4 | 11.3 |
| Max Thick | 3–21 | 11.1 | 4.2 | 2–15 | 7.7 | 2.4 | 2–18 | 7.4 | 3.8 | 2–29 | 8.0 | 3.8 |
| Mid-Pt Wid | 11–62 | 29.5 | 9.6 | 17–55 | 30.1 | 8.0 | 8–42 | 21.5 | 7.8 | 11–58 | 28.3 | 9.7 |
| Mid-Pt Thk | 3–20 | 9.7 | 4.1 | 3–14 | 6.8 | 2.4 | 1–5 | 5.8 | 2.9 | 1–16 | 6.1 | 3.0 |

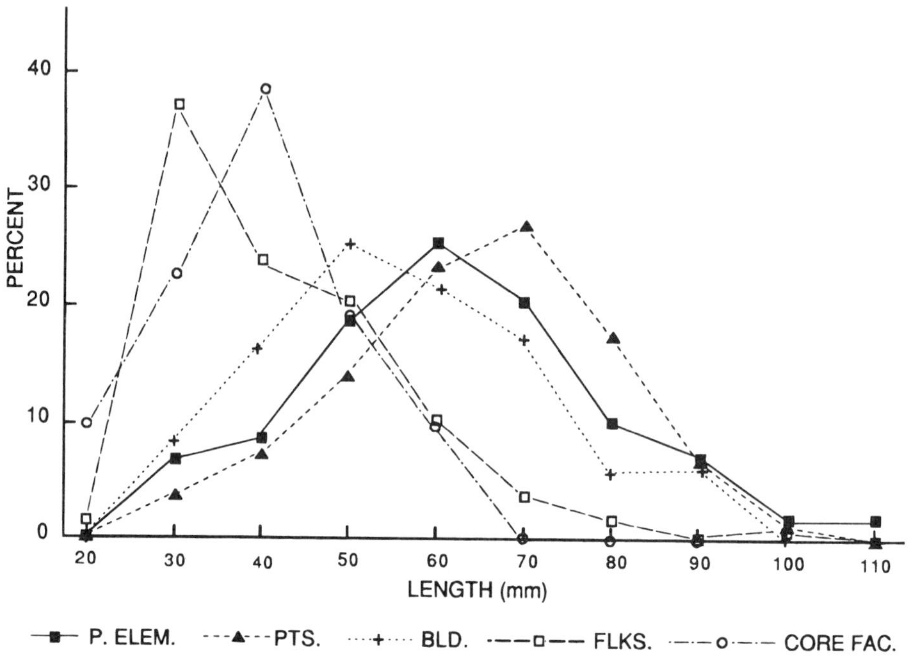


**Figure 5.8.** Graph showing the length distributions of various classes of debitage from Tor Faraj: primary elements (P.ELEM.), Levallois points (PTS.), blades (BLD.), flakes (FLKS.), and core facets (CORE FAC.). Note that flakes and core facets are much shorter than primary elements, points, and blades.

inance of cores on flakes (Table 5.6), coupled with their small dimensions, underscores the economizing measures taken by the occupants of the site in extending flake production. These typically display opposed platforms and resemble the truncated and facetted specimens initially described by Schroeder (1969) and subsequently classified as Type IV Nahr Ibrahim cores by Solecki and Solecki (1970). The split cobble cores are even smaller and most likely represent the ends of cobbles that were removed early in core preparation. They characteristically display a rounded, cortical covered face opposite a worked face. Amorphous cores are the smallest of the class and their lack of standardization in platform orientation and scar patterns is probably an expression of their extreme exhaustion.

*Techniques of Point Production.* An attribute study of Levallois points revealed two attribute clusters based upon patterned differences in platforms and scar patterns (Table 5.7). These attribute groupings also differ in their associated dimensions. Group 1 consists of those points with a V-shaped (*chapeau de gendarme*) platform and Y-arrete scar pattern (Figure 5.11:a–c). Group 2 includes points with a diamond or triangular shaped platform and a scar pattern formed by three or more converging ridges near the base (Figure 5.11:e,f). Group 1 points are also longer relative to their widths (average L:W ratio of 2.0) than the points of Group 2 (average L:W ratio of 1.5).

These two point varieties were tied to different production strategies. Cores for generation of Group 1 points, with their classic *chapeau de gendarme* bases, appear to have been prepared along the lines described by Boëda *et al.* (1990:56). Although they describe the *proceeds techniques* for preparing a core for flake production, the sequence seems to mesh well with the platform and butt characteristics of the Group 1 points from Tor Faraj. Initial platform preparation

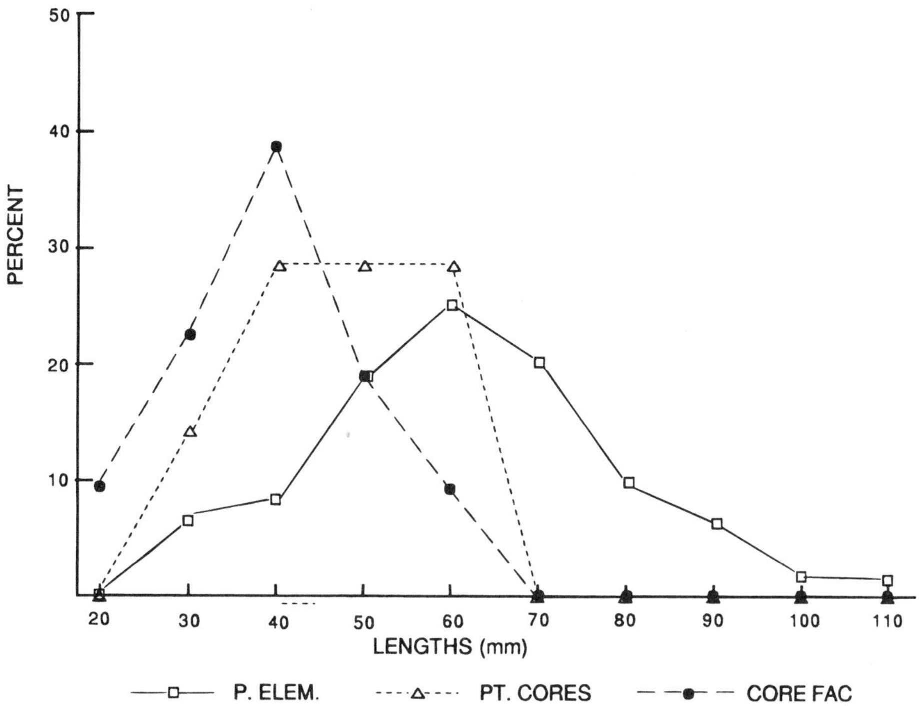


**Figure 5.9.** Graph showing the length distributions of various classes of debitage from Tor Faraj: primary elements (P.ELEM.), point cores (PT. CORES), and core facets (CORE FAC.).

**Table 5.6. Core Types and Dimensions from Tor Faraj, Layer Top C**

| Core type | N | % | $x_T$ | $x_N$ |
|---|---|---|---|---|
| Levallois point | 7 | 17.1 | 43.5 | 32.1 |
| Cores on flakes | 16 | 39.0 | 30.9 | 23.1 |
| Cores on split cobbles | 6 | 14.6 | 26.8 | 22.8 |
| Cores w/ opposed and 90° platforms | 6 | 14.6 | 35.8 | 21.9 |
| Amorphous | 6 | 14.6 | 26.6 | 15.9 |

involved isolating a prominence through the creation of two lateral notches. A noticeable characteristic of the Tor Faraj sample is the substantial depth of the notches that act to exaggerate the protuberance of the base. The platform preparation was linked to the face of the core through the removal of flakes in a manner intended to generate a Y-arrete pattern and to subsequently guide the delivery of a triangular point. About 75% of the Group 1 points show these removals originating from the base, whereas some 25% display bidirectional preparation. Subsequent to completing these removals, the edge of the platform was ground for strengthening and regularization. Almost all of the Group 1 points show extensive edge grinding.

The Group 2 points with their narrower bases and simpler dorsal scar patterns would have required less core preparation. Rather than using a protuberance isolated by lateral notches to receive the blow, a prominent ridge dividing relatively large platform facets served the same purpose. The narrow, convergent dorsal scar pattern of the Group 2 points is less standardized than

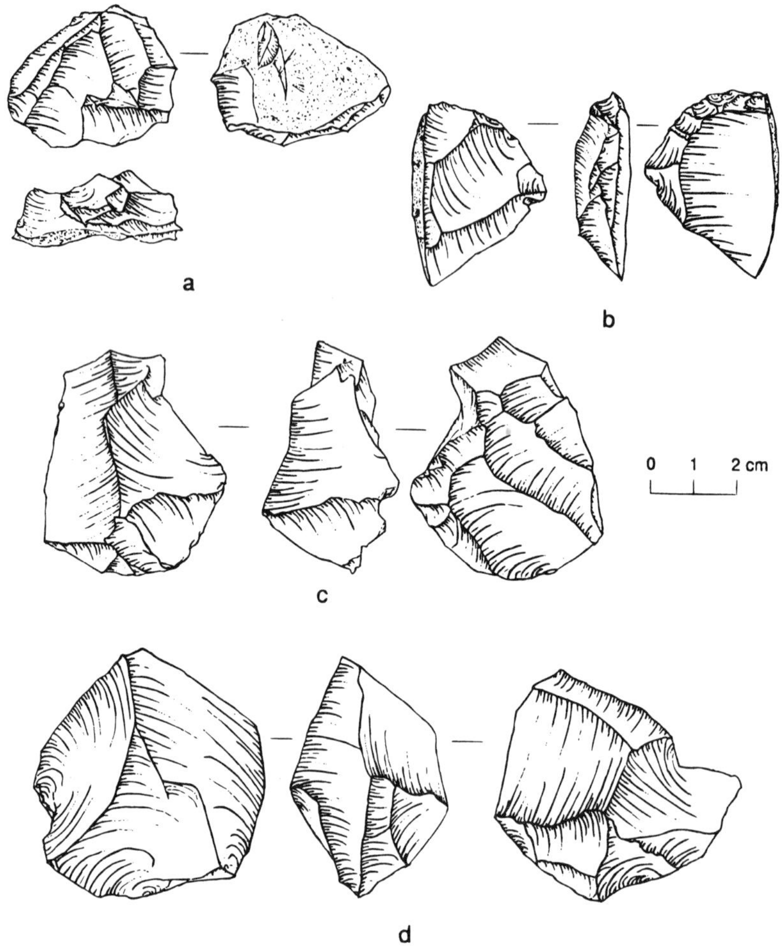


**Figure 5.10.** Illustrations of cores from Tor Faraj: (a) small point core on primary element, (b) small point core on flake, (c) amorphous core, (d) point core.

**Table 5.7. Comparison of the Dimensions of Group 1 and Group 2 Levallois Points from Tor Faraj**

| | Group 1 | Group 2 |
|---|---|---|
| × Length | 51.40 | 56.60 |
| × Width | 27.90 | 28.90 |
| × L/W Ratio | 1.84 | 1.96 |

the Y-arrete pattern of the Group 1 points, but nevertheless served to deliver triangular forms. In combination the structure of the platform and the scar pattern of the Group 2 points generated more elongated specimens. When compared to the wide-based Group 1 points, the Group 2 points would have been more economical to produce. Not only would they have employed less of a core's face for production, rejuvenation of core surfaces would have been much less complicated.

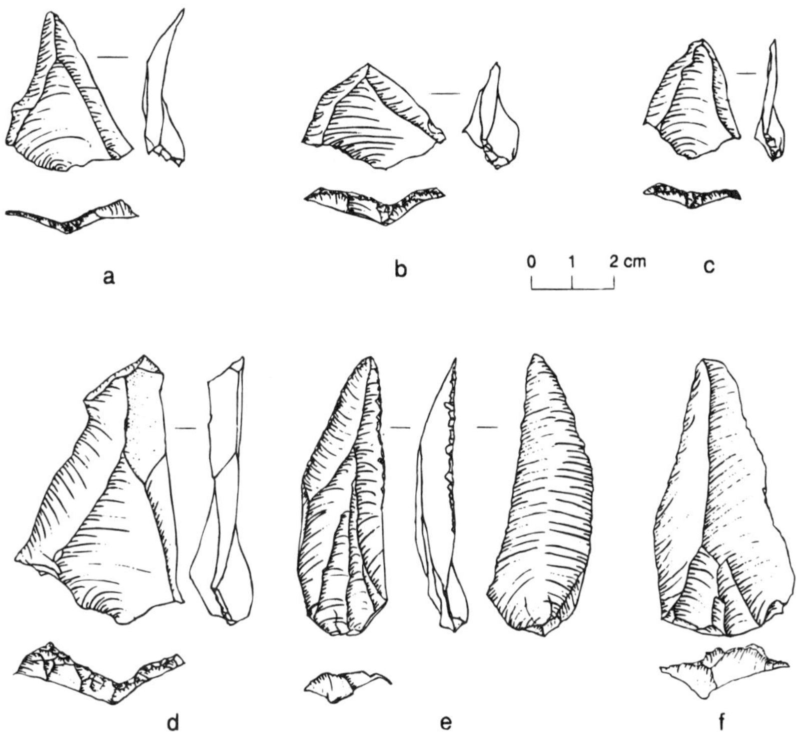


**Figure 5.11.** Illustrations of Levallois points from Tor Faraj: (a–c) small points with classic form, (d) either a mis-struck point or a product of reshaping the core's face after point removal, (e–f) elongated points lacking classic form.

Production of both varieties of points, however, occurred near the beginning of the reduction stream immediately following the removal of cortex. Primary elements reflect this early stage of core preparation in their elongation (compare dimensions of primary elements points, and blades in Table 5.5) and in their parallel and parallel-convergent scar patterns (Table 5.8). This closely resembles the early segment of the operational sequence (*chaîne opératoire*) as described for point production at Kebara (Boëda *et al.* 1990:67). And, as at Kebara and in most other Levantine Mousterian assemblages, unidirectional preparation is dominant (Crew 1976; Munday 1976; Meignen and Bar-Yosef 1989). Following the initial removal of cortical pieces at Tor Faraj, however, a significant amount of bi-directional core shaping took place as expressed in the scar patterns of blades (Table 5.8) and points (ranging from 9–42%).

*Loading.* In an attempt to trace the methods (i.e., hard-hammer percussion, soft-hammer percussion, and pressure flaking) that were employed for processing material throughout the reduction stream, bulb attributes were examined on primary elements, blades, flakes, points, and chips. An analysis of the sizes and the weights of chips (i.e., specimens less than 25 mm maximum dimension) was also undertaken, given that such metric data for microdebitage have been found to be diagnostic of specific methods of loading (Henry et al. 1976; Stahle and Dunn 1984; Fladmark 1982; Clark 1986; Ahler 1989).

The bulbs of primary elements, blades, flakes, and points are overwhelmingly diffuse with the great majority of these displaying bulbar scars; lipped bulbs occur in low proportions (Table 5.9). In contrast, about one-third of the chips exhibit lipped bulbs. Diffuse bulbs have been tra-

Table 5.8. Facet Patterns Observed on Debitage from Layer Top C, Tor Faraj

| | Prim elem | | Blade | | Flake | |
|---|---|---|---|---|---|---|
| | N | % | N | % | N | % |
| Unidirectional parallel | 23 | 47.9 | 34 | 42.5 | 81 | 48.2 |
| Unidirectional convergent | 9 | 18.8 | 10 | 12.5 | 19 | 11.3 |
| Centripedal | 14 | 29.1 | 2 | 2.5 | 11 | 6.5 |
| Parallel truncated | 2 | 4.2 | 9 | 11.3 | 30 | 17.9 |
| Bidirectional | — | — | 16 | 20.0 | 24 | 14.3 |
| Y-arete | — | — | 9 | 11.3 | 3 | 1.8 |
| | 48 | 100% | 78 | 100% | 168 | 100% |

Table 5.9. Bulb Attributes for Debitage and Chips from Tor Faraj

| | Prim elem | | Blade | | Flake | | Chip | |
|---|---|---|---|---|---|---|---|---|
| | N | % | N | % | N | % | N | % |
| Indeterminate | 1 | 1.7 | 12 | 12.2 | 3 | 1.5 | 4 | 2.1 |
| Lipped | 7 | 11.9 | 3 | 3.1 | 16 | 8.0 | 42 | 21.5 |
| Unlipped | 16 | 27.1 | 18 | 18.4 | 42 | 20.9 | 87 | 44.6 |
| Unlipped w/scar | 35 | 59.3 | 65 | 66.3 | 139 | 69.2 | 0 | 0.0 |
| Other | 0 | 0.0 | 0 | 0.0 | 1 | 0.4 | 62 | 31.0 |
| Total | 59 | 100% | 98 | 100% | 201 | 100% | 195 | 100% |

ditionally linked to hard-hammer percussion (Brezillon 1968:75–78; Bordes 1969:11–12, Bordaz 1970:25–27), whereas lipped bulbs have been associated with soft-hammer percussion (Brezillon 1968:75; Henry *et al.* 1976:57; Patterson 1982:55). This suggests that the occupants of Tor Sabiha relied predominantly upon hard-hammer percussion for core shaping and blank removal, but that some soft-hammer percussion was employed for final tool shaping, rejuvenation, and maintenance tasks.

Metrical attributes of the chips also support the argument that a small portion of the class was produced with soft-hammer percussion. Experimentally produced specimens that were size graded between 4 and 10 mm yielded mean thicknesses of 2.1 mm for hard-hammer percussion, 1.9 mm for soft-hammer percussion, and 1.3 mm for pressure flaking (Henry *et al.* 1976). When the chips from Tor Faraj were similarly analyzed they yielded a mean thickness of 2.3 mm with a unimodal distribution ranging from 0.5–5.0 mm (Figure 5.12). While the large majority of chips were produced by hard-hammer percussion, a minor portion was apparently delivered with soft-hammer or even pressure loading given the lower end of the distribution.

Direct evidence of the implements used in flaking are limited to four small spherical, sandstone hammerstones that range from 30–60 mm in diameter. The two largest specimens clearly show evidence of battering.

## Tools and Type Lists

Levallois points and retouched pieces are the two dominant tool classes followed by markedly lesser numbers of notches, burins, sidescrapers, and endscrapers (Table 5.10). Perforators, denticulates, truncations, and special tools are present, but uncommon.

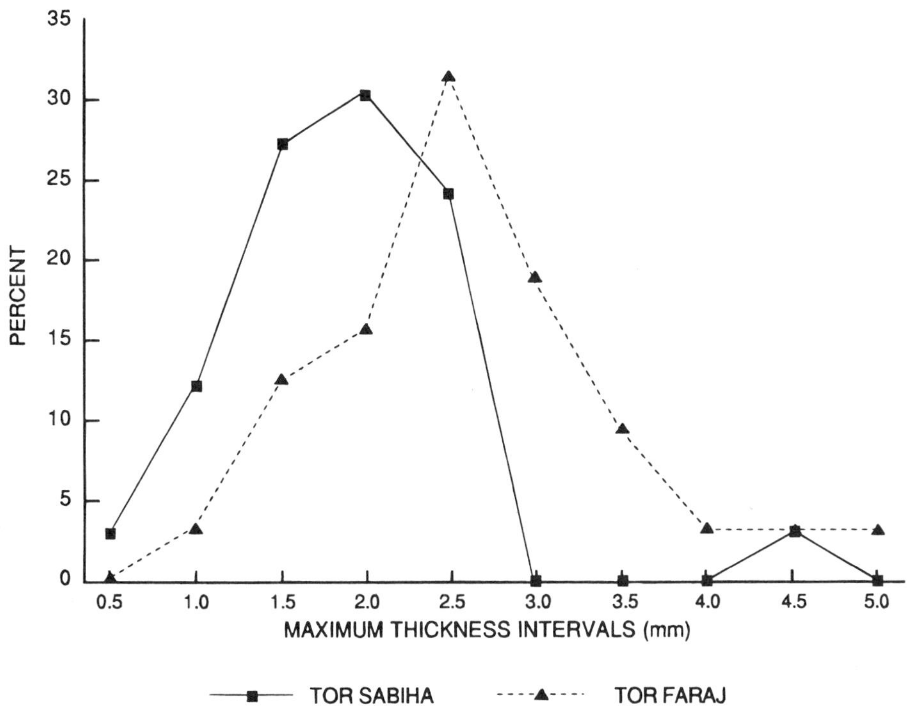


**Figure 5.12.** Graph showing the distribution for the maximum thicknesses of chips (micro-flakes) that have been size graded between 4–10 mm.

*Levallois Points.* Depending upon the layer, from 27–39% of the points have been retouched. As with other tool classes, a substantial proportion (30–54%) of these are inversely retouched. The retouch is typically light and marginal, although a few specimens display heavy, invasive scars. An examination of the location of retouch shows a clear emphasis on the proximal and medial segments of the points with the distal one-third rarely having been worked (Figure 5.13). Interestingly, this pattern is near identical for both obverse and inverse surfaces. The distributional pattern of the retouch with its proximal orientation, coupled with its general symmetry on surfaces and edges, suggests that it was related to hafting. The points may have been intentionally retouched to dull their edges for better bonding or they may have been accidentally retouched by pressures on their edges from their hafts. Regardless of the specific explanation of the retouch, its connection with the points having been hafted is in agreement with the findings of Shea (Chapter 6) in his wear-pattern study.

*Sidescrapers.* Those with inverse retouch represent the most common type within the class, followed by simple straight, simple convex, and double straight types (Figures 5.14:f; 5.15:f; 5.16:f). Although substantial, the retouch is rarely heavy and normally covers about two-thirds the length of the edge. Elongated flakes and blades typically represent the blank form.

*Endscrapers.* These are not common, nor well done. In addition to the simple endscrapers on flakes, other specimens include those with straight (displaying thicker and heavier retouch than truncations) and denticulated bits.

Table 5.10. Type Lists for Assemblages from Tor Faraj and Tor Sabiha

| Site/Assemblage | J430 Top C | J430 Up C | J430 Md C | J8 Sum C |
|---|---|---|---|---|
| A. Levallois points | 49.0% | 67.2% | 68.1% | 41.0% |
| 1. Unretouched | 29.2% | 52.7% | 52.2% | 26.3% |
| 2. Retouched | 19.8% | 14.5% | 15.9% | 14.7% |
| B. Sidescrapers | 3.7% | 1.8% | 0% | 3.1% |
| 1. Simple straight | 0.9% | 0% | 0% | 0.7% |
| 2. Simple convex | 0.9% | 0% | 0% | 0.7% |
| 3. Double straight | 0.4% | 0% | 0% | 0% |
| 4. Inverse | 1.4% | 1.8% | 0% | 1.5% |
| C. Endscrapers | 2.8% | 0% | 0% | 0% |
| 1. Simple on flake | 1.4% | 0% | 0% | 0% |
| 2. With straight bit | 0.9% | 0% | 0% | 0% |
| 3. With denticulated bit | 0.4% | 0% | 0% | 0% |
| D. Burins | 6.6% | 1.8% | 0% | 0.7% |
| 1. Angle on snap or old surface | 3.7% | 1.8% | 0% | 0% |
| 2. Carinated | 0.4% | 0% | 0% | 0% |
| 3. On straight truncation | 1.4% | 0% | 0% | 0.7% |
| 4. On obligue truncation | 0.4% | 0% | 0% | 0% |
| 5. Transverse | 0.4% | 0% | 0% | 0% |
| E. Perforator | 0.4% | 0% | 0% | 0% |
| 1. Typical | 0.4% | 0% | 0% | 0% |
| F. Truncations | 1.4% | 1.8% | 0% | 2.3% |
| 1. Blade | 1.4% | 1.8% | 0% | 2.3% |
| G. Notches | 3.7% | 5.4% | 11.3% | 7.7% |
| 1. On flake | 2.8% | 3.6% | 6.8% | 7.7% |
| 2. On blade | 0.9% | 1.8% | 4.5% | 0% |
| H. Denticulate | 2.3% | 1.8% | 0% | 10.0% |
| 1. Unilateral | 1.4% | 1.8% | 0% | 7.7% |
| 2. Bilateral | 0.9% | 0% | 0% | 2.3% |
| I. Retouched pieces | 26.4% | 16.3% | 15.9% | 31.0% |
| 1. Flake, continuous, obverse | 2.3% | 1.8% | 0% | 5.4% |
| 2. Blade, continuous, obverse | 4.7% | 0% | 2.2% | 3.1% |
| 3. Flake, continuous, inverse | 7.0% | 1.8% | 0% | 0% |
| 4. Blade, continuous, inverse | 2.3% | 0% | 2.2% | 3.1% |
| 5. Blade, continuous, bilateral | 0.9% | 0% | 0% | 1.5% |
| 6. Flake, discontinuous | 6.1% | 12.7% | 11.3% | 9.3% |
| 7. Blade, discontinuous | 2.8% | 0% | 0% | 8.5% |
| J. Special tools | 3.3% | 3.6% | 4.5% | 3.8% |
| 1. *Piece esquillee* | 0.4% | 0% | 0% | 0% |
| 2. Nahr Ibrahim retouch | 2.3% | 1.8% | 4.5% | 3.1% |
| 3. Varia | 0.4% | 1.8% | 0% | 0.7% |
| Total | 99.6% | 101.5% | 99.8% | 101.1% |

*Burins.* Angle burins on snaps or old surfaces dominate the class followed by truncation, carinated, and transverse types. Most have been formed by single-blows and are found on flakes.

*Perforator.* A single simple perforator created by lateral notches on a flake represents the class.

*Truncations.* These consist of straight and oblique forms created by light, marginal retouch on blades.

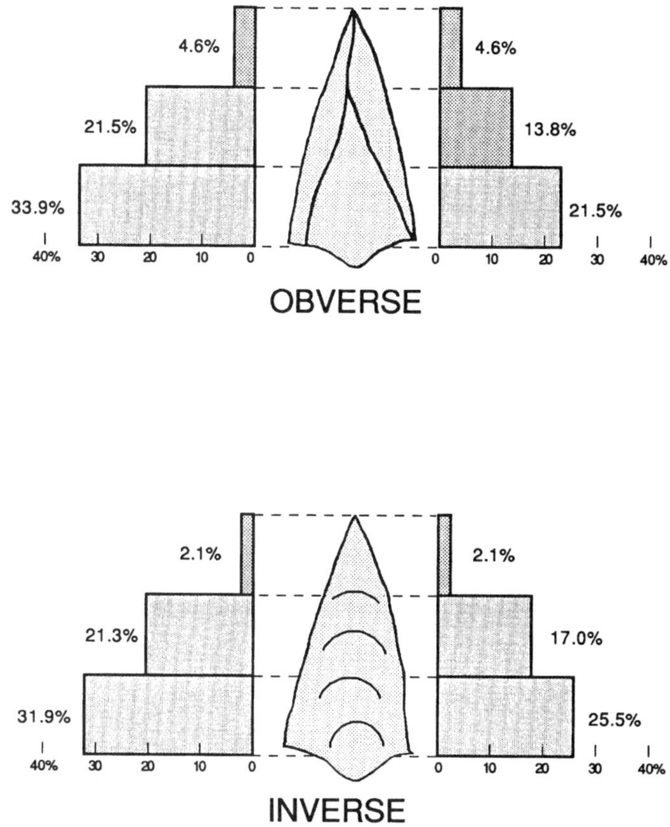


**Figure 5.13.** Schematic illustration of Levallois point showing the proportionate distributions of retouch for various locations on points recovered from Tor Faraj.

*Notches.* Most are single notches on flakes produced by steep, heavy retouch (Figure 5.17:f,g).

*Denticulates.* The class includes unilateral and bilateral types occurring for the most part on elongated flakes (Figure 5.15:d).

*Retouched Pieces.* Obverse and inverse retouched pieces are roughly balanced as is their distribution on blades and flakes (Figure 5.14:e; 5.16:a,c–e).

*Special Tools.* Flakes that have been truncated and faceted dominate the class (Figure 5.14:c,d). These have been extensively described for other Levantine Mousterian assemblages (Schroeder 1969; Solecki and Solecki 1970; Crew 1976; Goren-Inbar 1988). The modification generally has been attributed to thinning the specimens in order to enhance hafting. Crew (1976), however, was unable to detect any wear on such specimens in his microscopic study. In following the scheme of classification adopted for the Nahr Ibrahim assemblage (Solecki and Solecki 1970), types I (proximal truncation) and II (distal truncation) are about equally represented and all occur on the inverse surface. The retouch forming the truncations is composed of relatively large, generally uniform flake scars. In most cases the truncations served

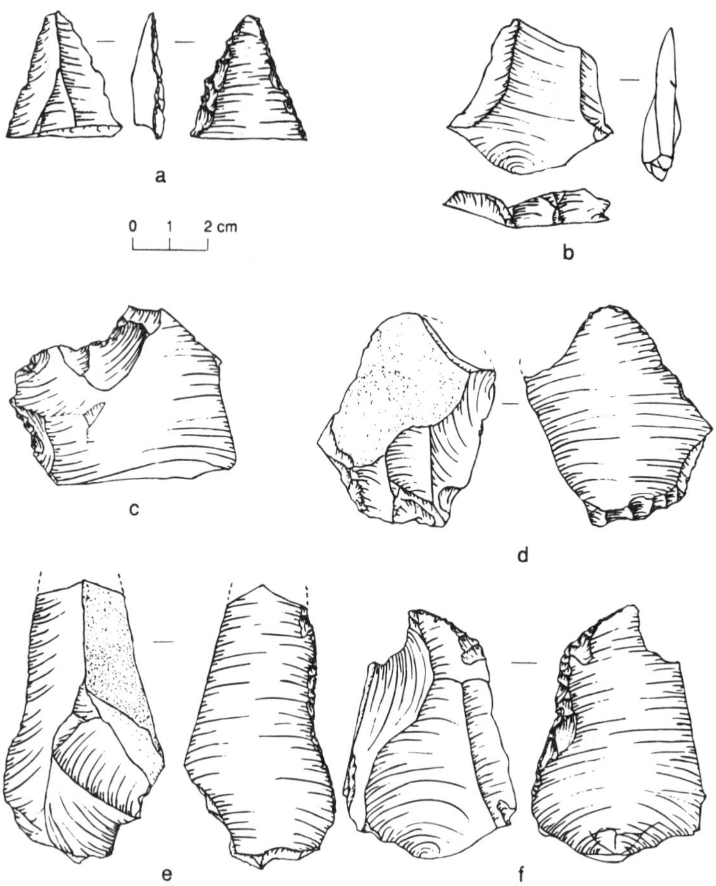


Figure 5.14. Illustrations of artifacts from Tor Faraj: (a) heavily retouched point tip, (b) point base, (c,d) truncated-faceted pieces, (e) inversely retouched blade, (f) inversely retouched sidescraper.

as platforms for removal of very small flakes. Those in which the truncation served as a platform for the removal of large flakes were classified as cores. Most of these exhibited opposed platforms.

## Retouch By Debitage Variety

An examination of the distribution of tools relative to the blanks upon which they were manufactured shows flakes and Levallois points to be the most common varieties, followed by blades and primary elements (Table 5.11). A comparison of this distribution with the proportionate representation of the varieties of unmodified blanks reveals that flakes and blades were selected for tools at about the same proportions as they occurred as blank forms. In other words, there appears to have been little preferential selection by the occupants of Tor Faraj for either of these forms as tools. In contrast, primary elements are proportionately underrepresented as blanks for tools while Levallois points are overrepresented (Table 5.11). This is consistent with the technological characteristics and reduction patterns which reflect an emphasis upon Levallois point production.

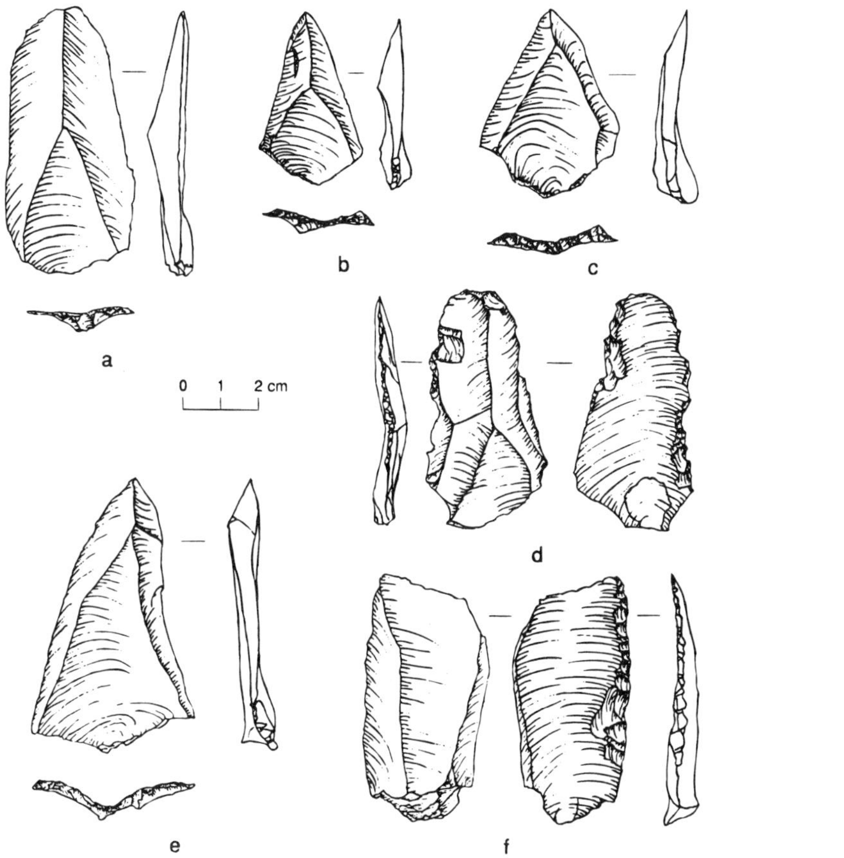


**Figure 5.15.** Illustrations of artifacts from Tor Faraj: (a) Levallois blade, (b,c,e) Levallois points, (d) inversely retouched denticulate, (f) inversely retouched sidescraper.

## Inverse Retouch: A Stylistic Indicator

Although obverse retouch is dominant, retouch present on the inverse surfaces of tools ranges from ca. 33–43% throughout the deposit (Table 5.12). The proportions of obverse and inverse retouch also remain relatively constant across tool classes. With a single exception (Kebara), the relative frequency of inverse retouch in other Levantine Mousterian assemblages is quite low, ranging from 0–4%. It appears beyond coincidence that the only other assemblage to register such a high proportion of inverse retouch comes from the nearby and partially synchronous site of Tor Sabiha.

## Tor Sabiha

The artifact assemblage from Tor Sabiha was recovered from a relatively thin (20–30 cm) culture bearing horizon in Layer C. Spatially this horizon was defined in a 13 $m^2$ block excavation and in two isolated units (8 and 21). The artifacts were concentrated in the northwestern portion of the block with over 60% of the assemblage coming from only two units (Figure 5.18).

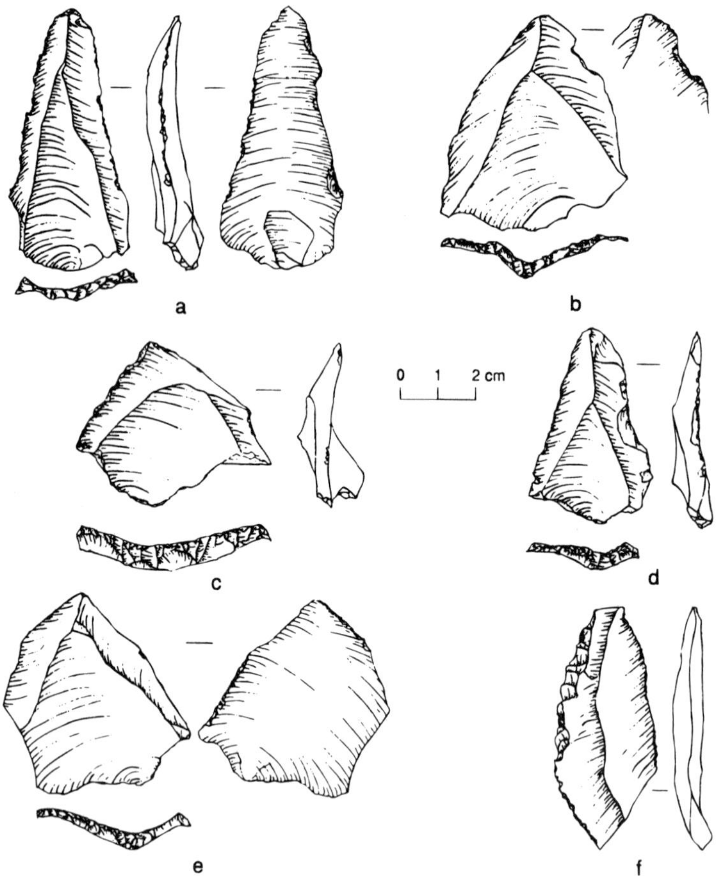


**Figure 5.16.** Illustrations of artifacts from Tor Faraj: (a,b,d) retouched Levallois points, (c,e) retouched rejuvenation flakes, (f) sidescraper on blade.

Chips account for the great majority of these and are principally responsible for the overall distributional pattern. In that their relative densities show a strong degree of spatial overlap with those of primary elements and cores, this small area of artifact concentration was most likely tied to initial core reduction.

The overall technotypologic profile of the assemblage, however, shows that it was final, not initial, processing which was emphasized by the occupants of the site. This observation is based upon the relatively low frequencies of cores and primary elements in the assemblage, coupled with relatively high proportions of flakes derived from secondary retouch. The little core preparation and blank production undertaken at the site focussed on the generation of Levallois points. Most of the points were apparently produced off the site (or at least outside the area of excavation) as indicated by a high core to point ratio. The importance of Levallois points as blanks finds expression in the toolkit in the dominance of unretouched and retouched forms. As at Tor Faraj, the points and other tools show a high proportion of inverse retouch.

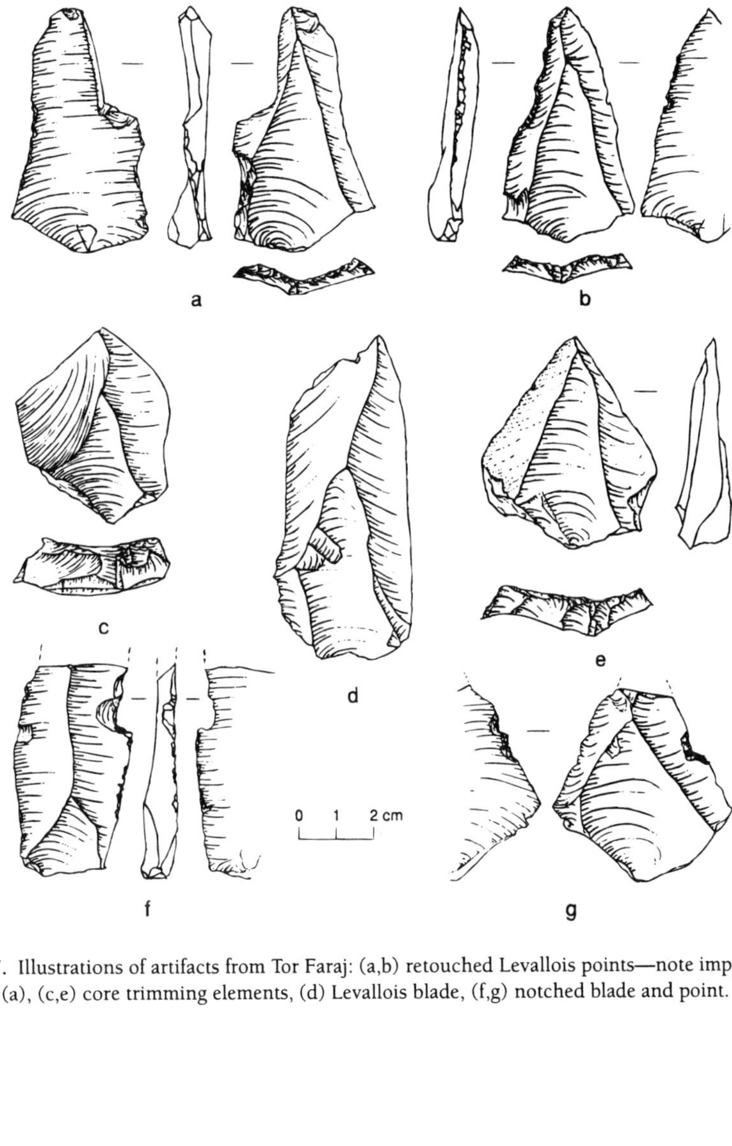


**Figure 5.17.** Illustrations of artifacts from Tor Faraj: (a,b) retouched Levallois points—note impact spall on (a), (c,e) core trimming elements, (d) Levallois blade, (f,g) notched blade and point.

**Table 5.11. Distribution of Retouch by Debitage Classes from Tor Faraj**

| | Top C | | Upper C | | Mid C | | Low C | |
|---|---|---|---|---|---|---|---|---|
| | N | % | N | % | N | % | N | % |
| Debitage class | | | | | | | | |
| Flake | 64 | 44.7 | 13 | 59.1 | 10 | 47.6 | 4 | 50.0 |
| Blade | 33 | 23.1 | 0 | 0.0 | 2 | 9.5 | 3 | 37.5 |
| Prim element | 6 | 4.2 | 2 | 9.1 | 1 | 4.8 | 0 | 0.0 |
| Lev point | 40 | 28.0 | 7 | 31.8 | 8 | 38.1 | 1 | 12.5 |
| Total | 143 | 100 % | 22 | 100 % | 21 | 100 % | 8 | 100 % |

Table 5.12. Distribution of Retouch on Surfaces of Retouched Tools from Tor Faraj

| | Top C | | Upper C | | Mid C | | Low C | |
|---|---|---|---|---|---|---|---|---|
| | N | % | N | % | N | % | N | % |
| Retouch | | | | | | | | |
| Obverse | 86 | 66.2 | 13 | 56.5 | 13 | 61.9 | 4 | 66.7 |
| Inverse | 44 | 33.8 | 10 | 43.5 | 8 | 38.1 | 2 | 33.3 |
| Total | 130 | 100% | 23 | 100% | 21 | 100% | 6 | 100% |

## Technology

*Raw Material.* Abundant chert sources are available within 1–2 km of the site in the limestone exposed along the 500 m high escarpment of the plateau and in the cobbles of the wadi beds on the floor of the Judayid Basin. Although over 90% of the chert identified within the assemblage came from the plateau and its drainages, distant sources (16–20 km) in the Rift Valley and near Jebel Humeima were also represented.

*Indices.* Despite the importance of Levallois point production, the poor representation of Levallois flakes and blades results in a low overall Levallois Index of 4.4 (Table 5.3). The assemblage yields a high Blade Index of 37.1 and an overall Facetting Index of 37.5. Points received the greatest attention in platform preparation followed by blades, flakes, and primary elements (Table 5.13).

*Reduction Stages, Cores, and Debitage Attributes.* The relatively low frequencies of cores and primary elements coupled with the high proportions of chips (flakes with maximum dimensions less than 20 mm) suggest that final over initial processing activities were emphasized at Tor Sabiha (Table 5.3). Given the close proximity of the site to abundant chert resources, it would be expected that a greater amount of core preparation and blank production would have been undertaken. As will be explored in Chapter 7, however, distance to raw material was only one of a complex set of factors that influenced decisions governing prehistoric stone tool technology.

The little core shaping and blank production that was undertaken appears to have been focussed on the generation of Levallois points. Four of the ten cores recovered from the site are point cores and most of the remaining cores, composed of truncated and facetted flakes and opposed platform varieties, are quite small and may well have served as point cores before their exhaustion (Table 5.14, Figure 5.19). Although the point cores display the largest facet dimensions of the core types, they appear to have been entirely exhausted when abandoned. Their mean maximum facet dimensions, for example, are some 23 mm shorter and 8 mm narrower than the mean Levallois point dimensions (Figure 5.20). In that primary elements are so small and poorly represented in the assemblage, it is surprising to find that three of the four point cores have substantial cortex on their backs. This certainly implies that cores were shaped for the most part away from the site.

The large dimensions of the points relative to primary elements and blades in conjunction with a point-to-total core ratio of 1:5.3 are consistent with many points having been produced off the site. A comparison of the dimensions of debitage shows points to be substantially larger than either primary elements or blades (Table 5.15; Figure 5.20). As discussed earlier, the preparation of a point core involved removing primary elements and blades with orientations that

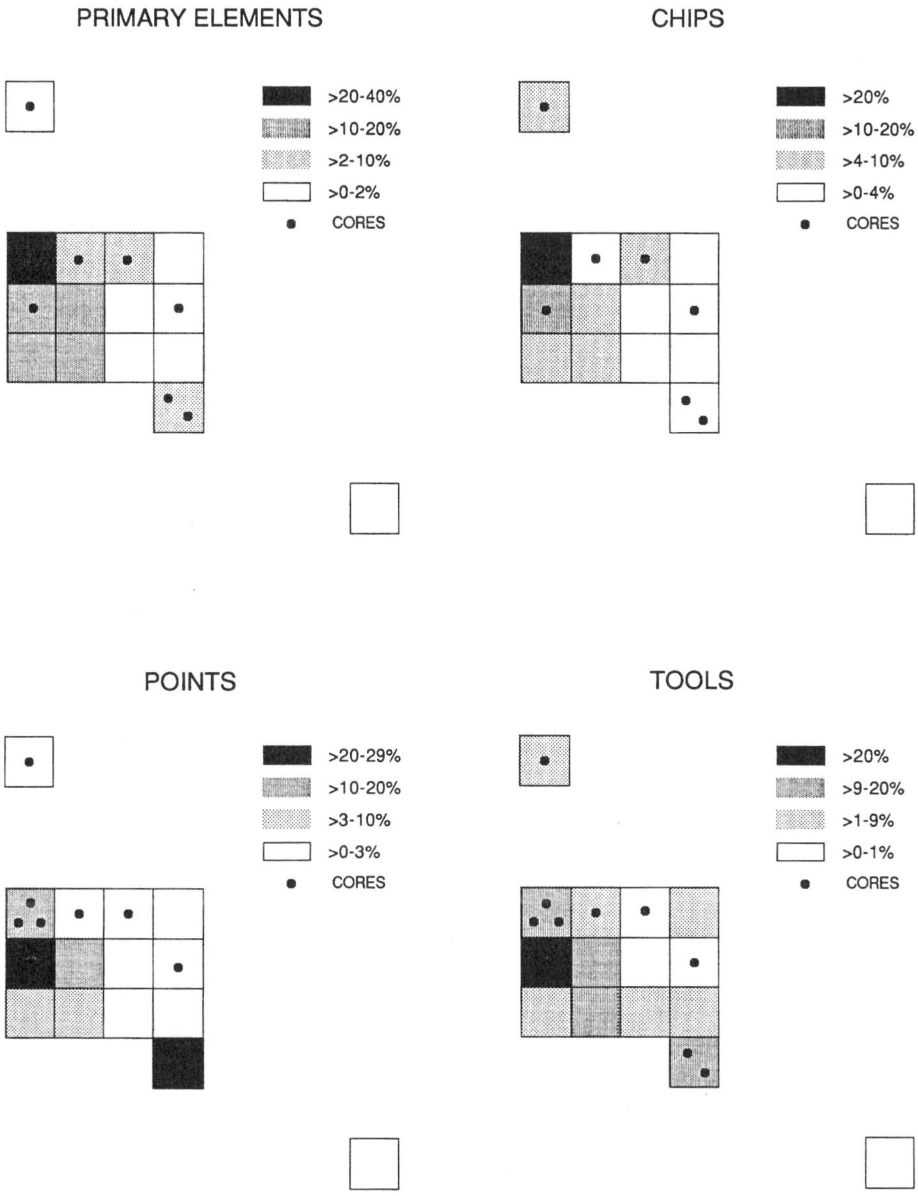


**Figure 5.18.** Plan of block excavation at Tor Sabiha showing the proportionate distributions of various artifacts.

yielded parallel or parallel-convergent scar patterns on its face for guiding the subsequent detachment of a point. The lengths of these removals would have commanded most of a core's face along the ultimate striking axis of the point removal, and thus one would expect the mean lengths of points, primary elements, and blades to be roughly similar if all of the remnants of this process were captured within an assemblage. However, the mean length of the points at Tor Sabiha exceeds the mean lengths of primary elements and blades by 14 mm and 18 mm, re-

Table 5.13. Platform Faceting by Debitage Class, Tor Sahiba

| | Prim elem | | Blade | | Flake | | Point | |
|---|---|---|---|---|---|---|---|---|
| Platform | *N* | % | *N* | % | *N* | % | *N* | % |
| Single | 11 | 47.8 | 37 | 39.4 | 47 | 47.0 | 10 | 20.1 |
| Dihedral | 2 | 8.7 | 14 | 14.9 | 22 | 22.0 | 6 | 12.5 |
| Multiple | 2 | 8.7 | 18 | 19.1 | 17 | 17.0 | 26 | 54.2 |
| Crushed | 5 | 21.7 | 14 | 14.9 | 12 | 12.0 | 2 | 4.2 |
| Cortex | 3 | 13.0 | 11 | 11.7 | 2 | 2.0 | 4 | 8.3 |
| Total | 23 | 100 % | 94 | 100 % | 100 | 100 % | 48 | 100 % |

Table 5.14. Core Types and Facet Dimensions, Tor Sabiha

| Core type | *N* | % | *x* | *x* |
|---|---|---|---|---|
| Levallois point | 4 | 40.0 | 32.4 | 20.6 |
| Cores on flakes | 3 | 30.0 | 24.5 | 15.1 |
| Cores w/ opposed and 90° platforms | 3 | 30.0 | 32.4 | 22.5 |
| Total | 10 | 100 % | | |

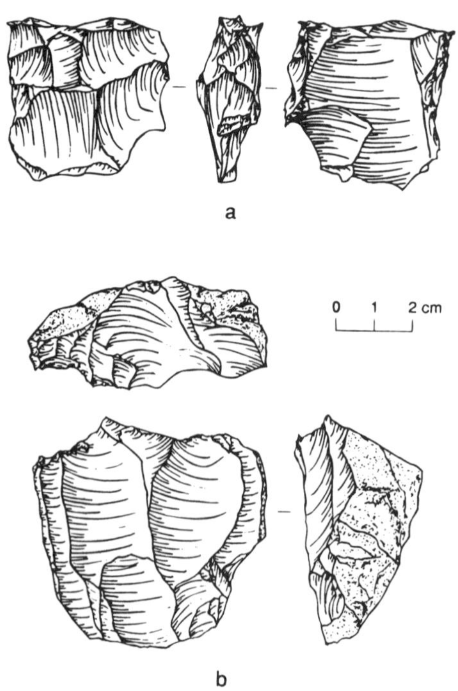


**Figure 5.19.** Illustrations of cores from Tor Sabiha: (a) core on flake, (b) unstruck point core.

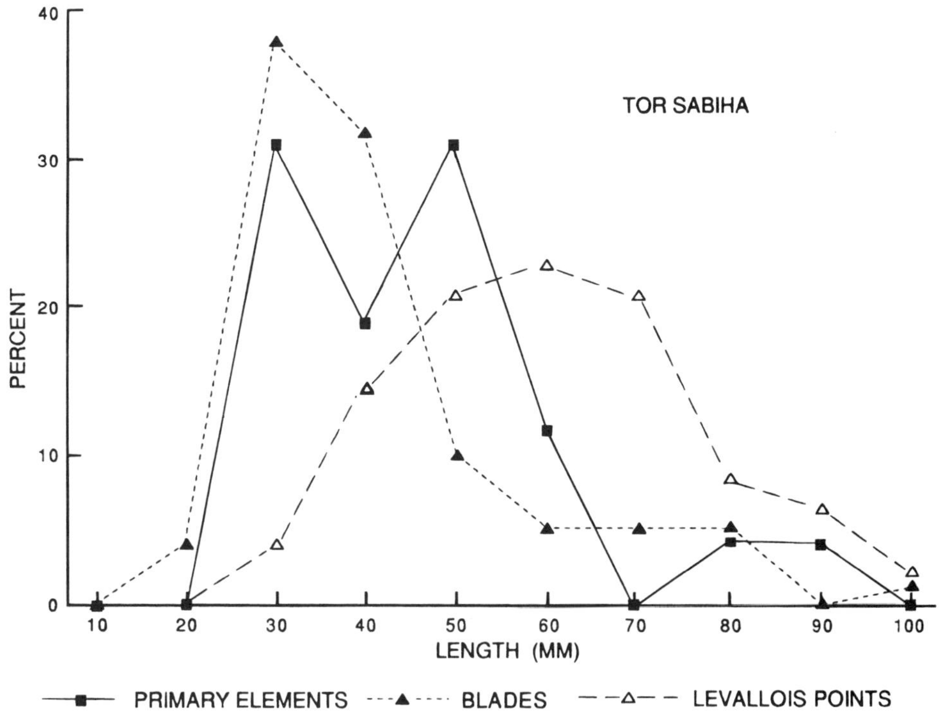


**Figure 5.20.** Graph showing the length distributions of primary elements, Levallois points and blades from Tor Sabiha. Note that the points are significantly longer than blades and primary elements.

**Table 5.15. Metric Data for Debitage from Tor Sahiba**

| | Prim Elem (N = 26) | | | Points (N = 46) | | | Blades (N = 100) | | | Flakes (N = 102) | | |
|---|---|---|---|---|---|---|---|---|---|---|---|---|
| | *r* | *x* | SD | *r* | *x* | SD | *r* | *x* | SD | *r* | *x* | SD |
| Max length | 25–85 | 41.3 | 15.3 | 24–98 | 54.9 | 16.5 | 19–91 | 36.9 | 15.2 | 15–72 | 28.7 | 9.9 |
| Max width | 12–49 | 26.8 | 2.2 | 15–52 | 28.6 | 7.5 | 7–40 | 18.6 | 7.5 | 8–48 | 25.5 | 7.5 |
| Max thick | 4–17 | 9.3 | 4.0 | 2–11 | 5.9 | 2.7 | 2–18 | 5.3 | 2.7 | 1–17 | 5.3 | 2.6 |
| Mid-pt wid | 11–46 | 23.9 | 11.4 | 11–37 | 23.7 | 5.8 | 5–35 | 16.0 | 6.9 | 7–45 | 23.6 | 7.2 |
| Mid-pt thk | 3–15 | 8.4 | 3.7 | 2–11 | 5.9 | 2.0 | 1–14 | 4.0 | 2.3 | 1–16 | 4.3 | 2.5 |

spectively. If the proportionate distribution of points, blades, and primary elements are compared by interval lengths, points clearly form a much larger portion of the distribution for lengths greater than 60 mm (Figure 5.20). In combination then, these data imply that while some points were produced at Tor Faraj, most were produced elsewhere and imported to the site.

*Loading.* The prevalence of diffuse bulbs, mostly with bulbar scars, on primary elements, blades, flakes, and points indicates a reliance upon hard-hammer percussion for their delivery (Table 5.16). The chips show a quite different pattern, however, for on chips, diffuse bulbs are only slightly more common than lipped ones. Given the linkage between lipped bulbs and soft-hammer percussion, this would imply that a substantial proportion of the chips were produced with a soft-hammer. Moreover, metrical analysis of the chips (following the procedure discussed under the artifacts of Tor Faraj) furnished a mean thickness of 1.6 mm with an essentially unimodal distribution ranging from 0.5–3.0 mm (Figure 5.12). One might recall that specimens which were experimentally produced by soft-hammer and pressure flaking furnished mean thicknesses of 1.9 mm and 1.3 mm, respectively. When coupled with the bulb characteristics of the chips, these data suggest that a considerable amount of soft-hammer percussion and even some pressure flaking were undertaken by the occupants of Tor Sabiha.

## Tools and Type List

The assemblage is dominated by Levallois points and retouched pieces followed by moderate frequencies of notches and denticulates (Table 5.10). Scrapers, burins, perforators, and truncations rarely occur.

*Levallois Points.* About 35% of the points have been retouched and, as within other retouched tool classes, a large proportion (58%) of these have been modified on their inverse surfaces (Figure 5.23:b). The retouch, typically light and marginal, is roughly balanced in its distribution on the lateral edges with a clear decline in frequency from the proximal to the distal ends of specimens (Figure 5.22). This pattern, common to both faces, is indicative of the points having been hafted. About 14% of the points display evidence of heat-treatment, whereas over 6% show large impact spalls.

Points exhibiting classic *chapeau de gendarme* and Y-arrete scar patterns (Figure 5.23:b) account for less than one-third of the specimens, while those with diamond/triangular butts and convergent scar patterns represent the large majority (i.e., 69%) of the points (Figure 5.21:d). Over half (51.1%) of the points were struck from bi-directionally prepared cores and tend to be elongated as reflected in an average length:width ratio of 1.9:1.

*Sidescrapers, Burins, and Truncations.* These classes are so poorly represented that typologic patterns are difficult to detect. The great majority of specimens are retouched inversely with blades and points providing the most common blank forms (Table 5.10; Figure 5.23:c,d).

*Notches and Denticulates.* These classes are well represented in the assemblage and show patterned differences in the distributions of retouch and blank selection. Notches were formed exclusively on flakes with obverse retouch. In contrast, denticulates occur about equally on

**Table 5.16. Bulb Attributes for Debitage and Debris from Tor Sabiha**

| | Prim elem | | Blade | | Flake | | Chip | |
|---|---|---|---|---|---|---|---|---|
| | N | % | N | % | N | % | N | % |
| Indeterminate | 2 | 7.7 | 2 | 2.0 | 3 | 2.9 | 1 | .5 |
| Lipped | 1 | 3.8 | 14 | 14.0 | 5 | 4.9 | 65 | 30.7 |
| Unlipped | 8 | 30.8 | 31 | 31.0 | 21 | 20.6 | 95 | 44.8 |
| Unlipped w/ scar | 13 | 50.0 | 53 | 53.0 | 73 | 71.6 | 0 | 0.0 |
| Other | 2 | 7.7 | 0 | 0.0 | 0 | 0.0 | 51 | 24.1 |
| Total | 26 | 100 % | 100 | 100 % | 102 | 100 % | 212 | 100 % |

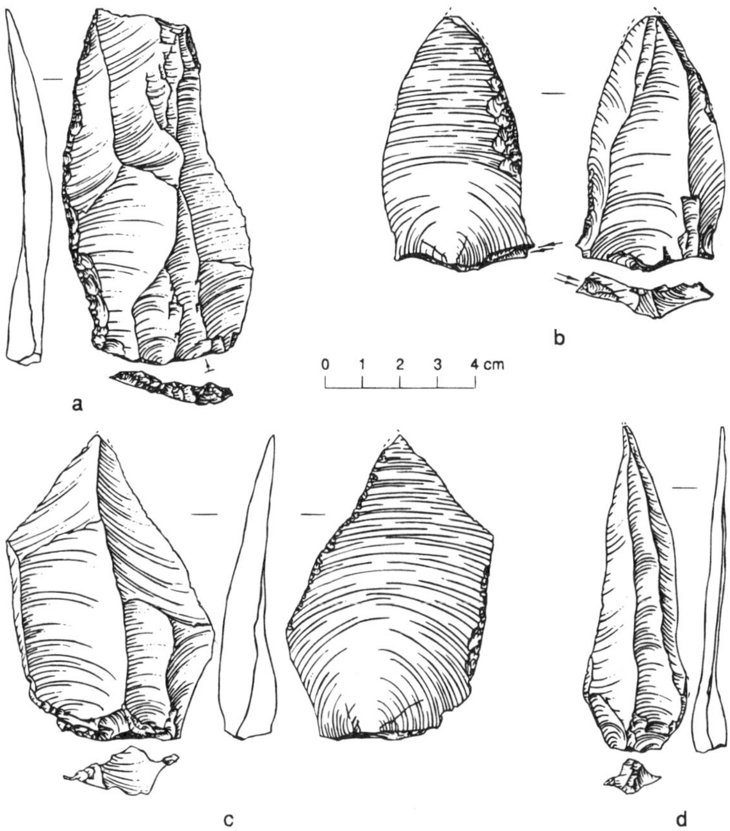


**Figure 5.21.** Illustrations of artifacts from Tor Sabiha: (a) sidescraper on Levallois blade, (b) inverse sidescraper on Levallois point, (c) inverse retouch on flake, (d) Levallois point.

flakes and blades/points with the retouch also being roughly balanced on obverse and inverse surfaces (Figure 5.23:a,e).

*Retouched Pieces.* Aside from points, this forms the largest class in the assemblage. The retouch is typically light and marginal (Figure 5.21:c). Unlike the distributional pattern of retouch displayed by most other tool classes, however, inverse retouch is relatively uncommon (i.e., 12.5%). As for blank selection, retouched pieces occur about equally on flakes and blades.

*Special Tools.* Flakes that have been truncated and facetted dominate the class. They are evenly distributed between proximal, distal, and laterally retouched varieties.

## Retouch By Debitage Class

When the retouched tools are examined relative to the debitage classes of the blanks upon which they were manufactured, flakes appear as the most commonly utilized blank followed by blades and points (Table 5.17). Not a single tool was fashioned from a primary element. A comparison of this distribution with that of unretouched blanks by debitage class indicates that blades and flakes were selected in about the same proportions as they occurred after production. On the other hand, primary elements were preferentially avoided for use as blanks for tools,

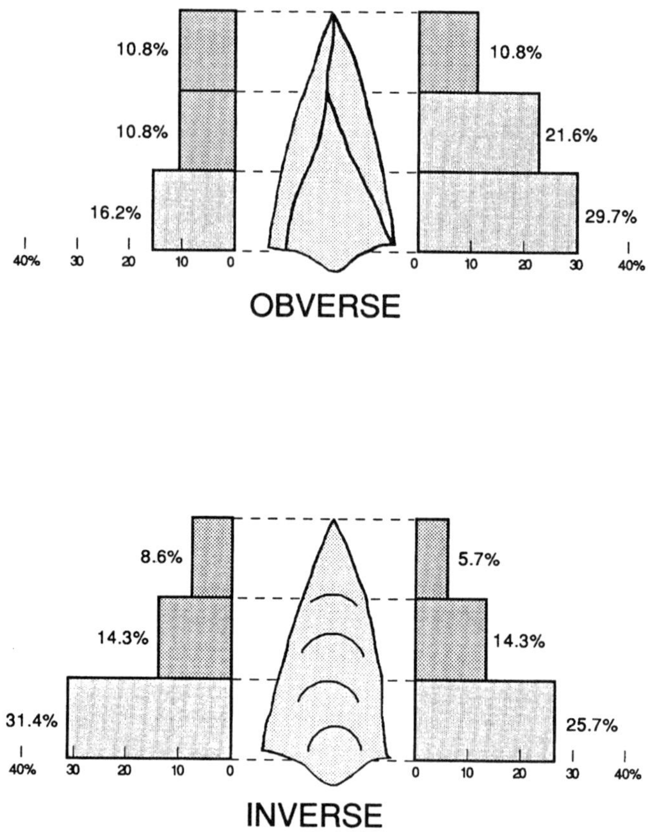


**Figure 5.22.** Schematic illustration of Levallois point showing the proportionate distributions of retouch for various locations on points recovered from Tor Sabiha.

while Levallois points were overwhelmingly preferred. Whereas Levallois points account for only 2% of the blanks, they constitute over 26% of the retouched tools.

### Inverse Retouch

A little over one-third of the retouched specimens from Tor Sabiha were modified on their inverse surfaces (Table 5.18). If the retouched piece class is excluded, leaving only those tool classes characterized by intensively worked edges (e.g., scrapers and denticulates), the relative proportion of inversely worked tools is 54%. As previously discussed, with the exception of those assemblages from Tor Faraj and Kebara, other Levantine Mousterian assemblages show very low proportions of inverse retouch.

## SUMMARY

The sheltered sites of Tor Faraj and Tor Sabiha appear to have been occupied ca. 70,000 years ago based upon amino acid racemization and uranium series dates of ostrich eggshell fragments. Paleoenvironmental data collected from their deposits indicate a generally arid climate,

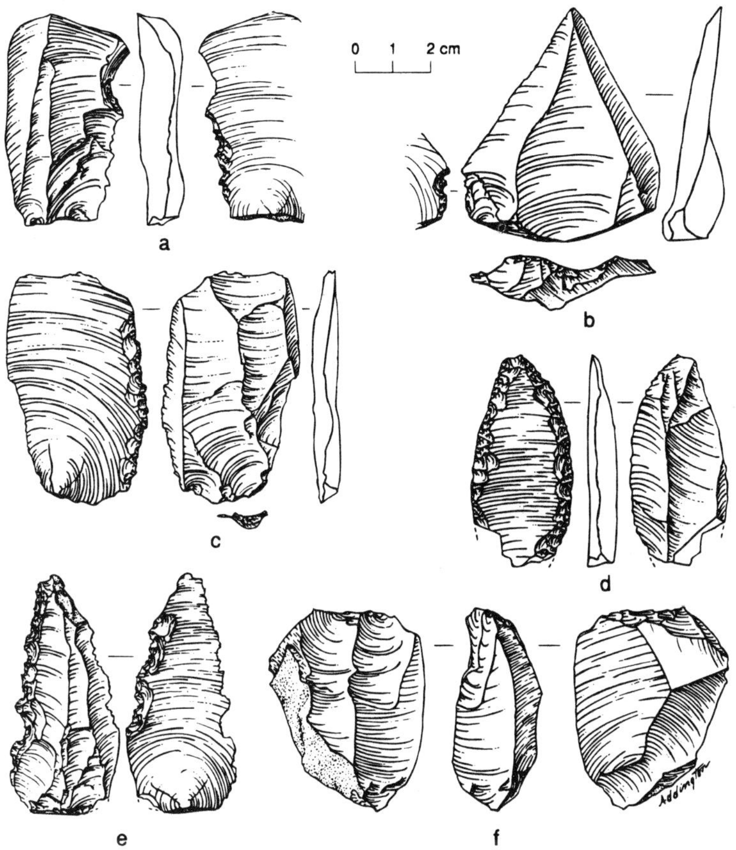


**Figure 5.23.** Illustrations of artifacts from Tor Sabiha: (a) denticulate, (b) Levallois point with inverse retouch, (c) inverse sidescraper, (d) inverse convergent sidescraper, (e) denticulate, (f) Levallois point core.

cooler and more moist than today. Low shrubs and grasses would have dominated the landscape with isolated stands of trees occurring along water courses and in the uplands.

Although the sites are only some 17 km apart, they have quite different local settings and resource catchments. With its eastern exposure, Tor Sabiha (ca. 400 m higher than Tor Faraj) is most likely to have served as a warm season camp. Depending on how early it was occupied in the season, standing water would have been available for its inhabitants within 0.5–3 km. They would also have been able to exploit numerous, high-quality chert sources that were exposed along the nearby escarpment of the plateau. In contrast, Tor Faraj would have provided an ideal cold-season camp spot with its southern exposure and lower elevation. Seasonal water would most likely have been available within the wadi beneath the shelter, but high-quality raw material would not have been found within the site's catchment.

The assemblages from the two sites reflect an emphasis upon the production of Levallois points from both unidirectionally and bidirectionally prepared cores. The majority of the points were generated using classic *chapeau de gendarme* basal preparation along with Y-arrete scar patterns, but a significant proportion of the points display compact diamond/triangular shaped butts and parallel convergent scars. Although the production of Levallois points was clearly a focus of the inhabitants of the two sites, the generation of other Levallois products was gener-

Table 5.17. Distribution of Retouch on Debitage from Tor Sahiba

| | Retouched | | As blanks | |
|---|---|---|---|---|
| | N | % | N | % |
| Debitage Class | | | | |
| Flake | 41 | 43.2 | 843 | 20.1 |
| Blade | 29 | 30.5 | 652 | 38.8 |
| Prim elem | 0 | 0.0 | 152 | 9.0 |
| Lev pt | 25 | 26.3 | 34 | 2.0 |
| Total | 95 | 100% | 1681 | 100% |

Table 5.18. Distribution of Retouch on Surfaces of Retouched Tools from Tor Sabiha

| | N | % |
|---|---|---|
| Retouch | | |
| Obverse | 57 | 66.3 |
| Inverse | 29 | 33.7 |
| Total | 86 | 100% |

ally ignored as reflected in the low Levallois Indices of the assemblages. The toolkits of the sites display roughly similar configurations when tool classes are compared. The differences that exist relate to the higher frequencies of burins at Tor Faraj and higher frequencies of notches and denticulates at Tor Sabiha. The presence of high proportions of inverse retouch serves to distinguish the assemblages of Tor Sabiha and Tor Faraj from most other Levantine Mousterian assemblages and may well represent a stylistic attribute.

The major differences in the assemblages are found in the stages of reduction they represent, not in techniques or styles of blank production and tool fabrication. While all stages of reduction are represented at the sites, much more core shaping and blank production was undertaken at Tor Faraj than at Tor Sabiha. Tool fabrication, maintenance, and rejuvenation were carried out at both sites, but these appear to have been the focal activities at Tor Sabiha.

In Chapter 7, the differences in the settings and the material residues of the two sites are explored more fully in an effort to better understand Middle Paleolithic cultural ecology and cognition. Logistical and opportunistic behaviors for provisioning the two sites are contrasted and placed within the context of an adaptive strategy that relied upon transhumance. These behaviors are then examined relative to different cognitive patterns as identified in developmental psychology.

# Chapter 6

# Lithic Microwear Analysis of Tor Faraj Rockshelter

JOHN J. SHEA

Whether in ethnographic or in archaeological contexts, the use of tools can be best understood in terms of "operational sequences" (Creswell 1976). Operational sequences involve four linked sets of activities; material procurement, manufacturing, utilization, and discard. Many analyses of Levantine Mousterian assemblages have attempted to reconstruct the adaptive context of stone tool production and use (Binford 1968; Jelinek 1982; Munday 1979; Marks 1988; Bar-Yosef and Meignen 1992). Yet, most of these reconstructions have emphasized only the initial stages of the operational sequence. Actualistic studies suggest that tool utilization and discard behavior have a profound effect on the structure of the archaeological record (see examples in Schiffer 1987). Thus, our ignorance of Levantine Mousterian tool use and discard patterns presents a major obstacle to a comprehensive explanation of Levantine Mousterian industrial variability.

In order to shed light on patterns of prehistoric stone tool use and discard behavior, archaeologists have increasingly turned to the analysis of microscopic wear patterns on the edges of stone tools (Keeley 1980). Microwear analysis has been performed on Middle Paleolithic collections from France (Beyries 1988; Anderson-Gerfaud 1990), Eastern Europe (Schelinski in Plisson 1988), and the Levant (Shea 1991). This chapter reports the results of a microwear analysis performed on the Levantine Mousterian lithic assemblage from Layer C of Tor Faraj rockshelter.

## THE TOR FARAJ SAMPLE AND METHODS OF ANALYSIS

The lithic assemblage from Tor Faraj rockshelter Layer C presents an ideal opportunity to examine patterns of Levantine Mousterian tool use. Level C consists largely of uncemented sediment and ash lenses, and the edges and surfaces of the tools from this level are generally well-preserved. The technotypological characteristics of the Tor Faraj assemblage are described in Henry (1992, Chapter 5).

All of the tools larger than 2.5 cm in any dimension were examined for lithic use-wear traces using a Nikon SMZ stereomicroscope at magnifications between 5–120×. At these magnifications, it is possible to observe the major features of the microfracturing, striations, abrasive rounding, and some aspects of the polish that together comprise lithic use-wear (Keeley

1980; Odell 1981). Selected tools were also examined using a binocular high-magnification (70–300×) microscope following the methods described by Keeley (1980).

Microwear was identified as use-wear on those parts of edges where overlapping concentrations of microfractures and abrasive wear exceeded the minor "background" wear that occurred on most stone tool edges. Once an edge had been identified as possibly worn from use, both the tool and the worn edge itself were registered and described in terms of a series of variables noting the provenance of the tool, its technological and typological characteristics, the morphology of the worn edge, and microwear phenomena (see Appendix 6.1). For each worn edge, microwear characteristics were recorded in terms of the values of nine use-wear variables. These variables noted the distribution, orientation, and morphology of microfractures, striations, abrasive rounding, and polish. In replicative experiments, systematic comparisons of the values of these nine variables, or "wearcodes," allowed wear patterns from different modes of stone tool use to be reliably discriminated from each other (Shea 1991:70–117). The functions of archaeological tools were reconstructed by comparing archaeological wearcodes to those recorded in an experimental reference collection. The data records for all of the worn edges identified in the Tor Faraj C assemblage are presented in Appendix 6.2.

The microwear evidence from Tor Faraj will be discussed below, first in terms of the inferred occurrence of use-wear on different artifact categories, and then in terms of the specific modes of use inferred from patterned variation among microwear traces.

## LOCATION OF USE

Of the 824 artifacts examined from Tor Faraj C, 93 exhibit microwear patterns referable to stone tool use (Table 6.1). Some of these tools are illustrated in Figure 6.1 accompanied by photomicrographs in Figures 6.2–6.9. About half of the artifacts in the use-wear sample ($n$ = 50 or 53.8%) exhibit only one worn edge, while at least 11 artifacts (12%) feature as many as four discrete concentrations of wear. A total of 169 worn edges were identified.

Most of the use-worn tools are largely-complete "whole" flakes. Only one "core" and eight flake fragments preserve use-wear traces. While oval flakes and blades considerably outnumber points among the "complete flakes" in the Level C assemblage, points and oval flakes are about equally common among the use-worn tools. As is the case in most Levantine Mousterian assemblages thus far examined (Shea 1991), pointed artifacts exhibit wear traces far more frequently than their numbers in an assemblage might suggest. Blades feature recognizable microwear traces about twice as often as do oval flakes. Curiously, however, this ratio between blades and flakes is reversed among the Levallois tools, where Levallois flakes exhibit microwear nearly three times more often (53%) than do Levallois blades (16%).

Levallois artifacts comprise only 16% of the Tor Faraj assemblage, but they represent 59% of all the worn tools. Although many Levallois tools were probably made in one place and discarded in another, the high rate at which Levallois debitage exhibits microwear could suggest that the greater outlay of flintknapping effort into the production of Levallois artifacts resulted in a higher incidence of Levallois debitage being selected for use. Among the worn Levallois tools, Levallois points ($n$ = 31) are far more common than either blades ($n$ = 6) or oval flakes ($n$ = 16), roughly paralleling the raw frequency of these artifact-types in the Tor Faraj assemblage.

All of the twenty-nine retouched tools in the Tor Faraj assemblage exhibit microwear traces. While microwear coincides with retouch on most of these, this is not always the case. Where retouched edges do not appear to be use-worn it is possible that the edge was modified to facilitate a more secure grip or to better lash the stone tool to a handle. It is also possible that many retouched edges may not have been used for a sufficient length of time and/or against hard enough material to create a detectable wear pattern.

Table 6.1. Occurrence of Microwear on different categories of Artifacts in the Tor Faraj C assemblage

| Artifact categories | Total *n* | Total % | Worn *n* | Worn % | % of worn |
|---|---|---|---|---|---|
| Total artifacts | 824 | 100.0 | 93 | 11.2 | 100.0 |
| Whole flakes | 590 | 71.6 | 84 | 14.2 | 90.3 |
| Points | 65 | 7.9 | 35 | 58.3 | 37.6 |
| Blades | 93 | 11.3 | 14 | 15.1 | 15.1 |
| Flakes | 432 | 52.4 | 35 | 8.1 | 37.6 |
| Levallois Debitage | 132 | 16.0 | 55 | 41.7 | 59.1 |
| Points | 65 | 7.9 | 31 | 47.7 | 33.3 |
| Blades | 37 | 4.5 | 6 | 16.2 | 6.4 |
| Flakes | 30 | 3.6 | 16 | 53.3 | 17.2 |
| Non-Levallois debitage | 458 | 55.6 | 29 | 6.3 | 31.2 |
| Retouched tools | 29 | 3.5 | 29 | 100.0 | 31.2 |
| Unretouched artifacts | 789 | 95.7 | 64 | 8.1 | 68.8 |
| Cortical flakes | 173 | 21.0 | 25 | 14.5 | 26.9 |
| Debris | 228 | 27.6 | 8 | 3.5 | 8.6 |
| Cores | 6 | 0.7 | 1 | 16.7 | 1.0 |
| Tools with | | | | | |
| one worn edge | | | 50 | | 53.8 |
| two worn edges | | | 21 | | 22.5 |
| three worn edges | | | 11 | | 11.8 |
| four worn edges | | | 11 | | 11.8 |

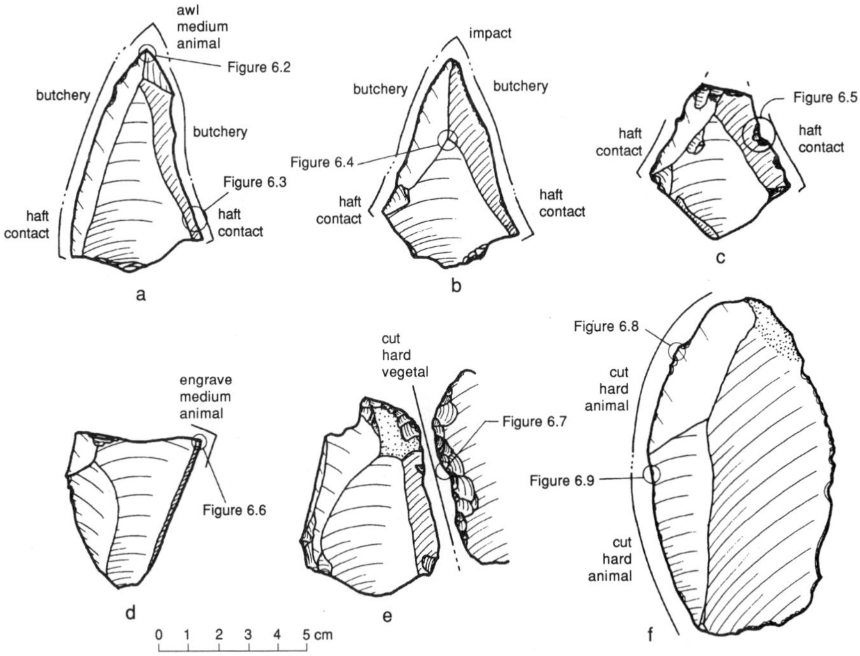


**Figure 6.1.** Selected worn artifacts from Tor Faraj Level C with notes on inferred modes of use: (a, b) Levallois points; proximal fragment of Levallois point; proximal flake fragment with burin facet; inversely-retouched sidescraper on Levallois flake; unretouched Levallois flake. The locations of photomicrographs appearing in Figures 6.2–6.9 are indicated.

Figure 6.2. Photomicrograph of microfractures on the tip of the Levallois point in Figure 6.1:a (25×). The oblique orientation of these microfractures suggests a twisting ("awling") motion during use.

Figure 6.3. Photomicrograph of a cluster of small microfractures on the proximal part of the right lateral edge of the Levallois point in Figure 6.1a (25×). The clustering of these scars suggest contact with lashing fibers.

**Figure 6.4.** Photomicrograph of abrasion ("bright polish") on the dorsal ridge of the Levallois point in Figure 1b (120×). The location of this abrasion suggests contact between the tools and a wooden handle.

**Figure 6.5.** Photomicrograph of a rather large cluster of microfractures on the left lateral edge of the Levallois point fragment in Figure 6.1c (8×). The restricted character of this wear, the presence of a nearly-identical scar cluster on the opposite edge, and the abrasion of the dorsal surface of the tool between these clusters suggests forceful contact with lashing fibers. The numerous large bending fractures near the distal part of this tool suggest it was possibly damaged from having been used as a spear point, but the lateral snap is in and of itself not diagnostic of this activity.

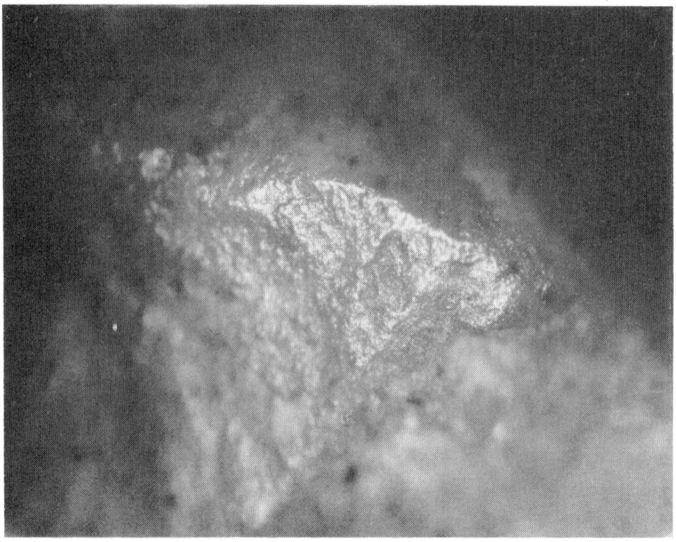

**Figure 6.6.** Photomicrograph of abrasion (matte polish) on the burin *biseau* of the flake fragment in Figure 6.1d (160×). Such wear is consistent with an engraving motion performed on a semirigid, nonsiliceous material, possibly leather.

**Figure 6.7.** Photomicrograph of "bright polish" on the ventral side of the retouched edge of the sidescraper in Figure 6.1e (120×). Such polish, with its sharply delimited border, suggests a rigid, moderately siliceous material, possibly some kind of hardwood.

**Figure 6.8.** Photomicrograph of extensive "matte polish" on the upper left lateral edge of the large Levallois flake in Figure 6.1f (160×). This abrasive wear is accompanied by numerous small bending fractures and is most plausibly referable to the cutting of a yielding or semirigid, nonsiliceous material. Here, such wear is attributed to butchery.

**Figure 6.9.** Photomicrograph of extensive "matte polish" on the lower left lateral edge of the large Levallois flake in Figure 6.1f (160×). The wear preserved on this edge is essentially the same as shown in Figure 6.8.

Much of our perception of Mousterian industrial variability is shaped by characterizations and relative frequency measurements of retouched tools and Levallois debitage (Rolland and Dibble 1990). In this regard, it is reassuring that both retouched tools and Levallois debitage exhibit microwear traces more often than do non-Levallois debitage and unretouched artifacts. It should be noted, however, that a substantial proportion of the tools from Tor Faraj exhibiting microwear traces are non-Levallois debitage (31.2%), and moreover, that most of the worn tools are unretouched (68.8%). Indeed, retouched edges ($n$ = 26) amount to only 15.4% of all the edges exhibiting microwear traces. Considered together, this suggests that technotypological characterizations of Middle Paleolithic variability that emphasize the retouched tool component (e.g., Bordes 1961) are unlikely to monitor the totality of functional variability, at least in Levantine Mousterian assemblages.

## INTERPRETATION OF WEAR PATTERNS

A total of 169 worn edges have been identified in the Tor Faraj assemblage. A selection of these worn edges are illustrated in Figure 6.1. Wear interpretations are based on a systematic comparison between the archaeological wear patterns and a reference collection of more than 1000 tools worn in projectile point use, butchery, hide-working, bone-carving, woodworking, and a variety of other activities, as well as "control samples" of tools worn from trampling.

### Reconstructing Tool Functions

For each worn edge, an attempt was made to reconstruct tool motion during use, the worked materials, and the kinds of activity in which the edges were likely to have been employed.

### Tool Motion

The frequencies with which different motions have been reconstructed for worn edges in the Tor Faraj C assemblage are listed in Table 6.2. Most of the worn edges (51.5%) exhibit bifacially-symmetrical microfracturing and abrasion, suggesting they had been used in a longitudinal cutting motion. A smaller number (9.5%) feature the asymmetrical microfracturing and abrasion that typically results from edge-transverse "scraping" motions. Large-scale microfracturing suggesting "adzing or chopping" motions occurs on two edges. Similarly-large bending fractures on the tips of twenty-seven pointed artifacts (16.0% of worn edges) suggest use in a forceful piercing motion. Although lateral snapping occurred frequently in projectile experiments (see also Odell and Cowan 1986), this kind of breakage pattern can result from many other kinds of wear vectors, including trampling and soil compaction; thus lateral snaps were not considered diagnostic of piercing motion. Four points and projections display the intensely-clustered microfractures and directional abrasive wear consistent with awling and engraving motions.

Clusters of small microfractures located symmetrically on opposite sides of the same tool and accompanied by roughening of adjacent ridges and surfaces of the tool suggest the kind of sliding contact that occurs between a stone tool and either lashing fibers or the handle to which it has been attached. These wear patterns were assessed as worn from "haft contact." Most of these wear patterns occur on the proximal ends of pointed artifacts, especially Levallois points, but there are also some truncated flakes that appear to feature wear from "haft contact."

### Worked Materials

Because high-magnification analysis of polishes was not performed on all of the worn edges, specific identifications of worked materials (e.g., wood, hide, bone, meat) were not attempted. Rather, a more-general approach was employed, one that utilized independent criteria for the

**Table 6.2. Frequency of Different Tool Motions Inferred for the Tor Faraj C Microwear Sample**

| | Worn edges | |
|---|---|---|
| Inferred tool motion | *n* | % |
| Cut | 87 | 51.5 |
| Scrape | 16 | 9.5 |
| Adze/chop | 2 | 1.2 |
| Pierce | 27 | 16.0 |
| Engrave | 2 | 1.2 |
| Awl | 2 | 1.2 |
| Haft contact | 31 | 18.3 |
| Unknown/indeterminate | 2 | 1.2 |

recognition of worked material resistance and its relative siliceousness. As a result, worked material identifications were made in terms of the categories described in Table 6.3.

Most of the materials that worked with stone tools can be characterized as having either yielding, semirigid, or rigid resistance properties. In experiments, the size and intensity of the microfracturing, and the degree to which abrasive wear (polish and striations) are more or less invasive than the microfracturing, have proven to be reliable indicators of worked material resistance (Odell 1981).

On edges used against yielding materials, such as soft plant matter or meat, microfractures are small (generally less than 1 mm from initiation to termination) and tend to be arrayed in a single row of nonoverlapping scars, while abrasive wear extends far beyond the microfractures onto the surface of the tool. On edges used against semirigid materials, such as tanned leather or wood, microfractures are somewhat larger (more than 2 mm from initiation to termination, although this varies with loading speed) and abrasion is about equally as invasive as these microfractures. On edges used against rigid materials, such as bone, microfractures are generally more than 3–4 mm from initiation to termination (although again this varies with loading speed), scars are arrayed in "stacks" of hinge- and step-terminated fractures, and abrasion does not usually extend onto the edge beyond these microfractures.

The majority of worn edges in the Tor Faraj C assemblage (55.0%) display wear patterns suggesting semirigid materials (Table 6.4). Equal numbers of worn edges are referable to yielding materials (8%) as are referable to rigid materials (8%). No specific identification of worked material resistance could be made for the tips of the pointed artifacts worn from piercing motions, or for the clusters of scars worn from haft contact, although materials of either rigid or semi-rigid resistance seem about equally likely.

Recent studies suggest the silica content of worked materials has a major effect on the quality of light reflected from the polished surfaces of flint tools (Fullagar 1991). When viewed at magnifications greater then 80–100×, polishes that result from the working of nonsiliceous animal tissues generally exhibit a dull or "matte" appearance. In contrast, polishes on edges used on moderately-siliceous wood and soft plant tissues exhibit a "bright" appearance. Truly "vitreous" polishes result primarily from cutting highly-siliceous grasses and from tilling soil, of which a silica is a major component (Keeley 1980). Although the differences between, "matte," "bright," and "vitreous" light reflection in the Tor Faraj sample were assessed visually, and are thus subjective perceptions of wear phenomena, preliminary studies suggest these differences can be measured and quantified (Keeley 1980:62–63).

Table 6.3. Worked Material Categories Obtained by "Nesting" Inferred Resistance and Siliceousness (each cell lists the materials referable to a particular category in the author's experiments)[a]

| | Inferred Silica Content | | |
|---|---|---|---|
| Inferred resistance | Moderately nonsiliceous (matte polish) | Siliceous (bright polish) | Highly siliceous (vitreous polish) |
| *Yielding*<br>(abrasion > microfracturing small feather terminated fractures | *Soft Animal*<br>fresh skin<br>meat<br>fat<br>hair | *Soft Vegetal*<br>reeds<br>roots & tubers<br>leaves<br>nonwoody plant parts | *Soft Inorganic*<br>tropical grasses<br>fresh bamboo<br>loose sand |
| *Semirigid*<br>(abrasion = microfracturing, medium [1–2 mm] fractures with mixed feather, step, and hinge terminations) | *Medium Animal*<br>dried skin<br>leather<br>dried or frozen meat<br>fish butchery<br>cartilage<br>soaked antler<br>boiled horn<br>weathered bone | *Medium Vegetal*<br>Various soft woods including birch, maple, oak, fir, pine, willow, acacia | *Medium Inorganic*<br>sand<br>silt<br>clay |
| *Rigid*<br>(abrasion < micro-fracturing, large step and hinge fractures) | *Hard Animal*<br>fresh bone<br>antler<br>horn<br>ivory<br>mollusc shell | *Hard Vegetal*<br>dried and seasoned wood such as oak, maple<br>some tropical hardwoods | *Hard Inorganic*<br>stone<br>ceramics<br>gravel |

[a]Tips of points and edges/surfaces worn from haft contact generally exhibit too little abrasive wear to be unambiguously referable to one of these categories.

Of the worn edges in the Tor Faraj sample, 67 (39.7%) exhibited matte polishes and 37 (22.0%) exhibited bright polishes (Table 6.4). No worn edges exhibited vitreous polishes. A clear determination of polish brightness could not be made for 65 or 38.5% of the worn edges. Most of the latter ($n$ = 58) were either the tips of points worn from a piercing motion or the edges and adjacent tool surfaces worn from "haft contact."

By dividing the worn edges assigned to "yielding," "semirigid," and "rigid" categories into subgroups on the basis of polish brightness, each worn edge could be assigned either to one of the nine worked material categories indicated in Table 6.4, or to an indeterminate category.

"Medium animal" is the combined resistance and relative siliceousness category to which the largest percentage of worn edges were assigned (32.0%). The next-most-common category was "medium vegetal" (20.7%). Edges worn from "haft contact" and "piercing," both of which usually occur on pointed artifacts, together amount to fully 34.4% of the worn edges.

## Reconstructing Activities

Some insights into the likely activities in which worn edges were employed can be gained by considering which combinations of tool motions and worked materials occur together. When these combinations are tabulated (Table 6.5), "cut/medium animal" is by far the most common

**Table 6.4. Worked materials inferences for the Tor Faraj C Microwear sample**

| Worked material characteristics | Worn edges | |
|---|---|---|
| | *n* | % |
| Resistance | | |
| Yielding (abrasion > microfracturing) | 8 | 4.7 |
| Semirigid (abrasion = microfracturing) | 93 | 55.0 |
| Rigid (abrasion < microfracturing) | 8 | 4.7 |
| Indeterminate | 60 | 35.5 |
| Polish brightness | | |
| Matte | 67 | 39.7 |
| Bright | 37 | 21.9 |
| Indeterminate | 65 | 38.5 |
| Inferred Worked Material Category | | |
| Soft animal | 6 | 3.6 |
| Soft vegetal | 1 | 0.6 |
| Soft indeterminate | 1 | 0.6 |
| Medium animal | 54 | 32.0 |
| Medium vegetal | 35 | 20.7 |
| Medium indeterminate | 4 | 2.4 |
| Hard animal | 7 | 4.1 |
| Hard vegetal | 1 | 0.6 |
| Unknown | 2 | 1.2 |
| Indeterminate (pierce, Haft Contact) | 58 | 34.3 |

**Table 6.5. Tool motion and worked materials combinations for Tor Faraj C.**

| Tool motion | Worked material | *n* EUs | % EUs | Inferred activity |
|---|---|---|---|---|
| Cut | Soft animal | 6 | 3.6 | Butchery |
| | Soft vegetal | 1 | 0.6 | Soft plant |
| | Medium animal | 49 | 29.0 | Butchery |
| | Medium vegetal | 22 | 13.0 | Woodworking |
| | Medium indeterminate | 3 | 1.8 | Unknown |
| | Hard animal | 5 | 3.0 | Bone contact |
| | Hard vegetal | 1 | 0.6 | Woodworking |
| Scrape | Medium animal | 3 | 1.8 | Hide working |
| | Medium vegetal | 10 | 5.9 | Woodworking |
| | Medium indeterminate | 1 | 0.6 | Unknown |
| | Hard animal | 2 | 1.2 | Bone contact |
| Adze/Chop | Medium vegetal | 2 | 1.2 | Woodworking |
| Grave | Soft indeterminate | 1 | 0.6 | Unknown |
| | Medium vegetal | 1 | 0.6 | Woodworking |
| Awl | Medium animal | 2 | 1.2 | Hide working |
| Pierce | Indeterminate | 27 | 16.0 | Impact |
| Sliding | Haft contact | 31 | 18.3 | Haft contact |
| Unknown | Unknown | 2 | 1.2 | Unknown |

(49 worn edges or 29.0%). "Piercing" and "haft contact" together account for an additional third of the worn edges. Edges reconstructed as worn from cutting or scraping "medium vegetal" materials together comprise less than one-fifth (18.9%) of the microwear sample.

Where possible, each of the inferred motion and worked material combinations listed in Table 6.5 has been assigned to one of seven "activities" (impact, butchery, hide-working, bone contact, plant processing, woodworking, and haft contact). These assignments were based on the ways in which these combinations occurred together in the experimental program. For example, all worn edges inferred to have been used against either "medium vegetal" or "hard vegetal" materials have been grouped together in the "woodworking" activity category. In experiments, the combinations, "cutting soft animal" and "cutting medium animal," occurred both in butchery and hide-working. Here, both of these combinations have been assigned to the "butchery" activity category. Thus, the "hide-working" category, as it is constituted here (including only edges worn from scraping or from awling), registers only the minimum number of worn edges that are plausibly referable to that activity.

It has been argued elsewhere (Shea 1988) that the wear referable to piercing motions and haft contact on pointed Levantine Mousterian artifacts results from the use of hafted stone spear points. Unfortunately, it has been proven impossible to differentiate the wear on the stone armatures of throwing spears from the wear on the tips of lances thrusting spears. Accordingly, all tips of points worn from piercing motions have been assigned to the "impact" activity category.

The somewhat ambiguous terms, "plant processing" and "bone contact," reflect the difficulty of assigning specific motion and worked material combinations to more specific activities. Plant processing could reflect food preparation or the cutting of brush either for fire or for fuel. Similarly, tools can come into forceful contact with "hard animal" materials either from carving bone tools or from heavy-duty butchery. In this case, however, the absence of carved bone tools from most Levantine Mousterian contexts suggests these tools are probably worn from some kind of butchery.

Butchery is the activity that is most frequently represented in the Tor Faraj sample (29.6%). If the worn edges referable to impact (16.0%), haft contact (18.3%) (which mainly occurs on impact-damaged points), hide-working (4.7%), and bone contact (4.7%) are added to those worn from butchery, then fully 73.3% of the microwear sample is referable to the procurement and processing of animal tissues. Woodworking and plant processing together account for only 21.3% of the worn edges from Tor Faraj Level C.

Table 6.6 compares the percentages of worn edges assigned to the most common "activities" in Tor Faraj C to those for other Levantine Mousterian assemblages analyzed using the same approach as that described here (Shea 1991). These assemblages include Units IX–XIII of Kebara Cave (Bar-Yosef *et al.* 1986), Tabun Cave Units I (Beds 18–26), II, and IX (Jelinek 1982), Hayonim Cave Level E (Bar-Yosef 1979); Qafzeh Cave Units XV, XVII–XXIV (Vandermeersch 1981).

The percentage of worn edges referable to impact and to hafting in Tor Faraj (34.3%) is higher than in all other Levantine Mousterian assemblages. Worn edges referable to butchery and to bone contact are also more common in the Tor Faraj microwear sample than they are in the other Levantine Mousterian samples. The percentage of worn edges referable to woodworking, in contrast, is lower in Tor Faraj than in any of the other Levantine Mousterian microwear samples.

## DISCUSSION

Without information about the use and discard of stone tools, our understanding of lithic "operational sequences" is incomplete. Lithic microwear analysis provides insights into the formation of the Tor Faraj assemblage and into some aspects of Levantine Mousterian industrial variability.

Table 6.6. Functional Comparison of Tor Faraj C to other Levantine Mousterian assemblages[a]

| Assemblage | Worn edges | Percent: Impact & half contact | Butchery & bone contact | Wood working | Other |
|---|---|---|---|---|---|
| Tor Faraj C | 169 | 34.3 | 34.3 | 21.3 | 10.1 |
| Kebara IX–XIII | 1591 | 24.7 | 26.6 | 35.1 | 13.6 |
| Tabun I/18–26 | 184 | 8.2 | 24.4 | 41.3 | 26.1 |
| Tabun II | 90 | 27.2 | 26.7 | 33.3 | 12.8 |
| Tabun IX | 338 | 20.2 | 29.3 | 36.1 | 14.4 |
| Hayonim E | 414 | 20.8 | 24.6 | 40.9 | 14.7 |
| Qafzeh XV | 486 | 31.1 | 24.1 | 27.8 | 17.0 |
| Qafzeh XVIII–XXIV | 206 | 11.7 | 20.4 | 41.7 | 26.2 |

[a]Source: Shea (1991).

At least in terms of the occurrence of wear, the Tor Faraj assemblage differs little from other Levantine Mousterian assemblages. Points, Levallois debitage, and retouched tools exhibit wear traces more often than one would predict from the raw frequency with which they occur in an assemblage. Where Tor Faraj differs from other Levantine Mousterian samples is in the patterned variation of the microwear traces from this site. When the microwear patterns observed in Tor Faraj are interpreted by formal analogy with wear from replicative experiments, the high percentages of wear referable to impact, butchery, bone contact and haft contact suggest that the deposition of the Tor Faraj assemblage coincided with a degree of predatory activity that is unusually high for Levantine Mousterian assemblages.

While it is tempting to equate the relative frequencies of different kinds of microwear traces with the actual frequency of the activities inferred from them, there is neither ethnographic nor experimental evidence to support such a straightforward equation. Moreover, there are good reasons to expect the lithic microwear record to be biased against the preservation and recognition of certain activities (Bamforth 1988; Shea and Klenck in press). If anything, however, one would expect the record to be biased *against* the preservation of wear from the kinds of relatively brief uses of tools against nonsiliceous materials that occur during hunting and butchery of animal carcasses. Yet wear traces referable to just these activities actually comprise the *majority* of the Tor Faraj microwear evidence. This could suggest that the microwear evidence from Tor Faraj actually *underrepresents* the degree of hominid predatory activity associated with the formation of the Tor Faraj assemblage.

The location of Tor Faraj in the highlands of the interior southern Levant suggests that when it was occupied by Levantine Mousterian hominids, it would have been surrounded by a more xeric and steppic landscape than was present around Kebara, Tabun, Hayonim, and Qafzeh caves. All of these other sites are located in the coastal lowlands, and even if they were occupied during relatively warm, dry periods (i.e. oxygen-isotope stages 5a, 5c, or 5e) they would likely have been situated within the Mediterranean oak-pistachio woodland ecozone. If this was the case, then hominids living around Tor Faraj would probably have had fewer plant and small animal food sources available to them, and may have experienced more acute seasonal foraging deficits, than those hominids living in the coastal woodlands and in the more northerly parts of the Transjordan Plateau. If the hominid groups living in this "marginal zone" were unable to move to higher-quality resource patches, one way in which hominids may have attempted to overcome nutritional shortfalls might have been to hunt large game more regularly.

If hunting was especially common near Tor Faraj, then it seems reasonable to expect a corresponding increase in both the incentive and the energetic "payoff" for hominids living there to devise more reliable hunting weapons, such as spears tipped with stone points (Bleed 1986). It also seems reasonable to expect lithic microwear referable to the procurement and processing of animal tissues to be more common than in assemblages from the woodland ecozone, an expectation that is met by the microwear evidence discussed here. The large number of pointed artifacts and impact-damaged points from Tor Faraj may reflect Levantine Mousterian hominids producing stone tools at strategic locations on the landscape and returning to these sites to repair and replace damaged hunting equipment.

Relatively high percentages of points occur in many of the Levantine Mousterian assemblages from the southern and interior Levant (Marks 1981; Jelinek 1982). If many of these tools were also used as hafted stone spear points, then one might reasonably infer that hunting large mammals was a more significant part of Levantine Mousterian subsistence adaptations in the southern and interior parts of the Levant than it was in those parts of the coastal lowlands from which Middle Paleolithic assemblages have been recovered.

Our understanding of Levantine Mousterian functional variability and its behavioral correlates is very rudimentary. While microwear analysis can provide insight into some of the factors that contribute to large-scale pattern of Levantine Mousterian industrial variability, it cannot in and of itself refute the many behavioral interpretations of Levantine Mousterian technotypological variability that have thus far been proposed. More detailed models of the hominid behavioral strategies that resulted in Levantine Mousterian industrial variability must relate the microwear evidence to other nonlithic aspects of the archaeological record.

## Appendix 6.1. Key to Variables Recorded

Each worn edge identified in the Levantine Mousterian assemblage from Tor Faraj C is listed in Appendix 6.2, where it is described in terms of its values for the variables discussed below.

*PLAN* (Excavation Plan Unit): Plan unit of excavation from which the artifact was recovered.

*ARTNO* (Artifact Number): Artifact registration number, as given by the excavator. If unnumbered, then alphabetic designations (e.g., A–Z, AA–ZZ, etc.) were substituted.

*LV* (Levallois): Presence or absence of Levallois preparation (facetted *chapeau de gendarme* striking platform, flake scars on dorsal from previous removals).

0. Non-Levallois or Indeterminate.
1. Levallois (*chapeau de gendarme* present, dorsal scars suggest prior preparatory flaking).

*BL* (Blank Shape): Shape of artifact in plan view.

1. Point—triangular shape.
2. Blade—laminar/elongate rectangular shape.
3. Oval Flake—oval, subrectangular shape.
4. Other—flake fragment, angular piece, core.
5. Cobble/pebble.

*BT* (Blank Technology) technological characteristics of the worn artifact.

1. Primary Cortical Flake—> 50% cortical surface on dorsal face.
2. Secondary Cortical Flake—< 50% cortical surface on dorsal face.
3. Tertiary Flake—no cortical surface on dorsal face.
4. Other Conchoidal Fracture Product—flake fragment, core, angular piece.
5. Cobble/pebble.

*MD* (Tool Modification): absence/presence and kind of modification/retouch on worn artifact.

0. Unmodified.
1. Unifacial Retouch only.
2. Bifacial Retouch only.
3. Both Unifacial and Bifacial retouch.
4. Burin, either by itself, or with other retouch.

*WE* (Number of Worn Edge): Sequential number of each worn edge beginning with worn edges located on the distal and proceeding clockwise around the perimeter of the tool.

*PCL* (Location of EU): those parts of an edge that exhibit continuous microwear traces.

Values are numeric, and indicate segments of an eight-point polar-coordinate grid numbered clockwise from the most distal point of the tool.

*EM* (Edge Modification): Notes the presence of retouch and other technological effects imposed on the edge, either prior to or during tool use.

0. Unmodified.
1. Unifacial retouch.
2. Bifacial retouch.
3. Both unifacial and bifacial retouch.

*ES* (Edge Shape in Plan View) notes the shape of the employed unit prior to use as viewed in plan view.

1. Convex edge.
2. Straight edge.
3. Concave edge.
4. Recurved edge.
5. Point (3-sided projection).
6. Projection (4-sided projection).
7. Convex surface.
8. Flat/planar surface.
9. Concave surface.

*SPA* (Spine Plane Angle): The angle (in degrees) formed by the intersection of the planes of the working edge/protection measured at points approximately 2 mm from the edge.

*WEARCODE*: The microwear characteristics of an EU recorded in terms of values for the following nine subvariables.

*Abrasion Location*: The location of abrasive wear traces (polish, striation, dulling) on the employed unit.

0. Absent, or no change from unworn condition.
1. Edge-unifacial.
2. Edge-bifacial.
3. Point/projection—one facet.
4. Point/projection—opposed facets.
5. Surface > 5 mm from edge.
6. Edge-bifacial asymmetrical.

*Dulling*: The degree of microtopographic flattening on abraded surfaces nearest to the edge.

0. Absent, or no change from unworn condition.
1. Roughened—isolated areas of smoothed microtopography.
2. Rounded—continuous areas of smoothed microtopography along edge.
3. Flattened—continuously-smoothed areas of microtopography > 1 mm in extent.

*Lustre/Polish Reflectivity*: The reflection of light from abraded surfaces.
0. Absent, or no change from unworn condition.
1. Matte—diffuse overall reflection of light.
2. Bright—focused reflection of light from specific areas.
3. Vitreous—brilliant reflection of light from extensively-rounded surfaces.
4. Additive Phenomena—adhesion of mineral particles obscures tool surfaces.

*Orientation of Striations*: The modal orientation of striations to the working edge.
0. Absent, or no change from unworn condition.
1. Parallel to edge, projection facet.
2. Perpendicular to edge, projection facet.
3. Oblique to edge, projection facet.
4. Sets of parallel striations > 5 mm from edge.
5. Randomly-oriented striations on surface > 5 mm from edge.

*Microfracturing Location*: The location of microfracturing traces on employed units.
0. Absent, or no change from unworn condition.
1. Edge-unifacial
2. Edge-bifacial.
3. One facet of point/projection.
4. Opposed facets of point/projection.
5. Surface > 5 mm from edge.
6. Edge-bifacial asymmetrical.

*Microfracturing Distribution*: The distribution of microfracturing traces on an employed unit.
0. Absent, or no change from unworn condition.
1. Continuous—larger microfractures overlap with each other along a worn edge.
2. Close—larger microfractures abut each other but generally do not overlap along a worn edge.
3. Uneven—larger microfractures are separated from each other by one or more microfracture-scar's width along a worn edge.
4. Clumped—larger microfractures are arrayed in one or more discrete clusters along a worn edge.
5. Alternating, bifacial, and clumped—larger microfractures are arrayed in multiple discrete clusters that alternate from one face of the edge to the other.
6. Macrobreak—a single large scar.

*Microfracture Termination*: The modal pattern of microfracture termination(s).
0. Absent, or no change from unworn condition.
1. Feather type 1—defined finial.
2. Feather type 2—indistinct finial.
3. Hinge—recurved termination.
4. Step—truncated termination.
5. Shear fracture/macrofracture/torsion snap.
6. Comminution—multiple, overlapping and incompletely-propagated shear fractures.
7. Feather (either kind) and hinge fractures in roughly even numbers.
8. Feather (either kind) and step fractures in roughly even numbers.
9. Hinge and step fractures in roughly even numbers.

*Size of Microfractures*: Modal length of microfractures measured from initiation to termination.
0. Absent, or no change from unworn condition.
1. Small—< 1 mm.
2. Medium—< 2 mm.
3. Large—> 2 mm, < 5 mm.
4. Macrofracture—> 5 mm.

*Trajectory of Microfracture Propagation*: Modal trajectory of microfracture propagation in relation to worn edge.

0. Absent, or no change from unworn condition.
1. Perpendicular to edge.
2. Oblique to edge.
3. Even mix of edge-perpendicular and edge-oblique propagations.

*TM* (Tool Motion): The inferred modal pattern of the movement of the tool through space during use. (For definitions, see Shea 1991.)

CT = Cut.
SH = Shave (here grouped together with Cut).
SC = Scrape.
PL = Plane.
AZ = Adze/Hoe/Pick.
CH = Chop (here grouped together with Adze/Hoe/Pick).
WD = Wedge.
PC = Pierce.
GR = Engrave.
AW = Awl/Drill.
AB = Abrade.
PD = Pound.
HF = Haft Contact.
UK = Unknown/Indeterminate Motion.

*WM* (Worked Material Category): The inferred worked materials category. See discussion in text.

SA = Soft Animal.
SV = Soft Vegetal.
SI = Soft Inorganic.
MA = Medium Animal.
MV = Medium Vegetal.
MI = Medium Inorganic.
HA = Hard Animal.
HV = Hard Vegetal.
HI = Hard Inorganic.
HF = Haft Contact.
IS = Indeterminate Yielding Resistance.
IM = Indeterminate Semirigid Resistance.
IH = Indeterminate Rigid Resistance.
UI = Impact Against Unknown Material.
UK = Indeterminate/Unknown Worked Material.

*IA* (Inferred Activity): The activity in which the employed unit is inferred to have been utilized.

PI = Projectile Point Impact.
BT = Butchery.
HD = Hide Working.
BC = Bone Carving.
SP = Soft Plant Processing.
HW = Heavy-Duty Woodworking.
LW = Light-Duty Woodworking.
(Here, both heavy- and light-duty woodworking are grouped together.)
DG = Digging
SC = Stone-Carving.
SK = Stone-Knapping.
HF = Haft Contact.
UK = Unknown/Indeterminate Activity.

Appendix 6.2. Observations on worn tools from Tor Faraj Layer C

| ARTN | LV | BL | BT | MD | WE | PCL | EM | ES | SPA | WEARCODE | TM | WM | IA |
|---|---|---|---|---|---|---|---|---|---|---|---|---|---|
| 06 | 0 | 3 | 3 | 0 | 1 | 670 | 0 | 1 | 20 | 212021822 | CT | MV | LW |
| 35 | 0 | 4 | 4 | 3 | 1 | 670 | 0 | 2 | 40 | 212021722 | CT | MV | LW |
| 42 | 1 | 1 | 3 | 0 | 1 | 810 | 0 | 5 | 20 | 000034521 | PC | UI | PI |
| 42 | 1 | 1 | 3 | 1 | 2 | 340 | 1 | 1 | 45 | 210024111 | HF | HF | HF |
| 42 | 1 | 1 | 3 | 1 | 3 | 670 | 1 | 1 | 30 | 221022112 | CT | SA | BT |
| 45 | 0 | 3 | 2 | 1 | 1 | 450 | 1 | 3 | 85 | 122011821 | SC | MV | LW |
| 45 | 0 | 3 | 2 | 1 | 2 | 700 | 1 | 2 | 65 | 222022822 | CT | MV | LW |
| 47 | 0 | 1 | 4 | 1 | 1 | 810 | 1 | 5 | 20 | 000034521 | PC | UI | PI |
| 47 | 0 | 1 | 4 | 1 | 2 | 230 | 1 | 8 | 40 | 211061822 | CT | MA | BT |
| 47 | 0 | 1 | 4 | 1 | 3 | 670 | 1 | 8 | 30 | 211061822 | CT | MA | BT |
| 48 | 0 | 2 | 1 | 1 | 1 | 400 | 1 | 1 | 40 | 121011421 | SC | MA | BC |
| 52 | 0 | 1 | 4 | 0 | 1 | 120 | 0 | 2 | 40 | 211021712 | CT | MA | BT |
| 52 | 0 | 2 | 3 | 1 | 1 | 234 | 0 | 2 | 35 | 221022712 | CT | MA | BT |
| 52 | 0 | 1 | 4 | 0 | 2 | 780 | 0 | 2 | 32 | 211022712 | CT | MA | BT |
| 54 | 1 | 1 | 3 | 0 | 1 | 810 | 0 | 5 | 50 | 000034521 | PC | UI | PI |
| 54 | 1 | 1 | 3 | 0 | 2 | 234 | 0 | 2 | 50 | 221021812 | CT | MA | BT |
| 54 | 1 | 1 | 3 | 0 | 3 | 670 | 0 | 2 | 50 | 221022712 | CT | MA | BT |
| 55 | 1 | 3 | 3 | 0 | 1 | 123 | 0 | 1 | 50 | 221021812 | CT | MA | BT |
| 56 | 1 | 1 | 3 | 0 | 1 | 123 | 0 | 1 | 10 | 221022712 | CT | MA | BT |
| 56 | 1 | 1 | 3 | 0 | 2 | 678 | 0 | 1 | 15 | 221022712 | CT | MA | BT |
| 57 | 1 | 1 | 3 | 0 | 1 | 123 | 0 | 4 | 35 | 211022222 | CT | MA | BT |
| 57 | 1 | 1 | 3 | 0 | 2 | 678 | 0 | 2 | 20 | 211022222 | CT | MA | BT |
| 57 | 1 | 1 | 3 | 0 | 3 | 450 | 0 | 3 | 50 | 000025811 | HF | HF | HF |
| 58 | 0 | 3 | 4 | 0 | 1 | 600 | 0 | 5 | 25 | 411041212 | AW | MA | HD |
| 59 | 0 | 3 | 2 | 0 | 1 | 120 | 0 | 3 | 25 | 612061222 | SH | MV | LW |
| 60 | 1 | 1 | 3 | 0 | 1 | 123 | 0 | 2 | 10 | 211021112 | CT | SA | HD |
| 60 | 1 | 1 | 3 | 0 | 2 | 450 | 0 | 3 | 55 | 000014121 | HF | HF | HF |
| 60 | 1 | 1 | 3 | 0 | 3 | 678 | 0 | 2 | 15 | 211021112 | CT | SA | HD |
| 61 | 1 | 1 | 3 | 0 | 1 | 810 | 0 | 5 | 20 | 000034521 | PC | UI | PI |
| 61 | 1 | 1 | 3 | 0 | 2 | 230 | 0 | 1 | 40 | 211023112 | CT | MA | BT |
| 61 | 1 | 1 | 3 | 0 | 3 | 450 | 0 | 3 | 70 | 000014811 | HF | HF | HF |
| 61 | 1 | 1 | 3 | 0 | 4 | 670 | 0 | 1 | 40 | 211023112 | CT | MA | BT |
| 64 | 1 | 1 | 3 | 1 | 1 | 810 | 0 | 5 | 20 | 000034533 | PC | UI | PI |
| 64 | 1 | 1 | 3 | 1 | 2 | 450 | 1 | 1 | 35 | 110024112 | HF | HF | HF |
| 64 | 1 | 1 | 3 | 1 | 3 | 230 | 0 | 2 | 25 | 211021812 | CT | MA | BT |
| 64 | 1 | 1 | 3 | 1 | 4 | 670 | 0 | 2 | 20 | 211021812 | CT | MA | BT |
| 67 | 1 | 1 | 3 | 0 | 1 | 810 | 0 | 5 | 15 | 000034521 | PC | UI | PI |
| 67 | 1 | 1 | 3 | 0 | 2 | 230 | 0 | 2 | 20 | 211022722 | CT | MA | BT |
| 67 | 1 | 1 | 3 | 0 | 3 | 450 | 0 | 3 | 50 | 000014111 | HF | HF | HF |
| 67 | 1 | 1 | 3 | 0 | 4 | 670 | 0 | 1 | 35 | 211022722 | CT | MA | BT |
| 68 | 0 | 4 | 1 | 1 | 1 | 812 | 1 | 3 | 30 | 222022822 | CT | MV | LW |
| 68 | 0 | 4 | 1 | 1 | 2 | 670 | 1 | 1 | 60 | 222022822 | CT | MV | LW |
| 69 | 1 | 1 | 3 | 0 | 1 | 810 | 0 | 5 | 45 | 000034531 | PC | UI | PI |
| 69 | 1 | 1 | 3 | 0 | 2 | 450 | 0 | 4 | 60 | 000014813 | HF | HF | HF |
| 70 | 0 | 1 | 3 | 1 | 1 | 810 | 1 | 5 | 35 | 000034521 | PC | UI | PI |
| 70 | 0 | 1 | 3 | 1 | 2 | 500 | 0 | 1 | 40 | 000014111 | HF | HF | HF |
| 71 | 1 | 1 | 3 | 0 | 1 | 810 | 0 | 5 | 15 | 000034521 | PC | UI | PI |
| 71 | 1 | 1 | 3 | 0 | 2 | 230 | 0 | 4 | 25 | 211061822 | CT | MA | BT |

**Appendix 6.2.** *(Continued)*

| ARTN | LV | BL | BT | MD | WE | PCL | EM | ES | SPA | WEARCODE | TM | WM | IA |
|---|---|---|---|---|---|---|---|---|---|---|---|---|---|
| 71 | 1 | 1 | 3 | 0 | 3 | 450 | 0 | 3 | 55 | 000014811 | HF | HF | HF |
| 71 | 1 | 1 | 3 | 0 | 4 | 670 | 0 | 4 | 30 | 211062712 | CT | MA | BT |
| 72 | 1 | 1 | 2 | 0 | 1 | 120 | 0 | 2 | 30 | 122012821 | SC | MV | LW |
| 73 | 0 | 4 | 4 | 0 | 1 | 810 | 0 | 5 | 20 | 000034521 | PC | UI | PI |
| 73 | 0 | 4 | 4 | 0 | 2 | 230 | 0 | 4 | 45 | 211021833 | CT | HA | BC |
| 73 | 0 | 4 | 4 | 0 | 3 | 670 | 0 | 4 | 40 | 211021833 | CT | HA | BC |
| 74 | 1 | 4 | 4 | 1 | 1 | 670 | 0 | 1 | 12 | 211022811 | CT | MA | BT |
| 76 | 1 | 1 | 3 | 0 | 1 | 123 | 0 | 2 | 30 | 211022122 | CT | MA | BT |
| 76 | 1 | 1 | 3 | 0 | 2 | 678 | 0 | 2 | 40 | 211022122 | CT | MA | BT |
| 77 | 1 | 3 | 3 | 0 | 1 | 230 | 0 | 1 | 25 | 211011821 | CT | MA | HD |
| 77 | 1 | 3 | 3 | 0 | 2 | 458 | 0 | 7 | 17 | 510014111 | HF | HF | HF |
| 78 | 1 | 1 | 3 | 0 | 1 | 123 | 0 | 2 | 15 | 211022122 | CT | MA | BT |
| 78 | 1 | 1 | 3 | 0 | 2 | 450 | 0 | 3 | 30 | 000024112 | HF | HF | HF |
| 78 | 1 | 1 | 3 | 0 | 3 | 678 | 0 | 1 | 25 | 211022112 | CT | MA | BT |
| 79 | 1 | 3 | 3 | 0 | 1 | 567 | 0 | 4 | 20 | 212021222 | CT | MV | LW |
| 80 | 0 | 2 | 2 | 0 | 1 | 780 | 0 | 2 | 30 | 610022112 | SH | SV | SP |
| 82 | 0 | 2 | 4 | 0 | 1 | 678 | 0 | 3 | 25 | 222021222 | CT | MV | LW |
| 83 | 1 | 3 | 3 | 0 | 1 | 123 | 0 | 4 | 35 | 211022122 | CT | SA | HD |
| 83 | 1 | 3 | 3 | 0 | 2 | 678 | 0 | 1 | 40 | 211022122 | CT | SA | HD |
| 87 | 1 | 1 | 3 | 0 | 1 | 810 | 0 | 5 | 15 | 000014531 | PC | UI | PI |
| 88 | 1 | 1 | 3 | 0 | 1 | 810 | 0 | 5 | 30 | 000034533 | PC | UI | PI |
| 88 | 1 | 1 | 3 | 0 | 2 | 230 | 0 | 1 | 20 | 211022822 | CT | MA | BT |
| 88 | 1 | 1 | 3 | 0 | 3 | 450 | 0 | 4 | 65 | 000014811 | HF | HF | HF |
| 88 | 1 | 1 | 3 | 0 | 4 | 670 | 7 | 1 | 20 | 221022822 | CT | MA | BT |
| 92 | 1 | 1 | 3 | 1 | 1 | 123 | 0 | 1 | 35 | 212021222 | CT | MV | LW |
| 92 | 1 | 1 | 3 | 1 | 2 | 560 | 0 | 1 | 25 | 212062812 | SH | MV | LW |
| 93 | 0 | 3 | 4 | 0 | 1 | 810 | 0 | 5 | 10 | 000034521 | PC | UI | PI |
| 96 | 1 | 1 | 3 | 1 | 1 | 810 | 0 | 5 | 30 | 000034531 | PC | UI | PI |
| 96 | 1 | 1 | 3 | 1 | 2 | 670 | 1 | 1 | 50 | 621011121 | SC | MA | BT |
| 96 | 1 | 1 | 3 | 1 | 3 | 450 | 0 | 3 | 60 | 000014113 | HF | HF | HF |
| 96 | 1 | 1 | 3 | 1 | 4 | 230 | 0 | 1 | 55 | 221022112 | CT | SA | BT |
| 97 | 1 | 1 | 3 | 0 | 1 | 810 | 0 | 5 | 20 | 000034531 | PC | UI | PI |
| 97 | 1 | 1 | 3 | 0 | 2 | 230 | 0 | 4 | 30 | 211021822 | CT | MA | BT |
| 97 | 1 | 1 | 3 | 0 | 3 | 450 | 0 | 2 | 45 | 000014121 | HF | HF | HF |
| 97 | 1 | 1 | 3 | 0 | 4 | 670 | 0 | 3 | 30 | 211021822 | CT | MA | BT |
| 98 | 1 | 1 | 3 | 0 | 1 | 810 | 0 | 5 | 15 | 000034521 | PC | UI | PI |
| 98 | 1 | 1 | 3 | 0 | 2 | 360 | 0 | 3 | 25 | 000024111 | HF | HF | HF |
| 105 | 1 | 1 | 3 | 0 | 1 | 123 | 0 | 4 | 30 | 211021822 | CT | MA | BT |
| 105 | 1 | 1 | 3 | 0 | 2 | 678 | 0 | 4 | 35 | 211021822 | CT | MA | BT |
| 106 | 1 | 1 | 3 | 0 | 1 | 810 | 0 | 5 | 20 | 000034521 | PC | UI | PI |
| 106 | 1 | 1 | 3 | 0 | 2 | 450 | 0 | 1 | 70 | 000024112 | HF | HF | HF |
| 107 | 1 | 1 | 3 | 0 | 1 | 810 | 0 | 5 | 30 | 000034512 | PC | UI | PI |
| 107 | 1 | 1 | 3 | 0 | 2 | 230 | 0 | 1 | 45 | 211021812 | CT | MA | BT |
| 107 | 1 | 1 | 3 | 0 | 3 | 450 | 0 | 3 | 55 | 000014811 | HF | HF | HF |
| 107 | 1 | 1 | 3 | 0 | 4 | 670 | 0 | 1 | 35 | 211021812 | CT | MA | BT |
| 110 | 1 | 1 | 3 | 0 | 1 | 810 | 0 | 5 | 30 | 000034521 | PC | UI | PI |
| 110 | 1 | 1 | 3 | 0 | 2 | 230 | 0 | 4 | 40 | 211022822 | CT | MA | BT |
| 110 | 1 | 1 | 3 | 0 | 3 | 450 | 0 | 3 | 55 | 000025113 | HF | HF | HF |

Appendix 6.2. *(Continued)*

| ARTN | LV | BL | BT | MD | WE | PCL | EM | ES | SPA | WEARCODE | TM | WM | IA |
|---|---|---|---|---|---|---|---|---|---|---|---|---|---|
| 110 | 1 | 1 | 3 | 0 | 4 | 670 | 0 | 1 | 35 | 211022812 | CT | MA | BT |
| 113 | 0 | 3 | 4 | 0 | 1 | 810 | 0 | 5 | 15 | 000034531 | PC | UI | PI |
| 113 | 0 | 3 | 4 | 0 | 2 | 230 | 0 | 2 | 20 | 220021823 | CT | IM | UK |
| 113 | 0 | 3 | 4 | 0 | 3 | 670 | 0 | 2 | 20 | 220021823 | CT | IM | UK |
| 116 | 1 | 1 | 3 | 1 | 1 | 230 | 0 | 2 | 30 | 212021822 | CT | MV | LW |
| 116 | 1 | 1 | 3 | 1 | 2 | 810 | 1 | 1 | 70 | 222022722 | CT | MV | LW |
| 117 | 1 | 3 | 3 | 1 | 1 | 123 | 1 | 7 | 60 | 121011431 | SC | HA | BC |
| 121 | 1 | 1 | 3 | 0 | 1 | 810 | 0 | 5 | 20 | 000034533 | PC | UI | PI |
| 121 | 1 | 1 | 3 | 0 | 2 | 450 | 0 | 2 | 30 | 000014121 | HF | HF | HF |
| 123 | 1 | 1 | 3 | 0 | 1 | 810 | 0 | 5 | 20 | 000034512 | PC | UI | PI |
| 123 | 1 | 1 | 3 | 0 | 2 | 230 | 0 | 1 | 30 | 221022712 | CT | MA | BT |
| 123 | 1 | 1 | 3 | 0 | 3 | 450 | 0 | 3 | 50 | 000025111 | HF | HF | HF |
| 123 | 1 | 1 | 3 | 0 | 4 | 670 | 0 | 1 | 30 | 221022712 | CT | MA | BT |
| 124 | 1 | 3 | 3 | 0 | 1 | 230 | 0 | 4 | 20 | 212021822 | CT | MV | LW |
| 125 | 1 | 3 | 3 | 0 | 1 | 120 | 0 | 1 | 40 | 222021822 | CT | MV | LW |
| 601 | 1 | 2 | 3 | 1 | 1 | 810 | 0 | 5 | 18 | 000014531 | PC | UI | PI |
| 601 | 1 | 2 | 3 | 1 | 2 | 450 | 1 | 3 | 29 | 000025122 | HF | HF | HF |
| 644 | 1 | 3 | 4 | 0 | 1 | 340 | 0 | 3 | 45 | 000014811 | HF | HF | HF |
| A | 0 | 3 | 1 | 2 | 1 | 230 | 0 | 2 | 67 | 122011821 | SC | MV | LW |
| B | 0 | 1 | 4 | 1 | 1 | 678 | 0 | 2 | 40 | 622012822 | SH | MV | LW |
| C | 1 | 3 | 3 | 0 | 1 | 234 | 1 | 2 | 45 | 212021822 | CT | MV | LW |
| C | 1 | 3 | 3 | 0 | 2 | 780 | 1 | 1 | 25 | 112011821 | SC | MV | LW |
| D | 0 | 3 | 2 | 1 | 1 | 810 | 1 | 3 | 70 | 220000000 | UK | UK | UK |
| D | 0 | 3 | 2 | 1 | 2 | 234 | 0 | 2 | 10 | 211022712 | CT | MA | BT |
| E | 0 | 2 | 2 | 0 | 1 | 560 | 0 | 2 | 20 | 212022822 | CT | MV | LW |
| F | 1 | 1 | 3 | 0 | 1 | 810 | 0 | 5 | 20 | 000034522 | PC | UI | PI |
| F | 1 | 1 | 3 | 0 | 2 | 450 | 0 | 3 | 65 | 000014813 | HF | HF | HF |
| G | 0 | 4 | 4 | 0 | 1 | 810 | 0 | 5 | 20 | 000034521 | PC | UI | PI |
| H | 1 | 1 | 3 | 1 | 1 | 120 | 0 | 4 | 35 | 222021822 | CT | MV | LW |
| I | 1 | 3 | 2 | 0 | 1 | 560 | 0 | 1 | 40 | 221021933 | CT | HA | BC |
| I | 1 | 3 | 2 | 0 | 2 | 780 | 0 | 1 | 40 | 221021933 | CT | HA | BC |
| J | 1 | 4 | 4 | 0 | 1 | 360 | 0 | 3 | 50 | 000024821 | HF | HF | HF |
| K | 1 | 3 | 2 | 7 | 1 | 781 | 7 | 1 | 70 | 211014811 | SC | MA | HD |
| K | 1 | 3 | 2 | 7 | 2 | 360 | 1 | 3 | 70 | 000014821 | HF | HF | HF |
| L | 1 | 2 | 3 | 0 | 1 | 123 | 0 | 1 | 40 | 211022822 | CT | MA | BT |
| L | 1 | 2 | 3 | 0 | 2 | 450 | 0 | 3 | 55 | 000024111 | HF | HF | HF |
| L | 1 | 2 | 3 | 0 | 3 | 678 | 0 | 2 | 25 | 211022822 | CT | MA | BT |
| M | 0 | 3 | 2 | 0 | 1 | 780 | 0 | 5 | 30 | 312031821 | GV | MV | LW |
| N | 0 | 3 | 2 | 8 | 1 | 678 | 1 | 3 | 35 | 622012822 | SH | MV | LW |
| O | 0 | 4 | 4 | 8 | 1 | 670 | 0 | 2 | 40 | 000024821 | HF | HF | HF |
| P | 0 | 3 | 2 | 0 | 1 | 200 | 0 | 1 | 15 | 112011222 | SC | IM | UK |
| Q | 0 | 4 | 4 | 8 | 1 | 123 | 0 | 4 | 20 | 622061731 | AZ | MV | HW |
| R | 0 | 4 | 4 | 8 | 1 | 670 | 0 | 1 | 50 | 000024111 | HF | HF | HF |
| S | 0 | 3 | 2 | 1 | 1 | 100 | 1 | 5 | 25 | 411041212 | AW | MA | HD |
| T | 1 | 2 | 3 | 0 | 1 | 123 | 0 | 1 | 15 | 211022812 | CT | MA | BT |
| U | 1 | 2 | 3 | 0 | 1 | 678 | 0 | 4 | 20 | 211022812 | CT | MA | BT |
| V | 0 | 3 | 1 | 1 | 1 | 230 | 1 | 2 | 50 | 122011821 | SC | MV | LW |
| W | 0 | 2 | 2 | 0 | 1 | 123 | 0 | 1 | 35 | 612062822 | SH | MV | LW |

Appendix 6.2. *(Continued)*

| ARTN | LV | BL | BT | MD | WE | PCL | EM | ES | SPA | WEARCODE | TM | WM | IA |
|---|---|---|---|---|---|---|---|---|---|---|---|---|---|
| X | 0 | 3 | 2 | 0 | 1 | 678 | 0 | 2 | 10 | 112011821 | SC | MV | LW |
| Y | 0 | 3 | 2 | 7 | 1 | 700 | 0 | 3 | 25 | 222021822 | CT | MV | LW |
| Z | 0 | 3 | 4 | 1 | 1 | 810 | 1 | 5 | 40 | 000034531 | PC | UI | PI |
| Z | 0 | 3 | 4 | 1 | 2 | 230 | 0 | 1 | 40 | 211022812 | CT | MA | BT |
| Z | 0 | 3 | 4 | 1 | 3 | 670 | 0 | 2 | 40 | 211022812 | CT | MA | BT |
| AA | 0 | 3 | 2 | 0 | 1 | 781 | 0 | 1 | 25 | 221021933 | CT | HA | BC |
| AB | 1 | 3 | 3 | 0 | 1 | 360 | 0 | 2 | 30 | 000025121 | HF | HF | HF |
| AC | 1 | 3 | 4 | 0 | 1 | 120 | 0 | 6 | 65 | 321032111 | GV | IS | UK |
| AD | 1 | 3 | 2 | 0 | 1 | 234 | 0 | 2 | 45 | 211022822 | CT | MA | BT |
| AE | 1 | 2 | 2 | 0 | 1 | 340 | 0 | 1 | 10 | 222021822 | CT | MV | LW |
| AF | 1 | 3 | 3 | 0 | 1 | 810 | 0 | 1 | 25 | 111011431 | SC | HA | BC |
| AG | 0 | 2 | 3 | 1 | 1 | 120 | 0 | 3 | 35 | 112011221 | SC | MV | LW |
| AG | 0 | 2 | 3 | 1 | 2 | 340 | 0 | 1 | 30 | 112011221 | SC | MV | LW |
| AH | 1 | 2 | 3 | 0 | 1 | 810 | 0 | 5 | 10 | 000034522 | PC | UI | PI |
| AH | 1 | 2 | 3 | 0 | 2 | 200 | 0 | 2 | 40 | 211021823 | CT | MA | UK |
| AH | 1 | 2 | 3 | 0 | 3 | 360 | 0 | 4 | 70 | 000025811 | HF | HF | HF |
| AH | 1 | 2 | 3 | 0 | 4 | 700 | 0 | 1 | 45 | 211021812 | CT | MA | BT |
| AI | 0 | 3 | 3 | 0 | 1 | 810 | 0 | 1 | 45 | 122011821 | SC | MV | LW |
| AJ | 0 | 2 | 2 | 0 | 1 | 678 | 0 | 1 | 30 | 210021822 | CT | HV | LW |
| AK | 1 | 1 | 3 | 0 | 1 | 123 | 0 | 2 | 20 | 211025812 | CT | MA | BT |
| AK | 1 | 1 | 3 | 0 | 2 | 450 | 0 | 3 | 60 | 000024111 | HF | HF | HF |
| AK | 1 | 1 | 3 | 0 | 3 | 678 | 0 | 1 | 40 | 210023823 | CT | IM | UK |
| 71A | 0 | 3 | 2 | 8 | 1 | 230 | 0 | 3 | 50 | 112012821 | SC | MV | LW |
| 71A | 0 | 3 | 2 | 8 | 2 | 810 | 0 | 6 | 90 | 220000000 | UK | UK | UK |
| 83A | 1 | 3 | 2 | 0 | 1 | 340 | 0 | 1 | 50 | 612061842 | CH | MV | HW |

# Chapter 7

# Late Levantine Mousterian Patterns of Adaptation and Cognition

For most of this century one of the central concerns of paleoanthropology has been with the origin of culture. Through investigations of fossils and artifacts archaeologists have attempted to identify the contexts in which our early hominid ancestors became increasingly dependent upon cultural adaptation. This progressive shift from organic to behavioral adaptation was accompanied by extraordinarily rapid rates in both biologic and cultural evolution, but over the last 100,000 years or so the tempo of cultural evolution has greatly exceeded that within the organic domain. As reflected in the changes seen in human material culture and economic systems, cultural evolution has steadily accelerated to its presently dizzying pace. At some point within this long, 3–4 million year evolutionary sweep hominids came to rely more upon behavioral responses than biologic ones in coping with environmental demands. These learned, patterned behaviors and their material consequences have come to distinguish the human species, for we are now and have been for some time dependent upon them for our survival.

Within the last decade, scholars have become increasingly interested in identifying when in prehistory we crossed this adaptive threshold. Perhaps more importantly, they want to understand the factors that contributed to our making this significant evolutionary step. In terms of timing, the late Pleistocene has been targeted by many scholars as the interval which encompassed the transition from the "protocultural" to the "cultural" behaviors that we associate with contemporary societies. This is a time when hominids are thought to have emerged into a "fully human ecological niche" (Sackett 1988:44) coinciding with the biological (i.e., *Homo sapiens neanderthalensis* to *H. sapiens*) and macrocultural (i.e., Middle to Upper Paleolithic) transitions that are most clearly defined within southwestern Europe (Mellars 1973, 1989; Chase and Dibble 1987; White 1982; Marshack 1988).

## MODERN CULTURE AND COGNITION

In addressing these questions, researchers have emphasized different prehistoric behavioral expressions (e.g., symbols, tool specificity, regionalism) of the emergence of modern culture, but they essentially share the notion that a new kind of cognition prompted these changes in behavior. While agreeing that differences in mental faculties may have allowed for the emergence of culture as we know it, Lewis Binford (1989:36) argues that:

> culture develops in response to evolutionary processes. It does not simply "bloom" in response to somatically based behavioral potentials or realities.

With this in mind, he has led the way in stressing the importance of isolating the archaeological residuals of anticipation and planning depth as a means of distinguishing archaic from modern strategies of decision making (1979, 1985, 1987, 1989).

In following this lead at an analytic level, researchers focussing on the late Pleistocene have attempted to measure anticipation and planning depth through the examination of procurement and settlement strategies and intrasite organization. Procurement and transport of raw materials (Geneste 1988; Meignen 1988; Roebucks *et al.* 1988), provisioning strategies (Munday 1976, 1979; Marks 1988; Kuhn 1990; Henry 1992), artifact uses (Shea 1989, 1991), and inter/intrasite artifact patterns (Hietala and Marks 1981; Marks and Friedel 1977; Gilead and Grigson 1984; Goren-Inbar 1990b; Saragusti and Goren-Inbar 1990; Meignen *et al.* 1989) have been studied in an effort to reconstruct prehistoric behaviors. Direct studies of cognition have also been undertaken in relation to measuring the complexity of manufacturing steps (Gowlett 1984), modes of selection, and standardization of form in tool production (Wynn 1979, 1981; Goren-Inbar 1990a; Watanabe and Kuchikura 1973).

The results of these studies have been mixed relative to their definition of cognitively linked behavioral differences between Middle Paleolithic and succeeding populations. Interestingly, the results seem to break primarily along geographic lines. It is the Near Eastern, especially Levantine, evidence that contrasts most strongly with the prevailing notion of a late emergence of modern behaviors coinciding with the appearance of anatomically modern humans and Upper Paleolithic industries as seen from Europe. Perhaps not by coincidence, the cultural and biologic successions of the Levant also contrast markedly with those defined for Europe. Not only do we find the remains of Neanderthals and anatomically modern humans within Middle Paleolithic horizons, but they are even associated with the same industry level taxon, i.e., Levantine Mousterian C type (Bar-Yosef 1989). Unlike the Middle to Upper Paleolithic transition of Europe, Middle and Upper Paleolithic industries of the Levant are weakly demarcated. Both typological (e.g., burins) and technological (e.g., blade production) elements that are traditionally taken as markers of the Upper Paleolithic in Europe are consistently present within Levantine Mousterian assemblages. Thus in the Levant, both biological and archaeological successions appear earlier than those of Europe. If this is indeed the case, should we also expect the evolution of human adaptive strategies to have followed a similar pace?

## THE IMPORTANCE OF TOR FARAJ AND TOR SABIHA

The occupations of Tor Faraj and Tor Sabiha offer an opportunity for significantly refining our understanding of the settlement and procurement patterns followed by Late Levantine Mousterian populations of the Levant some 70,000 years ago. This should indirectly furnish additional insights of the cognitive level of these populations. The importance of the two sites rests in their apparent contemporaneity, their close proximity, and their different natural settings.

### Contemporaneity

Amino acid racemization analysis of ostrich eggshell fragments yielded dates of 69,000 ± 6,000 B.P. for both Tor Sabiha and Tor Faraj. Uranium series dates, also on ostrich eggshell, were less consistent. While a date of 62,400 ± 14,000 B.P. was obtained from the top of Layer C at Tor Faraj, the other U-series dates of ca. 31,000 B.P. from Tor Faraj and ca. 29,000 B.P. from Tor Sabiha are clearly too young. The general similarities of the technotypologic profiles of the assemblages from the two sites also suggest synchroneity as does the unusually high proportions of inverse retouch, a likely stylistic attribute. Finally, paleoenvironmental evidence from the two sites points to a dry relatively cool setting that is consistent with the paleoclimatic reconstruction for Oxygen Isotope stage 4, dated to ca. 74–50,000 B.P. (Bar-Yosef 1989) or 70–62,000 B.P. (Farrand 1979; Goldberg 1986; Martinson *et al.* 1987), again in agreement with the AAR and oldest U-series dates.

The contemporaneity of the two nearby occupations enables them to be examined as part of a common settlement-procurement system. While such studies of Middle Paleolithic settlement and procurement practices have been undertaken in the Levant (S. Binford 1968; Marks and Friedel 1977; Munday 1976, 1979; Coinman *et al.* 1988) and Europe (Geneste 1988: Meignen 1988; Roebucks 1988; Kuhn 1990; Jelinek 1991), the scale by which the occupations have been judged synchronous has generally been quite large. In the absence of absolute dates and other refined chronometric evidence, the synchroneity of occupations has often been assumed on the basis of common membership in large archaeological or geoclimatic intervals. Dependence upon such coarse grained chronologies raises the question of how accurate such studies have been in tracing Mousterian settlement-procurement patterns.

## Natural Settings and Planning Depth

Resting at the upper edge of the piedmont with an eastern exposure, Tor Sabiha would have represented a poor camp during the winter wet season. The cave itself is small and the surrounding cliff face with little to no overhang provides only limited protection. In contrast, the lower setting of the large Tor Faraj with its southern exposure and extensive overhang would have made a fine winter camp-spot.

As for water availability, the Wadi Aghar running beneath Tor Faraj is likely to have been charged during the winter wet season, but may not have flowed in the dry season. Tor Sabiha's catchment, on the other hand, would have commanded year round water sources in the form of springs on the plateau as well as seasonal sources provided by streams on the floor of the basin and bedrock pools on Jebel el Jill. The catchments of the two sites also differ markedly with regard to the availability of chert. While the catchment of Tor Faraj is poor in chert, Tor Sabiha's catchment contains an abundance of fine quality cherts for stone tool production.

The natural settings of the sites thus provide an opportunity for seeing how Late Levantine Mousterian groups responded to differences in the seasonal and spatial distributions of their resources. The important question here centers on the degree to which these responses resembled those of modern humans. In a modern context they involve adjustments in the sizes and the mobility levels of groups. They also entail changes in the way groups organize their technology for accomplishing tasks.

In general, the size and mobility of prehistoric foraging groups is reflected in the intensity of an occupation as measured by the density, distribution, and contexts of artifacts. Elucidating the shifts in technological organization that prehistoric foragers employed to accommodate differences in the distributions of resources is more complex. The significance of understanding the technological reorganization that accompanied a change in the procurement of resources, however, may be at the very heart of identifying when prehistoric foragers crossed the threshold to a dependence upon behavioral adaptation. As previously discussed, a substantial degree of anticipation or planning depth used in organizing a technology to cope with the environment is a hallmark of modern culture. As Binford (1989:37) recently has noted:

> we will not grasp the transition unless we begin to worry about how to measure variables such as planning depth, mobility, group size, and compositional variability and then proceed to see how these properties vary with environmental conditions as a clue to the ecology of ancient populations.

Thus given the differences in environmental settings, a comparison of the behavioral residue from the archaeologically synchronous deposits of Tor Faraj and Tor Sabiha should be helpful in accessing the degree of cultural modernity achieved by the Middle Paleolithic occupants of the sites.

## TOOLKITS AND PROVISIONING STRATEGIES

Studies of the technological organization of prehistoric foragers tend to focus upon tools (design or toolkit composition) or upon those procedures related to raw material acquisition, core shaping, and blank production. Although there is an analytic distinction between these bodies of data, within the domain of prehistoric behavior they were undoubtedly interwoven and responsive to the same conditions.

### Tool Design and Toolkit Configurations

Variation in tool design and the compositions of toolkits have been found to be linked to one of the fundamental organizational features of foragers—that is, varying levels of residential mobility. The size and weight of tools and the diversity of toolkits appear to be inversely related to the degree of residential mobility.

Although portability is an obvious factor relating the size and weight of implements to mobility (Shott 1986), it probably is not the only factor. Bleed (1986), for example, distinguishes between "reliable" and "maintainable" design systems and ties their adoption to different functional contexts in the ethnographic and archaeological records. In the main, reliable designs prevail in settings where the time and duration of use of implements can be reasonably predicted, whereas maintainable designs dominate those settings in which the schedule of use is unpredictable, but need is continuous. We generally find that those settings in which a long, but predictable "downtime" is acceptable are also settings of low residential mobility. Those settings which require implements designed for a limited downtime are settings accompanied by high levels of mobility. Moreover, because of the objectives inherent in the design, reliable designs are typically larger, heavier, and bulkier than maintainable designs.

The linkage between the diversity of toolkits and levels of residential mobility also has received considerable attention. While researchers consistently find toolkit diversity and residential mobility to be inversely related, they point to various factors as being responsible for such a linkage. Perhaps the most commonly held notion is based upon an assumption that the number of functional units (expressed variously as tool classes, wear-pattern indices, etc.) within a tool assemblage co-varies directly with the number of exploited resources and that these, in turn, increase with the duration of residence. This last connection assumes that the longer a setting is occupied the more likely it is that its catchment will be enlarged and that other resources will come into phase, both of which increase the number and diversity of resources. Beyond this argument, others hold that as the cost of transport increases with higher levels of mobility more flexible, multi-purpose tools are adopted to reduce the numbers of tools that have to be portaged (Shott 1986).

### Provisioning Strategies

In a recent effort to examine the "nature and extent of anticipatory organization within a technological system," Steven Kuhn (1990:69–76) has introduced a model of "technological provisioning." In synthesizing Binford's (1973, 1977, 1979) concepts of curation and expediency, Kuhn's provisioning model is based upon the assumption that a principle objective of a technological system is to "make tools available when and where they are needed." To meet this objective, the organization of a technological system is dependent upon several other considerations including a group's level of mobility, the distribution of resources, and the nature of the tasks for which the tools are intended. Given our interest in the cultural ecology of Middle Paleolithic groups, it is their anticipation of these relationships and the decisions that they made governing the organization of their technology that is so important. It is precisely those differences in the level of anticipation by which the technology was organized that Kuhn's model reveals.

Kuhn recognizes three levels of provisioning that have to do with organizing technologies for activities, individuals, and/or places. The *provisioning of activities* requires the least amount of planning depth in that in this context needs are anticipated only as they occur. Making tools "expediently" as an immediate need arises, and abandoning them when that need ceases, is only efficient, however, in those contexts where raw material is immediately available and where manufacturing time does not interfere with meeting the goals of the task. In those situations where raw material and time are not available, anticipation or planning depth plays a stronger role in organizing the technology. The *provisioning of individuals* with "personal gear" frees individuals and groups from the constraints imposed by raw material availability and enables them to engage their technology more or less immediately when a need arises. This requires a considerable amount of anticipation or planning ahead in organizing the technology with the fundamental constraint being that of portability. Depending upon the specificity by which needs can be anticipated, individuals may be provisioned with specialized tools, generalized tools, or even raw materials for making tools. The *provisioning of places* is an alternative strategy for organizing a technology which also depends on a substantial anticipation of needs. In contrast to the provisioning of activities or individuals, the provisioning of places is not constrained by the distribution of raw materials, manufacturing time, or portability. The major constraint in organizing a technology dependent upon the provisioning of a place is the duration of residence. As the duration of an occupation increases, there is an attendant reduction in the importance of raw material availability, manufacturing time, and portability of implements in organizing a technology.

Application of Kuhn's technological provisioning model to real-world artifact assemblages allows for assessing the relative importance of raw material availability, manufacturing time, portability, and residential mobility in shaping technological systems. But perhaps more importantly, it provides a means of measuring the degree of planning depth or anticipation associated with the organization of the technology.

## PROVISIONING STRATEGIES AT TOR FARAJ AND TOR SABIHA

Although each of the three provisioning strategies probably contributed to the accumulation of artifacts at both sites, the occupation of Tor Faraj appears to have been provisioned primarily as a place, whereas the technology of Tor Sabiha was organized mainly around the provisioning of activities and, to a lesser extent, individuals. This interpretation is based upon patterned differences between the assemblages of the two sites relative to their portability, raw material utilization, manufacturing time, and toolkit diversity. The patterned differences in these factors also imply that the technological organizations responsible for the assemblages were shaped by different levels of residential mobility.

### Manufacturing Time

The number of stages of production represented by an assemblage is a reasonable measure of its associated manufacturing time, as well as a probable indirect expression of the degree of time-stress under which it was produced. Approaches to tracing the stages involved in manufacturing chipped stone tools have focussed principally upon certain morphological characteristics of an assemblage (Henry 1973; Collins 1975; Bradley 1975; Hassan 1976; Geneste 1985). Although they may vary in detail most approaches recognize four major stages: (1) raw material acquisition, (2) primary processing, (3) tool manufacture, and (4) tool use, rejuvenation, and maintenance. These stages normally are identified on the basis of the relative proportions of cortical material, cores, blanks (flakes, blades, etc.), and retouched tools. Although conventional morphological analyses allow for a general reconstruction of reduction strategies, they ignore an important source of information, i.e., the inexorable trend of diminution during the course of an item's fabrication and use-life. Through measuring and comparing the dimensions of core facets, blanks, and tools, one may gain a more precise understanding of the reduction stream within each of the stages as defined conventionally (Henry 1989b).

In following an analytical approach initially developed by Andre Leroi-Gourhan (1964), Geneste (1985) has presented a means of tracing the operational sequence (*chaîne opératoire*) manifested in Mousterian assemblages. Geneste's analysis is based upon the proportionate representation of 26 "types" broken into six "phases" and a *divers* category (1985). The types and phases are arranged sequentially following a reduction stream from the testing of raw material (Phase 0) to retouching blanks into tools (Phase 3). A comparison of the assemblages from Tor Sabiha and Tor Faraj along the lines suggested by Geneste shows patterned differences between the two sites (Figure 7.1). The Tor Sabiha assemblage, weakly represented in the earliest *chaine* phase, is substantially represented in the final phase. The assemblages from Tor Faraj show a reversal of this pattern, but with less variability between the initial and final phases.

Further examination of the reduction streams of the assemblages shows an even stronger emphasis on final processing at Tor Sabiha. The degree to which the second *chaine* phase was emphasized is defined essentially by the proportion of blanks (Levallois and ordinary) within the assemblage. The problem is that the analysis does not tell us whether the blanks were pro-

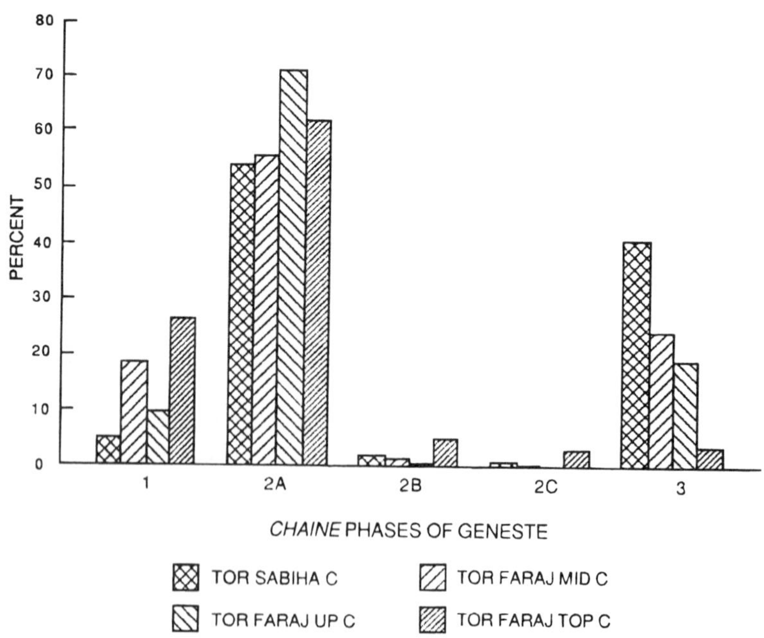


**Figure 7.1.** Comparison of the assemblages recovered from Tor Faraj and Tor Sabiha according to the *Chaine* Phases of Geneste. The reduction stream from initial to terminal processing flows from left to right.

duced on-site or imported. This bit of knowledge is crucial to establishing the time and energy that was devoted to manufacturing implements within the occupations. The relative dimensions of the debitage and ratios for certain artifacts assist in resolving this problem.

The production of Levallois points involved the rather precise shaping of cores through the removals of elongated primary elements and blades. In that these removals created the parallel convergent and Y-arrete scar patterns that guided the subsequent detachment of points, the lengths of these removals should be as great or greater than those of the points. With this in mind and in recalling the dimensional comparisons of Levallois points with primary elements from Tor Sabiha (Figure 5.20), only the smallest of the points (i.e., those less than 60 mm) were likely to have been consistently produced at the site. Over 60% of the points have lengths falling between 60–100 mm, while less than 20% of the primary elements fall within this range. Most of the larger points then must have been imported to the site. Similar comparisons of the lengths of primary elements and points from Tor Faraj (Figure 5.9) reveal a quite different relationship and indicate that most of the points could well have been manufactured within the shelter. The ratios of points to cores are also consistent with the patterns revealed in the dimensional studies. The point to core ratio of 2.5:1 at Tor Faraj is in general agreement with that observed at Boker Tachtit where core reconstructions demonstrated the production of 1.9 points per core (Volkman 1983). With some 5.3 points per core, the Tor Sabiha assemblage must include points that were imported to the site having been produced elsewhere.

The proportionate differences in loading within the chips also indicates a greater degree of final processing at Tor Sabiha than Tor Faraj. Samples of chips from the two sites, size graded to between 4 and 10 mm, show dissimilar weight distributions (Figure 7.2). The Tor Faraj sample peaks strongly at 0.20 g with few chips falling under 0.15 g, whereas the Tor Sabiha sample is much more strongly represented by chips weighing 0.15gm or less. These contrasting distributions become important when compared to experimentally produced samples which show

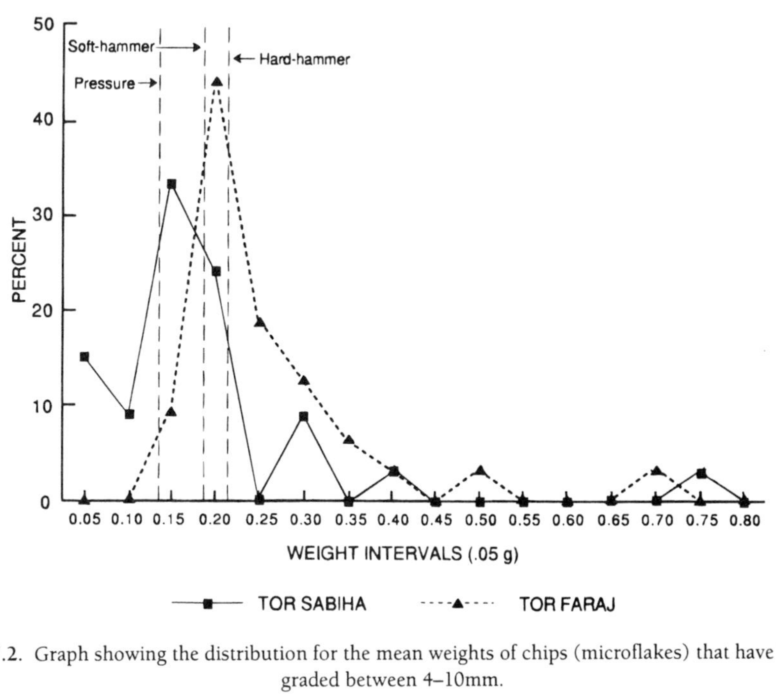


Figure 7.2. Graph showing the distribution for the mean weights of chips (microflakes) that have been size graded between 4–10mm.

the mean chip weights of pressure flaking, soft-hammer percussion, and hard-hammer percussion to be 0.13, 0.19, and 0.21, respectively (Henry *et al.* 1976). This would indicate that pressure and soft-hammer percussion were employed to a greater extent (relative to hard-hammer percussion) at Tor Sabiha than at Tor Faraj.

Several lines of evidence thus suggest that while at Tor Faraj, groups engaged in a full stream of reduction activities initiated by core shaping, blank production, tool manufacture, use, and rejuvenation. While each of these activities were present to some degree at Tor Sabiha, the final stages of reduction were the ones that were emphasized. The low proportions of primary elements and cores imply that little core shaping and blank production was undertaken at the site. Moreover, the dimensional comparisons of debitage suggest that a large proportion of the points must have been imported to the site.

Taken together then, these data trace alternative strategies for provisioning and organizing the technologies by the occupants of the sites. The substantial degree of initial reduction at Tor Faraj (some 17 km from the most utilized raw material) followed by tool manufacture, use, and renewal, implies a long manufacturing interval which was responsive to very little time-stress. The shelter clearly was provisioned as a place. In contrast, the manufacturing time enjoyed by the occupants of Tor Sabiha was typically short, encompassing the manufacture, repair, and rejuvenation of tools and blanks imported to the site. These imports would have represented the personal gear of individuals as well as residual items manufactured in the provisioning of activities within the nearby, chert-rich portions of the site's catchment.

## Portability

The lithic artifacts from Tor Faraj are consistently larger and heavier than those from Tor Sabiha (Table 7.1). The greater emphasis upon initial processing at Tor Faraj (as expressed in larger/heavier debitage) accounts in part for the differences in portability between the assemblages. But Levallois points and other retouched tools, which would have been produced at the same stage in both reduction sequences and were likely to have been carried about, are also larger and heavier at Tor Faraj. While these differences are in agreement with the expected differences in the portability of artifacts associated with the provisioning of places as opposed to individuals, one might question how the distribution of chert sources may have influenced the relative size/weight distributions of the assemblages. One might expect, for example, that artifacts would become smaller with increasing distance from their raw material sources, that is, a kind of "distance decay" relationship (Munday 1976, 1977; Geneste 1988; Henry 1989b). What is so surprising with respect to the Tor Faraj and Tor Sabiha assemblages is that the largest/heaviest assemblage is located the greatest distance from the chert sources. This, I think, underscores the fact that quite different technological organizations and provisioning strategies were employed by the occupants of the two sites.

## Raw Material Utilization

An examination of the distribution of chert varieties in the assemblages and the sources from where they were acquired also reveals differences in the way the two sites were provisioned. Consistent with provisioning Tor Faraj as a place, its occupants appear to have acquired most of their chert from specific (although variable by occupational level) targeted sources in bulk from well outside (17–20 km) their catchment. In contrast, the artifact accumulation at Tor Sabiha was derived primarily through the provisioning of individuals and activities from chert sources within the site's catchment. Chert was introduced to the site principally in the form of blanks or tools that were most likely produced at or near sources which had been selected rather indiscriminately from within the catchment.

Table 7.1. Weights (in grams) for Artifacts from Tor Sabiha and Tor Faraj

| | Tor Sabiha | | | Tor Faraj | | | | | | | | |
|---|---|---|---|---|---|---|---|---|---|---|---|---|
| | C | | | Mid C | | | Up C | | | Top C | | |
| | *r* | *x* | sd | *r* | *x* | sd | *r* | *x* | sd | *r* | *x* | sd |
| Points | 2–28 | 9.9 | 7.3 | 2–26 | 13.4 | 7.9 | 1–38 | 12.9 | 8.5 | 3–48 | 13.9 | 8.4 |
| Tools | 2–70 | 17.8 | 14.2 | — | — | — | — | — | — | 4–90 | 19.2 | 12.6 |
| Cores | 12–176 | 64.9 | 45.6 | — | — | — | — | — | — | 10–236 | 44.8 | 44.2 |
| Debitage | | | | | | | | | | | | |
| Blades | — | 3.5 | — | — | — | — | — | — | — | — | 5.9 | — |
| Flakes | — | 3.3 | — | — | — | — | — | — | — | — | 10.8 | — |
| P. Elem | — | 4.8 | — | — | — | — | — | — | — | — | 15.8 | — |
| L. Flk | — | 4.5 | — | — | — | — | — | — | — | — | 19.6 | — |
| L. Bld | — | 9.0 | — | — | — | — | — | — | — | — | 18.2 | — |

## Field Study

In 1985 a field study was undertaken to identify the distribution of chert sources within the region. *In situ* cherts are confined mainly to Late Cretaceous limestones and chalks that form the plateau and its flanks (Figure 7.3). The lower piedmont and lowlands fronting the plateau lack cherts as they are composed of Lower Cretaceous sandstones and Cambrian/Precambrian granites. Derived or gravel chert sources are also found in the beds of tributaries draining the plateau, but large, knappable-sized cobbles are confined to near the headwaters of these drainages within some 2–3 km from the edge of the plateau. Exceptions to this general distribution include a source confined to a small hill about 1 km northeast of Jebel Humeima and a source associated with Cenomanian limestones exposed over a larger area along the eastern flank of the Rift Valley some 20 km northwest of Tor Faraj. It should also be noted that aside from the plateau two other locations of Cenomanian limestones are plotted on geologic maps. One is located about 8 km southwest of Tor Faraj and another some 20 km southwest of the site on the edge of the Rift Valley. Examination of the nearest of these limestones, however, indicated that it does not contain chert.

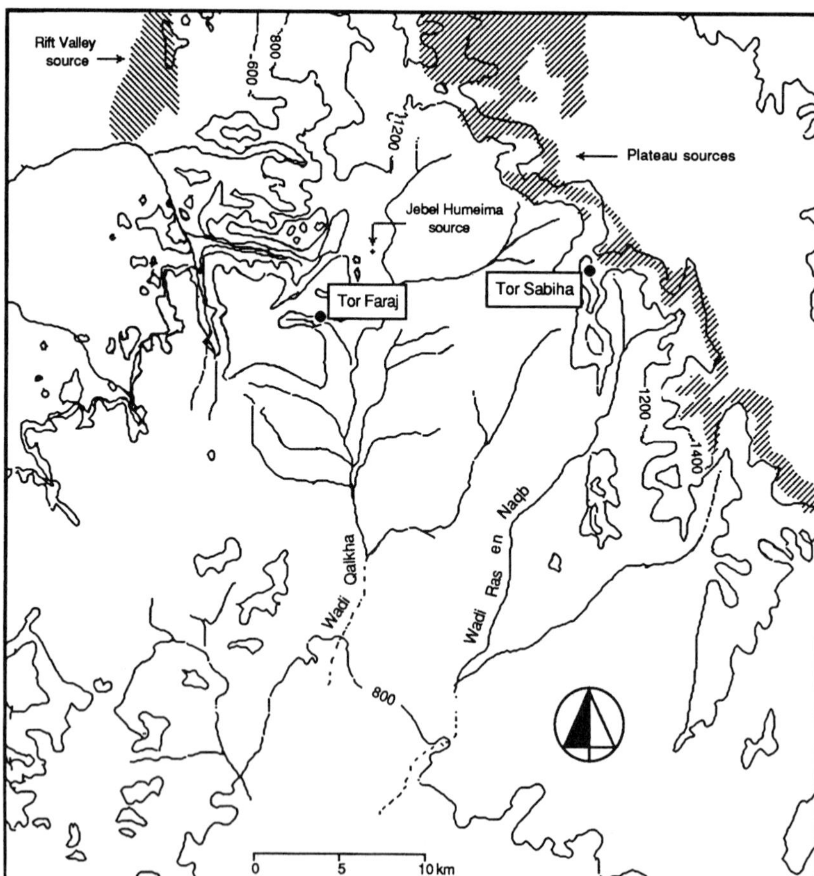


Figure 7.3. Map showing the locations of Tor Faraj and Tor Sabiha relative to the distributions of chert sources for the region as established from geological maps (Bender 1974; Jordanian Geographic Centre 1988).

The initial plan of the study called for spatially sampling specific Late Cretaceous stratigraphic units, as identified on geologic maps developed by Bender (1974) and others, for cherts. These samples were then to be classified as to chert varieties and subsequently compared to chert varieties identified in the artifact assemblages. This plan failed to take into account the abundance and great diversity of cherts exposed on the plateau. For example, along the escarpment near Ras en Naqb dozens of chert veins occur stratified within a 400 m thick section. They display striking variation across a range of attributes and occur in veins from a few centimeters to over a meter in thickness. Confronted with this overwhelming mass of information the plan of study was modified to first classifying the chert varieties contained within the artifact assemblages and then searching for these sources within the field.

### Classification of Cherts

In order to standardize the classification and to objectively define variations in the cherts, an attribute study was conducted for each assemblage. The study focussed on seven major attributes (Table 7.2) with four analysts cross-checking results until a high degree of standardization was achieved. Ultimately the study resulted in the identification of seven varieties and an additional catchall category that included "all other varieties."

Chert varieties 1 through 5 were found to come from Upper Cenomanian and Turonian limestones on the plateau, whereas varieties 7 and 8 came from the small hill near Jebel Humeima and the Rift Valley, respectively. Variety 6 was assigned to the catchall variety and in all likelihood the sources for these poorly represented cherts are on the plateau given that the chert bearing limestones are concentrated there. The great majority of plateau cherts identified within the assemblages displayed limestone in contrast to gravel cortex, thus indicating that *in situ* as opposed to derived cherts provided the bulk of the material.

### Patterns of Utilization

In classifying the cherts of primary elements, two important patterns of utilization stand out. First, the assemblages from both sites were composed mainly of cherts from the plateau (Table 7.3). With the exception of the assemblage from Upper Layer C of Tor Faraj, the gross patterns of utilization look very similar. When the exploitation of specific varieties of chert are examined, however, the assemblages of Tor Faraj show a quite different pattern of utilization from that of Tor Sabiha (Figure 7.4). While chert Variety 1 was the most intensively exploited by the occupants of Tor Faraj, varieties 3 and 5 were the most commonly used cherts at Tor Sabiha. Beyond the differences in utilization of specific varieties of chert, the patterns of chert use suggest quite different approaches to its acquisition. The dominance of Variety 1 at Tor Faraj indicates that a specific distant source was targeted for direct bulk procurement. The greater diversity and better balance between the chert varieties at Tor Sabiha point to a more casual, opportunistic exploitation of cherts within the site's catchment. This observation is further underscored by the substantial differences in the contribution of Variety 6 (the "all other" category) to the assemblages of the two sites. While Variety 6 represents only 0–2.6% of the Tor Faraj assemblages, it accounts for 13.7% of the Tor Sabiha assemblage.

## Toolkit Diversity

### Morphological Typology

The typologically defined toolkit from Tor Sabiha and Tor Faraj exhibit generally similar configurations with only subtle differences occurring in burin, notch, and denticulate classes (Table 5.3, Figure 7.5). The assemblages are dominated by Levallois points which account for

Table 7.2. Comparison of Attributes of the Seven Major Chert Varieties Identified Within the Tor Sabiha and Tor Faraj Assemblages (Note that Type 7 is the "catchall" variety)

| Type | 1 | 2 | 3 | 4 | 5 | 7 | 8 |
|---|---|---|---|---|---|---|---|
| Color Pattern and Inclusions | grayish brown (10 Yr 5/2) banding, small white specks | very dark gray (5 Yr 3/1) small light specks | brown-dark brown (7.5 Yr 5/2/2—4/2), few small white specks | reddish-dark reddish brown (5 Yr 5/3—3/3), wavy or sswirl pattern | dusky-dark dusky red (2.5 Yr 3/2—2.5/2), wavy with light blotches | light gray–grayish brown (10 Yr 7.2—5.2) uniform with light areas | very dark gray–black 5 Yr 3/1—2.5/1) uniform, small white inclusions |
| Cortex | gravel, diffuse boundary | limestone, well defined boundary | limestone, well defined boundary | limestone, well defined boundary | limestone, gravel | limestone, distinct | limestone, well degined |
| Patina | none | none | none | none | light gray | none | none |
| Luster | non-lustrous | low luster | low luster | high luster | high luster | low luster | non-lustrous |
| Grain size | coarse | moderate | moderate | fine | fine | fine | moderate |
| Translucence | low | low | low | high | high | low–moderate | low |

**Table 7.3.** Proportionate Utilization of Different Chert Sources as Indicated for the Tor Sabiha and Tor Faraj Assemblages

| | Tor Sabiha | Tor Faraj | | |
|---|---|---|---|---|
| | C | Mid C | Up C | Top C |
| Chert source | % | % | % | % |
| Plateau | 88 | 87 | 51 | 75 |
| Humeima | 8 | 5 | 29 | 9 |
| Rift valley | 4 | 8 | 20 | 16 |

some 50–68% of the toolkit. From a typological perspective, the proportionate dominance of points coupled with the fact that they served as multipurpose implements may well mask the functional variability that existed between the toolkit.

## Wear Pattern Studies

In his wear pattern study of the Tor Faraj assemblage, Shea (1991, Chapter 6) found that about 48% of the Levallois points exhibited use-wear. In a similar study focussing on points from Tor Faraj and Tor Sabiha, Lee (1987) found points to have been used in performing eight different activities with an average of about two "functional units" per point. Points in both assemblages were used most extensively on soft materials (representing meat, fresh hide, tendon, ligament) with less emphasis on working materials with medium resistance, those principally associated with woods (Table 7.4). Lee's (1987) study revealed statistically significant differences between the points from the two sites relative to the materials upon which they had been

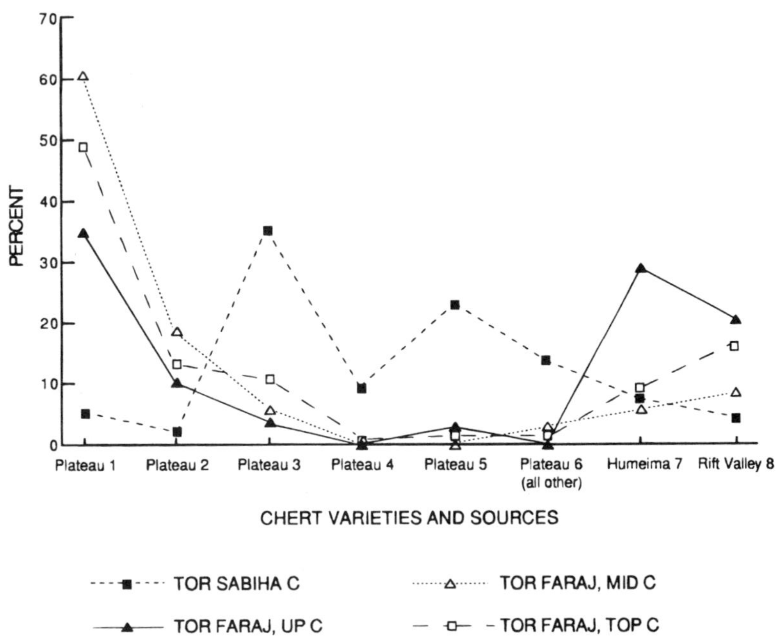


**Figure 7.4.** Histogram of the relative proportions of different cherts represented in the assemblages from Tor Faraj and Tor Sabiha.

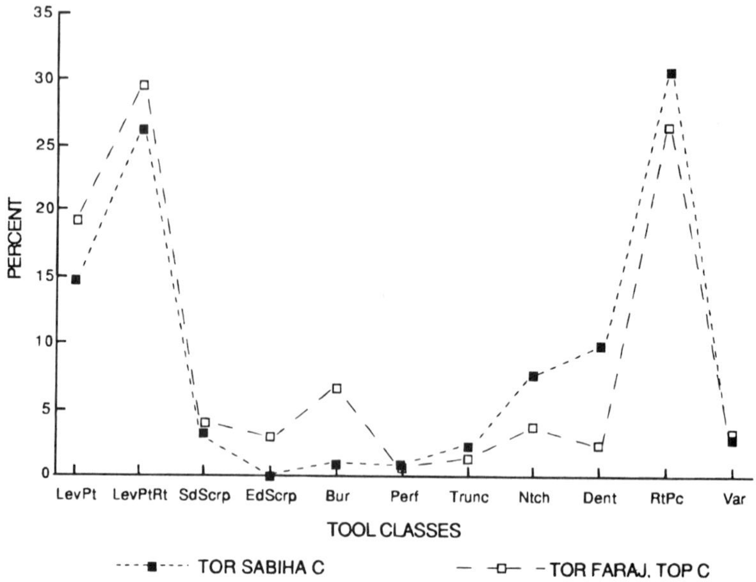


**Figure 7.5.** Histogram comparing the tool classes of assemblages from Tor Faraj and Tor Sabiha.

used, however. He found the Tor Sabiha sample to have been used to a greater extent on soft materials (animal), whereas the Tor Faraj sample was used more extensively on medium materials (wood). In the motion component of the study, "cutting" dominated the points from both sites, but was significantly less common in the Tor Faraj sample which also showed a much greater frequency of "scraping."

Shea's (1991) analysis of samples of all artifacts from Tor Faraj also identified cutting and scraping along with piercing as major motions for the assemblage as a whole. He identified butchery, light-duty woodworking, and hunting as principal inferred activities, although some hide-working and soft plant processing were noted.

## Breakage Patterns

Although Levallois points served a wide range of functions overlapping in part with those of formal tools, their breakage patterns differ markedly from other implements (Table 7.5). The paucity of broken points in the assemblages is surprising given that various lines of evidence (i.e., basally concentrated retouch, impact spalls, haft wear) confirm that points did serve as

**Table 7.4. The Degree of Resistance of Materials Worked as Indicated by Worn Edges of Levallois Points (modified after Lee 1987)**

| Resistance | Tor Sabiha | Tor Faraj |
|---|---|---|
| Soft | 69.5% | 63.2% |
| Medium | 30.6% | 36.8% |
| Hard | 0.0% | 0.0% |
| No. of functional units | 90 | 114 |

**Table 7.5. Breakage Patterns of Tools and Points (Comp = Complete, Brok = Broken)**

| | Tor Sabiha | | Tor Faraj | | | | | |
|---|---|---|---|---|---|---|---|---|
| | C | | Mid C | | Up C | | Top C | |
| | Comp | Brok | Comp | Brok | Comp | Brok | Comp | Brok |
| Tools | (N = 96) | | (N = 19) | | (N = 17) | | (N = 101) | |
| | 53.5 | 46.5 | 42.1 | 57.9 | 29.4 | 70.6 | 62.4 | 37.6 |
| Levallois points | (N = 53) | | (N = 35) | | (N = 36) | | (N = 102) | |
| | 81.1 | 19.9 | 94.3 | 5.7 | 89.5 | 19.5 | 94.1 | 5.9 |

projectiles, most likely as spear heads, in conjunction with their other uses. In having been employed as an impact implement, their rate of breakage should in fact exceed that of most other tools. This is what we see for large projectiles in other contexts where some 40–80% are broken with the large majority of these occurring as bases.

The unexpected breakage pattern of the points from Tor Sabiha and Tor Faraj may provide a clue as to how they were employed and perhaps, more importantly, how this influenced provisioning strategies. The large size and heaviness of the points indicate that it is unlikely that they were projected with the aid of an atlatl or throwing board. Ethnographic and archaeological studies of such dart points show their mean weights to range from 4.5–9 g (Fenenga 1953; Thomas 1978; Patterson 1985), whereas the points from Tor Sabiha and Tor Faraj range from about 10–14 g.

The distinction between spears thrown or thrusted by hand and those projected with the aid of a throwing board is important in that hand-thrown spears are generally formed from a uniform shaft while spears projected with a throwing board are typically composed of segmented shafts, consisting of a main- and a fore-shaft (Oswalt 1973:49–51). The use of a fore-shaft protects the breakage of the main-shaft on impact, prevents the loss of a spear impaled in an animal that gets away, and allows for the rapid rearming of a spear with spare fore-shafts.

In what must have been an early stage in the evolution of stone-tipped shaft weaponry, Levantine Mousterian groups may well have achieved some of the advantages accompanying the use of a fore-shaft by loosely mounting points to a unified shaft. If the points had been mounted in such a way that when they were thrust or thrown into prey, they would have separated easily from the shafts, then the shafts would have been less likely to have been broken or lost. Whether thrust or thrown, shafts could have been rearmed rapidly by casually securing points in a notch with mastic (Figure 7.6), a hafting procedure which, perhaps not coincidentally, would have been facilitated by the basal shape of points created by the Y-arrete scar pattern. Such a technology would explain the extraordinarily low rates of breakage observed on the points recovered from Tor Sabiha and Tor Faraj, for there would have been little reason for returning them to camp. Since broken points securely mounted in fore-shafts would have been returned to camp for recovery of the fore-shafts and possible rejuvenation of the points with retouch, neither of these technologies may have been associated with Levallois points.

## Residential Mobility and Group Size

Decisions related to manufacturing time, portability, raw material utilization, and toolkit configuration must have strongly influenced the adoption of certain provisioning strategies by groups of prehistoric foragers. The varying degrees of residential mobility experienced by such groups, however, may well have been the single most important factor governing the overall

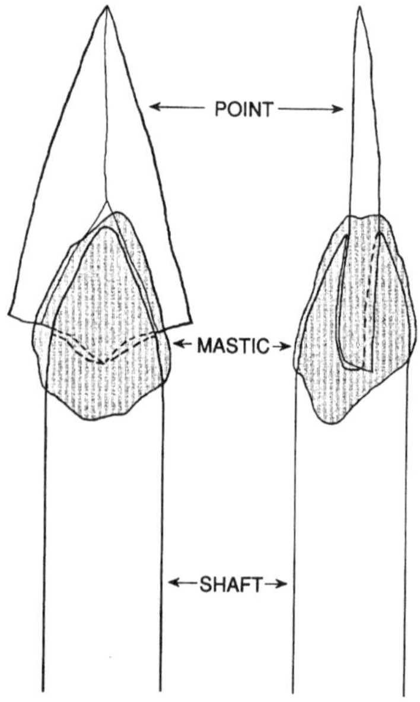


Figure 7.6. Schematic illustration of Levallois point mounted to spear shaft.

organization of their technology. This is essentially because foragers fundamentally depend upon movement, coupled with adjustments in group size, for meshing population levels with available resources. Given this adaptive linkage combined with the correlations researchers have found between certain mobility levels and artifact configurations, the reconstruction of settlement patterns provides perhaps the most direct means of understanding prehistoric adaptation.

Various lines of evidence are consistent in suggesting that Tor Faraj was more intensively occupied than Tor Sabiha. Most directly this occupational intensity can be translated to mean that this larger site was inhabited for longer intervals by larger groups. The relative influence of numbers and length of residence on artifact accumulations, however, is often difficult to ascertain.

## Group Size

Overall, Tor Faraj is some 100 m$^2$ larger than Tor Sabiha with most of this greater occupation area protected by an overhang. The artifact densities of Tor Faraj, depending upon occupation horizon, range from slightly lower to significantly higher than those of Tor Sabiha (Table 7.6). If discard rates were held constant at the two sites, this implies that there were indeed a greater number of occupants at Tor Faraj. If the number of residents and their artifact discard rates were the same at the two sites, then one would expect artifact densities to be significantly lower (i.e., ca. 47% or 78 artifacts per m$^3$) at the larger Tor Faraj. Estimates as to the number of occupants at each site can be made by using equations, developed from Yellen's (1977) ethnographic data, that define the correlation between the occupied areas and the number of occupants in contemporary !Kung Bushmen camps. Wiessner (1974, 1979), Casteel (1979), Read (1978), and Hassan (1981) have presented various equations that describe this relationship mathematically. Although differing in approach, both power and exponential curves yield high correlation coefficients (Hassan 1981) and similar population estimates at the range of 100–200 m$^2$. When applied to the occupation areas of Tor Sabiha and Tor Faraj, the resulting population estimates are 17 and 23 occupants, respectively.

**Table 7.6. Comparison of Variables Related to the Intensities of the Occupations of Tor Sabiha and Tor Faraj**

| | | Tor Faraj | | |
|---|---|---|---|---|
| | Tor Sabiha | MD C | UP C | TP C |
| Area ($m^2$) | | | | |
| Protected | —28— | — | 136 | — |
| Overall | —116— | — | 216 | — |
| Artifact density (0.1 $m^3$) | | | | |
| | 148 | 205 | 123 | — |
| Reduction Stream | | | | |
| Prim Elem: Debitage | 1:11.5 | 3.9 | 7.8 | 6.1 |
| Core: Tool | 1:12.9 | 4.5 | 26.5 | 5.0 |
| Core: Debitage | 1:174 | 43 | 160 | 19 |
| Tool: Debitage | 1:13.6 | 9.5 | 6.1 | 4.0 |
| Debitage: Chip | 1:2.3 | 1.6 | 1.2 | 0.1 |
| Tool: Chip | 1:31.3 | 15.4 | 7.3 | 0.4 |
| Core: Point | 1:5.3 | — | — | 1:2.5 |
| *Chaine* Phases (Rank order, High–Low, 1–4) | | | | |
| 1 | 4 | 2 | 3 | 1 |
| 2A | 4 | 3 | 1 | 2 |
| 2B | 2 | 3 | 4 | 1 |
| 2C | 2 | 3 | 4 | 1 |
| 3 | 1 | 2 | 3 | 4 |

## Seasonality

Differences in the seasons and durations of the occupations are also indicated by their natural settings, features, and artifact inventories. As previously described, the high elevation and eastern exposure of Tor Sabiha would have made it an unlikely choice for a cold season camp. More specifically, the intrasite distribution of artifacts reflects a concentration of activities along the western edge of the excavation block (Figure 5.19) at the base of a steep cliff. This distribution implies that the occupants of Tor Sabiha may well have positioned their activities in order to take advantage of the shade offered by the cliff, thus suggesting a warm season camp. The setting of Tor Faraj at a lower elevation with a southern exposure and a much larger protected area is more consistent with a winter camp. The Wadi Aghar running beneath the shelter also was more likely to have been charged during the winter wet season than during the summer and early fall. The rarity of phytoliths of flowering plant parts also points to the shelter having been occupied outside of the major growth and flowering seasons of the spring and summer.

## Periodicity and Duration of Use

Aside from the seasonal differences in the occupations, Tor Faraj appears to have been visited more regularly and occupied for longer periods than Tor Sabiha. The hearths and ash lenses of the Middle and Upper horizons of Layer C reflect two such episodes within the upper 100 cm of the deposit. Given the artifacts cemented to the backwall of the shelter some 100 cm above the floor in conjunction with the recovery of a Levallois point from the wall of a Bedouin storehouse some 3 m below the floor, Tor Faraj contains at least four meters of cultural deposit. In contrast, Tor Sabiha's cultural deposit is only 30 cm thick and the distributional patterns of artifacts indicate that it may well represent only a single occupational episode.

The longer duration of the occupations at Tor Faraj is indicated by longer reduction streams emphasizing initial processing, the direct bulk procurement of distant chert, heavier/larger artifact inventories, higher proportions of retouched tools, and less curation of Levallois points. The presence of hearths also may signal relatively long encampments at the shelter, or they could merely reflect the greater use of fire in winter encampments.

As previously discussed, several workers have noted the direct relationship between the length of a reduction stream expressed by an assemblage and the length of residence associated with its locus of recovery. The relatively high proportions of primary elements and cores in conjunction with dimensional comparisons that confirm the on-site production of blanks (especially Levallois points) underscore the importance that was placed upon initial production activities at the shelter. Studies of the tool typology, chips, and wear patterns substantiate a wide range of end-of-stream activities including tool manufacture, use, and maintenance.

The higher proportion of retouched tools in the Tor Faraj assemblages may also indicate longer periods of residence in the shelter. While the degree to which tools have been retouched has been proposed as a direct expression of the intensity of Mousterian occupations in Europe (Rolland 1981; Dibble 1984), such intensive retouch and edge modification is unusual for the Levantine Mousterian. As a parallel to this argument, however, one might suggest that the probability that blanks will be retouched at a site increases directly with the duration of residence at that site, especially if the site is in a chert-poor area. The nature of chert acquisition, in fact, provides the most convincing argument for extended occupations at the shelter. The evidence for initial processing coupled with the patterns of chert use shows that the occupants of Tor Faraj targeted specific chert sources on the plateau some 16 km away for direct procurement in bulk. They were clearly following a logistical pattern in provisioning a place.

In contrast, Tor Sabiha appears to have served as an ephemeral campsite at which final processing activities (i.e., tool fabrication, maintenance, and rejuvenation) were emphasized. In the main, initial processing (i.e., core shaping and blank production) appears to have been undertaken off the site, most likely centered around activities which were expediently provisioned within the chert-rich portions of the site's catchment. Other implements imported to the site would have included those manufactured in anticipation of those activities subject to time stress (e.g., hunting) and used in provisioning individuals. Levallois points, perhaps the clearest example of such implements, appear to have been produced mostly off-site. This is confirmed by the points being larger than those debitage classes (primary elements and blades) which would have necessarily preceded point production in a continuous reduction stream and by a core-to-point ratio (1:5.3) that substantially exceeds the anticipated production of around two points per core. The pattern of chert utilization within the Tor Sabiha assemblage is also consistent with an expedient, opportunistic procurement strategy within the nearby chert-rich portions of the site's catchment. Unlike the pattern of Tor Faraj in which one chert variety was dominant, chert selection at Tor Sabiha was much more heterogeneous.

The ephemeral nature of the occupation at Tor Sabiha is also indicated by an artifact distribution which suggests that a wide range of activities was concentrated in the same locus confined to 2 $m^2$ in the northwest corner of the excavation block (Figure 5.19). The spatially overlapping concentrations of primary elements, cores, points, tools, and chips indicate that when undertaken, initial processing activities were most likely followed by final processing, tool use, and maintenance—all in a continuous, probably rather brief interval. A similar "specialized" artifact distribution pattern, recognized in Level 1 of the terminal Levantine Mousterian occupation of Boker Tachtit, has also been associated with an ephemeral camp (Hietala and Marks 1981). More "generalized" patterns as seen from early (Rosh Ein Mor) and late (Far'ah II) Levantine Mousterian sites have been associated with longer term occupations or base camps (Marks and Friedel 1977; Marks 1989; Gilead and Grigson 1984).

While the durations of the occupations at Tor Faraj appear longer than those of Tor Sabiha, they were still relatively brief, at least to the extent that fragile hearth shadows and ash lenses remained reasonably undisturbed. In many ways the shallow hearths and ash lenses are reminiscent of those associated with the seasonal shepherd camps of contemporary Bedouin. After a few weeks of use they are abandoned for a year or more leaving the floors to be sealed by wind-borne sand.

## Settlement Pattern

The settings and artifact patterns of Tor Faraj and Tor Sabiha show a striking dichotomy which is suggestive of different segments of a settlement system. Given the proximity of the sites, they undoubtedly represent only a portion of such a system, however. Late Levantine Mousterian groups may well have followed an annual settlement pattern that encompassed a much larger territory stretching from the uplands on the plateau to the Rift Valley. Although such a territory would have provided an elevational range of over 1,700 m accompanied by substantial resource diversity, it could have been traversed over linear distances measuring no more than 70 km. In following this transhumant pattern Late Levantine Mousterian groups may have occupied a series of winter camps stretching from the Rift Valley to the lower piedmont followed by the occupation of the higher elevations of the piedmont and plateau during the warm, dry season. These winter camps may have been inhabited longer and by larger groups than their warm season counterparts because of the limited number of protected settings such as Tor Faraj. The availability of predictable water sources near such sites would have sustained large groups for extended periods. Groups may well have recognized the benefits of pooling their efforts and returns in the collection of fuels during the cold season in what must have been a fuel-poor steppe environment.

The Wadi Aghar serves as a natural avenue between the western plain of the broad Hisma Valley and the floor of the Rift Valley some 25 km away and 1,000 m lower elevation. It reaches its headwaters at a low drainage divide about 5 km upstream from Tor Faraj and from this point a branch of the Wadi Aheimir falls rapidly to the Rift Valley about 20 km distant. Today Bedouin herders regularly follow this route in moving camels between the Hisma and the rift. The substantial proportions (10–22%) of chert from the Rift Valley in the Tor Faraj assemblages indicate that Levantine Mousterian groups probably used the same passageway. Beyond resting along this natural route, Tor Faraj also provides a commanding view of the plain forming the western end of the Hisma. In fact, one can see all the way to the plateau and its outliers, including Jebel el Jill. On the opposite face of the jebel, as seen from Tor Faraj, the cave of Tor Sabiha offers another extensive view, but of the Judayid Basin and the escarpment of the Ma'an Plateau. Like Tor Faraj, Tor Sabiha also rests along a natural access route, in this case between the Hisma and the uplands on the plateau. The route follows a ridge formed by an extension of Jebel el Jill and another outlier which connects to the plateau proper near Ras en Naqb. This ridge provides the only gentle ascent to the southern edge of the plateau as denoted by ancient roads and trails. Given their relative locations then, it is not surprising that these two settings would contain evidence for the movement of Levantine Mousterian groups between winter-lowland and summer-upland settings.

In the most comprehensive settlement pattern study of the Levantine Mousterian, Marks and his colleagues have suggested an evolution from a "radiating" to a "circulating" pattern in the Highland Negev (Marks and Friedel 1977; Marks 1989). The radiating pattern is thought to have consisted of large basecamps, apparently occupied over long intervals, which were tied to ephemeral exploitation camps and quarry sites. While the base camps were provisioned as places using a logistical strategy, the provisioning of individuals and activities appears to have accounted for the artifact accumulations at the ephemeral camps. With a single exception, the nine ephemeral sites fall within 10 m of a chert outcrop, whereas the four base camps are each situated within 600 m from a chert source. Chert was introduced to the base camps and those ephemeral camps distant from raw material mostly in the form of blanks and shaped cores (Munday 1976, 1977; Marks 1988). A circulating pattern of settlement is thought to have emerged during the terminal Levantine Mousterian and came to characterize subsequent Upper Paleolithic and Epipaleolithic settlement systems in the Negev. In contrast to the marked variability between the site types of the radiating pattern, the site types of the circulating pattern are relatively uniform in being composed of ephemerally occupied multipurpose camps. The technologies associated with such camps would have largely involved the provisioning of activities and individuals.

The settlement pattern associated with the two south Jordan sites differs from those proposed for the Highland Negev in two ways. First, rather than strictly matching a radial or a circulating pattern, the data from Tor Faraj and Tor Sabiha are more consistent with a settlement system in which group size and residential permanence were adjusted seasonally to accommodate fluctuations in resources, creature comforts, and social needs. Secondly, the acquisition of raw material in bulk from distant sources by the occupants of Tor Faraj differs from the Negev strategy in which raw material appears to have been imported to sites primarily in the form of cores and blanks.

Perhaps in recognizing the rigidity of the idealized radiating and circulating patterns, Marks and Friedel (1977) pointed to ethnographic analogies that were typically composites of the two. Settlement patterns of ethnographic foragers characteristically include large, long duration base camps supported in part by ephemeral special purpose camps during periods of resource abundance followed by the dispersal of the population into smaller, more transitory, multipurpose camps for the remaining portion of an annual cycle While this cycle of coalescence and dispersal is typical of most foragers, the relative degree of mobility among ethnographic foragers ranges from near permanent settlements to almost full nomadism (Kelly 1983). In viewing the evidence from the Negev from an evolutionary perspective, Marks and Friedel (1977) argued strongly that:

> the settlement patterns of the survey zone are actually accurate reflections of total settlement systems found within this part of the Levant and that the gradual trend from a relatively sedentary to a relatively nomadic lifestyle. . . . is a general diachronic pattern for the area.

Their notion of a trend of increasing settlement mobility finds support in the data from the south Jordan sites. In dating to ca. 69,000 B.P., Tor Faraj and Tor Sabiha fall between the Early Levantine Mousterian sites of Rosh Ein Mor and Nahal Aqev (ca. 87,000 B.P.) and the transitional occupation at Boker Tachtit, Level 1 (ca. 45,000 B.P.). Speculation that the adoption of a more mobile settlement system was triggered by the onset of arid conditions is also consistent with the paleoenvironmental evidence retrieved from the south Jordan deposits. In contrast to the pluvial conditions associated with the Early Levantine Mousterian, Tor Sabiha and Tor Faraj appear to have been occupied under relatively dry conditions.

The south Jordan Late Levantine Mousterian settlement system appears to have differed from that of the Early Levantine Mousterian of the Negev relative to the specific pattern of settlement and the overall level of settlement mobility. Some interesting continuities as well as differences can be seen in the provisioning strategies, however. The Early Levantine Mousterian strategy of provisioning individuals with prepared cores and blanks appears to have continued to be employed during Late Levantine Mousterian times as evidenced in the ephemeral occupation of Tor Sabiha (and level 1, Boker Tachtit). The provisioning of longer-term occupations at base camps as places through a logistical strategy also appears to have persisted as seen from the occupation horizons at Tor Faraj.

But unlike the base camps of the Early Levantine Mousterian that were provisioned principally through the importation of prepared cores and blanks, the occupants of Tor Faraj relied upon importing chert in bulk directly from distant sources. Although the chert nodules that were collected from sources some 16–20 km away appear to have undergone some trimming, this did not approach the degree of core shaping identified at the Early Levantine Mousterian basecamps of the Negev. This is underscored by the high proportions (18.2%) and large dimensions (X lengths = 57.4 mm) of primary elements at Tor Faraj when compared to similar data for Rosh Ein Mor (14.6%; 47.7 mm) and Nahal Aqev (8.5%; length—no data), even though each of the Negev sites is separated by no more than 600 m from its chert source. Why would the residents of Tor Faraj consistently have decided against extensively reducing chert at its source for improved portability, especially given the long transport distance back to the site? The answer to this question may have to do with whether or not the groups/individuals were stressed for time during the process of acquiring raw material. The degree of time-stress that they experienced is likely to have been governed by the location of the chert source relative to the limits of their site catchment. If sources fell within a site catchment (i.e., ca. 10 km radius), raw materials could have been acquired in an expedient, opportunistic fashion often accompanying other procurement activities. Binford (1979) has described this as an embedded strategy. Chert would often have been needed for fabricating implements to be used in other activities and these other activities would normally have produced resources that would also have required portaging. Both of these requirements would have favored core shaping at or near source areas. In resting within a site catchment, raw materials would have been imported in this reduced form continuously, probably daily, during the occupation of a site.

For sites like Tor Faraj that utilized chert sources well outside their catchments, procurement of chert could not have accompanied other activities, nor could it have been undertaken in an opportunistic, expedient fashion. Distant chert sources resting outside the site catchment would have been targeted and time would have been devoted to collecting and importing cherts from these sources. Although the devotion of time to this specialized activity would have allowed for collecting chert periodically in volume from a great distance, it also would have greatly reduced the time available for a group's other support activities (e.g., collecting food and fuel) while away from a base camp. Thus the time available for procurement and transport of raw materials from great distances would have been stressed to the extent that extensive reduction and core shaping would have reduced the overall efficiency of the process. The setting and season of occupation of Tor Faraj may have even heightened the time-stress placed upon the procurement of chert from the plateau. In resting between ca. 1,400–1,700 m most of the locations of chert sources would have been uncomfortably cold and lacking most food resources during the winter.

## COGNITION IN THE LEVANTINE MOUSTERIAN

For whatever reason, it is clear that the residents of Tor Faraj engaged in the bulk acquisition of specific chert varieties from sources at least 16 km distant. This implies that in organizing their technology, a logistical strategy of provisioning a place was employed and that this strategy required a significant degree of planning depth and foresight.

### Planning Depth and Flexibility

As discussed earlier, limited planning depth has been proposed as a salient characteristic of Mousterian cognition and a feature distinguishing archaic from fully modern thought processes. While evidence from Europe is certainly consistent in showing little utilization of distant raw materials during the Middle Paleolithic, this was not the case for the Late Levantine Mousterian inhabitants of Tor Faraj. Beyond the issue of planning depth, the variability in Levantine Mousterian mobility and group size coupled with adjustments in the organization of technology fails to fit the rigid, routinized patterns postulated for Middle Paleolithic behaviors. Along these lines, Binford (1989:35) argues that:

> the greatest contrast between the Mousterian and the culturally organized systems of modern hunter-gatherers is in the relationships among mobility, environment, and technological organization. Modern hunter-gatherer adaptations achieve their stability and security by highly flexible patterns of movement within their environments and flexible group size and composition.

The seasonal shift from the large, long-duration, logistically organized base camp of Tor Faraj to the smaller, more transitory, expediently provisioned camp at Tor Sabiha appears to reflect an adaptive pattern that more closely resembles that of modern foragers than the postulated Mousterian pattern.

## Memory and Conservation Logic

The differences in the ways the residents of Tor Faraj and Tor Sabiha organized their technologies also provides another clue as to their patterns of cognition. The knowledge of where specific, distant chert sources were situated and the procurement of chert in quantities sufficient for future use both imply a significant degree of anticipation. From another perspective, such anticipation had to have been grounded in memory. That is to say a cognitive map of chert sources and a knowledge that chert would be needed in certain amounts for sustaining the technology for a certain length of time would both have depended upon accurately recalling past experiences. A second area of cognition upon which the organization of the technology would have depended is that of "conservation logic." This involves the ability to compare accurately and convert between different units of measure such as dimensions of length, number, mass, area, and volume. A decision of whether to use a logistical or an opportunistic procurement strategy would have required many such comparisons and conversions.

An examination of memory and related cognitive changes in the ontogeny of modern humans may shed some light on evolutionary developments. This is not to imply that the cognitive ontogeny of modern humans necessarily mirrors our hominid phylogenic development. But a general understanding of the cognitive development of modern humans may provide inspiration for model building in much the same way that studies of contemporary foragers have increased our understanding of their prehistoric predecessors.

### The Piaget Model

Much of the thinking in developmental psychology is based upon the theories of Jean Piaget in which he argued that cognition develops in a series of discrete stages. Although generally accepted, Piaget's stages have been questioned as to whether they represent distinct steps in a cognitive succession or merely segments along a continuum of change (Flavell 1978; Carey 1985) and as to whether they occur naturally in the lock-step way he described (Fisher 1980; Case 1985). There has also been concern about the cross-cultural applicability of Piaget's findings, but these appear to stem more from the use of inappropriate measures than true differences (Price-Williams 1961; Ember 1977). In sum, there is a general consensus that within the ontogeny of our species, levels of cognition undergo fundamental changes in a rather uniform way.

In returning to the question of cognitive evolution during the late Pleistocene, what concerns us here are the differences in thinking that are related to opportunistic and logistical behaviors. Perhaps not by coincidence, these behaviors are based upon several cognitive patterns that distinguish Piaget's preoperational and operational stages. In fact, conservation logic, upon which a logistical strategy depends, is not only diagnostic of the Piaget's operational stage but was viewed by him as the basis for all rational thinking (Bjorklund 1989:24).

Memory would also have played an important role in supporting the behaviors accompanying a logistical procurement strategy. A precise knowledge of where distant high-quality chert sources were located in relation to a base camp, topographic features, and other critical resources (e.g., food and fuel) would have been required. Given the heavy relief of the area between Tor Faraj and the chert sources of the plateau, a detailed cognitive map would have been a necessity for efficient procurement. Deep, vertically walled canyons and sheer escarpments create numerous barriers to straight-line travel. During most of the year, water and fuel are hard to find.

The kind of memory and recall related to building cognitive maps is termed "constructive" by developmental psychologists. Such memory is based upon what we already know about the world and our recall of events as shaped by previous knowledge (Bjorklund 1989:158). As with general cognitive development, Piaget and his colleagues also investigated the development of memory and found a marked improvement in constructive memory to coincide with the operational stage (Piaget and Inhelder 1973). Recent research has shown, however, that constructive memory may be directly linked to how well information is initially encoded (Liben 1981). From a developmental perspective, encoding at the operational stage is more reflective of reality than at earlier stages.

In using the findings of developmental psychology and specifically Piaget's stages as means of measuring the necessary cognitive levels of those behaviors associated with a logistical procurement strategy, it is apparent that a mature operational level, as seen in that of contemporary humans, must have been attained by the residents of Tor Faraj some 69,000 years ago.

### Other Applications of Piaget

Piaget's findings also have been applied to stone tool fabrication in an effort to measure the cognitive level of prehistoric knappers (Wynn 1979, 1985). Wynn (1985) has stressed the importance of the concept of "reversibility" and the recognition of true Euclidean relationships to certain lithic technologies. Both of these thought processes are diagnostic of Piaget's operational stage. Wynn (1985) argues that an ability to conceive of an action and its inverse simultaneously (concept of reversibility) would have been necessary even for the production of bifacially symmetric handaxes. This would certainly have been a cognitive necessity in the rather precise shaping of cores for Levallois point removal. If Wynn is correct, the cognitive level of modern humans was achieved well before the Middle Paleolithic and we have to look beyond cognitive changes to understand the behavioral evolution of the Late Pleistocene.

## SUMMARY AND CONCLUSIONS

Although representing only fragments of a 70,000-year-old adaptive system, the archaeologically synchronous occupations at Tor Faraj and Tor Sabiha suggest that the Late Levantine Mousterian residents of the sites employed a transhumant strategy to cope with seasonal fluctuations in resources and creature comfort. Tor Faraj is most likely to have served as a large, long-term, winter encampment that was revisited regularly. It was logistically organized for the acquisition of resources, most notably that of specific chert varieties that were procured in bulk from distant (>16 km) sources. In contrast, Tor Sabiha represents a small, high elevation, transitory camp most likely to have been inhabited during the summer. An expedient, opportunistic procurement strategy was followed in which chert was acquired indiscriminately from sources within the site's catchment through the provisioning of individuals and activities.

The alternative ways the occupants of the two sites organized their technology shows that Levantine Mousterian groups commanded significant planning depth as well as flexibility in structuring their technology, both characteristics of fully modern foragers. From a strictly cognitive perspective, these groups appear to have had the mental faculties associated with mature, operational levels of modern humans as defined by Piaget and others. These characteristics of significant planning depth and flexibility in organizing technologies coupled with advanced cognitive levels are also seen from other areas of the Levant. Aside from the previously discussed evidence from the Highland Negev (Marks and Friedel 1977; Munday 1977), marked differences in the intensity by which sites were occupied are noted from the El-Kowm Basin of Syria and the Carmel–Galilee area of northern Israel (Bar-Yosef 1989). Intrasite artifact patterns defined within Levantine Mousterian sites also denote differences in the organization of technology and associated activities. A comparison of such patterns as seen from Rosh Ein Mor and Boker Tachtit (Marks 1988, 1989), Fa'ra II (Gilead and Grigson 1984), Quneitra (Goren-Inbar 1990b), and Kebara Cave (Bar-Yosef *et al.* 1992) shows that the complexity and spatial segregation of activities increased with the duration of encampments. This again is consistent with the ways in which modern foragers order their activities.

Symbolic activities, an alternative means for accessing cognition, are expressed by the intentional inhumations exposed at the Levantine Mousterian cave sites of Skhul, Qafzeh, Amud, and Kebara. Although rare, grave offerings include a wild boar skull with Skhul V and a deer antler accompanying a child burial at Qafzeh (Bar-Yosef 1989). Such symbolic activities are again in keeping with the other lines of evidence that point to essentially modern behaviors for Levantine Mousterian groups, in this case shared by early moderns and Neanderthals.

Taken together then, diverse lines of evidence suggest that Levantine Mousterian groups had come to occupy a fully human ecological niche and were using culture as their extrasomatic means of adaptation in much the same way as modern humans. If this was indeed the case, then the variability that we see in the behavioral residuals of these populations through time and space is likely to have been produced more by differences in human ecology than cognition.

The traditionally held diachronic pattern for the Levantine Mousterian is that evidenced by the technotypologic succession of Tabun Cave in which Type D, C, and B variants (labelled for their stratigraphic layer of recovery) were initially defined (Jelinek 1981). For some time now the long sequence at Tabun has been viewed as a reference column by which the Levantine Mousterian might be ordered, but this has not gone without controversy. The definition of regionally specific successions has prompted the pan-Levantine application of the Tabun sequence to be questioned. Moreover, biostratigraphic sequences and recent chronometric dates from nearby Levantine Mousterian deposits also fail to agree with those of Tabun (Tchernov 1989; Bar-Yosef and Meighnen 1992). The apparent lack of concordance between the Tabun sequence and others centers on the relative placement of Tabun D. For Lebanon, Copeland (1981) and others have argued that assemblages of the C-Type pre-date those of Tabun D, while in the Negev D-Type assemblages are thought to have persisted until transitional Upper Paleolithic times (Marks 1992).

Although the diachronic patterning of the industries of the Levantine Mousterian remains problematic, the spatial patterning of these industries is clearer and perhaps even more informative. Those sites containing D-Type assemblages are found in the arid zone and along the coastal plain as far north as Sidon, while sites with C- and/or B-Type assemblages are confined to the Mediterranean woodlands of northern Israel, Lebanon, and western Syria (Figure 7.7).

In that the contrasting spatial distributions of the industries correspond so strongly with

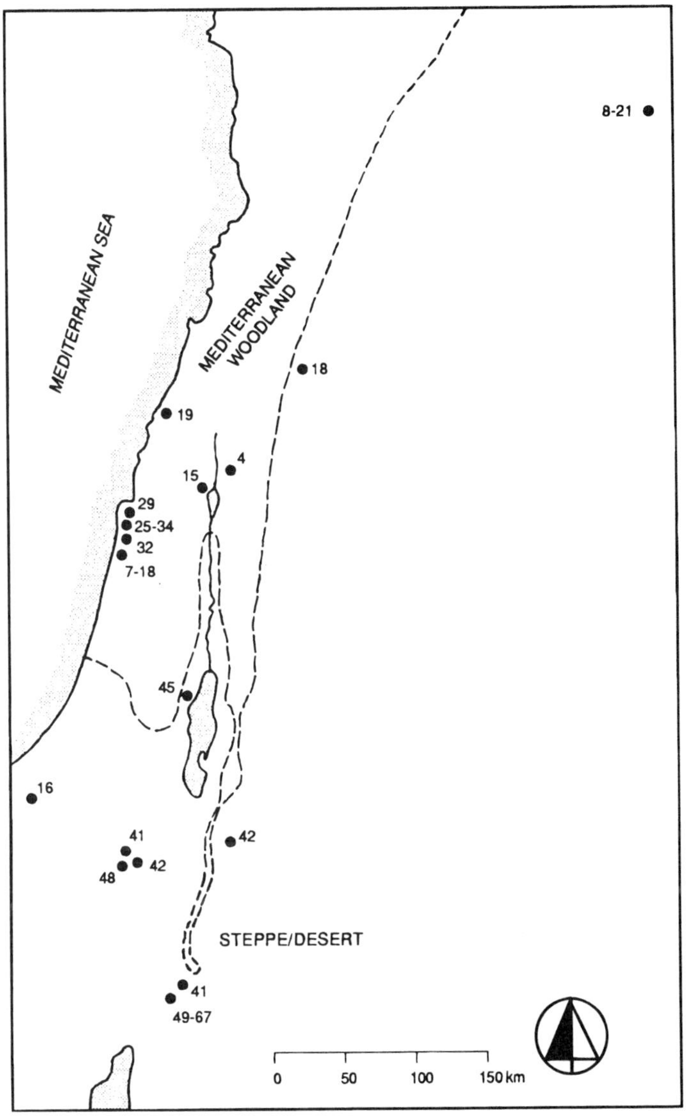


**Figure** 7.7. Map of the Levant showing the relationship of the relative proportions of Levallois points reported for assemblages and the regions biotic zones. Note that assemblages with higher proportions of points are

differences in environment, the distinction between D- and C-Type industries may merely reflect behavioral responses to different environments regardless of their relative chronological placement. The three sites in which the two industries are known to overlap would have rested along environmental boundaries separating woodland from steppe. This may seem to be a surprising interpretation for the coastal settings of Tabun and Adlun, but one has to consider that the coastal plain would have been substantially wider during a period of Mediterranean regression as is associated with the D-Type occupations (Jelinek 1981; Bar-Yosef 1981).

The broad differences required of subsisting in steppe as opposed to better watered woodland settings finds expression in the distinctive lithic characteristics of the industries. While D-

Type assemblages are characterized by relatively high frequencies of elongated Levallois points and blades, C-Type assemblages are typified by low frequencies of points and the production of broad oval Levallois flakes. The emphasis upon point and blade production is consistent with the importance of hunting and the high degree of residential mobility that would have been required for occupation of steppe environments. In his wear pattern study, Shea (1990, Chapter 6) also found the D-Type assemblages to be associated with proportionately greater amounts of extractive use (hunting) as opposed to the higher proportions of maintenance activities for the C-Type assemblages. An even more fundamental difference between the two industries relates to the amount of edge that was produced on blanks given their distinctive lithic technologies (Henry 1986b). The point and blade oriented technology of the D-Type Industry would have produced substantially more edge per standard volume of chert than the broad oval flake technology of the C-Type Industry. The greater edge production efficiency of the D-Type assemblages would have supported the higher mobility levels accompanying occupation of an open, arid environment through increased portability and conservation of raw materials.

We have a relatively good understanding of the way in which those groups responsible for D-Type assemblages went about coping with environmental challenges behaviorally. This understanding suggests behaviors that are very much like those of modern foragers which for 70,000–100,000 years ago is considerably earlier than what is generally suspected. Thus there is a certain irony in the fact that we have yet to identify these groups biologically. While both the remains of Neanderthals and early modern humans have been found associated with C-Type assemblages and Neanderthals with B-Type, fossil remains have yet to be recovered with D-Type artifacts. The environmental distribution of D-Type sites may offer a clue as to the biological identify of their occupants, however. Given their known range in Europe and the Near East, Neanderthals appear to have occupied predominantly woodland environments. As for early modern humans, Foley (1984) and others have pointed to the adaptive significance of a more gracile, elongated body form in enabling such populations to occupy open, arid environments. Interestingly, as with the behavioral (technologic) adaptations, the biologic adaptations appear to have been geared primarily toward enhancing mobility in arid environments. This is not to say that early modern humans were incapable of inhabiting woodlands, for the remains from Skhul and Qafzeh rest within the woodland zone. But it does seem that they were better adapted physically and, perhaps even more importantly if my hunch is right, behaviorally for life in arid environments.

# Chapter 8

# The Upper Paleolithic Sites

NANCY R. COINMAN AND DONALD O. HENRY

Ironically, our understanding of the Levantine Upper Paleolithic is not as well developed as that of the preceding Middle Paleolithic. This largely stems from the rather substantial geographic distribution of Upper Paleolithic sites that stretch from the southern Sinai Peninsula to coastal Lebanon and inland to the eastern desert of Jordan. The presence of such a high density of sites in core areas of the arid zone suggests that Upper Paleolithic populations were better able to exploit the extremely dry settings of the steppe-desert zone than were their Levantine Mousterian predecessors. An amelioration in climate also may have allowed for a more intensive settlement of the arid zone at this time, but even during the more pronounced pluvial conditions of the early Levantine Mousterian, groups appear not to have inhabited such dry settings. At least the surveys of large portions of Sinai and eastern Jordan have yet to define a strong Levantine Mousterian presence. The other major problem in understanding the variability of the Levantine Upper Paleolithic is tied to a chronology developed from a largely inadequate number of radiocarbon dates. The uneven geographic distribution of the dates further complicates the chronology for most of the dates come from sites in the Negev and Sinai and are confined to a single industry.

With the discovery of Upper Paleolithic sites in the arid zone, researchers came to recognize that the unileal developmental sequence as initially proposed by Garrod (1951, 1954) and Neuville (1934, 1951) needed restructuring. This prompted the partitioning of Upper Paleolithic assemblages into two large, partially synchronous complexes: the Ahmarian and the Levantine Aurignacian (Gilead 1981; Marks 1981). Radiocarobon dates for the Ahmarian Complex cluster at the beginning and end of the sequence. Early Ahmarian assemblages date from ca. 38,000–30,000 B.P. and Late assemblages date to ca. 20,000 B.P. Although not well dated, Levantine Aurignacian assemblages appear to fall somewhat later with dates ranging from ca. 32,000–17,000 B.P. (Marks and Ferring 1988). The Ahmarian is characterized by a technology related to the production of relatively narrow, thin blades (and bladelets) from well prepared cores, whereas the Levantine Aurignacian technology is typically described as one in which flakes and thick blades were produced from simply prepared, predominantly single platform cores (Gilead 1981; Marks and Ferring 1988; Gilead and Bar-Yosef 1993). The initial steps of blank production also appear to distinguish the two complexes. Whereas a *lame à crête* technique was employed to initiate the finely controlled removal of blades in the Ahmarian, more opportunistic approaches were followed in the initial blank removals by Levantine Aurignacian knappers. The typological contrasts of the complexes have been described as revolving around a dominance of finely retouched blades, especially el Wad points, in the Early Ahmarian, small Ouchtata bladelets in the Late Ahmarian, and a dominance of thick endscrapers and burins in the Levantine Aurignacian. It should be noted, however, that Ahmarian assemblages evince much more technotypological homogeneity than do those of the Levantine Aurignacian.

The biogeographic distribution of Ahmarian and Levantine Aurignacian assemblages offers an important clue as to one source of variability between the complexes. Whereas Ahmarian assemblages are centered in the core of the arid zone, those of the Levantine Aurignacian are found principally within the Mediterranean woodlands. Interestingly, the emphasis upon the production of a "light weight" toolkit often dominated by points and based upon blade and bladelet production resembles the general configuration of the D-Type Levantine Mousterian assemblages that were also distributed principally within the arid zone; although not deep desert. In contrast, the "heavy-durable" toolkits of the Levantine Aurignacian are based upon the production of thick blades/flakes and contain few points. Again, this parallels the configurations of the C- and B-Type assemblages of the Levantine Mousterian that are confined to the Mediterranean woodlands.

It seems unreasonable to attribute to coincidence these dichotomies in technologies that remain correlated with certain environmental settings for some 70,000–80,000 years. As previously discussed (Chapter 7), the blade dominated assemblages would have provided for a more efficient technology given their greater portability and more economic use of chert. In the arid zone, the portability offered by a stone tool technology would have been especially important because of the high mobility levels that groups would have experienced in moving between widely spaced, low-density resources. Also, unlike the near ubiquitous availability of chert sources in the limestones of the Mediterranean woodlands, chert sources are relatively patchy in the arid-zone as limestones give way to sandstones and granites. In such a setting, technological systems that produced a greater amount of working edge per weight/volume of raw material would have had a selective advantage.

Indirectly, the adaptive differences between those populations associated with the Ahmarian and Levantine Aurignacian technologies may give us some insights as to their origins. The Ahmarian appears to represent an indigenous population that expanded upon the capacity of Late Levantine Mousterian groups in exploiting the arid zone. This apparent unilineal development within the steppe-desert region is supported by the remarkable sequence revealed at the Highland Negev site of Boker Tachtit. Here we see the transition from the terminal phases of the Levantine Mousterian into the early Ahmarian (Marks 1983b,c). The Levantine Aurignacian is more problematic and suggests a more complex cultural history. Bar-Yosef and Gilead (1993) recently have suggested that it represents a population intrusive to the region; perhaps originating in the Balkans. While this notion of a European connection is consistent with the general technotypological characteristics of the Levantine Aurignacian lithic assemblages and modest worked bone industry, much more detailed comparative work is needed for confirmation. Alternatively, the Levantine Aurignacian may well represent an indigenous development within the Mediterranean woodland zone. From an ecologic perspective, the population responsible for the Levantine Mousterian B-Type Industry represents a logical precursor.

## THE SOUTHERN JORDAN SITES

Upper Paleolithic occupations were discovered at six sites, all of which were found within the Jebel Qalkha study area (Figure 8.1). Initial investigations at the sites were confined to test excavations designed to gather fundamental information on the nature and extent of the cultural deposits. Although these probes were undertaken in anticipation of larger excavations at each of the sites, constraints on funding and time limited subsequent work to a single site, that of Tor Hamar (J431). Even though largely limited to test excavations, the research furnishes important evidence on the artifactual configurations of assemblages for a poorly understood archaeological period from a virtually unknown region.

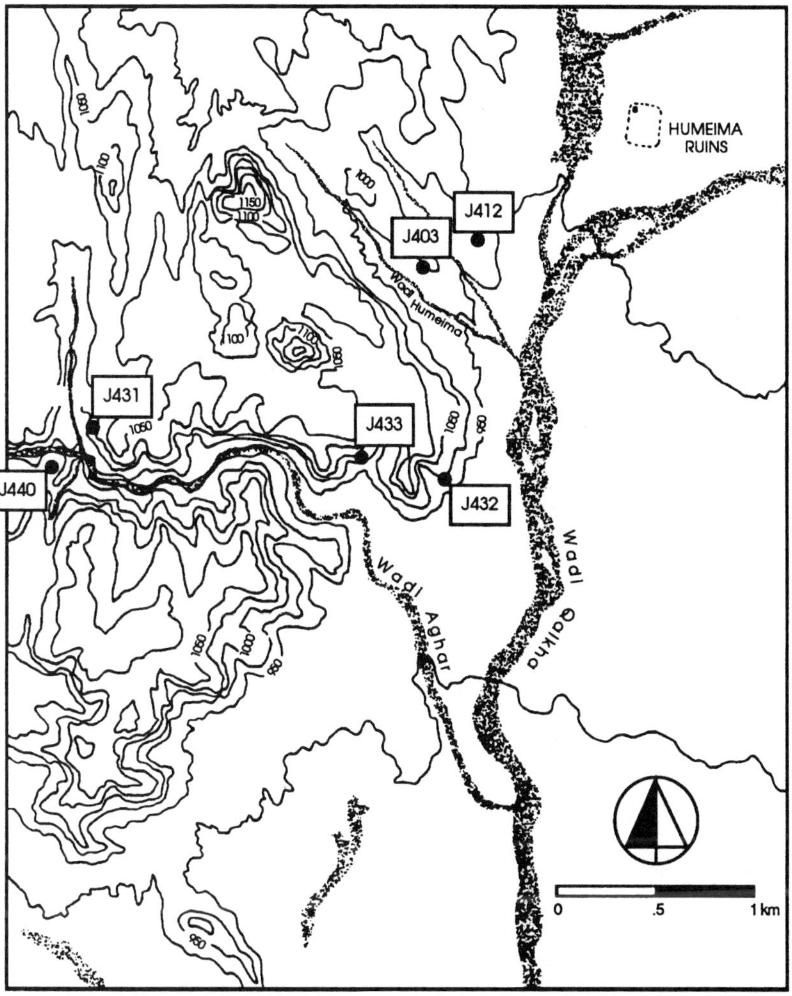


**Figure 8.1.** Jebel Qalkha area with site settings.

Buried occupations were discovered at each of the sites, but only the deep deposit of Tor Hamar (J431) revealed other components. These consisted of overlying Epipaleolithic and underlying Levantine Mousterian horizons. While not confirmed, the sites of Tor Aeid (J432) and Wadi Aghar (J433) are suspected of containing Levantine Mousterian horizons. From a cultural historic perspective, the occupations represent the Early Ahmarian (J431 and J432) and a non-Ahmarian complex, possibly the Levantine Aurignacian (J403 and J412).

## SITE SETTINGS AND RESOURCES

All of the sites are tied to south- or west-facing rockshelters that are situated along the steep walls of canyons that drain into the Wadi Qalkha. They fall within an elevational belt ranging from ca. 940–980 masl. Typically, the shelters display shallow overhangs of 2–5 m that extend some 15–20 m along the cliffs. Artifacts on the surface appear concentrated in these protected areas and also continue down-slope to varying distances beyond the drip-lines of the shelters.

### Site Areas

The surface concentrations of artifacts show considerable variability in size, ranging from 156–1064 $m^2$ (Table 8.1). The sizes and configurations of the concentrations are likely to reflect the cumulative patterns of utilization of the shelters as well as the degree to which the artifacts have been dispersed through natural actions. In general, one would expect the inflation of the artifact distributions by erosion and natural dispersion to be the greatest in those sites with a high degree of slope. A review of Table 8.1 shows that the site with the highest degree of slope (J431), however, has one of the smallest areas of artifact concentration. This may be explained in part by artifacts actually being removed or greatly dispersed from settings with extreme degrees of slope, whereas in settings with a moderate slope areas of concentration would merely be extended downslope. Such a site setting of moderate slope is exhibited at site J403 where the largest artifact concentration is recorded. In short, this would imply that the artifact concentrations of sites J412 and J432 rather accurately define the cultural limits of the sites, while the concentrations at J403 and J432 over- and under-represent the cultural limits, respectively.

### Site Catchments

Given the close proximities of the sites, their catchments are near identical. These include the steep-walled sandstone canyons and flat-topped *jebels* of the area west of the Wadi Qalkha, as well as the plain and shallow drainages that form the western end of the broad valley of the Wadi Hisma. Today, surface water is available in the *wadis* as runoff during the winter wet season and in bedrock pools with the canyons until late spring. Although not flowing today, a seep or spring was discovered on the south wall of the canyon across from site J403. Chert sources (described in detail in Chapter 7) within the catchments consist of those from a small hill formed by a limestone remnant near Jebel Humeima about 3–10 km walking distance from the sites and small stream cobbles from the bed of the Wadi Qalkha. All other chert sources are located beyond the catchments of the sites at distances of some 16–20 km.

## STRATIGRAPHIES AND PALEOENVIRONMENTS

With the exception of the excavation at Tor Hamar, the test excavations exposed relatively shallow deposits, ranging from 20–90 cm beneath the surface. None of these reached culturally sterile sediments, however. By necessity the excavations were terminated because the 1 × 1 m or 1 × 2 m test units would not accommodate greater depths (J403, J412, J432, J432) or because the floors of units had become choked off by rocks (J433). The maximum thickness of cultural

**Table 8.1. Comparative Data for Upper Paleolithic Sites and Artifact Contexts**

| Site | Protected area ($m^2$) | Area of artifact concentration ($m^2$) | Thickness cultural deposit (m) | Slope |
|---|---|---|---|---|
| J403 | 102 | 1,064 | 0.7 | 25° |
| J412 | 55 | 550 | 0.8 | 13° |
| J431 | 113 | 307 | 0.9 | 50° |
| J432 | 36 | 540 | 0.8 | 10° |

deposits could be defined only at Tor Hamar where a 90 cm thick horizon was exposed beneath some 1.3 m of Epipaleolithic occupations.

### Areal Stratigraphy

A common depositional unit composed of a pink-light brown silt was identified at four of the sites (J403, J412, J431, J432). The silt is widespread in the area with exposures showing it to underlie red sands containing Epipaleolithic occupations (Qalkan and Early Hamran assemblages) and to overlie red sandy silts. Although archaeological evidence has not been recovered from the earlier red silts exposed in these sections, one might recall that the nearby Levantine Mousterian occupation of Tor Faraj (J430) rests in a red sandy silt. Also, the Upper Paleolithic succession within the Tor Hamar (J431) deposit shows Layer F (reddish brown silt) to overlie a red sandy silt in Layer G. The only other Upper Paleolithic occupation to come from a red sandy silt is that of J433. The shallow deposit (20 cm) makes it difficult to determine if this is truly reflective of the occupation's regional stratigraphic placement, however.

### Paleoenvironments

The palynological study (Chapter 16) indicates a relatively moist setting for the Early Ahmarian occupations. The presence of alder, elm, oak, and abundant grasses indicates a much more moist setting than exists for the Wadi Aghar today. Moreover, the presence of spores of ferns and tetrades of pollen of rushes point to standing water. Moist conditions also are indicated by pollen from the Levantine Aurignacian deposits, although a rise in Chenopodiaceae (type *Noaea*) implies the on-set of a warmer, drier setting during the deposition of the upper part of the deposit at Tor Fawaz.

Faunal remains were recovered only from the Tor Hamar deposit and these were of insufficient numbers to allow for a meaningful paleoenvironmental reconstruction (Chapter 20).

## SITE PLANS AND STRATIGRAPHIES

### Tor Fawaz (J403)

The site rests near the north rim of a sandstone canyon that forms the southern branch of the Wadi Humeima. The shallow, south-facing shelter is situated about 15 m above the wadi bed and provides a commanding view of the plain that stretches east of the Wadi Qalkha. Two walls of sandstone blocks accompanied by thick dung layers point to the contemporary use of the shelter by Bedouin herders.

A high density of artifacts with a milky white patination covers an area of over 1,000 $m^2$ and extends from the back wall of the shelter to near the edge of the wadi (Figure 8.2). Four test units were placed along the fall line of the slope just beyond the edge of the shelter, and a fifth unit was excavated behind the drip-line. The units excavated on the slope revealed shallow deposits (10–20 cm deep) with abraded, patinated artifacts contained within a rubble scree. The unit within the shelter exposed a much thicker cultural deposit (>70 cm) consisting of a compact light brown silt (7.5 YR 6/4) with a layer of angular roof fall.

### Jebel Humeima (J412)

This shallow, southwest-facing rockshelter emerges near the eastern tip of a low sandstone ridge that ends just west of the Roman townsite of Humeima. Artifacts are concentrated over an area of ca. 550 $m^2$ on the shelter's gently sloping terrace (Figure 8.3). Their fresh edges and lack of patination suggest that most of the occupational horizon has been exposed only recently.

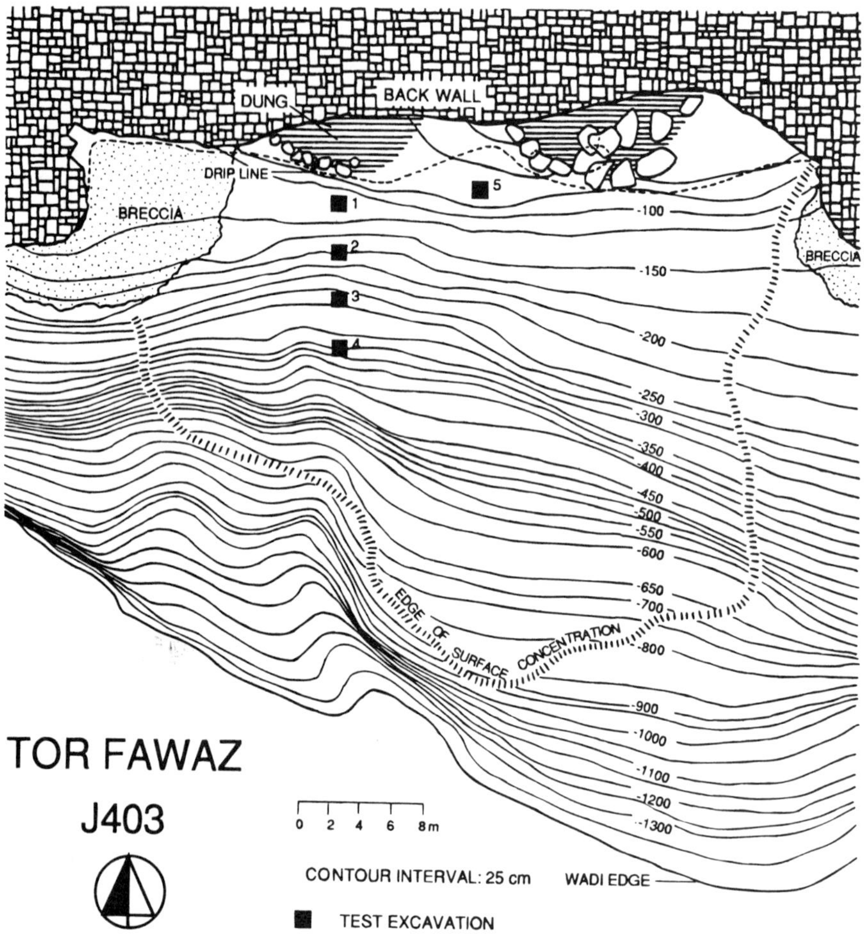


**Figure 8.2.** Excavation plan for Tor Fawaz (J403).

Two test units were excavated on either side of the shelter's drip-line. In Unit 1 the cultural deposit was exposed to a depth of 90 cm, at which point bedrock or a large rock closed off most of the unit. Unit 2 was excavated to 60 cm without reaching culturally sterile sediments or bedrock. Both units revealed a single depositional unit composed of a very compact light brown silt (7.5 YR 6/4) interspersed with a few angular sandstone fragments.

## Tor Hamar (J431)

The site consists of a small rockshelter and narrow terrace perched on the north wall of the Wadi Aghar about 3 km into the canyon. The canyon widens considerably at this point to accommodate the merging of lateral streams with the main channel. Near this confluence several large pools have been formed in the sandstone bed of the wadi. These not only hold water during the winter rainy months, but well into the spring. Near the end of April following an unusually dry winter in 1984, two of the largest pools still contained water. The wadi also serves as a principal route connecting the Hisma to the Rift Valley some 20 km to the west and 1,000 m lower in elevation.

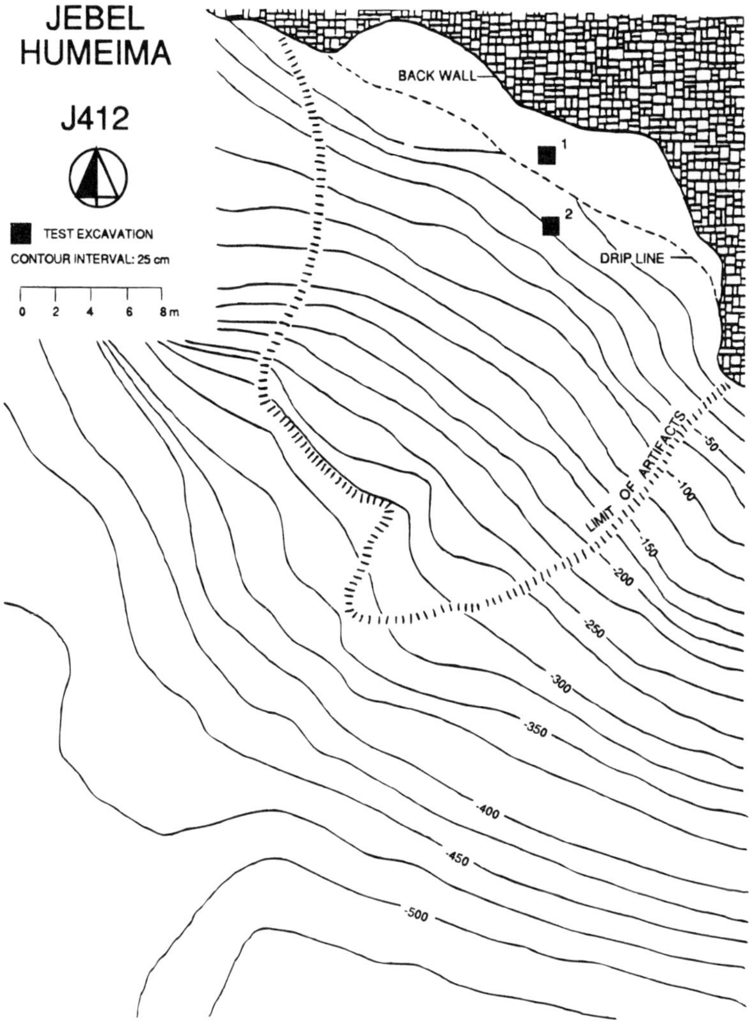


**Figure 8.3.** Excavation plan for Jebel Humeima (J412).

## Site Plan

The shelter, exposed to the southwest, is formed by the lowest of a series of overhangs that project from a steep cliff. At least a part of the overhang was truncated subsequent to the prehistoric occupation as evidenced by the presence of artifacts in the cemented sediments that underlie the roof-fall at its northwestern end (Figure 8.4). Three low stone walls accompanied by deposits of dung were located in the central portion of the site at the time of discovery. The two southernmost of these Bedouin windbreaks were removed prior to the 1988 excavation season.

Chert artifacts and bone occur in high densities within the shelter and on the adjacent slope in association with an ashy deposit. A deep conical mortar fashioned in a large sandstone block was also found in one of the Bedouin walls. During the first season at the site in the winter of 1983/84, a *sondage* (Units 1 and 2) was excavated to a depth of 1.3 m, exposing an Epipaleolithic

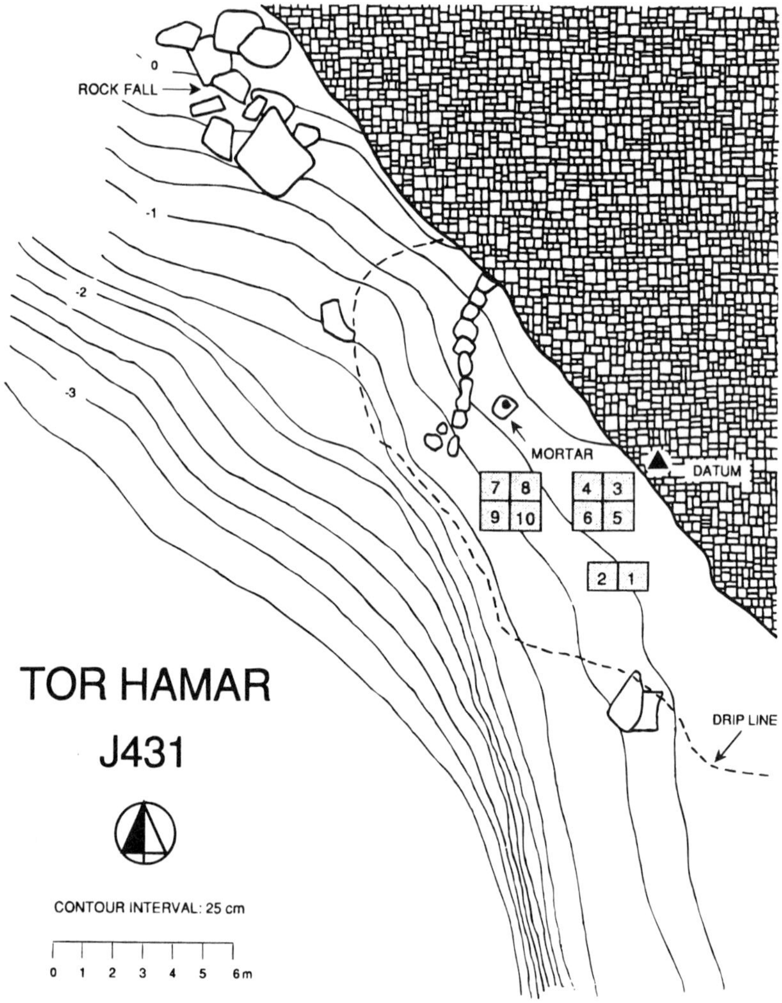


**Figure 8.4.** Excavation plan for Tor Hamar (J431).

(Mushabian Complex, Madamaghan Industry) horizon (Henry and Garrard 1988). Eight additional units, formed by two 2 m² blocks, were excavated to a depth of 2.2 m in the following season during the summer of 1988.

## Stratigraphy

The excavation exposed eight relatively level-bedded strata (A–H) within 22 arbitrary levels (Figure 8.5). The upper 1.8 m of the deposit contained Epipaleolithic artifacts (Layers A–D—Mushabian; Layer E—Qalkhan), whereas Upper Paleolithic materials were found in Layers F–G. Very little of Layer H was removed and few artifacts were recovered.

*Layer A* consists of a fine, powdery light grey ash. The upper 5–10 cm (A1) contain dung, twigs, grass seed, and other organic material of recent origin. The lower part of the layer (A2) is free of such material.

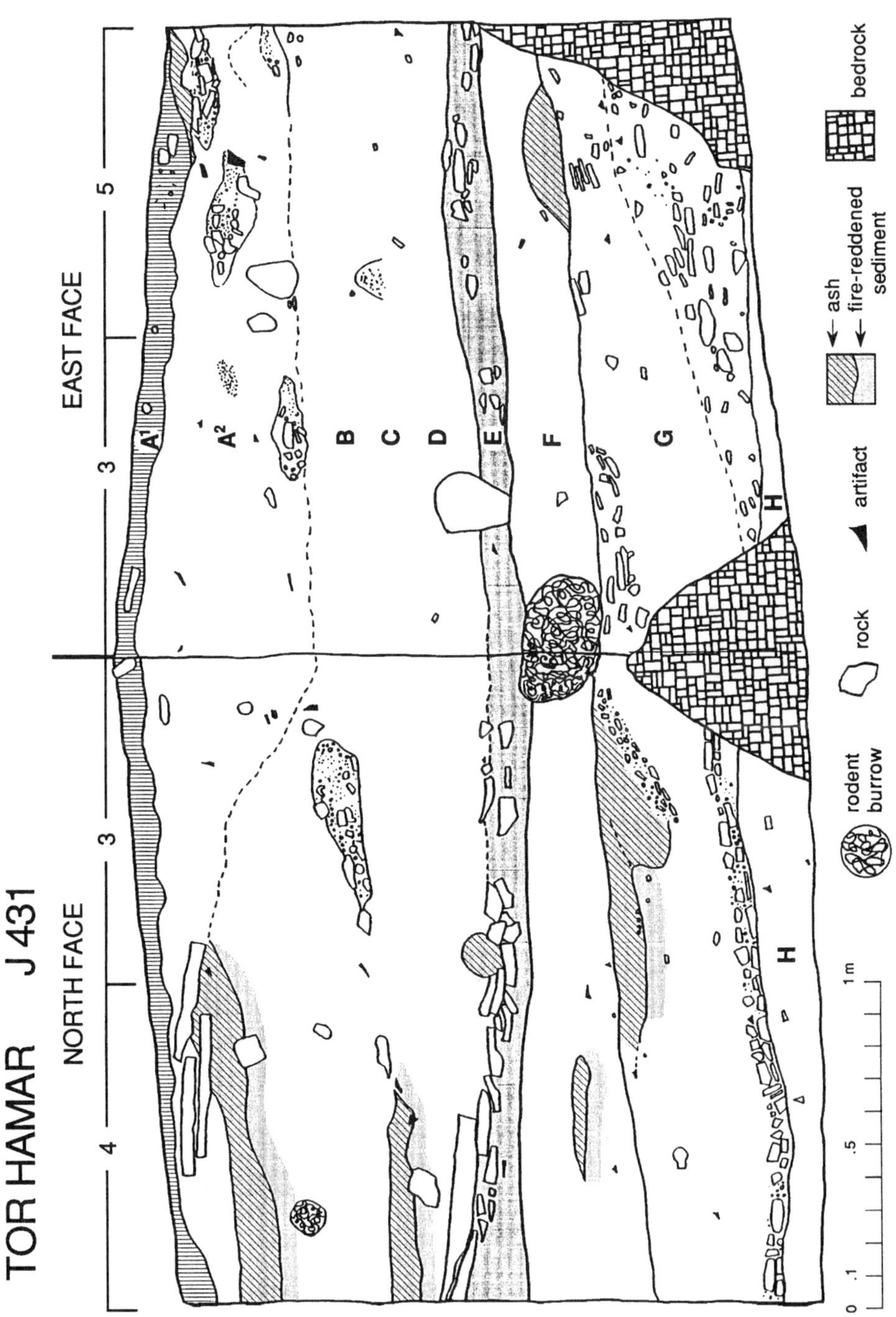


Figure 8.5. Stratigraphy at Tor Hamar (J431).

*Layers B, C, and D* are differentiated by varying proportions of ash and sand. Ash lenses point to shallow hearths underlain by fire-reddened sediments that denote a relatively level-bedded stratigraphy. While these layers could be distinguished during excavation, they were virtually impossible to identify in the sections after the sediments had dried. Spraying the sections with a fine water mist also failed to reveal their contacts. Layer B differs from Layer A in having a higher yellow-red sand content and in being more compact—less powdery. Layer B grades into Layer C at ca. 110–120 cm BD (below datum) with the latter being composed of a predominantly yellow-red sand. Layer C, in turn, grades into Layer D (more compact and ashier) at about 140 cm BD.

Mushabian artifacts and animal bones were found throughout layers A–D. Two radiocarbon dates were obtained from charcoal recovered from upper Layer C: 12,680 ± 320 (SMU 1399); 12,320 ± 90 (ETH 806).

*Layer E* is composed of a 20 cm thick sandstone rubble zone associated with dark grey ash and charcoal. The rubble, having fallen from the ceiling of the shelter, appears to have been associated with a large fire which was confined to the easternmost excavation block. The layer displays sharp contacts with contiguous strata and in units 4 and 6 is capped by a large 50 cm thick block of roof-fall. Qalkhan artifacts and animal bones were recovered throughout the layer.

*Layer F* consists of a compact light brown silt containing ash lenses underlain by fire-reddened sediment. Upper Paleolithic artifacts along with animal bones increase in density with depth through the layer.

*Layer G* is formed by a very compact yellowish red silt with angular sandstone rubble that increases with depth. Zones of calcium carbonate are also present. Artifact density is very low in the upper 30 cm of the layer, but increases sharply in the lower 10 cm of the stratum in association with the rubble. The artifact assemblage from the layer is Upper Paleolithic.

*Layer H* consists of a soft yellowish red sandy silt void of sandstone rubble and exhibiting a very low artifact density. The layer rests on a sandstone rubble cemented with calcium carbonate. Excavation was terminated at the rubble contact.

## Tor Aeid (J432)

The site rests on a gently sloping terrace fronting a steep cliff just north of the mouth of the Wadi Aghar. To the east the site overlooks the Wadi Qalkha and the plain beyond, while immediately to the south the terrace falls away to a minor wadi which drains the Jebel Humeima. This small wadi not only provides access to the top of the cliff, but contains a number of seasonal pools. Perhaps inspired by these, Nabateans constructed a large cistern in the wadi just above the site, and this is still used by Bedouin as a source of water.

A concentration of Upper Paleolithic artifacts on the surface of the terrace extends from a shallow overhang to where the slope begins to drop off steeply (Figure 8.6). This covers an area of ca. 540 $m^2$. Although quite shallow, the overhang serves as a storage area and seasonal corral for Bedouin as evidenced by a rock wall, dung, and ash. The terrace has been lightly cultivated with a shallow plow zone. Levallois points and point cores were found eroding from the terrace along the 3–3.5 m contour.

A 1 × 2 m *sondage*, excavated in the northeast corner of the artifact concentration, revealed an 80 cm thick cultural deposit in which three strata were defined. These consisted of: Layer A—a loose silt and ash layer corresponding to a plow zone that extended some 10–15 cm from the surface; Layer B—a compact light brown silt with a high density of large, angular sandstone rocks (possibly roof-fall) accompanied by charcoal flecks; Layer C—continuation of the light brown silt with greater compaction and fewer rocks. Ahmarian artifacts were found throughout the excavated deposit.

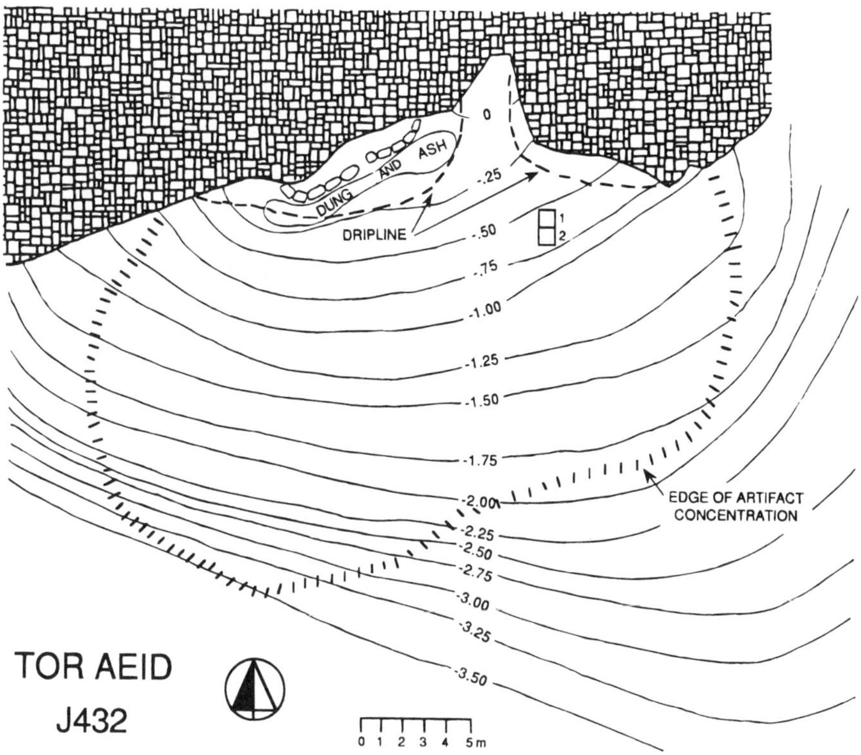


Figure 8.6. Excavation plan for Tor Aeid (J432).

## Site J440

The site is located across the Wadi Aghar from Tor Hamar (J431) at about the same elevation on the north side of a small canyon. It consists of artifacts distributed over ca. 600 $m^2$ of a slope that extends from a shallow rockshelter to the edge of a wadi some 6 m below. The slope has been lightly cultivated and terraced with low stone walls that form three plots.

The distribution was surface collected, but no excavation was undertaken. The surface collection consisted of six roughly 2 $m^2$ units arranged as an east-west transect through the distribution along the fall-line of the terrace from the shelter to the edge of the wadi.

## Wadi Aghar (J433)

The site is situated at the base of the sandstone cliff that forms the northern wall of a large embayment at the mouth of the Wadi Aghar. The cliff furnishes a shallow overhang and a southern exposure. Alluvial fans extend from the cliff to the floor of the embayment and the wadi channel. On one of these fans artifacts were found in high densities extending from the head of the fan to its foot where it merged with level ground. Upper Paleolithic artifacts as well as Levallois points and cores were recovered from the slope of the fan.

Three 1 $m^2$ units were excavated on the near level surface at the head of the fan, positioned between a large (1 × 3 m) sandstone block and the cliff. Sediments beneath the block have been exposed by erosion to a depth of 20–30 cm and these contain *in situ* artifacts. The *sondage* reached bedrock or perhaps a large rock at a depth of only 35 cm. Within this shallow cultural deposit, three strata and a feature were encountered, however. These included: Layer A—a powdery grayish tan sand; Layer B—an unconsolidated light reddish brown sandy silt; and Layer C—compact, cemented pinkish sand. A small rock-lined hearth containing burned sediment and ash was exposed at the base of Layer B in the northwest corner of the test unit.

## LITHIC ARTIFACTS

### Tor Fawaz (J403)

#### Introduction

The lithics from Tor Fawaz are somewhat unique in that they exhibit more post-depositional damage than any of the other Upper Paleolithic assemblages in the study area. All lithics recovered from the five excavation units at the site exhibit unusually high amounts of abrasion and damage to edges, but there is substantial variation in the degree of abrasion and intactness among unit assemblages within the site. Not surprisingly, lithics from the shallow derived deposits of Units 1–4 along the fall line of the slope exhibited higher rates of breakage than those recovered from Unit 5, and these rates, discussed below, are higher than those observed for other Upper Paleolithic assemblages in this study.

Although the entire assemblage from the site appears to be impacted differentially by post-depositional processes, it is similar in composition to other Upper Paleolithic assemblages, with all major stages of reduction represented in the debitage and debris (Appendix 8.1). Typologically, the retouched tool assemblage is atypical. The latter may be explained, in part, its derived nature. Typical Upper Paleolithic tools are less frequent relative to notches and retouched flakes and blades. Most of these exhibit very informal, discontinuous, and minimal retouch. In most respects, however, the Tor Fawaz assemblage resembles the Upper Paleolithic assemblage at Jebel Humeima (J412).

#### Raw Material

Cores recovered from Tor Fawaz are predominantly rolled cobbles with gravel cortex, followed by 21.7% limestone nodules (Table 8.2). Only 17.4% are reduced to the point where no outer surface remains. Within the site, there are major color differences in the raw material as a result of differential patination. Units 1–4, representing surface collections extending down the slope, are characterized by brown flints with limited patination, while lithics recovered from buried strata in Unit 5 are heavily patinated with a milky white patina. The spatial separation of Unit 5, stratigraphic depth, and the differences in patination may be evidence for temporal differences between Units 1–4 and Unit 5. Alternatively, the differences may be accounted for primarily by post-depositional processes acting differentially on lithic materials across the site.

#### Technology

Cores from Tor Fawaz illustrate a dominant emphasis on single platform cores (56.5%) with a limited number of cores with opposed platforms (17.4%) or change-of-orientation cores (17.4%) (Table 8.2). Only 6% of the debitage exhibits bidirectional, opposed dorsal facets that are associated with core reduction using opposed platforms (Appendix 8.2). The majority of dorsal facets on primary elements and debitage blades are parallel facets indicating unidirectional blank removal from single platforms. Less than 3% of the debitage blades exhibit truncated, intercepted facet patterns that might be suggestive of a change in the orientation of platforms on cores corresponding to the 17% change-of-orientation cores.

Table 8.2. Classification of Cores by Raw Material and Core Class from Tor Fawaz (J403)

| | Unit 1 | Unit 2 | Unit 3 | Unit 4 | Unit 5 | Total *n* | Total % |
|---|---|---|---|---|---|---|---|
| A. Raw Material | | | | | | | |
| Rolled Cobble | 5 | 3 | 1 | | 5 | 14 | 60.9 |
| Limestone Cobble | | 1 | 2 | | 2 | 5 | 21.7 |
| Indeterminate | 1 | 2 | | | 1 | 4 | 17.4 |
| Total | 6 | 6 | 3 | | 8 | 23 | 100.0 |
| B. Core Class | | | | | | | |
| Single Platform | 2 | 2 | 3 | | 6 | 13 | 56.5 |
| Opposed Platforms | | | | | | 4 | 17.4 |
| Opposite | 1 | 1 | | | 1 | | |
| Same | | | | | | | |
| Twisted | | 1 | | | | | |
| Change of Orientation | 2 | 1 | | | 1 | 4 | 17.4 |
| Others | | | | | | 2 | 8.7 |
| Discoidal | 1 | | | | | | |
| Other | 1 | | | | | | |
| Total | 7 | 5 | 3 | | 8 | 23 | 100.0 |

Platforms on debitage pieces conform closely to the simple, unfaceted platforms on cores (Appendix 8.2). Approximately 70% of both the primary elements and debitage blades exhibit single faceted platforms coinciding with the relatively unprepared platform cores. Faceted platforms are rare within the debitage. Crushed platforms occur infrequently on primary elements (7.1%) but are far more frequent on debitage blades (23.5%).

Core platforms were rejuvenated most often using the core tablet technique. Core trimming elements are relatively frequent at 2.1% of the debitage with a CTE:core ratio of 1:0.66 (Appendix 8.1). The core tablets are not the typical elongated tablet type. Most tend to be wider than they are long, removing only the front half of the platform—a shape that may be dictated by wider cores and the overall configuration of the raw material. The occurrence of core trimming elements as blanks for tools increases their overall frequency in the assemblage with the core tablet technique representing 50% of the core rejuvenations (Table 8.3). In addition to core tablets, another important technique involves the removal of "platform blades" (37.5%). These elements are blade-like pieces that represent removals along one side of the platform. The characteristic edge of the platform forms a portion of the dorsal ridge. Other core trimming elements consist of atypical elements and fragments, two of which are unusual. One is a huge overpassed blade which was initiated from one of two opposed platforms of a core and effectively renewed most of the face of the core, including a large portion of the opposite platform. The other is a large piece that was struck from the distal back section of a core, removing a large portion of the core's face and platform.

## Reduction Mechanics

Hard hammer percussion, indicated by unlipped platforms, is characteristic of 66.6% of the primary elements, with 9.5% of these exhibiting *fractures a languette* (Appendix 8.2). Some 31% of the primary elements show lipping that indicates the supplementary use of soft hammer for primary removals. Although the high frequency of crushed platforms discussed above implies soft hammer percussion, a surprising 77.8% of the debitage blades exhibit unlipped plat-

Table 8.3. Core Trimming Elements in the Debitage and Retouched Tools at Tor Fawaz (J403)

| | | Core tablets | Platform blades | Others |
|---|---|---|---|---|
| Debitage | *n* | 14 | 15 | 6 |
| | % | 40.0 | 42.8 | 17.2 |
| Tools | *n* | 10 | 3 | — |
| | % | 2.3 | 0.7 | |
| Total CTEs | | 24 | 18 | 6 |
| % of Total CTEs | | 50.0 | 37.5 | 12.5 |

forms, suggesting that hard hammer prevailed throughout most of the secondary removals. Only 11% of the debitage blades show lipping.

Lateral profiles of debitage elements are possible indicators of percussion modes. Curved and twisted profiles should indicate soft hammer mode in removal, while flat profiles are indicative of heavier loading and more perpendicular loading angles associated with hard hammer modes. In this assemblage, there is evidence for mixed modes of percussion and loading angles. Primary elements tend to be curved or flat, whereas debitage blades exhibit flat (54.0%) and twisted (38.0%) profiles (Appendix 8.2).

Evidence from debitage terminations shows that the delivery of loading force was fairly consistent throughout primary and secondary removals. Primary elements exhibit 74.6% feathered terminations, whereas 86.8% of the debitage blades have feathered distal profiles. Primary pieces exhibit a much higher percentage of underpassed (16.9%) or overpassed (2.8%) terminations suggestive of errors in loading force, while more controlled force reduced errors to only 9% of the debitage blade sample.

## Reduction Activity

Evidence for a complete reduction sequence is present at J403, with expected proportions of debitage, debris, and tools representing the major stages (Appendix 8.1). Primary processing of cores at the site is supported by the fact that the frequency of pieces with complete cortex or natural outer surface is high, at 22% of the primary pieces (Appendix 8.2). Cores, however, are relatively infrequent at the site with a core:debitage ratio of 1:71, while primary elements make up 9.8% of the debitage—moderately high among the sites in the study area. The ratio of cores:primary elements is 1:7.1, a high ratio because of the low frequency of cores. Debris represents 50% of the assemblage, which is a moderate percentage, but the debitage:chips ratio of 1:0.75 is fairly low. The tool component is 8.4% of the assemblage, probably somewhat inflated because of the informal nature of many of the retouched pieces, discussed below.

There is no clear conclusion on the goals of production if only the frequencies of debitage blades and flakes are compared (see Appendix 8.1) since they occur in similar frequencies in the debitage. A higher frequency of tools made on blade blanks (55.7%), however, argues for some importance in the production of elongated blanks (Table 8.6). Cross sections of debitage blades indicate an almost equal production of 1st and 2nd order blades/bladelets. In the retouched tools, all stages of the reduction sequence are represented by 1st and 2nd order primary blade tools (10%), by primary and secondary flake tools (10.3%), and by the highest percentages associated with 1st, 2nd, and 3rd order plain blades (45.7%) and plain flakes (28.4%).

Dimensional data used to reconstruct a complete reduction sequence in the Tor Fawaz assemblage (Tables 8.4 and 8.5, Figure 8.7) suggest that primary removals tended to be flakes

Table 8.4. Dimensional Data (mm) for Cores and Complete Debitage from Tor Fawaz (J403)

| A. Cores (*n* = 19) | Length | Width[a] | Width[b] | Weight (g) | CSI[c] |
|---|---|---|---|---|---|
| Mean | 53.6 | 36.3 | 39.5 | 56.2 | 19.7 |
| sd | 10.2 | 9.8 | 10.3 | 34.8 | 8.1 |
| min | 34.1 | 22.0 | 24.3 | 24.5 | 11.6 |
| max | 77.1 | 58.0 | 59.3 | 168.6 | 42.3 |
| **B. Primary Elements (*n* = 26)** | **Length** | **Width[a]** | **Thickness** | **L: W Ratio** | **Size[d]** |
| Mean | 39.2 | 25.9 | 8.6 | 1.7 | 107.3 |
| sd | 12.0 | 10.4 | 3.9 | 0.6 | 91.4 |
| min | 17.8 | 8.4 | 2.5 | 0.7 | 5.3 |
| max | 63.1 | 51.5 | 16.6 | 3.3 | 315.5 |
| **C. Debitage Blades (*n* = 56)** | **Length** | **Width[a]** | **Thickness** | **L: W Ratio** | **Size[d]** |
| Mean | 42.6 | 16.6 | 6.8 | 2.6 | 87.4 |
| sd | 19.3 | 8.3 | 4.3 | 0.5 | 139.4 |
| min | 18.2 | 6.6 | 1.3 | 1.9 | 1.9 |
| max | 89.7 | 42.1 | 18.9 | 4.4 | 708.2 |
| **D. Core Facets (*n* = 19)** | **Length** | **Width[a]** | | | |
| Mean | 33.4 | 15.2 | | | |
| sd | 10.9 | 5.2 | | | |
| min | 21.2 | 6.3 | | | |
| max | 52.0 | 28.9 | | | |

[a]Width at midpoint.
[b]Maximum width.
[c]Core Size Index (length × width[a]/100).
[d]Size (length × width[a] × thickness/100).

with wider and thicker proportions than secondary debitage, which is indicated by a mean length:width ratio of 1.7 and a mean width:thickness ratio of 3.3 (Table 8.4). Comparisons of the metric distributions show that primary elements and debitage have fairly similar distributions in length, but mean widths and thicknesses vary, as we might expect (Figure 8.8). The mean length of primary elements (39.2 mm) is only slightly shorter than the mean length of debitage blades (42.6 mm), but the modal length interval for primary elements is 30–40 mm, while it is 20–30 mm for debitage blades (Figure 8.8a).

The mean dimensions of major tool classes illustrate that overall, the tool assemblage consists of large tools and that different classes of tools exhibit more or less restricted dimensional parameters (Table 8.5 and Figure 8.7). The mean length of complete endscrapers, represented by a small sample of complete tools (*n* = 4), is 68.3 mm. Endscrapers tend to be much larger than burins, displaying an overall size index of 438.0 mm, compared to 192.0 mm for burins. The latter generally occur on smaller, thinner blanks and are not the thick carinated types found at other Upper Paleolithic sites in the study area (e.g., J432). Major differences in blank forms between burins and endscrapers are highlighted by their mean length:width ratios; for burins it is 2.1, in contrast to 1.6 for endscrapers.

The dimensions of other complete tools indicate they occupy larger, less restricted dimensions (Table 8.5, Figure 8.7). Complete tools made on flakes and blades range from 13 mm to 81 mm in length with means of 31.8 mm for flake tools and 47.0 mm for blade tools. Although the range in tool length is extensive, flake and blade tools are quite different and tend to be con-

Table 8.5. Dimensional Data (mm) for Retouched Tools from Tor Fawaz (J403)

| | Length | Width | Thickness | Nose thickness | Bit width | Size[a] | L: W Ratio |
|---|---|---|---|---|---|---|---|
| Endscrapers | | | | | | | |
| Complete ($n = 4$) | | | | | | | |
| mean | 68.3 | 42.5 | 14.8 | 14.9 | NA | 438.0 | 1.6 |
| sd | 14.1 | 1.6 | 3.3 | 3.1 | | 178.1 | 0.3 |
| min | 55.2 | 41.1 | 11.8 | 11.6 | | 334.2 | 1.3 |
| max | 88.3 | 44.8 | 18.8 | 17.6 | | 703.9 | 2.1 |
| Incomplete ($n = 15$) | | | | | | | |
| mean | 42.4 | 28.1 | 11.4 | 10.1 | NA | 142.3 | 1.7 |
| sd | 10.7 | 10.8 | 4.7 | 4.7 | | 88.1 | 0.7 |
| min | 21.8 | 15.6 | 4.9 | 3.6 | | 17.1 | 0.8 |
| max | 61.3 | 52.8 | 22.7 | 18.8 | | 345.2 | 3.4 |
| Burins | | | | | | | |
| Complete ($n = 2$) | | | | | | | |
| mean | 59.3 | 28.6 | 11.3 | NA | 10.7 | 192.0 | 2.1 |
| sd | 1.6 | 0.4 | 0.6 | | 4.5 | 8.5 | 0.1 |
| min | 58.1 | 28.3 | 10.9 | | 7.5 | 185.9 | 2.0 |
| max | 60.4 | 28.9 | 11.8 | | 13.9 | 198.1 | 2.1 |
| Incomplete ($n = 12$) | | | | | | | |
| mean | 42.1 | 25.4 | 10.2 | NA | 6.5 | 132.7 | 1.8 |
| sd | 16.4 | 9.4 | 3.7 | | 2.8 | 108.5 | 0.7 |
| min | 24.6 | 10.1 | 5.1 | | 3.5 | 15.2 | 1.1 |
| max | 81.2 | 42.8 | 15.9 | | 13.0 | 368.8 | 3.2 |
| Other Tools | | | | | | | |
| Complete Blades ($n = 22$) | | | | | | | |
| mean | 47.0 | 20.8 | 7.6 | NA | NA | 105.0 | 2.3 |
| sd | 16.3 | 8.0 | 4.0 | | | 109.7 | 0.4 |
| min | 19.9 | 11.4 | 2.6 | | | 9.2 | 1.5 |
| max | 81.7 | 35.5 | 18.7 | | | 386.9 | 3.0 |
| Complete Flakes ($n = 47$) | | | | | | | |
| mean | 31.8 | 25.3 | 6.0 | NA | NA | 65.3 | 1.3 |
| sd | 12.6 | 9.7 | 2.6 | | | 80.6 | 0.3 |
| min | 13.2 | 13.7 | 2.3 | | | 6.4 | 0.5 |
| max | 66.8 | 56.5 | 14.8 | | | 344.9 | 2.0 |

[a]Size = length × width × thickness/100.

Table 8.6. Percentages of Different Reduction Stages Represented in the Retouched Tools from Tor Fawaz (J403)

| | Primary Blades | | Plain Blades | | | Flakes | | | |
|---|---|---|---|---|---|---|---|---|---|
| | 1 | 2 | 1 | 2 | 3 | 1 | 2 | 3 | |
| | Cortical | | | | | | Cortical | | Others |
| Retouched Tools ($n = 333$) | 5.3 | 4.7 | 17.2 | 21.6 | 6.9 | 3.7 | 6.6 | 28.4 | 5.6 |

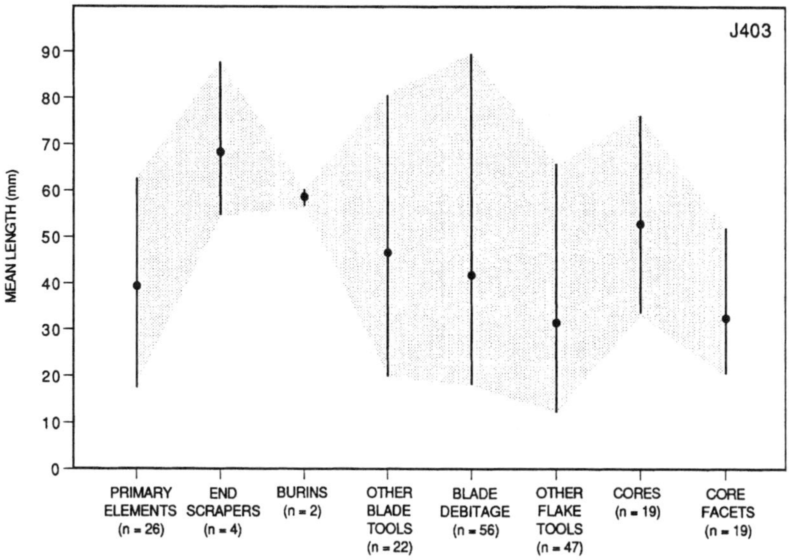


**Figure 8.7.** Mean lengths and ranges of major debitage and tool classes from Tor Fawaz (J403).

siderably smaller when mean blank size is compared to mean blank size indices for endscrapers and burins.

The large tool dimensions together with the fact that 20.3% of the tools are cortical pieces (see Table 8.6) indicates that many tool blanks were selected from large debitage produced early in the reduction sequence after initial primary flakes were removed.

The dimensions of cores at abandonment suggest cores were far from exhausted (Figures 8.7, 8.8 and Table 8.4). The Core Size Index (CSI) of 19.7 makes the J403 cores among the largest of the Upper Paleolithic assemblages in the study area, exceeded only by those at J433 with a CSI of 31.2. Mean core length is 53.6 mm, which is substantially longer than the average length of blade debitage, indicating that the cores were still useable when abandoned. The length of core facets, however, implies that the final removals were shorter than the average blade but well within the dimensional range of bladelets.

A number of Levallois elements and features occur in this Upper Paleolithic assemblage, consisting of a single discoidal core, a small number of faceted platforms, and three Levallois points.

## Tool Typology

The retouched tools comprise 8.4% of the assemblage from Tor Fawaz, the highest percentage for the Upper Paleolithic sites in the study area (Figures 8.9, 8.10, 8.11; Appendix 8.1). As mentioned above, a major characteristic of the J403 assemblage is the high frequency of breakage. For tools, the rate of intactness is only 22.5%, with similar rates for all tool groups except burins, for which it is even lower (11.1%) (Table 8.7). The high frequency of breakage in the tools is consistent with the overall derived nature of the assemblage. Within the site, Unit 5 with the deepest cultural deposits exhibits the highest proportion of intact retouched pieces (30.3%), while Unit 1 exhibits the most breakage with 92.9% incomplete pieces (Table 8.7). When only tools other than endscrapers or burins are considered, the rate of breakage is twice as high among blade tools (86.6%) as tools made on flakes (41%) (Table 8.7).

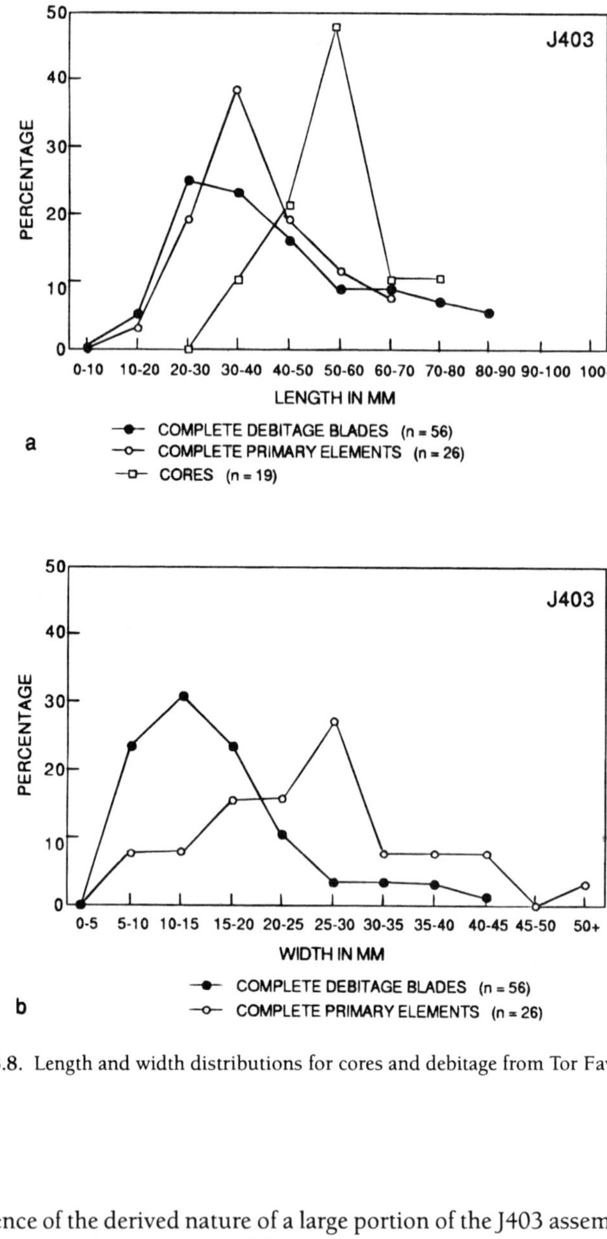


**Figure 8.8.** Length and width distributions for cores and debitage from Tor Fawaz (J403).

A consequence of the derived nature of a large portion of the J403 assemblage may be what appears to be the rather informal nature of the tools. The typological composition may be skewed because of the relatively high frequencies of less formal tool types, such as retouched flakes, blades, and a variety of notches (see Appendices 8.1 and 8.3). The overall character of most of these retouched pieces is their *lack* of definitive patterns of retouching. This may be due to the difficulty of discerning heavily abraded retouch from post-depositional edge damage and abrasion. Relative frequencies of the more typical Upper Paleolithic tool types (i.e., endscrapers, burins) are necessarily low, although their absolute numbers are not unusual. Therefore, it is unlikely that the proportions of the general tool groups can be correlated directly with possible site activities and function.

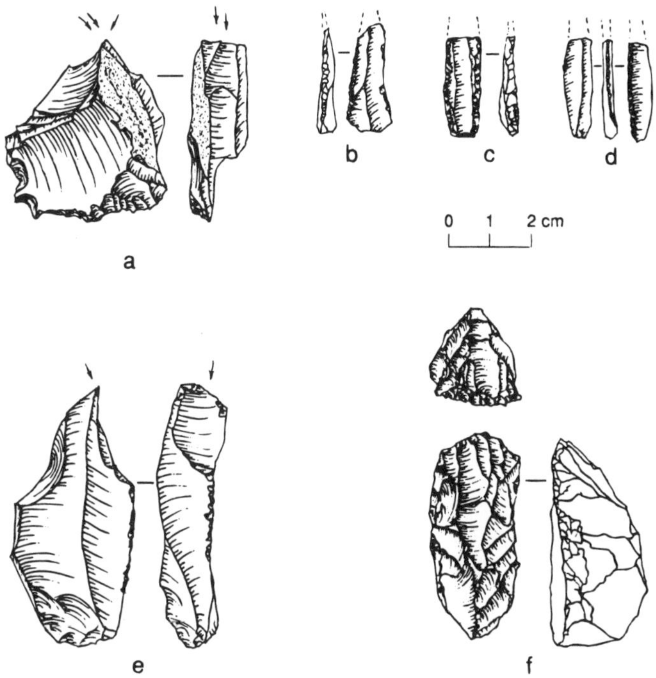


**Figure 8.9.** Tools from Tor Fawaz (J403): (a) burin on a retouched flake, (b,c) retouched bladelets, (d) bladelet with fine inverse retouch, (e) burin, (f) carinated endscraper.

Together, retouched pieces comprise 57.6% of the tools, with various retouched flakes accounting for 27.3%, and blades and bladelets making up 23.1% and 7.2%, respectively. Partial retouch, indicative of less formal tools, occurred most frequently on blades and flakes, while obverse retouch was the most characteristic of bladelets. One bladelet exhibiting very fine, marginal Ouchtata retouch was recovered from Unit 5 (Level 3). Ouchtata bladelets are characteristic of Late Ahmarian assemblages in which bladelets comprise a large portion of the retouched tools. Such a bladelet is unusual in this assemblage which is characterized by large blades and flakes and lacks a significant bladelet component.

Notches represent almost 16% of the tools, varying from small fragments with limited retouch to large tools with well-defined notches (Figure 8.10:b,d). Two of the three "massive" tools are pieces with notches. Endscrapers and truncations make up the most numerous tools after the notches. Endscrapers occur almost equally on flakes and blades (Figure 8.10:a,c) with only one example of a carinated endscraper (Figure 8.9:f). Truncations occur most often on blades with only two examples of bitruncated pieces. Angle burins made on old surfaces or breaks are the most typical type of burin in this assemblage. Others make up a variety of types, including dihedral burins (Figure 8.9:a), *burins plan*, carinated burins, and burins on truncations. Multiple or composite tools are numerous, three of which are combinations of lateral or sidescraper-notch tools. In conjunction with these typical Upper Paleolithic tools, three examples of Levallois points were recovered from three separate units. One atypical example is retouched with a notch and shows subsequent retouching.

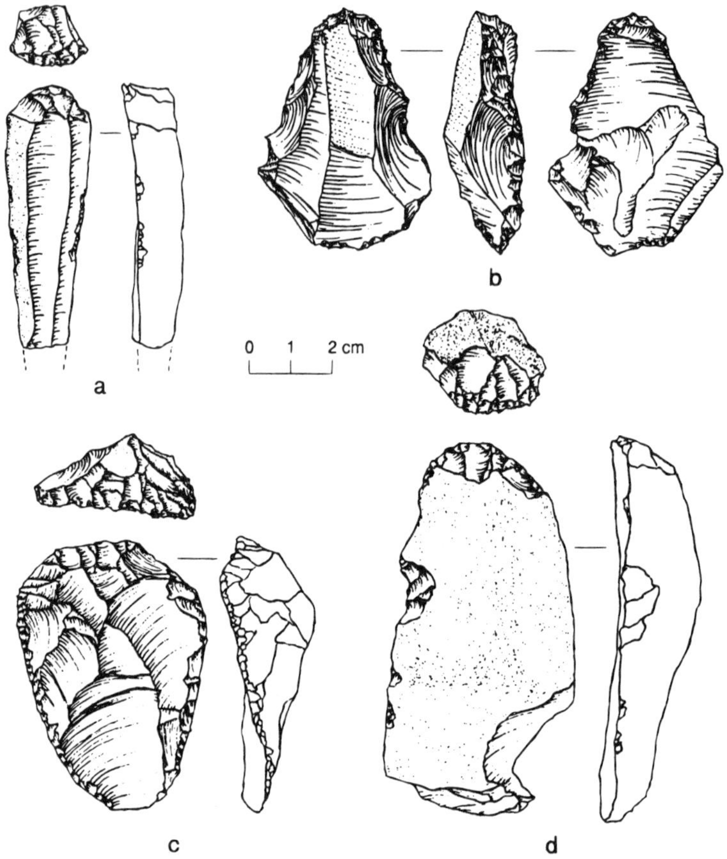


Figure 8.10. Tools from Tor Fawaz (J403): (a) endscraper on a blade, (b) retouched flake with notch, (c) steep scraper on a retouched flake, (d) endscraper on a cortical blade.

## Jebel Humeima (J412)

### Introduction

The lithic assemblage from Jebel Humeima (J412) is comprised of all lithics recovered from two, 2 m$^2$ test units during the 1980 field season and described previously by Henry (1982). Lithic materials in both units represent a single depositional unit and, as such, are treated analytically together (Appendix 8.4). Interlevel and interunit evaluations identified few significant differences among arbitrary levels or between units that would suggest otherwise. The assemblage represents an undated Upper Paleolithic blade technology focused on the production of flakes and large blades with little evidence for small bladelet debitage or tools. The assemblage has the strongest similarity to current descriptions of the Levantine Aurignacian and represents an increasing number of Upper Paleolithic assemblages from the southern Levant that vary significantly from those identified in the northern Levant.

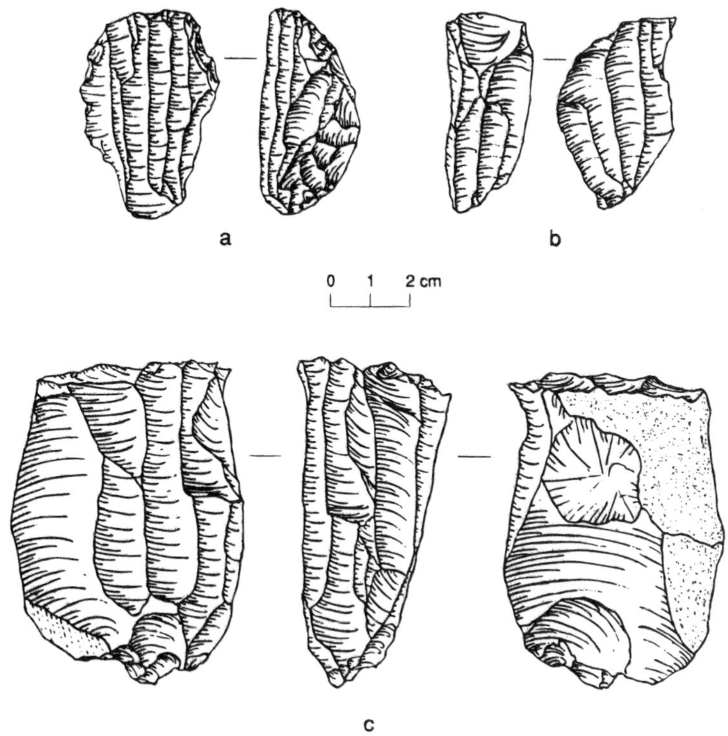


**Figure 8.11.** Cores from Tor Fawaz (J403): (a) bladelet core, (b) bladelet core with opposed platforms, (c) blade core.

## Raw Material

Raw material represented in cores illustrates a slight selection preference for rolled cobbles (48.6%) over limestone nodules with cortical rind (34.3%) (Table 8.8). Within the site, these two categories of raw material are represented differentially. In Unit 2, rolled cobbles occurred most frequently, while in Unit 1, they are equally represented. Small, heavily reduced cores with no remaining natural surfaces represent 17.1% of the core sample.

## Technology

Some aspects of the core technology can be extracted from the 35 cores recovered from the site. The majority are simple, single platform cores (60%), variations of subpyramidal cores with facets that suggest a focus on elongated blanks rather than flakes (Table 8.8). An important proportion of the cores (28.6%) have double or multiple platforms. Opposed platform cores represent 23%, with most of these exhibiting opposed platforms on opposite sides of the core. Dorsal facet patterns, which should be correlated with platform types on cores, vary between primary pieces and blade debitage (Appendix 8.5). Only 12.5% of the primary elements exhibit bidirectional removals that would be associated with reduction of opposed platform cores in

Table 8.7. Breakdown of the Condition of Tools by Assemblage, Excavation Unit, and Blank Form from Tor Fawaz (J403)

| A. Assemblage | (*n*) | Complete % | Incomplete % |
|---|---|---|---|
| Endscrapers | (20) | 20.0 | 80.0 |
| Burins | (18) | 11.1 | 88.9 |
| Other Tools[a] | (295) | 23.7 | 76.3 |
| Total | (333) | 22.8 | 77.2 |

| B. Excavation Unit | (*n*) | Complete % | Incomplete % |
|---|---|---|---|
| 1 | (42) | 7.1 | 92.9 |
| 2 | (70) | 18.6 | 81.4 |
| 3 | (74) | 24.3 | 75.7 |
| 4 | (48) | 25.0 | 75.0 |
| 5 | (99) | 30.3 | 69.7 |
| | (333) | 22.8 | 77.2 |

| C. Blank Form | (*n*) | Complete % | Incomplete % |
|---|---|---|---|
| Blades | (164) | 13.4 | 86.6 |
| Flakes | (81) | 59.0 | 41.0 |

[a]Based on tools other than endscrapers, burins, and very small fragments.

Table 8.8. Classification of Cores by Raw Material and Core Class from Jebel Humeima (J412)

| | Unit 1 | Unit 2 | Total *n* | Total % |
|---|---|---|---|---|
| A. Raw Material | | | | |
| Rolled Cobble | 9 | 8 | 17 | 48.6 |
| Limestone Cobble | 9 | 3 | 12 | 34.3 |
| Indeterminate | 6 | — | 6 | 17.1 |
| Total | 24 | 11 | 35 | 100.0 |
| B. Core Class | | | | |
| Single Platform | 12 | 9 | 21 | 60.0 |
| Opposed Platforms | | | 8 | 22.9 |
| Opposite | 4 | 2 | | |
| Same | 1 | | | |
| Twisted | 1 | | | |
| Change of Orientation | 2 | | 2 | 5.7 |
| Others | | | 4 | 11.4 |
| Discoidal | 1 | | | |
| Other | 3 | | | |
| Total | 24 | 11 | 35 | 100.0 |

the initial reduction stages, whereas interior blade debitage exhibits no bidirectional scars. Dorsal scar patterns on blade debitage are predominantly unidirectional with 87% parallel scars and converging scars on 11% of the sample. Truncated scars, indicative of cores in which there has been a change in the orientation and use of platforms, occur only in significant numbers on primary elements (17.5%). Given the number of opposed platform cores, representing an advanced or terminal stage in core reduction, we might expect interior debitage to exhibit evidence of bidirectional removals. This sample of debitage, however, indicates bidirectional removals restricted to primary elements and the initial processing of cores.

Platforms on cores are simple, unfaceted surfaces, exhibiting little formal preparation of a platform. Corresponding platform characteristics on debitage are similarly simple and unfaceted (Appendix 8.5). The majority of both primary elements and blade debitage are unfaceted, single platforms (53.5% and 67.9%, respectively). Cortical platforms are absent in the blade debitage, but their pronounced frequency in the primary elements (26.7%) corresponds well with the strong evidence for primary core reduction at the site, since they can be interpreted as diagnostic features of the earliest core removals. In contrast, the platforms of interior debitage exhibit a similar proportion of crushed platforms (21.4%), indicative of controlled percussion directed at increasingly narrowed platforms.

Techniques for maintenance and rejuvenation of core platforms are reflected in the core trimming elements, which represent only a small proportion (1.9%) of the debitage (Table 8.9, Appendix 8.4). Core tablets make up 20.7% (6) of all core trimming elements and 0.6% or 2 of the tools. Other core trimming elements account for the majority, indicating the core tablet technique for rejuvenating a core platform and improving the platform angle played a minor role in core reduction. More typical are blade-like elements that exhibit small remnants of a core's platform edge along the dorsal ridges. These elements represent removals along only part of a core's platform and were initiated from the front or the back of a core, a fact that suggests they may be associated with a change in platform orientation and establishing a new platform.

Bulb and platform characteristics of debitage indicate that most primary and secondary debitage was removed from cores using hard hammer percussion. For primary pieces this is indicated by unlipped pieces (69.3%), of which 12.5% had characteristic *fractures a languette* on the bulbs. Lipped primary pieces accounted for 29.5%, indicating that soft hammer techniques were used for removing some cortical pieces as well. Hard hammer percussion appears to have prevailed in the production of most debitage with 59% of the blades and bladelets being unlipped. Only 4% showed lipping indicative of soft hammer removals.

Lateral profiles in the debitage provide mixed evidence on loading force and the angle of force (Appendix 8.5). Heavy loading force and more perpendicular angles of force should be associated with larger cores and with hard hammer percussion and produce flat profiles. Soft hammer and smaller cores require less loading force and because of size, the angle of force is often more acute to the platform. The latter should result in higher frequencies of curved and twisted

**Table 8.9. Core Trimming Elements in the Debitage and Retouched Tools from Jebel Humeima (J412)**

| | | Core tablets | Platform blades | Others |
|---|---|---|---|---|
| Debitage | *n* | 4 | 17 | 5 |
| | % | 15.4 | 65.4 | 19.2 |
| Tools | *n* | 2 | 1 | — |
| | % | 66.7 | 33.3 | |
| Total CTEs | | 6 | 18 | 5 |
| % of Total CTEs | | 20.7 | 62.1 | 17.2 |

profiles. The evidence in this assemblage, however, shows that most primary elements are curved (46.9%) with a significant number of flat profiles (24.4%), suggesting early reliance on soft hammer to remove primary pieces. Interior blade debitage, on the other hand, is predominantly flat (47%) or twisted (36%). Evidence from lateral profiles and platform characteristics discussed above indicate that both hard and soft hammer modes were used interchangeably during the reduction sequence, rather than following a prescribed strategy of hard hammer followed by soft hammer.

Indications of control in loading in the delivery of force during core reduction can be interpreted from termination attributes on blanks. Feathered terminations were most frequent in both primary elements (64.3%) and blade debitage (70.6%) with hinged or underpassed terminations the next most frequent. These results and the lack of substantial numbers of overpassed terminations would seem to indicate relatively effective loading techniques applied to cores recovered from this site.

## Reduction Activity

Evidence for primary reduction activities at Jebel Humeima is provided in the unusually high proportion of primary elements (Appendix 8.4), particularly in Unit 1 where the proportion is 29.2% and the ratio of primary elements to debitage is 1:2.4. Debris associated with reduction activity is also proportionately higher, averaging 80.7% of the total site assemblage with a high ratio of debitage to chips (1:3.3). The percentage of cortex (or outer surface) on primary elements should be an additional indication of the amount of initial core preparation that took place at the site. Only 8.5% of the primary elements are initial removals from the core with complete cortical coverage, while most (91.5%) have varying amounts of cortex greater than 33% (Appendix 8.5). Additional evidence for initial on-site core reduction, discussed above, is represented by the high frequency of cortical platforms on primary elements (26.7%) (Appendix 8.5). The high number of primary elements with varying amounts of cortical coverage indicates on-site primary reduction of cores with some amount of initial reduction done off-site.

Relative frequencies of blank types provide the most direct means of assessing production goals. Reduction appears to have been directed at the production of flake debitage (48.2%), although 51.2% of the blanks selected for tools exhibit blade characteristics (Appendix 8.4). The combined percentage of blade blanks in debitage and tools is 27%, indicating the Jebel Humeima assemblage is low on a *lamellar* scale. Examination of reduction stages represented in the retouched tools only partially supports the implicit assumption that core reduction was directed at, and blank selection focused on, interior, non-cortical pieces (Table 8.12). Instead, evidence from this assemblage indicates that cortical blades and flakes represent a significant proportion of the retouched tools, and all stages of the reduction sequence are represented in the tool assemblage. A surprising 27.9% of the tools exhibit some amount of cortex, with 12.4% classified as 1st or 2nd order primary blades and 15.5% classified as primary or secondary flakes. Nevertheless, the majority of tools were made on plain blanks.

A complete reduction sequence can be reconstructed using dimensional data on cores, debitage, and tools (Figures 8.12, 8.13 and Tables 8.10, 8.11). Metric comparisons between primary elements and other debitage indicate that primary reduction of cores was achieved by the removal of flake-like primary pieces that were considerably shorter and wider than blade debitage (Figure 8.13). Mean length of primary elements is 31.7 mm with a mean length:width ratio of 1.5, while that of blade debitage is 41.8 mm with a length:width ratio of 2.7 (Table 8.10). The notable difference in mean lengths is diminished somewhat by the fact that primary elements and blade debitage exhibit very similar distributions for length, showing partially overlapping modal intervals (Figure 8.13a). But the predominantly flake-like character of the primary

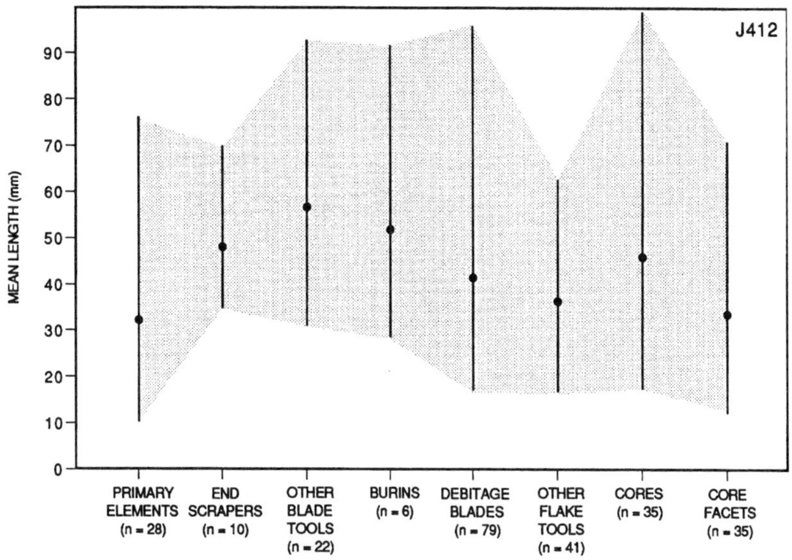


**Figure 8.12.** Mean lengths and ranges of major debitage and tool classes from Jebel Humeima (J412).

elements is well illustrated and contrasted with the interior blade debitage by the greater widths of primary blanks and significantly different modal intervals (Figure 8.13b).

Comparisons of the dimensions of tools with the debitage allows evaluating dimensional factors that may have been important in the selection of debitage blanks for retouching. The great range in the length of debitage blades from 18.2 mm to 96.0 mm encompasses the lengths of all the retouched tools (Figure 8.12, Tables 8.10, 8.11). The dimensions of specialized tools, such as endscrapers and burins, do not exhibit unique dimensional parameters. Burins appear to be somewhat more restricted in length, but endscrapers and burins fall easily within the parameters suggested by the pool of debitage blanks. The comparisons suggest that the debitage, at abandonment, represented a range of blanks of sufficient length that would have been suitable for retouched tools. However, the shorter mean length of the debitage blades would imply that, for the most part, the debitage pool was limited in blanks suitable for retouching into burins, endscrapers, or other blade tools. Furthermore, if overall blank size is considered, the remaining debitage is inadequate for most tool needs. The mean size of debitage blanks (69.1 mm) (Table 8.10) is significantly smaller than the blank sizes of any of the tools (Table 8.11). The complete pieces that make up the debitage would be more suitable for bladelet tools, but bladelet tools are not an important class of tools in this assemblage.

Discussion of the dimensional data has emphasized the smaller dimensions of primary elements and suggested that initial removals were cortical flakes. The data provide little information on the possibility that primary core reduction may also have produced large cortical *blades*. Indirect evidence that large primary blades were produced and selected out early in the reduction sequence to be made into tools may reside in the fact that 27.9% of the tools are cortical pieces (Table 8.12). Reconstruction of this portion of the reduction sequence requires the specific dimensions of all tools made on cortical blade blanks.

Cores make up a significant 2.5% of the debitage with a core:debitage ratio of 1:38.6. In comparison with other Upper Paleolithic sites in the study area, cores are somewhat more frequent in this assemblage. In mean length, however, the cores from J412 are the smallest (46.1

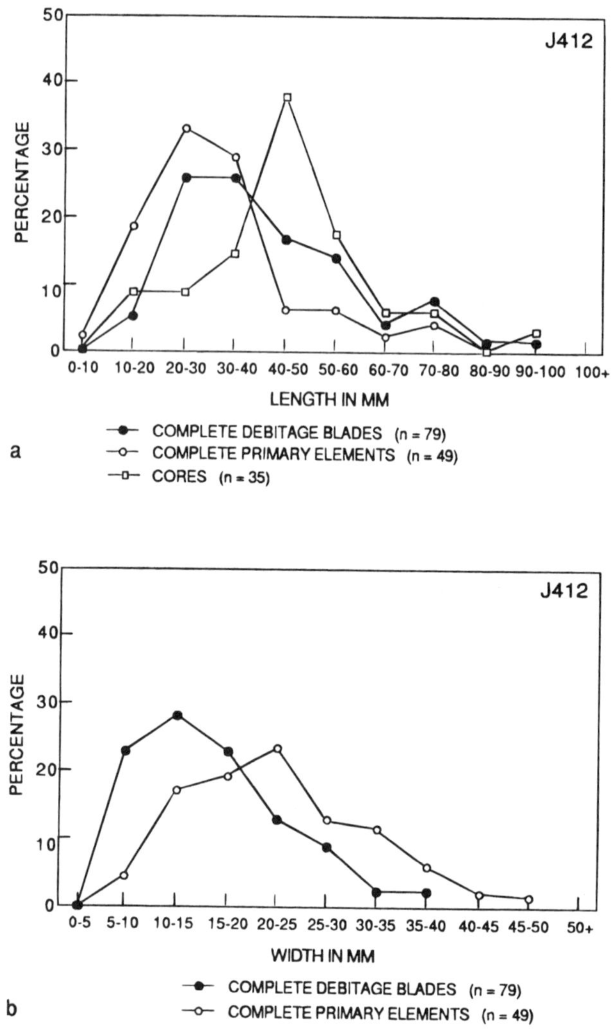


**Figure 8.13.** Length and width distributions for cores and debitage from Jebel Humeima (J412).

mm), but their overall size, indicated by the Core Size Index (15.2 mm), suggests they are quite similar to other Upper Paleolithic cores and exhibit typical average weights (Table 8.10). Although there appears to have been a great deal of primary processing taking place on site with some cores greatly reduced, cores do not appear to have been reduced more intensively than those observed in other assemblages. That is, there are no indications that raw material, in general, was scarce and that economizing behavior was responsible for intensive reduction at the site. Comparison of the mean lengths of cores and core facets with mean tool lengths suggests cores were no longer suitable for such tools as endscrapers and burins but useable for small bladelets. Again, however, the latter do not appear to have been important reduction goals, and cores were probably discarded because they were incompatible with tool requirements and/or exhibited irremediable technological errors.

**Table 8.10. Dimensional Data (mm) for Cores and Complete Debitage from Jebel Humeima (J412)**

| A. Cores (*n* = 35) | Length | Width[a] | Width[b] | Weight (g) | CSI[c] |
|---|---|---|---|---|---|
| Mean | 46.1 | 31.6 | 37.7 | 59.7 | 15.2 |
| sd | 17.2 | 11.5 | 11.6 | 53.1 | 10.0 |
| min | 16.8 | 13.9 | 19.1 | 8.0 | 2.8 |
| max | 99.5 | 61.6 | 70.0 | 227.0 | 53.5 |
| **B. Primary Elements (*n* = 49)** | **Length** | **Width[a]** | **Thickness** | **L : W Ratio** | **Size[d]** |
| Mean | 31.7 | 23.7 | 8.7 | 1.5 | 109.7 |
| sd | 15.1 | 11.8 | 5.7 | .7 | 174.6 |
| min | 9.9 | 8.4 | 1.7 | .6 | 1.9 |
| max | 76.5 | 68.9 | 23.7 | 3.2 | 738.5 |
| **C. Debitage Blades (*n* = 79)** | **Length** | **Width[a]** | **Thickness** | **L : W Ratio** | **Size[d]** |
| Mean | 41.8 | 16.1 | 6.1 | 2.7 | 69.1 |
| sd | 17.2 | 7.5 | 4.0 | .8 | 126.5 |
| min | 18.2 | 6.5 | 1.5 | 1.5 | 2.9 |
| max | 96.0 | 37.3 | 27.1 | 7.7 | 970.4 |
| **D. Core Facets (*n* = 35)** | **Length** | **Width[a]** | | | |
| Mean | 34.1 | 18.6 | | | |
| sd | 13.7 | 7.6 | | | |
| min | 12.5 | 9.3 | | | |
| max | 70.8 | 37.3 | | | |

[a]Width at midpoint.
[b]Maximum width.
[c]Core Size Index (length × width[a] / 100).
[d]Size (length × width[a] × thickness /100).

## Typology

The retouched tool assemblage at Jebel Humeima represents 3.5% of the total excavated sample ($n = 8858$) (Appendix 8.4). Typologically, the tool assemblage is dominated by the more informal classes of tools, such as notches, retouched blades, and retouched flakes (Appendix 8.6). Retouched bladelets comprise a small proportion (4.1%), but taken together with retouched blades, occur almost as frequently as retouched flakes. Notches represent the third largest class of tools, occurring most often on flakes (Figure 8.14d). Typical Upper Paleolithic tools, such as endscrapers and burins, represent relatively small proportions—8.7% and 7.6%, respectively. Proportionately more endscrapers were located in Unit 2 than in Unit 1, whereas the overall diversity of tools in Unit 2 is lower than it is in Unit 1.

Endscrapers as a class are quite variable, occurring equally on flake and blade blanks (e.g., Figures 8.14:a and 8.15:c). A significant number of endscrapers may be classified as carinated because of the thickness of the blank and/or the steepness of the nose. Mean nose thickness of endscrapers in this assemblage is 10.4 mm (Table 8.11), which is thick when compared to endscrapers from Upper Paleolithic assemblages at Tor Hamar or Tor Aeid. In contrast, burins from Jebel Humeima do not tend to be the thick, carinated types. Only one burin is typical of what might more appropriately be called a "burin-core" because of its thick blank, wide bit, and multiple spall scars (Figure 8.14:e). Burins in this assemblage occur on blade blanks more frequently

Table 8.11. Dimensional Data (mm) for Retouched Tools from Jebel Humeima (J412)

| | Length | Width | Thickness | Nose thickness | Bit width | Size[a] | L: W Ratio |
|---|---|---|---|---|---|---|---|
| Endscrapers | | | | | | | |
| Complete ($n$ = 10) | | | | | | | |
| mean | 56.3 | 36.5 | 10.8 | 10.4 | NA | 243.2 | 1.6 |
| sd | 17.1 | 8.5 | 4.2 | 6.7 | | 189.0 | 0.6 |
| min | 31.5 | 21.6 | 7.2 | 5.6 | | 63.2 | 0.8 |
| max | 93.2 | 51.4 | 21.6 | 27.9 | | 734.5 | 2.5 |
| Incomplete ($n$ = 19) | | | | | | | |
| mean | 40.2 | 31.5 | 11.4 | 10.2 | NA | 182.7 | 1.4 |
| sd | 10.9 | 8.9 | 4.3 | 3.6 | | 145.6 | 0.4 |
| min | 19.2 | 19.2 | 3.6 | 4.2 | | 22.0 | .9 |
| max | 60.8 | 48.0 | 19.7 | 17.2 | | 564.8 | 2.4 |
| Burins | | | | | | | |
| Complete ($n$ = 6) | | | | | | | |
| mean | 48.7 | 28.6 | 8.3 | NA | 4.2 | 131.1 | 1.7 |
| sd | 15.5 | 8.1 | 2.0 | | 2.1 | 96.7 | 0.5 |
| min | 34.8 | 19.1 | 6.0 | | 1.8 | 46.5 | 1.1 |
| max | 70.4 | 39.0 | 11.5 | | 7.4 | 275.4 | 2.4 |
| Incomplete ($n$ = 16) | | | | | | | |
| mean | 43.8 | 24.9 | 9.4 | NA | 5.9 | 119.4 | 1.9 |
| sd | 12.6 | 7.0 | 4.0 | | 3.5 | 90.5 | 0.8 |
| min | 25.2 | 9.5 | 4.2 | | 1.7 | 16.2 | 1.0 |
| max | 65.3 | 36.1 | 16.1 | | 13.3 | 261.5 | 4.3 |
| Other Tools | | | | | | | |
| Complete Blades ($n$ = 22) | | | | | | | |
| mean | 52.9 | 22.5 | 7.9 | NA | NA | 116.5 | 2.5 |
| sd | 14.7 | 8.3 | 2.7 | | | 114.6 | 0.8 |
| min | 28.4 | 10.0 | 3.5 | | | 19.6 | 1.4 |
| max | 92.3 | 48.1 | 132 | | | 536.1 | 5.1 |
| Complete Flakes ($n$ = 41) | | | | | | | |
| mean | 35.9 | 29.0 | 7.6 | NA | NA | 100.5 | 1.3 |
| sd | 11.5 | 9.5 | 3.3 | | | 97.1 | 0.3 |
| min | 17.1 | 9.9 | 2.0 | | | 3.5 | 0.6 |
| max | 63.0 | 55.9 | 16.8 | | | 434.3 | 1.9 |

[a]Size = Length × width × thickness /100.

than on flake blanks. Burins on snaps or old surfaces are frequent, and there are a few examples of transverse burins and *burins plan*. Truncations, denticulates, and perforators together make up 10.3% of the tools, occurring equally on flake and blade blanks.

The Jebel Humeima tool assemblage lacks distinctive Upper Paleolithic points, such as el-Wad points or pointed Ouchtata bladelets. Instead, the assemblage contains two complete Levallois points, considered to be intrusive in this otherwise Upper Paleolithic assemblage.

## Tor Hamar (J431): Layers F–G

### Introduction

The Upper Paleolithic materials in Layers F–G were excavated in two 2 m² units, subdivided into 1 m² quadrants (Figure 8.4). Excavation by 10 cm arbitrary levels in Layer F extended from 180–200 cm BD, whereas Layer G extended from 200–260 cm BD. The Upper Paleolithic

**Table 8.12. Percentages of Different Reduction Stages Represented in the Retouched Tools from Jebel Humeima (J412)**

| | Primary Blades | | Plain Blades | | | Flakes | | | |
|---|---|---|---|---|---|---|---|---|---|
| | 1 | 2 | 1 | 2 | 3 | 1 | 2 | 3 | |
| | Cortical | | | | | | Cortical | | Others |
| Retouched Tools ($n$ = 322) | 3.4 | 9.0 | 12.7 | 20.8 | 5.3 | 5.6 | 9.9 | 31.7 | 1.5 |

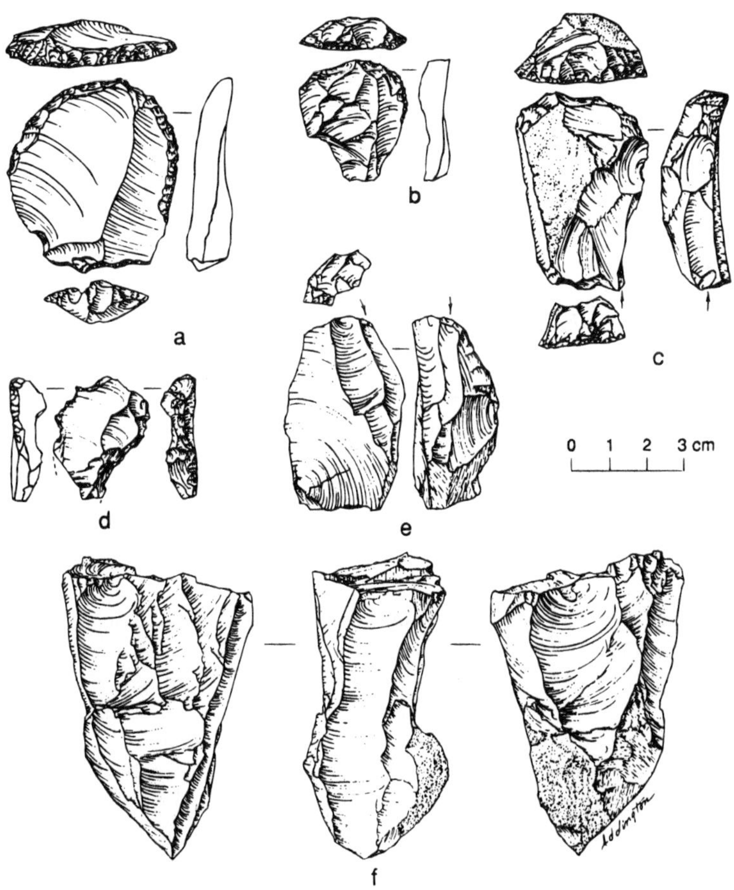


**Figure 8.14.** Cores and tools from Jebel Humeima (J412): (a,b) endscrapers, (c) endscraper-burin, (d) perforator, (e) burin on a truncation, (f) blade core.

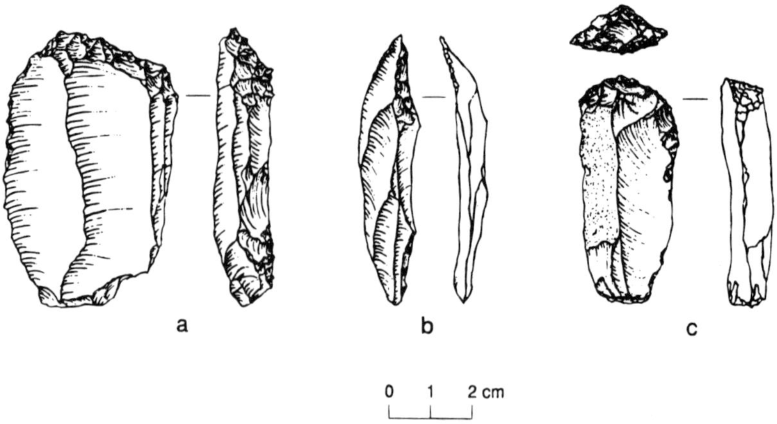


Figure 8.15. Tools from Jebel Humeima (J412): (a) atypical endscraper, (b) retouched blade on a core trimming element, (c) endscraper.

materials are quite consistent in assemblage composition throughout the 80 cm of cultural deposit, although the density of material varies between the two layers. Typologically, there are some differences between Layers F and G which may be due to a certain amount of mixing from subsequent layers. The following description and discussion is limited to lithic debitage and retouched tools recovered from Block I, Units 3–6 (Appendix 8.7). The cultural deposits from Layers F and G are undated and separated geologically from later Epipaleolithic Qalkhan deposits by rockfall representing a time period of unknown length. The hiatus must be of considerable duration since the cultural material in Layers F and G is most similar to Early Ahmarian assemblages from the Negev and Sinai, radiometrically dated to ca. 30,000–38,000 B.P. (Marks 1983b:37; Marks and Ferring 1988:Table 3; Phillips 1988:186).

## Raw Material

A breakdown of the forms of raw material utilized as cores in Layers F and G indicates that half of the cores represent rolled cobbles with gravel cortex (Table 8.13). Nodules with limestone rind or cortex account for 35.7% of the cores. Those lacking any original outer surface represent 14.3% of the cores.

## Technology

The collection of cores from Layers F and G is relatively small, representing only 0.9% of the debitage (Appendix 8.7). The largest proportion of cores is associated with Layer F. The majority are single platform cores (42.8%), while there are limited numbers of multiple platform cores (Table 8.13). These consist of cores with opposed platforms on the same side or those with opposed platforms twisted to the side. There are three examples of cores that exhibit a change in the orientation of successive platforms. Patterns of dorsal faceting on debitage is consistent with single platform core reduction since approximately 60% of primary elements and blade debitage exhibit unidirectional parallel facets (Appendix 8.8).

The platforms of cores are simple and unprepared, corresponding with unfaceted single platforms on primary elements (75.7%) and on 50% of the debitage blades. An additional 43.9% of the debitage blades have crushed platforms (Appendix 8.8). Dihedral and multi-faceted platforms occur infrequently.

**Table 8.13. Classification of Cores by Raw Material and Core Class for Tor Hamar, Layers F and G (J431)**

| | Layer F | Layer G | Total *n* | Total % |
|---|---|---|---|---|
| A. Raw material | | | | |
| Rolled Cobble | 5 | 2 | 7 | 50.0 |
| Limestone Cobble | 4 | 1 | 5 | 35.7 |
| Indeterminate | 1 | 1 | 2 | 14.3 |
| Total | 10 | 4 | 14 | 100.0 |
| B. Core class | | | | |
| Single Platform | 3 | 3 | 6 | 42.8 |
| Opposed Platforms | | | 3 | 21.4 |
| Opposite | | | | |
| Same | 2 | | | |
| Twisted | 1 | | | |
| Change of Orientation | 2 | 1 | 3 | 21.4 |
| Others | 2 | | 2 | 14.3 |
| Total | 10 | 4 | 14 | 99.9 |

Rejuvenation of platforms on cores was accomplished primarily with the core tablet technique (Table 8.14). The CTE:Core ratio is 1:.5 with an additional four core tablets identified as tool blanks. As in other Upper Paleolithic assemblages in the study area, an alternative technique for platform maintenance involved the removal of smaller portions of the platform, resulting in a sizeable proportion of core trimming elements identified as "platform blades" (see Tor Fawaz for a description). In addition, a unique combination of these two techniques occurs in four examples in which the proximal end of the core trimming element is a core tablet, but the medial and dorsal sections consist of portions of another platform. These would be consistent with change-of-orientation cores in which double platforms were present but not perpendicular to each other as in 90 degree cores.

## Reduction Mechanics

Indicators of percussion, loading, and relative efficiency and control in blank removals include platform characteristics and distal and lateral profiles (Appendix 8.8). The attributes indicate that hard and soft hammer modes were operative during both primary and secondary processing. The bulk of the primary elements were removed using hard hammer percussion, indicated by the presence of unlipped platforms (62.1%). Approximately 37.8% of the primary pieces, however, are lipped, suggesting that they were using soft hammers in the early stages as well. Lipped primary elements may be the smaller, cortical pieces resulting from primary core shaping and maintenance. Similarly, 52.1% of the debitage blades are unlipped, and 41.1% are lipped, implying that a comparable strategy incorporating both hard and soft hammers was used to produce interior or secondary products.

The delivery of force (loading) can be inferred from the distal profiles of blanks. Terminations tend to be feathered in both primary elements and interior blades. Loading errors resulting in underpassed or hinged terminations were more frequent in the earlier cortical pieces (25.4%) than on interior blades (13.0%).

Variation in the lateral profiles is significant and probably a function of the angle of force associated with different modes of percussion. In this assemblage, primary elements generally tend to be flat or curved (66.6%), a consequence of more perpendicular loading in a hard ham-

Table 8.14. Core Trimming Elements in the Debitage and Retouched Tools from Tor Hamar, Layers F and G (J431)

| | | Core tablets | Platform blades | Others |
|---|---|---|---|---|
| Debitage | *n* | 15 | 12 | 4 |
| | % | 48.4 | 38.7 | 12.9 |
| Tools | *n* | 4 | — | — |
| | % | 2.6 | | |
| Total CTEs | | 19 | 12 | 4 |
| % of Total CTEs | | 54.3 | 34.3 | 11.4 |

mer mode. In contrast, 79.7% of the interior blades are curved and/or twisted—profiles resulting from more acute loading angles associated with soft hammer.

## Reduction Activity

All major stages of the reduction sequence are represented in the debitage, debris, and retouched tools (Appendix 8.7). Debris accounts for a larger proportion than debitage, and tools account for a relatively small proportion (4.2%). The fairly high proportion of 18% primary pieces with complete cortical coverage (or natural surface) indicates primary processing activities at the site. This, however, is contradicted to some extent by evidence from core and debitage ratios. Ratios of primary elements to debitage and cores to debitage are high, indicating a low level of on-site initial processing of raw material (Appendix 8.7). Secondary processing of cores at the site, however, is indicated by the low ratios of core trimming elements to cores and debitage to chips.

General blank morphology of debitage and tool blanks provides an initial indication of reduction goals. In this case, the ratios of flakes to blades in both the debitage and tools are quite similar and indicative of production goals (see Appendix 8.7 and Table 8.17). Tools with blade proportions comprise 60.6% of the tools, whereas blade debitage is represented by the same proportion, suggesting reduction was directed toward elongated blanks. Preferred blanks for tools tended to be non-cortical blanks in 78% of the cases (Table 8.17).

Comparisons of dimensional data from cores, debitage, and tools provide evidence on the nature of reduction activity that took place at the site (Figures 8.16, 8.17 and Tables 8.15, 8.16). Core reduction at Tor Hamar involved the initial removal of primary elements that were relatively shorter and smaller than most of the subsequent debitage and tool blanks, with a mean length of 38 mm, a length:width ratio of 1.9, and a mean blank size of 77.2. Debitage blades, in contrast, represent a different class of reduction elements. When compared with primary elements, debitage blades appear initially to be quite similar; they exhibit a mean length only slightly longer at 39 mm and follow a parallel distribution for length (Figure 8.17a). Debitage blades, however, tend to be more elongated with a length:width ratio of 3.4 and a mean blank size of 17.3. The morphological differences between primary elements and debitage blades are well illustrated by blank widths, illustrated in Figure 8.17b. Mean blank width of primary elements is 22.4 mm, compared to 12.0 mm for debitage blades.

Blank size, calculated on the basis of length, width, and thickness (Table 8.15), is an effective means of comparing the overall size and 3-dimensional quality of reduction products for the purpose of placing them in a sequence of reduction (Figure 8.16). After the initial removal of rather small primary flakes with a mean size of 77.2 mm, blanks for the most robust tools were detached: (1) burins (mean size 196.9); (2) scrapers (mean size 193.2); and (3) other large flake tools (mean size 180.5). These were followed by the production of medium sized

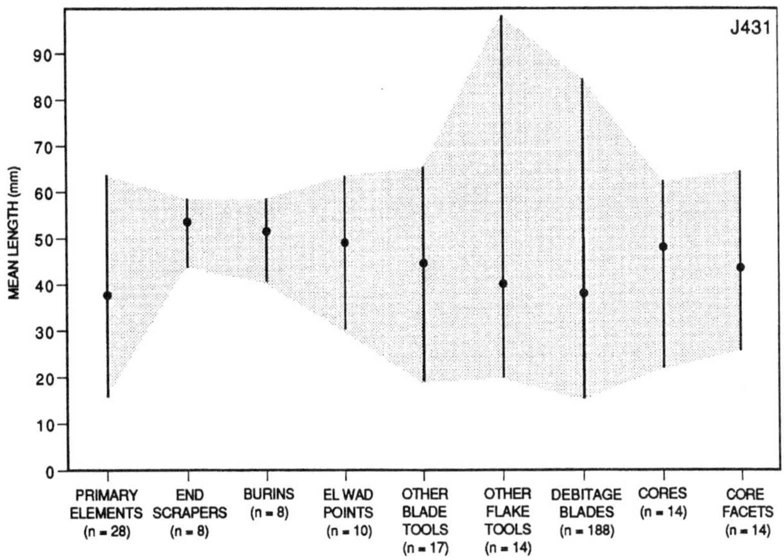


**Figure 8.16.** Mean lengths and ranges of major debitage and tool classes from Tor Hamar (J431).

debitage, a reduction group not documented in the assemblage—debitage in the size range of 40–180. It is possible that blanks from this group were made into tools that were taken out of the site. The remainder of the reduction elements represent other tools made on blades (mean size 38.6), a pool of residual debitage (mean size 24.3), and el-Wad points (mean size 19.3).

Endscrapers and burins were made on the largest blanks with mean lengths of approximately 53 mm. Length:width ratios for the larger tools are at or near 2:1. More specialized tools, such as el-Wad points, were made on more select blanks with substantially smaller sizes. These blanks are extremely elongated with a mean length of 49.8 mm and a mean length:width ratio of 4.6.

The end of the reduction stream is represented by the cores. In Figure 8.16a, the modal interval for core lengths, between 60 and 70 mm, implies a sizeable proportion of cores with ample remaining length. The mean dimensions of cores (48.6 mm) and core facets (44.3 mm) (Table 8.15 and Figure 8.16), however, indicate that the discarded cores were largely incompatible with the blank requirements of most of the specialized tools that dominate the tool assemblage.

## Tool Typology

The retouched tools in Layers F and G at Tor Hamar are Upper Paleolithic, but typologically, somewhat uncharacteristic and inconsistent with other early Ahmarian assemblages in the southern Levant (Appendices 8.7 and 8.9). One problem concerns a number of intrusive artifacts from the overlying Qalkhan layers which occur, with one exception, in the uppermost layer, F. Of greater importance, however, is a variety of inconsistent tool types which makes this assemblage unusual.

The tool assemblage comprises only 4.2% of the total lithic assemblage for Layers F and G, with 69% of the tools originating from the upper layer, F. The most prevalent tool classes include typical Upper Paleolithic tools—burins, endscrapers, and retouched blades and bladelets. Somewhat surprising are the large numbers of retouched flakes and other tools made on flake

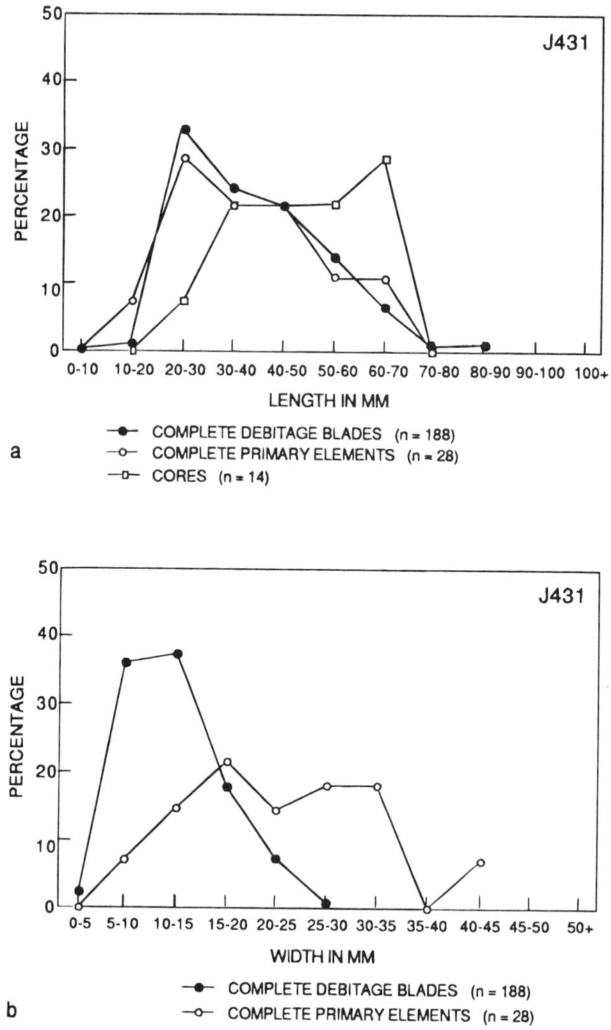


**Figure 8.17.** Length and width distributions for cores and debitage from Tor Hamar (J431).

blanks, especially tools we might expect to occur more often on blades (e.g., endscrapers) (see Appendix 8.9).

And yet, the most distinctive tools in this assemblage are the el-Wad points, identifying the assemblage with other Early Ahmarian assemblages. The points exhibit a number of consistent technological and typological characteristics (Figure 8.18: c,d,f). They occur on more elongated debitage blanks ranging from 31.4 mm to 64.2 mm with a mean length of 49.8 mm with the majority of points occurring in a modal interval between 47 and 57 mm (Table 8.16). Widths, ranging from 8.3 to 13.0 mm, are distributed in a modal interval of 11.5 mm and exhibit a mean of 10.8 mm. These dimensions are comparable to the ranges and modal distributions for lengths and widths of el-Wad points described for Lagama VII (Bar-Yosef and Phillips 1977:45, Figure 16), in which most points had lengths between 50 and 55 mm and an overall range of approximately 25–70 mm. Widths of the Lagama VII points ranged between 6 and 14

**Table 8.15. Dimensional Data (mm) for Cores and Complete Debitage from Tor Hamar, Layers F and G (J431)**

| A. Cores ($n$ = 14) | Length | Width[a] | Width[b] | Weight (g) | CSI[c] |
|---|---|---|---|---|---|
| Mean | 48.6 | 30.9 | 33.9 | 63.7 | 15.3 |
| sd | 12.8 | 8.8 | 8.3 | 38.3 | 6.8 |
| min | 22.5 | 21.3 | 23.5 | 22.1 | 5.3 |
| max | 63.4 | 45.3 | 48.8 | 164.6 | 28.7 |
| **B. Primary Elements ($n$ = 28)** | **Length** | **Width[a]** | **Thickness** | **L: W Ratio** | **Size[d]** |
| Mean | 38.0 | 22.4 | 6.8 | 1.9 | 77.2 |
| sd | 13.8 | 9.2 | 4.2 | .9 | 87.4 |
| min | 15.9 | 6.3 | 2.1 | .6 | 2.8 |
| max | 63.6 | 41.5 | 19.9 | 4.4 | 354.9 |
| **C. Debitage Blades ($n$ = 188)** | **Length** | **Width[a]** | **Thickness** | **L: W Ratio** | **Size[a]** |
| Mean | 39.0 | 12.0 | 3.7 | 3.4 | 24.3 |
| sd | 13.4 | 4.7 | 2.0 | .9 | 28.3 |
| min | 16.6 | 4.4 | 1.1 | 2.0 | .9 |
| max | 85.5 | 30.0 | 11.8 | 7.5 | 157.1 |
| **D. Core Facets ($n$ = 14)** | **Length** | **Width[a]** | | | |
| Mean | 44.3 | 20.4 | | | |
| sd | 12.6 | 7.2 | | | |
| min | 25.8 | 7.4 | | | |
| max | 64.7 | 35.2 | | | |

[a]Width at midpoint.
[b]Maximum width.
[c]Core Size Index (length × width[a]/100).
[d]Size (length × width[a] × thickness/100).

mm with a modal width between 8 and 10 mm. The dimensions of el-Wad points at Tor Hamar vary considerably from those described for K'sar Akil (Bergman 1981:326). El-Wad points from K'sar Akil, Level XI, are much shorter with most points between 30 and 40 mm and widths between 8 and 11 mm. Those from Level X at K'sar Akil tend to be only slightly longer (35–40 mm) and slightly narrower (7–9 mm).

The lateral profiles of el-Wad points from Tor Hamar are variable, ranging from flat ($n$ = 4) to incurvate ($n$ = 1) to curved and twisted ($n$ = 9). Retouching also varies, ranging from fine to semi-abrupt to abrupt, but it invariably occurs on obverse edges. Retouching was used to shape proximal, medial, and distal portions of the el-Wad points, in many cases to form pointed tips; in other cases, it is used to shape edges or proximal bases. The points from Tor Hamar are most similar to el-Wad point varieties at Boker A in which retouch is restricted to obverse edges or consists of backing (Marks 1977; Jones *et al.* 1983:299). They contrast, however, with another distinct variety of el-Wad point type at Boker A which features alternately retouched edges. None of the el-Wad points at Tor Hamar exhibit inverse retouch, and there is limited evidence of inverse retouch present in other retouched pieces (see Appendix 8.9).

Burins are the most numerous tools with a variety of types represented. Dihedral burins (Figure 8.19:b,d) and angle burins on a snap or an unaltered surface are the most prevalent. Three examples of carinated burins are thick burins with strong similarities to small bladelet cores (Figure 8.20). Complete burins are quite long, averaging 52.2 mm and exhibiting thick,

Table 8.16. Dimensional Data (mm) for Retouched Tools from Layers F and G at Tor Hamar (J431)

| | Length | Width | Thickness | Nose thickness | Bit width | Size[a] | L : W Ratio |
|---|---|---|---|---|---|---|---|
| Endscrapers | | | | | | | |
| Complete ($n$ = 8) | | | | | | | |
| mean | 54.0 | 37.8 | 8.7 | 7.4 | NA | 193.2 | 1.7 |
| sd | 6.2 | 14.9 | 3.4 | 3.2 | | 113.4 | 0.8 |
| min | 44.1 | 16.7 | 3.4 | 3.1 | | 29.0 | 0.8 |
| max | 59.2 | 60.7 | 13.3 | 14.1 | | 381.8 | 3.4 |
| Incomplete ($n$ = 8) | | | | | | | |
| mean | 50.1 | 29.2 | 8.4 | 8.7 | NA | 134.8 | 1.8 |
| sd | 15.8 | 8.7 | 3.2 | 3.9 | | 97.3 | 0.7 |
| min | 22.3 | 22.5 | 4.1 | 3.9 | | 25.9 | 0.9 |
| max | 67.5 | 49.2 | 12.5 | 12.1 | | 324.1 | 2.9 |
| Burins | | | | | | | |
| Complete ($n$ = 8) | | | | | | | |
| mean | 52.2 | 31.3 | 11.6 | NA | 12.1 | 196.9 | 2.0 |
| sd | 5.9 | 10.8 | 4.3 | | 7.9 | 116.1 | 1.1 |
| min | 41.1 | 12.2 | 7.2 | | 3.7 | 48.9 | 1.0 |
| max | 58.8 | 45.7 | 18.7 | | 25.3 | 409.0 | 4.5 |
| Incomplete ($n$ = 18) | | | | | | | |
| mean | 43.9 | 22.4 | 9.4 | NA | 7.6 | 109.0 | 2.0 |
| sd | 10.4 | 6.7 | 4.5 | | 3.9 | 115.0 | 0.5 |
| min | 24.4 | 13.3 | 4.2 | | 2.5 | 14.2 | 1.4 |
| max | 72.4 | 41.4 | 21.6 | | 15.4 | 525.0 | 3.1 |
| Other Tools | | | | | | | |
| Complete Blades ($n$ = 17) | | | | | | | |
| mean | 45.0 | 15.1 | 4.2 | NA | NA | 38.6 | 3.3 |
| sd | 13.5 | 7.7 | 1.9 | | | 47.7 | 1.0 |
| min | 19.8 | 4.7 | 1.3 | | | 1.6 | 1.7 |
| max | 65.9 | 39.1 | 8.2 | | | 207.5 | 5.2 |
| Complete Flakes ($n$ = 14) | | | | | | | |
| mean | 41.3 | 33.4 | 9.1 | NA | NA | 180.5 | 1.4 |
| sd | 19.2 | 16.0 | 4.1 | | | 247.5 | 0.5 |
| min | 20.3 | 13.8 | 2.7 | | | 8.4 | 0.7 |
| max | 98.6 | 60.5 | 20.0 | | | 963.5 | 2.1 |
| El Wad Points | | | | | | | |
| Complete ($n$ = 10) | | | | | | | |
| mean | 49.8 | 10.8 | 3.3 | NA | NA | 19.3 | 4.6 |
| sd | 10.4 | 1.8 | 1.1 | | | 11.2 | 0.6 |
| min | 31.4 | 8.3 | 2.2 | | | 5.7 | 3.7 |
| max | 64.2 | 13.0 | 5.7 | | | 37.7 | 5.8 |

[a]Size = length × width × thickness /100.

multifaceted bits (Figure 8.18:h). Burins also occur as multiple tools, one in conjunction with an endscraper, the other with a notch.

Scrapers are extremely heterogeneous in this assemblage with wide-ranging types, blank sizes, and retouch forming the scraping edge (compare Figures 8.18:j,k and 8.19:c,e). An unexpected type in this Early Ahmarian assemblage is an endscraper on a blade with Aurignacian retouch. Similarly, one sidescraper is present (Figure 8.19:a), and multiple notches are fairly numerous.

Table 8.17. Percentages of Different Reduction Stages Represented in the Retouched Tools from Tor Hamar, Layers F and G (J431)

| | Primary blades | | Plain blades | | | Flakes | | | |
|---|---|---|---|---|---|---|---|---|---|
| | 1 | 2 | 1 | 2 | 3 | 1 | 2 | 3 | |
| | Cortical | | | | | | Cortical | | Others |
| Retouched Tools ($n$ = 107) | 0.6 | 5.8 | 19.4 | 24.5 | 10.3 | 5.2 | 6.4 | 23.9 | 3.9 |

Retouched blades and bladelets together barely outnumber the retouched flakes, which is unusual in an Ahmarian assemblage. However, most retouched flakes are fragments and, in general, lack the formality of other tools. Retouched bladelets are not common and include both finely retouched pieces and pieces with abrupt backing, of which most are retouched unilaterally. One example of very fine Ouchtata retouch, more typical of later Ahmarian assemblages, is present. Inverse retouch is rare. The backed pieces appear to be intrusive Epipaleolithic microliths from Layer E, along with the La Mouillah point. One unexpected tool is an example of a *lamelle Dufour* (Figure 8.18:b), which has been suggested to be a by-product of carinated tool manufacture (Marks 1976b:238, 1976c:381–382). This example exhibits bipolar or *sur enclume* retouch along the left lateral edge and is not made of flint but obsidian, the only example known in the south Jordan assemblages.

## Tor Aeid (J432)

### Introduction.

Excavations in two units (Figure 8.6) produced a large lithic assemblage in which few differences among levels or between units were detected. The lithic assemblage exemplifies a blade technology with an emphasis on blades and less so on smaller bladelets. In both technology and typology, the Tor Aeid assemblage is most similar to Tor Hamar, Layers F–G, but there are some interesting variations. The Tor Aeid assemblage, in turn, is most analogous to Early Ahmarian assemblages, such as Abu Noshra I and II from South Sinai (Phillips 1988) and Boker A and BE in the Central Highlands of the Negev (Jones *et al.* 1983).

### Raw Material

A significant 28.6% of the cores recovered from Tor Aeid are reduced to the point at which no remaining outer surface is present that would allow identifying raw material forms (Table 8.18). Rolled cobbles make up 41.1% of the raw material utilized as cores, while the remaining cores represent limestone nodules.

### Technology

The majority of cores from Tor Aeid are single platform cores with simple, unfaceted platforms (Table 8.18). A significant portion of reduction, however, employed double opposed platforms. Changing the orientation of platforms, usually 90 degrees, also appears to have been an important option during the course of core reduction. The discarded cores correlate only loosely with observable features on debitage in the Tor Aeid assemblage. Cores tend to be subpyramidal in shape with limited evidence for distal and lateral trimming. Converging dorsal scars, compatible with pyramidal and subpyramidal core shapes, are present in only 14.9% of the blade debitage and 4.9% of the primary elements (Appendix 8.11). Dorsal facet patterns on debitage

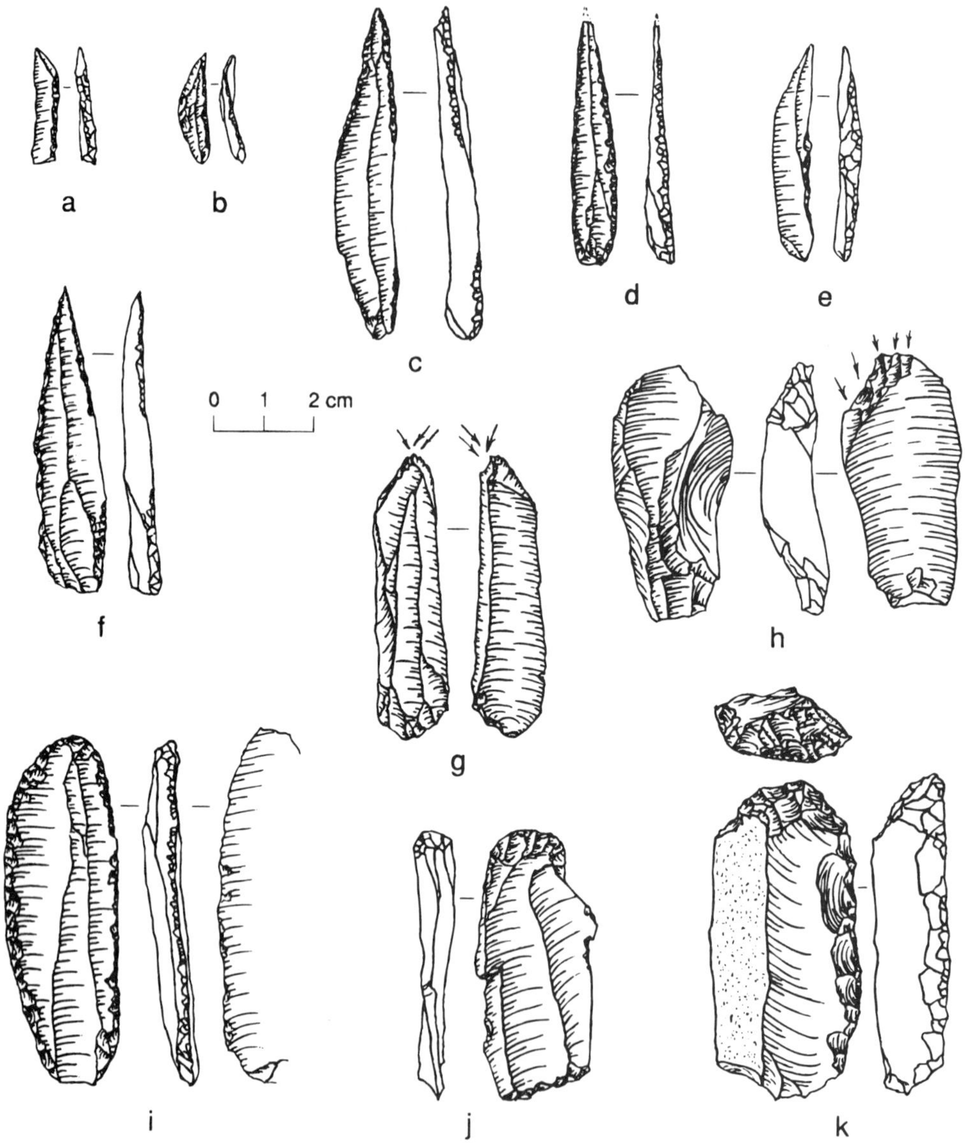


**Figure 8.18.** Tools from Tor Hamar (J431): (a,e) backed bladelets, (b) *lamelle Dufour*, (c,d,f) el-Wad points, (g) dihedral burin, (h) burin *plan* on a retouched blade, (i) endscraper on a retouched blade, (j) endscraper on a truncated blade with previous burin spalls, (k) endscraper on a retouched blade.

reflect a predominance of parallel scars—a pattern more consistent with prismatic cores. Opposed dorsal facets were observed most frequently on primary elements but only rarely (2.8%) on the blade debitage. Given the 21.4% discarded cores with opposed platforms, we might expect a higher occurrence of opposed facets on interior blade debitage, which would reflect an advanced stage in core reduction.

Platforms on cores are simple, unfaceted, and relatively unprepared except for regularizing platform edges. They correlate closely with platforms on debitage (Appendix 8.11). Single,

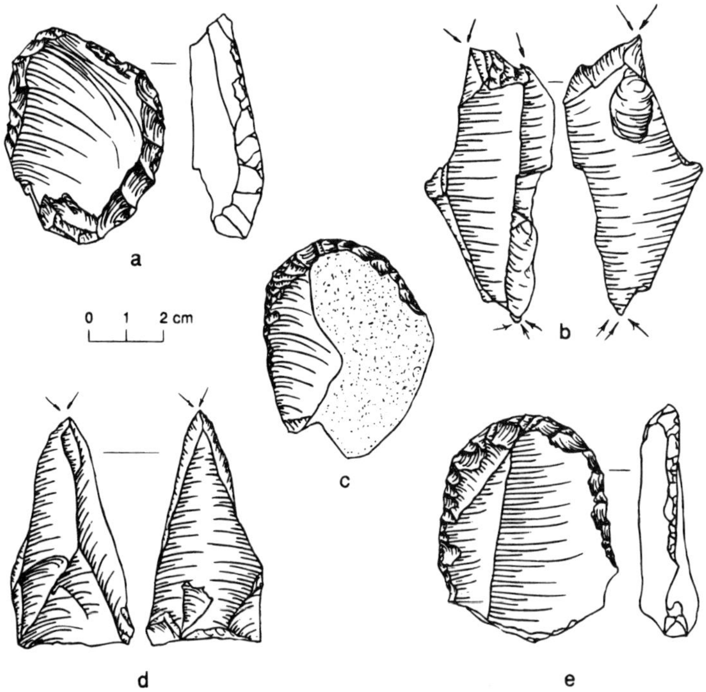


**Figure 8.19.** Tools from Tor Hamar (J431): (a) scraper, (b) multiple dihedral burin, (c) endscraper, (d) dihedral burin, (e) endscraper on a retouched flake.

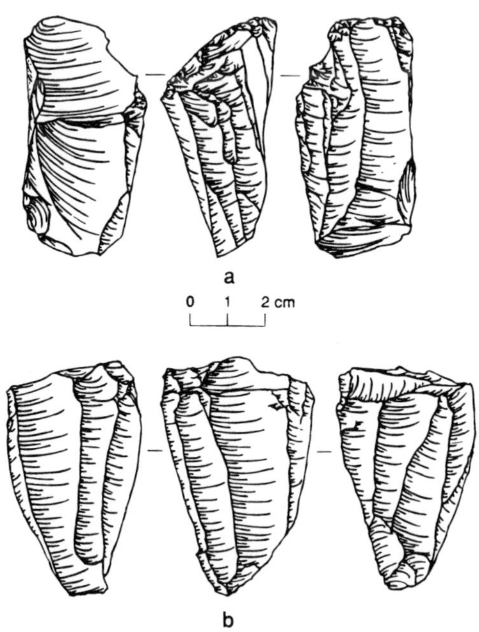


**Figure 8.20.** Cores from Tor Hamar (J431): (a,b) single platform bladelet cores.

Table 8.18. Classification of Cores by Raw Material and Core Class from Tor Aeid (J432)

| | Unit 1 | Unit 2 | Total | |
|---|---|---|---|---|
| | | | *n* | % |
| A. Raw Material | | | | |
| Rolled Cobble | 6 | 17 | 23 | 41.1 |
| Limestone Cobble | 7 | 10 | 17 | 30.3 |
| Indeterminate | 12 | 4 | 16. | 28.6 |
| Total | 25 | 31 | 56 | 100.0 |
| B. Core Class | | | | |
| Single Platform | 16 | 17 | 33 | 58.9 |
| Opposed Platforms | | | 12 | 21.4 |
| Opposite | | 7 | | |
| Same | 1 | | | |
| Twisted | 3 | 1 | | |
| Change of Orientation | 5 | 3 | 8 | 14.3 |
| Others | | | 3 | 5.4 |
| Discoidal | 1 | 1 | | |
| Other | 1 | | | |
| Total | 27 | 29 | 56 | 100.0 |

unfaceted platforms account for 54.8% of the primary elements, whereas 71.4% of blade debitage exhibits single faceted platforms. Differences are explained by cortical platforms that occur on primary elements (15.5%). Double and multifaceted platforms are infrequent, representing 9.5% of the primary elements and only 3.8% of the blades. In this assemblage, crushed platforms, resulting from a narrowed loading zone, occur at an unusually high rate on primary elements (20.2%), which is only slightly lower than the 24.8% on interior debitage blades.

Core trimming elements are frequent in the J432 assemblage, comprising 2.6% of the debitage and 5.5% of the blanks on which tools were made (Table 8.19 and Appendix 8.10). They are only slightly more frequent than the cores, resulting in a CTE:Core ratio of 1:0.9. Rejuvenation of core platforms was accomplished using two main techniques: (1) core tablet (52.8%), and (2) "platform blade" (38.6%) (Table 8.19). The latter are usually pointed and blade-like with characteristic edges of a core's platform forming part of the dorsal ridge of the piece. Core tablets were selected most often over other core trimming elements for retouching into tools (Table 8.19). An unusual example of a retouched tool on a core trimming element is illustrated in Figure 8.24:c.

## Reduction Mechanics

Information on the mechanics of debitage production is provided in the presence or absence of lipping on interior platform edges. Primary elements tend to be unlipped (65.5%) with 23.8% of these pieces exhibiting *fractures a languette* on the bulbs, suggesting that hard hammer percussion was the primary technique for initial core reduction (Appendix 8.11). Subsequent interior blade debitage exhibits a similar proportion of unlipped platforms, with only 18.7% showing lipped platforms, indicative of a soft hammer mode. An important 33.3% of the primary pieces show lipping, which suggests that both modes of percussion were used, somewhat interchangeably in the primary and secondary stages of reduction.

Loading efficiency was examined using blank terminations as indicators of loading success or errors. Most distal profiles show feathered terminations on both primary elements (78.3%) and blade debitage (82%). Loading errors in the form of overpassed or underpassed termina-

Table 8.19. Core Trimming Elements in the Debitage and Retouched Tools at Tor Aeid (J432)

| | | Core tablets | Platform blades | Others |
|---|---|---|---|---|
| Debitage | | | | |
| | *n* | 30 | 26 | 6 |
| | % | 48.4 | 41.9 | 9.7 |
| Tools | | | | |
| | *n* | 7 | 1 | |
| | % | 87.5 | 12.5 | — |
| Total CTEs | | | | |
| % of Total CTEs | | 37 | 27 | 6 |
| | | 52.8 | 38.6 | 8.6 |

tions occur at similar rates—18.6% of primary elements and 15% of the debitage blades—indicating errors in percussive force.

The lateral profiles of debitage blanks provide additional evidence for relative proficiency in the delivery of force. Most primary elements exhibit curved profiles (62.2%), while only 27% of the debitage blades are curved (Appendix 8.11). What distinguishes the blade debitage is the marked prevalence of twisted profiles (38.7%), a characteristic that is infrequent in primary elements (11.1%).

## Reduction Activity

All major stages of reduction are represented in the debitage, debris, and tools of the Tor Aeid assemblage. Although near equal proportions of debitage and debris are present, the tool component represents a fairly low percentage of the overall assemblage at 2.7%, resulting in a tool:debitage ratio of 1:15.8 (Appendix 8.10). Primary core processing at the site is reflected in the number of cores and the core tablets, discussed above, and in the substantial percentage of primary elements (8.5%).

The goals of reduction may be inferred on the basis of proportional differences in various classes of debitage and tools. Debitage production focused on a higher percentage of elongated pieces (53.5%) over those with flake dimensions (Appendix 8.10), and a similar proportion of the tools were made on blade blanks (Table 8.22). Based on the cross-sections of the sample of debitage blades, the largest proportion of blades (ca. 66%) are 2nd and 3rd order blades; that is, at or near the end of core reduction (Appendix 8.11). Tools made on cortical pieces comprise 22.3% of the tool assemblage—1st and 2nd order primary blades and primary and secondary flakes (Table 8.22). Nevertheless, the majority of tools (63.8%) were interior products representing the middle to end of the reduction sequence—1st, 2nd, and 3rd order plain blades and plain flakes.

Dimensional data for major classes of debitage and tools provide a means of reconstructing the stages of core reduction that took place at the site (Figures 8.21, 8.22 and Tables 8.20, 8.21). A comparison of the ranges of lengths in Figure 8.21 illustrates a great deal of overlap among some groups and fairly specific and restricted ranges for others. Mean lengths and ranges of primary elements, blade debitage, and other tools made on blades are quite similar. Primary elements removed early in the reduction sequence exhibit a distribution that is essentially the same as that for interior blade debitage and the same modal interval of 30–40 mm (Figure 8.22a). They differ only slightly in mean length (Table 8.20). Other tools made on blades mirror quite accurately the debitage from which they were selected. Primary elements, however, tend to be shorter and wider than the subsequent blade debitage, displaying a general blank form that is reflected in vastly different width distributions (Figure 8.22b) and a length:width ratio of 1.3.

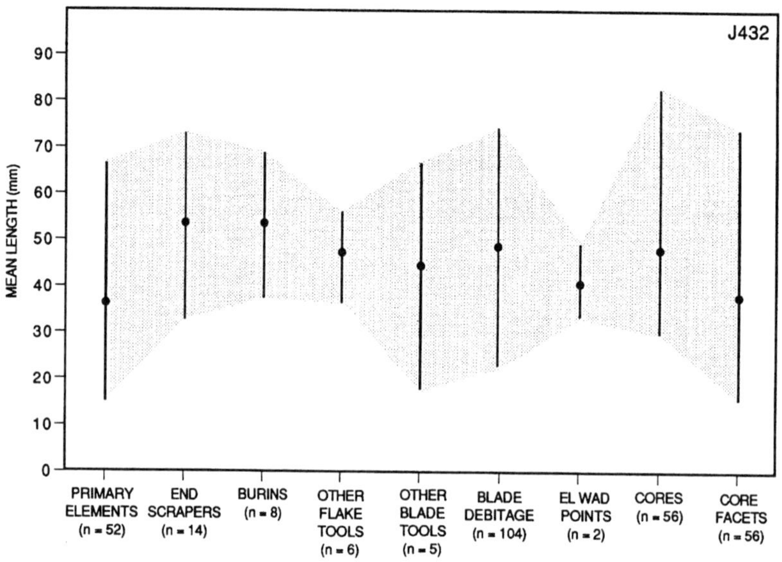


**Figure 8.21.** Mean lengths and ranges of major debitage and tool classes from Tor Aeid (J432).

These dimensions can be compared to the length:width ratios of 3.2 and 2.9 for blade debitage and other blade tools, respectively.

The metric characteristics of more specialized tools, such as endscrapers, burins, and especially el-Wad points, suggest more restricted dimensional parameters that should correlate closely with the requirements of specific tools. The most restricted sizes are those of the el-Wad points, but since the sample of complete points is so small ($n = 2$), it is difficult to generalize. Mean width and thickness of the remaining 14 fragments, however, indicate dimensions and ranges that are slightly smaller than those of the complete examples, indicating the two complete points are probably acceptable predictors for this assemblage. Endscrapers and burins are surprisingly similar in length, exhibiting similar ranges and mean lengths of approximately 54 mm (Figure 8.21, Table 8.21). Burins, however, are considerably larger in overall size than endscrapers, with a size index of 252.6 compared to 188.0 for endscrapers. Most of this difference is due to the thick blanks that characterize the burins in this assemblage. Therefore, length parameters may be similar, but endscrapers and burins represent specific and different dimensional products in core reduction.

A comparison of core dimensions and core facets with other classes of debitage and tools indicates that the final removals from the cores averaged 37.9 mm in length—smaller, on average, than the lengths of tools and most of the debitage. The fact that the range of facet lengths is as great as the range in core size might suggest that many cores were still suitable for the production of some kinds of debitage, particularly smaller bladelet blanks but incompatible with the requirements of tools with more restricted parameters.

## Typology

The tool component from Tor Aeid makes up only a small proportion of the overall assemblage (2.7%), implying tool manufacture, maintenance, and/or use were not important activities at the site (Appendices 8.10, 8.12). The tool component itself, however, is interesting and, in some ways, rather unique.

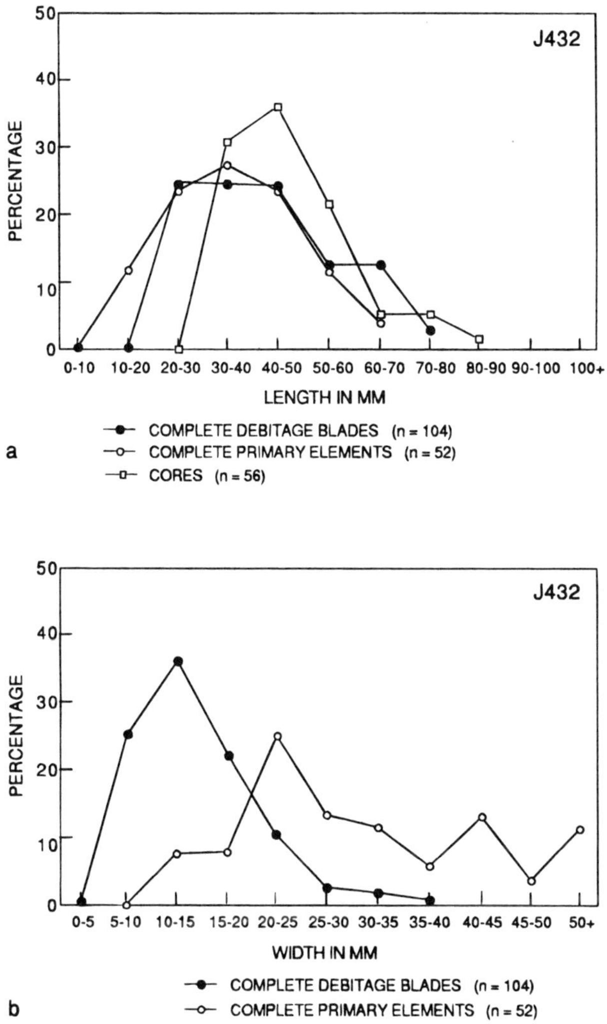


**Figure 8.22.** Length and width distributions for cores and debitage from Tor Aeid (J432).

The most diagnostic retouched pieces in this assemblage are the pointed tools. They are represented by 16 el-Wad points, of which only two are complete (Figure 8.23:b,c,d). Of the fragments, eight are proximal and six are distal, making it difficult to argue persuasively either for or against on-site provisioning of hafted toolkits, which would require relatively high frequencies of proximal point fragments. The el-Wad points occur on both flat and incurvate blanks, but rarely are the profiles twisted, in contrast to the frequency of twisted profiles in the Tor Hamar el-Wad points (see discussion of typology from Tor Hamar). Retouch characteristics of the el-Wad points are consistent in featuring semi-abrupt to fine retouch, often graded, on the right obverse sides of the points. Retouch appears to be used to shape the proximal ends, straighten the sides of medial sections, and shape the distal ends into points. Retouch to the left sides is less frequent, but when it does occur, it is used to shape the distal section. None of the points features inverse retouch, which is also uncharacteristic of the points described for Tor

Table 8.20. Dimensional Data (mm) for Cores and Complete Debitage from Tor Aeid (J432)

| A. Cores ($n$ = 56) | Length | Width[a] | Width[b] | Weight (g) | CSI[c] |
|---|---|---|---|---|---|
| Mean | 47.5 | 33.1 | 36.0 | 56.0 | 15.8 |
| sd | 11.6 | 7.9 | 6.9 | 32.3 | 5.9 |
| min | 30.5 | 13.2 | 21.8 | 20.6 | 7.4 |
| max | 82.7 | 49.7 | 51.5 | 164.4 | 36.4 |
| **B. Primary Elements ($n$ = 52)** | **Length** | **Width[a]** | **Thickness** | **L: W Ratio** | **Size[d]** |
| Mean | 36.2 | 32.3 | 10.3 | 1.3 | 161.6 |
| sd | 13.3 | 15.2 | 6.3 | .6 | 214.6 |
| min | 15.4 | 11.2 | 2.5 | .4 | 4.8 |
| max | 66.7 | 76.4 | 38.1 | 2.6 | 1334.6 |
| **C. Debitage Blades ($n$ = 104)** | **Length** | **Width[a]** | **Thickness** | **L:W Ratio** | **Size[d]** |
| Mean | 42.8 | 14.3 | 5.1 | 3.2 | 46.5 |
| sd | 13.8 | 6.1 | 3.7 | 0.8 | 59.1 |
| min | 20.8 | 4.5 | 1.4 | 2.0 | 1.7 |
| max | 73.8 | 35.2 | 24.6 | 6.1 | 267.0 |
| **D. Core Facets ($n$ = 56)** | **Length** | **Width[a]** | | | |
| Mean | 37.9 | 17.7 | | | |
| sd | 12.5 | 9.5 | | | |
| min | 16.6 | 5.6 | | | |
| max | 77.3 | 43.9 | | | |

[a]Width at midpoint.
[b]Maximum width.
[c]Core Size Index (length × width[a]/100).
[d]Size (length × width[a] × thickness/100).

Hamar. In this respect, the el-Wad points from these two south Jordan sites are different from el-Wad varieties described elsewhere that feature alternately retouched edges (e.g., Boker A [Jones *et al.* 1983:299]).

Although making comparisons based on the dimensions of a small sample of complete points from Tor Aeid is limited, some similarities and differences may be described. The dimensions of the el-Wad points at Tor Aeid provide the strongest contrast with those from Tor Hamar (Tables 8.21 and 8.16). The complete el-Wad points from Tor Aeid are shorter, having a mean length of 41.5 mm, compared to 49.8 mm for the Tor Hamar points. Overall size indices indicate that the Tor Aeid blanks are smaller (14.8), in contrast to 19.3 for those from Tor Hamar. Length:width ratios show the Tor Hamar points to be made on much more elongated blanks with a mean ratio of 4.6, while those from Tor Aeid exhibit a length:width ratio of 3.6. Comparisons with other el-Wad points described in the literature indicate the Tor Aeid examples are smaller but most similar to those identified at Lagama VII (Bar-Yosef and Phillips 1977:45) rather than the el-Wad points from K'sar Akil described by Bergman (1981) (see the edition on **Typology** for Tor Hamar).

Unexpected features of the Tor Aeid tool assemblage are the high frequencies of scrapers and burins, relative to blade and bladelet tools, which are typologically more typical of Levantine Aurignacian assemblages, as they are currently described. Whereas most of the scrapers are end-scrapers on blades (e.g., Figure 8.24:b), at least 30% were made on flakes (Figures 8.23: f,g and 8.24:a), and another 17% of them were made on core tablets or on thick blanks and classified as carinated (Appendix 8.12). Three are classified as sidescrapers.

Table 8.21. Dimensional Data (mm) of Retouched Tools from Tor Aeid (J432)

| | Length | Width | Thickness | Nose thickness | Bit width | Size[a] | L : W Ratio |
|---|---|---|---|---|---|---|---|
| Endscrapers | | | | | | | |
| Complete ($n$ = 14) | | | | | | | |
| mean | 54.1 | 30.9 | 10.5 | 8.2 | NA | 188.0 | 1.9 |
| sd | 11.7 | 7.7 | 3.7 | 4.1 | | 101.0 | 0.7 |
| min | 33.1 | 12.5 | 5.4 | 2.2 | | 35.7 | 1.2 |
| max | 73.3 | 43.1 | 16.2 | 17.2 | | 345.2 | 4.0 |
| Incomplete ($n$ = 19) | | | | | | | |
| mean | 50.1 | 25.6 | 10.7 | 8.4 | NA | 147.4 | 2.1 |
| sd | 10.8 | 7.0 | 5.3 | 5.1 | | 89.0 | 0.8 |
| min | 24.3 | 15.6 | 4.9 | 2.7 | | 22.3 | 0.9 |
| max | 66.5 | 44.5 | 19.9 | 19.5 | | 306.1 | 4.1 |
| Burins | | | | | | | |
| Complete ($n$ = 8) | | | | | | | |
| mean | 54.8 | 32.1 | 13.6 | NA | 13.4 | 252.6 | 1.8 |
| sd | 10.5 | 5.5 | 5.0 | | 6.6 | 121.2 | 0.5 |
| min | 38.2 | 19.6 | 4.8 | | 3.5 | 49.6 | 1.0 |
| max | 69.9 | 37.4 | 19.5 | | 23.6 | 379.7 | 2.6 |
| Incomplete ($n$ = 28) | | | | | | | |
| mean | 47.8 | 28.8 | 12.0 | NA | 9.9 | 178.7 | 1.8 |
| sd | 7.6 | 8.9 | 4.6 | | 4.9 | 116.0 | 0.5 |
| min | 27.8 | 17.5 | 4.1 | | 3.2 | 20.0 | 0.9 |
| max | 63.4 | 56.4 | 19.1 | | 19.4 | 472.9 | 3.3 |
| Other Tools | | | | | | | |
| Complete Blades ($n$ = 5) | | | | | | | |
| mean | 43.1 | 17.6 | 5.6 | NA | NA | 64.3 | 2.9 |
| sd | 20.2 | 8.8 | 3.5 | | | 53.2 | 1.3 |
| min | 17.8 | 3.7 | 0.7 | | | 0.4 | 1.7 |
| max | 66.8 | 25.7 | 8.8 | | | 121.2 | 4.9 |
| Complete Flakes ($n$ = 6) | | | | | | | |
| mean | 47.2 | 44.7 | 11.1 | NA | NA | 229.2 | 1.1 |
| sd | 8.0 | 12.6 | 2.8 | | | 74.3 | 0.3 |
| min | 37.6 | 35.6 | 8.5 | | | 148.8 | 0.8 |
| max | 56.2 | 69.0 | 16.2 | | | 336.6 | 1.6 |
| El Wad Points | | | | | | | |
| Complete ($n$ = 2) | | | | | | | |
| mean | 41.5 | 11.4 | 2.9 | NA | NA | 14.8 | 3.6 |
| sd | 11.1 | 2.8 | 0.7 | | | 10.4 | 0.1 |
| min | 33.6 | 9.4 | 2.3 | | | 7.4 | 3.6 |
| max | 49.4 | 13.4 | 3.4 | | | 22.2 | 3.7 |

[a]Size = length × width × thickness/100.

The burins are even more atypical of an Ahmarian assemblage (Appendix 8.12). Burins from Tor Aeid are unique among the Upper Paleolithic assemblages in the study area in that 48.9% of the burins occur on uncommon blanks, such as very thick core-like blanks, making them quite large with characteristically thick bits and multiple spall scars. Those that were made on blades (28.1%) are typically dihedral or angle burins. Burins on flakes account for the remaining 25%.

Retouched blades and bladelets comprise 16.5% of the tools (Appendix 8.12). Retouched blades consist mainly of pieces retouched unilaterally on either the obverse or inverse surfaces.

Table 8.22. Percentages of Different Reduction Stages Represented in the Retouched Tools from Tor Aeid (J432)

| | Primary blades | | Plain blades | | | Flakes | | | |
|---|---|---|---|---|---|---|---|---|---|
| | 1 | 2 Cortical | 1 | 2 | 3 | 1 | 2 Cortical | 3 | Others |
| Retouched Tools ($n$ = 145) | 6.5 | 5.9 | 16.4 | 20.4 | 5.3 | 4.6 | 5.3 | 21.7 | 13.8 |

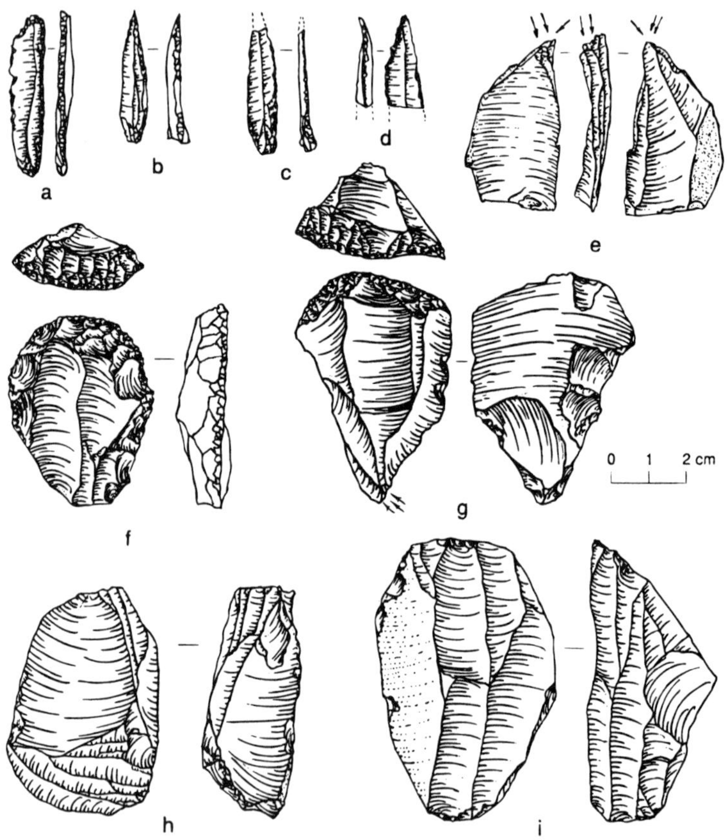


Figure 8.23. Cores and tools from Tor Aeid (J432): (a) backed bladelet, (b,c) el-Wad points, (d) fragment of a pointed blade, (e) dihedral burin, (f) endscraper on a retouched flake, (g) endscraper on a dihedral burin with notch, (h) 90-degree core, (i) opposed platform core.

One example of abrupt retouch is present in the blades (Figure 8.23:a). Retouched bladelets make up 6.2% of the tools with most bladelets exhibiting unilateral retouch on the obverse sides. Two backed pieces and one bladelet with very fine retouch were recovered. The retouch on the bladelets represent a range of retouching techniques, supporting the general position that the technology and typology represented in the Tor Aeid assemblage are somewhat unique for Early Ahmarian and Upper Paleolithic assemblages.

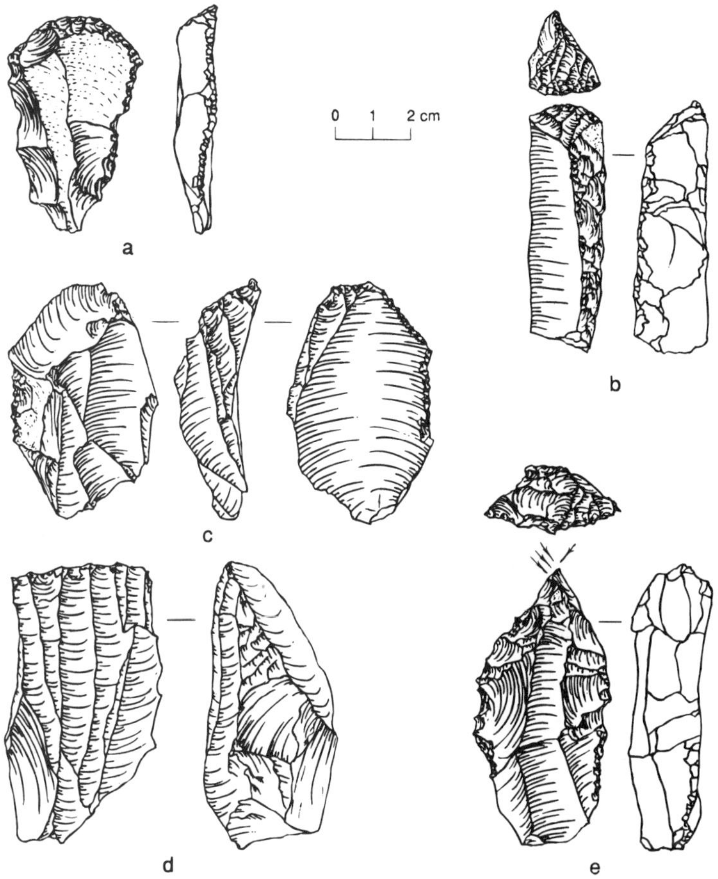


Figure 8.24. Cores and tools from Tor Aeid (J432): (a) endscraper and denticulate on a retouched piece, (b) thick endscraper on a backed blade, (c) core trimming element, (d) bladelet core, (e) thick burin on a retouched blade.

## Site J440

### Introduction

Surface collections at Site J440 provide a nonrandom sample of lithic artifacts representing an Upper Paleolithic assemblage with technological and typological features that are both similar to and different from those at other Upper Paleolithic sites in the area. In some respects, the technology is more sophisticated than that at Jebel Humeima or at Tor Fawaz, but it lacks important features that are typical of Early Ahmarian assemblages at Tor Hamar (F–G) and Tor Aeid. Some of the difficulty in associating this assemblage with other Upper Paleolithic assemblages may be attributed to the limitations of recovering a representative sample of the cultural material in the surface collection.

### Technology

Some features of core reduction and tool manufacture are not represented in the classes of lithic debitage and tools recovered from the surface of Site J440 (Appendix 8.13). Cores and primary elements are present in high proportions, but the associated debris that results from reducing cores and manufacturing or maintaining tools is absent. The debitage and tools, however, provide a fairly accurate description of some of the more general technological characteristics of the assemblage.

A wide variety of cores are present, only some of which are typical Upper Paleolithic cores. Of the seventeen cores recovered, 10 are single platform cores. Most of these have fairly simple platforms with some amount of platform regularization and abrasion and exhibit final facets suggesting bladelet production. Three cores have opposed platforms. Also present are very small cores that are greatly reduced with irregular platform edges and step fractures. One large core retains evidence for a variety of reduction strategies. In its final form, it has opposed platforms with large facets and irregular, scalloped platform edges. Prior to this orientation, it featured a simple platform with a regularized edge which was then re-oriented with a second platform at 90 degrees to the original face. In addition to these cores, one very small core is a thick piece of debitage that has been worked as a discoidal core. Two single platform cores exhibit large facets on their platforms, but otherwise resemble blade and bladelet cores. Five of the cores exhibit extremely fine waxy texture, suggesting the possibility of heat treatment or alteration. In three of these, the limestone cortex is also altered and appears harder and less porous with a slightly bluish coloration.

The core trimming elements indicate that platform rejuvenation was accomplished most often using the core tablet technique, but evidence for other methods of trimming cores in general is also present. One example originated from a change-of-orientation core and combines a small core tablet with a large portion of another platform and core face.

The largest proportion of the debitage consists of blades and bladelets with unfaceted platforms, ranging from small and narrow to fairly large and triangular in shape. Crushed platforms are present but less frequent. Blade and bladelet platforms are both lipped and unlipped. Blades are generally narrow and thin and indicate a fairly well-developed blade technology, corresponding to a number of the finer blade cores. Bladelets are not small and delicate but tend toward the larger size ranges or the small blade category.

Flake debitage is generally large and thick with sizable platforms, some of which have lipped platforms. Small trimming flakes are infrequent.

## Typology

Most of the classes of tools found at Upper Paleolithic sites are represented in the tool assemblage at J440 (Appendix 8.14). Perforators are represented by a single example on a multiple use tool. Scrapers, however, are less numerous than expected, and burins make up the largest class, followed by a relatively large group of retouched flakes. Pointed tools, such as el-Wad points, are absent.

The most distinctive tools in this surface collection are the burins. They occur most frequently on thick, core-like blanks, making them quite large, bulky, and rather unusual tools (e.g., Figure 8.25:a,b,e). Burins often are formed on an old surface or a snap, with a few made on truncations. One example exhibits spalls that twist around the bit, which subsequently was truncated (Figure 8.25:b). Most burins exhibit multiple burin scars and, because the blanks are bulky, the bits tend to be wide.

Endscrapers on flakes are more typical than scrapers on blades (Figure 8.25:d) with one example of a large bilateral sidescraper. Retouched blades and bladelets lack any degree of standardized retouch, consisting for the most part of limited inverse and obverse retouch. Retouched flakes are more numerous but quite heterogeneous and suggest a rather informal group of tools and fragments.

In summary, the evidence from the cores, debitage, and the tools is inconclusive in determining an overall technology or industrial affiliation for this assemblage. It appears to be dis-

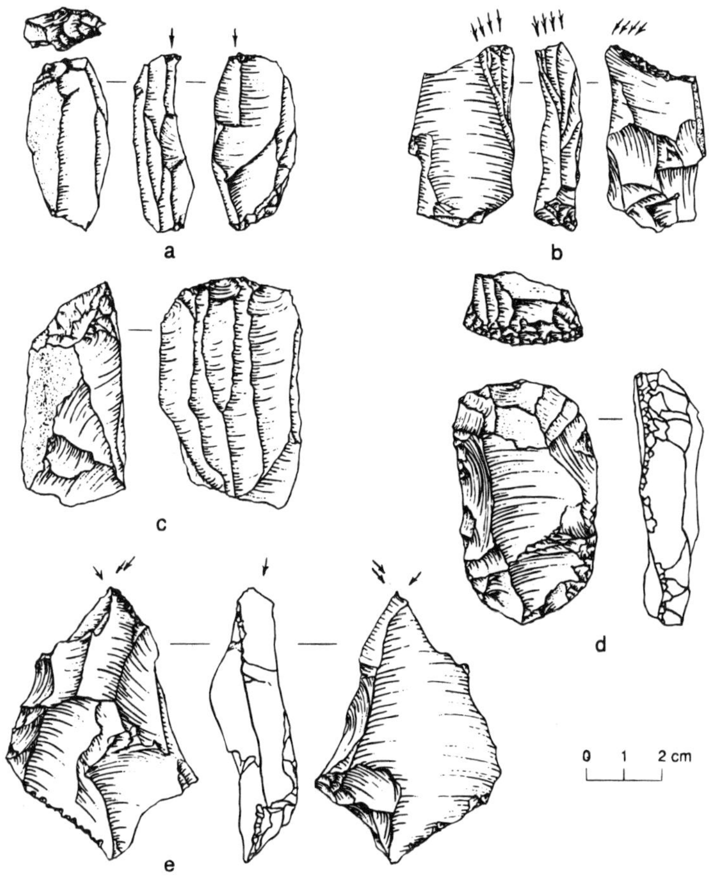


Figure 8.25. Cores and tools from Site J440: (a) burin on a previous endscraper, (b) burin on a truncation and core tablet, (c) single platform bladelet core, (d) thick endscraper, (e) dihedral burin.

tinctly Upper Paleolithic, but bears similarities to both Ahmarian and Levantine Aurignacian assemblages in the study area. The cores are mixed, with examples of blade/bladelet cores as well as more atypical examples, some with faceted platforms. The debitage, on the other hand, contains elements that suggest a fairly well-developed and sophisticated blade technology focused on blades rather than bladelets, making the assemblage similar to the Early Ahmarian technologies at Tor Hamar and Tor Aeid. Pointed tools, however, are not present and blade tools are less frequent than large tools on thick blanks and flake tools, a feature we might expect in a less developed blade technology characteristic of Levantine Aurignacian assemblages. Instead, the tool assemblage is dominated by a somewhat heterogeneous group of bulky burins similar to those from Tor Aeid (J432). It is possible to argue that the surface collection has merely sampled a functionally distinct area of the site organized around burin tools, although the function is itself difficult to define on the basis of the variety of burins. It is likely that the surface collection is unrepresentative of the range of activities of the site, and the high frequency of burins represents an overly broad, poorly-defined class of tools.

## Wadi Aghar (J433)

### Introduction

The lithic artifacts from Wadi Aghar (J433) exhibit a number of characteristics that suggest the assemblage represents an early Upper Paleolithic assemblage that is transitional between local Middle and Upper Paleolithic technologies. The geomorphological context of this assemblage in a red sandy silt, similar to Middle Paleolithic cultural deposits at Tor Faraj (J430), provides some stratigraphic support for the early Upper Paleolithic timing of these cultural deposits.

Three test excavation units produced relatively sparse cultural materials (Appendix 8.15) with little indication of an *in situ* technological transition. Comparisons with other Upper Paleolithic assemblages in the study area revealed the Wadi Aghar material to be different, exhibiting distinct as well as mixed technological attributes.

The test excavations reflect limited reduction activity at the site and suggest an occupation of short duration, but because testing was restricted to three test units, the results may not be representative of potentially different occupation periods and site activities. Sample size is small, and some patterns and conclusions may be strongly affected by small sample sizes. Nevertheless, it is apparent that this assemblage is unique among the South Jordan sites and provides an initial set of impressions about early Upper Paleolithic technology in this area of the Levant.

### Raw Material

The sixteen cores recovered at Wadi Aghar reflect a predominant selection of rolled cobbles with gravel cortex and limited use of limestone nodules (18.7%) as raw material (Table 8.23).

### Technology

The technology of the Wadi Aghar assemblage centered on the production of relatively large debitage. Cores and primary elements are larger than those from Upper Paleolithic assemblages in the study area. Dorsal scar patterning and platform faceting are indicative of some

**Table 8.23. Classification of Cores by Raw Material and Core Class from Wadi Aghar (J433)**

| | Unit 1 | Unit 2 | Unit 3 | Feature 1 | Total *n* | Total % |
|---|---|---|---|---|---|---|
| A. Raw Material | | | | | | |
| Rolled Cobble | 2 | 4 | 3 | 1 | 10 | 62.5 |
| Limestone Cobble | 2 | | 1 | | 3 | 18.7 |
| Indeterminate | | 1 | 1 | 1 | 3 | 18.7 |
| Total | 4 | 5 | 5 | 2 | 16 | 99.9 |
| B. Core Class | | | | | | |
| Single Platform | 3 | 4 | 2 | | 9 | 56.2 |
| Opposed Platforms | | | | | 2 | 12.5 |
| Opposite | | | 1 | 1 | | |
| Same | | | | | | |
| Twisted | | | | | | |
| Change of Orientation | 1 | 2 | | | 3 | 18.8 |
| Others | | | | | 2 | 12.5 |
| Discoidal | | | | 1 | | |
| Other | | 1 | | | | |
| Total | 4 | 7 | 3 | 2 | 16 | 100.0 |

late Middle Paleolithic characteristics and distinguish a considerable portion of the assemblage. A mixture of Upper and Middle Paleolithic attributes co-occur in the remainder of the assemblage, reflecting techniques of core preparation, reduction, and tool making that are atypical of the Upper Paleolithic.

Given an overall assemblage numbering only 325 pieces, cores are relatively abundant, yielding a core:debitage ratio of 1:10.9 (Appendix 8.15). Single platform cores make up 56% of the cores, while only two examples of opposed platforms are present. The cores, however, do not represent a well-defined, consistent core reduction strategy. Rather, there are indications of a number of strategies, as well as some mixture of strategies. Between the cores and the debitage, however, a somewhat dominant core reduction strategy can be characterized.

In general, there is little formal preparation of cores, and limited evidence for intensive use of the cores themselves. Cores vary in size, including small, rather informal thick flakes or blades, moderately reduced cobbles, and some very large cores exhibiting a limited number of facets or scars (Figures 8.27:f, 8.28:a, 8.29:a). Mean core length is 60.6 mm, ranging from 44 mm to 90.4 mm with weights ranging from 34.1 g to 335.7 g (Table 8.26). Some of the largest cores provide the most direct information on core reduction, exemplifying a number of rather atypical characteristics of the technology. These can be matched with corresponding features in the debitage and tools.

Core platforms include both single, unfaceted platforms and a number with two or more facets, but in general, the platform facets are quite large and somewhat irregular. The most conspicuous feature of the cores is the absence of platform edge regularization. Only one core, a small sub-pyramidal blade/bladelet core, exhibits any edge abrasion, and this core is atypical of this assemblage (Figure 8.28:a). The remainder of the cores have irregular, scalloped edges along the platform from which debitage with similarly large, scalloped platforms was removed (Figure 8.29:a). While there is ample evidence for relatively simple platforms with single or dihedral faceting on debitage, it also is evident that more complex multifaceted platforms comprise an important subgroup within the debitage. An examination of all platforms present on debitage and tools indicates that 63% of the platforms are single, while double and multifaceted platforms are present on 26.1% of the pieces (Table 8.24). These are relative proportions not present in any of the other Upper Paleolithic assemblages. Within the group with unfaceted platforms, the majority are large platforms with conspicuous scalloped edges, although some small, narrow platforms are also present. Dihedral platforms are generally unprepared, whereas the examples of multifaceted platforms range from three facets to heavy faceting and edge preparation. The latter do not have counterparts in the platforms of cores recovered from the site.

Strategies for platform rejuvenation are not well defined or standardized, although seven core trimming elements have been identified. Only one is a clear illustration of core alteration, which may have been unintentional. Originating from the distal end of the core face, the core

**Table 8.24. Types of Platforms on Debitage and Tools from Wadi Aghar (J433)**

| | Platform Types | | | | | |
|---|---|---|---|---|---|---|
| | Single | Double | Multiple | Crushed | Cortical | Total |
| Debitage | | | | | | |
| Blades/Bladelets | 52 | 4 | 9 | 5 | 3 | 73 |
| Flakes | 16 | 5 | 7 | 1 | — | 29 |
| Tools | 7 | — | 6 | 4 | — | 17 |
| Total | 75 | 9 | 22 | 10 | 3 | 119 |
| % of Total | 63.0 | 7.6 | 18.5 | 8.4 | 2.5 | 100.0 |

trimming element removed a large portion of the front of a core having a scalloped platform edge. The remaining core trimming elements are irregular, four of which are most similar to core tablets but lack the classic characteristics of tablets associated with blade cores. Instead, the platform removals are irregular in shape with limited platform landmarks but tablet-like thicknesses.

### Reduction Mechanics

The proportions of large, unlipped platforms on both primary removals and subsequent interior debitage blades implies a heavy reliance on hard hammer percussion in core reduction (Appendix 8.16). Indications that a limited amount of soft hammer percussion was used in primary and secondary reduction is apparent by the less frequent occurrence of lipped platforms.

Loading or delivery errors seem to have been restricted to underpassed or hinged terminations and an unusually high frequency of stepped distal profiles on primary elements (Appendix 8.16). Most of the debitage exhibit feathered terminations. Primary elements tend to have curved profiles with only a limited number that are twisted or convex. The lateral profiles of interior blade debitage, however, are quite variable, the largest proportion being incurvate.

### Reduction Activity

The composition of the assemblage from Wadi Aghar suggests that a limited amount of core reduction and tool making took place on site. Debris resulting from core reduction and tool manufacturing and maintenance is relatively low and scarce in absolute numbers as well (Appendix 8.15). Debitage ratios are low and retouched pieces limited to 27 tools. Although the sample is small, production goals appear to have favored more elongated blanks with blade proportions since the flake:blade ratio is 1:1.8 (Appendix 8.15), and the mean length:width ratio in the debitage blades is 3.1, a ratio that is high relative to the size of the blanks (see Table 8.26). The debitage and tool assemblages exhibit dimensional characteristics that illustrate the unique character of the overall assemblage (Tables 8.26, 8.27, and 8.28).

Complete primary elements are significantly larger than their counterparts in other assemblages. Mean length is 58.5 mm, and the overall size index of primary blanks is 260.8 (Table 8.26)—more than twice as large as most primary elements in other Upper Paleolithic assemblages. A comparison of length distributions for primary elements with other Upper Paleolithic assemblages illustrates the differences (Figure 8.26a). Similarly, the interior debitage blades from the J433 assemblage are longer than those from other assemblages. Mean length is 55.2 mm, but overall size and elongation indicates they are long but relatively narrow and thin when compared to primary elements and debitage blades from other assemblages (Table 8.26). A comparison of debitage blades in Figure 8.26b illustrates their atypical length. In the other assemblages, modal intervals for length of blades range from 20–50 mm, while the modal interval for the J433 blades extends from 50–70 mm.

The exceptional length of the debitage blades appears to be associated with other technological attributes. There is a general trend toward elongated blanks that are pointed with unidirectional converging dorsal scar patterns, even though the overall percentage of converging dorsal scar patterns for debitage blades is only about 14% (see Appendix 8.16). In tools and debitage combined, there are 16 complete elongated pointed blades with converging dorsal facets. A two-way table of platform and dorsal scar attributes illustrates the associated characteristics (Table 8.25). Single, unfaceted platforms are typical of most of the pointed blades (81.3%), the majority of which have large, scalloped platform edges. They co-occur most frequently on blanks with unidirectional converging dorsal facets. Blades of this type would have been the products of large subpyramidal cores with converging facets, such as the core illustrated in Figure 8.29a. Only one of the pointed blades exhibits a multifaceted platform, but

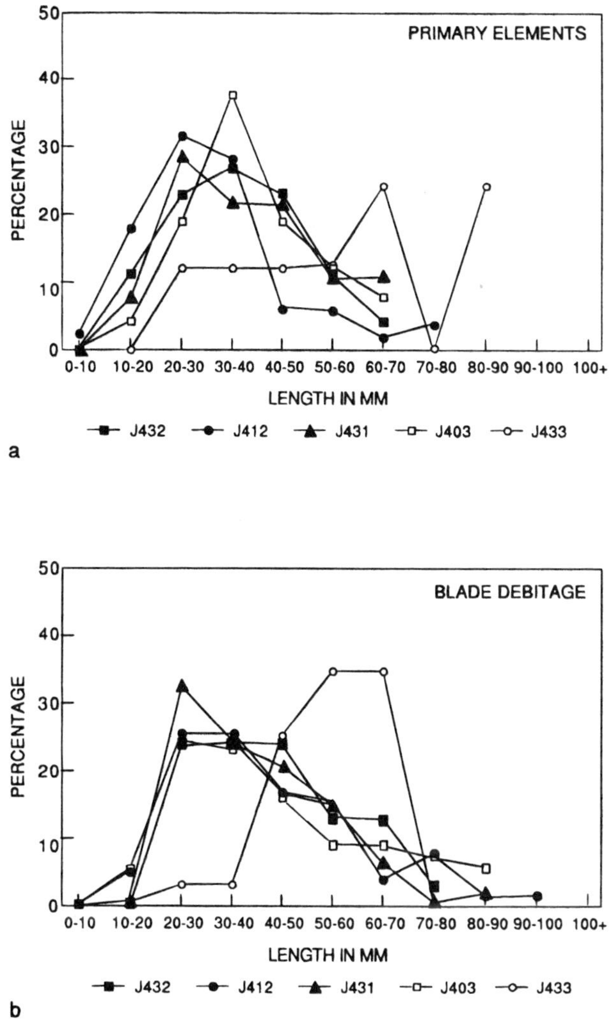


**Figure 8.26.** Comparisons of length distributions for primary elements and blade debitage from Wadi Aghar (J433) and other Upper Paleolithic assemblages.

rather than occurring on a blank with bidirectional facets, it also has unidirectional converging facets. The two-way table establishes a strong association between converging facets and single platforms, but at the same time, it illustrates the general lack of technological consistency in debitage characteristics.

As a subpopulation which includes both tools and debitage, the pointed blades exhibit dimensional characteristics that suggest they are only slightly longer than other blades but represent a discrete sub-set of blades. This is shown in Table 8.27 by the more restricted ranges for all dimensions, greater elongation, and larger overall blank size when they are compared to primary elements and the general population of blade debitage (compare Tables 8.26 and 8.27).

The dimensions of retouched tools provide additional evidence for an emphasis on long blanks in this assemblage (Table 8.28). Complete endscrapers and burins exhibit mean lengths

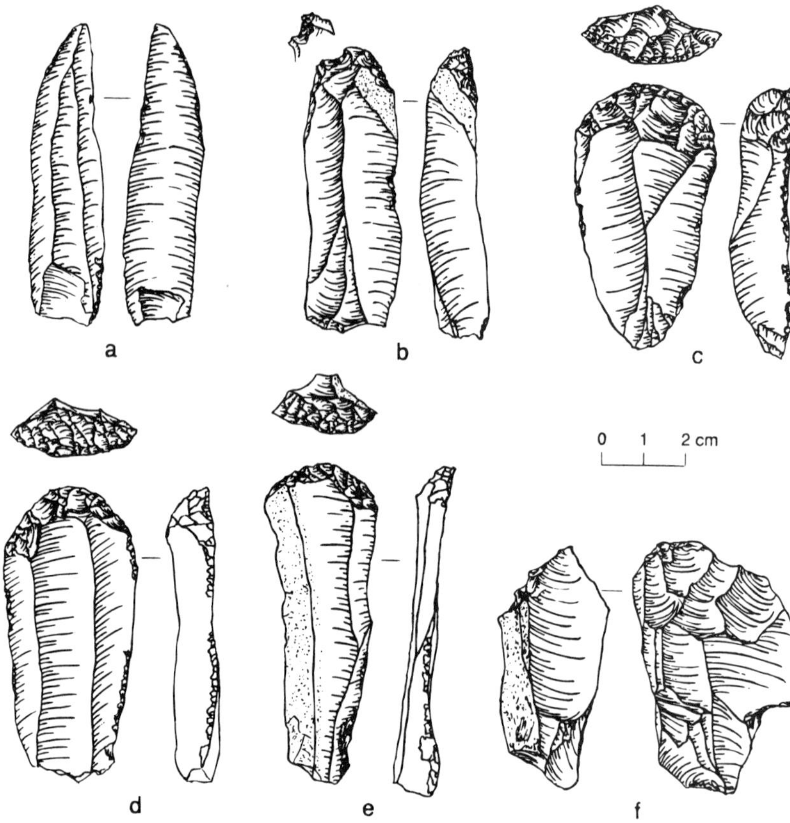


Figure 8.27. Cores and tools from Wadi Aghar (J433): (a) elongated point with retouch, (b) endscraper with notch, (c) endscraper, (d,e) endscrapers on retouched blades, (f) burin core.

over 50 mm, which is not unusual for Upper Paleolithic tools in these classes, but the mean length:width ratios of all tools from Wadi Aghar are greater. In contrast, the incomplete tools in this assemblage make up an unusually bulky group of tools, being larger in all respects than the complete tools (Table 8.28). In general, the length:width ratios of the retouched tools indicate a definite selection for elongated blanks with all ratios over 2.0.

## Typology

Representing 8.3% of the assemblage, retouched tools are relatively frequent but lack the diversity normally represented in Upper Paleolithic tool assemblages (Appendix 8.17). Endscrapers make up the largest group (40.7%) with examples of well-executed endscrapers on plain and retouched blades (e.g., Figure 8.27:d,e). Nose thickness is quite variable (see Table 8.28), ranging from relatively flat to extremely thick.

Burins are less frequent, occurring as angle burins on old surfaces, as multiple mixed burins, or burins on multiple use tools with endscrapers. Bit widths are not particularly thick on the burins in this assemblage, with overall size smaller than burins in the other assemblages (Table 8.28). This is due to the lack of large, thick, bulky burins, which are typical of some of the Upper Paleolithic assemblages.

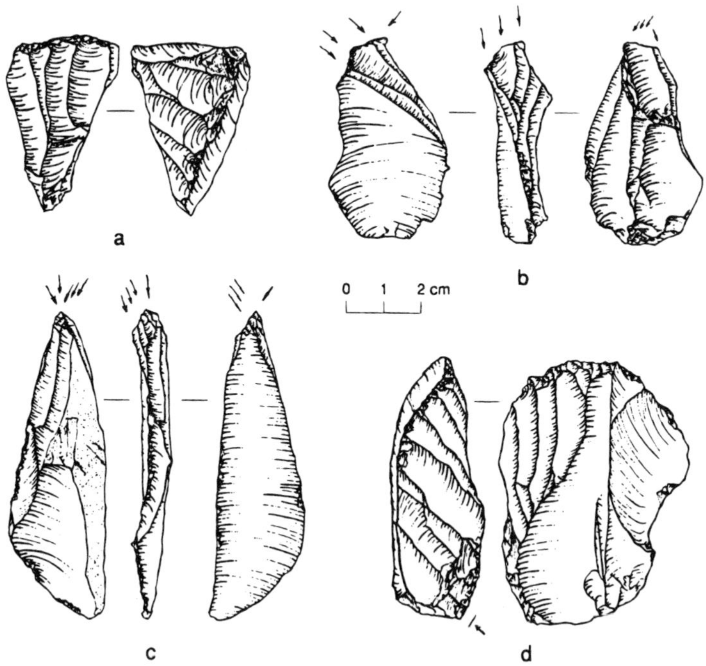


**Figure 8.28.** Cores and tools from Wadi Aghar (J433): (a) twisted opposed platform core, (b) burin *plan*, (c) dihedral burin on the proximal end, (d) core-scraper.

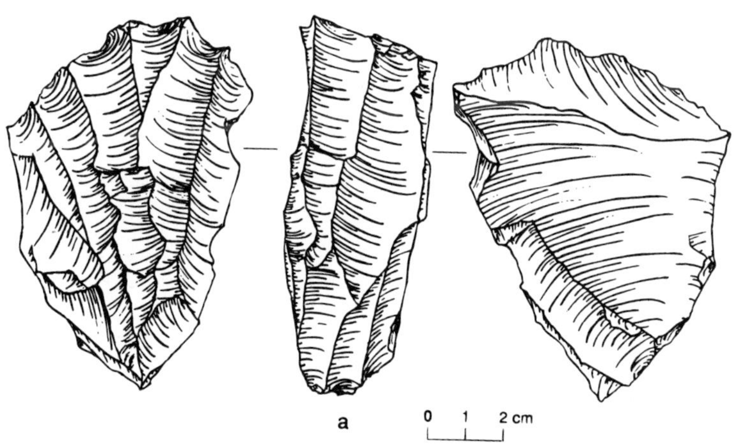


**Figure 8.29.** Core from Wadi Aghar (J433): (a) twisted opposed platform core.

Retouched blades and notches are unremarkable. Bladelets are infrequent in the debitage and absent from the retouched tool assemblage. Pointed blades were not generally retouched, and only one example is present in the retouched tools (Figure 8.27:a). It exhibits alternating retouch to the left distal edges, as well as obverse retouch on the right proximal section.

Table 8.25. Two-Way Table for Platform Type and Dorsal Scar Patterns from Wadi Aghar (J433)

| Platform Type | | Dorsal Facets: Unidirectional Parallel | Bidirectional Opposed | Unidirectional Converging | Row Total |
|---|---|---|---|---|---|
| Indeterminate | *n* | 1 | | | 1 |
| | % | 100.0 | | | 6.3 |
| Single | *n* | 3 | 2 | 8 | 13 |
| | % | 23.1 | 15.4 | 61.5 | 81.3 |
| Multifaceted | *n* | | | 1 | 1 |
| | % | | | 100.0 | 6.3 |
| Crushed | *n* | | 1 | | 1 |
| | % | | 100.0 | | 6.3 |
| Column Total | *n* | 4 | 3 | 9 | 16 |
| | % | 25.0 | 18.8 | 56.3 | 100.0 |

Table 8.26. Dimensional Data (mm) for Cores and Complete Debitage from Wadi Aghar (J433)

| A. Cores (*n* = 16) | Length | Width[a] | Width[b] | Weight (g) | CSI[c] |
|---|---|---|---|---|---|
| Mean | 60.6 | 48.8 | 52.8 | 103.9 | 31.2 |
| sd | 14.5 | 17.7 | 16.9 | 75.1 | 18.5 |
| min | 44.0 | 25.7 | 28.7 | 34.1 | 11.5 |
| max | 90.4 | 99.8 | 100.4 | 335.7 | 85.1 |
| **B. Primary Elements (*n* = 8)** | **Length** | **Width[a]** | **Thickness** | **L: W Ratio** | **Size[d]** |
| Mean | 58.5 | 34.0 | 12.0 | 1.9 | 260.8 |
| sd | 20.5 | 14.2 | 4.7 | 0.9 | 156.1 |
| min | 29.8 | 20.5 | 5.9 | 0.7 | 36.3 |
| max | 85.5 | 64.4 | 18.5 | 3.4 | 457.8 |
| **C. Debitage Blades (*n* = 32)** | **Length** | **Width[a]** | **Thickness** | **L:W Ratio** | **Size[d]** |
| Mean | 55.2 | 18.7 | 6.2 | 3.1 | 67.4 |
| sd | 9.5 | 4.7 | 2.4 | 0.7 | 36.8 |
| min | 29.5 | 9.4 | 2.7 | 1.9 | 11.3 |
| max | 69.1 | 28.7 | 13.5 | 4.7 | 163.1 |
| **D. Core Facets (*n* = 16)** | **Length** | **Width[a]** | | | |
| Mean | 40.7 | 21.7 | | | |
| sd | 14.4 | 8.9 | | | |
| min | 20.1 | 12.0 | | | |
| max | 69.1 | 48.9 | | | |

[a]Width at midpoint.
[b]Maximum width.
[c]Core Size Index (length × width[a]/100).
[d]Size (length × width[a] × thickness/100).

Table 8.27. Dimensional Data (mm) for Pointed Blades from Wadi Aghar (J433)

| | Length | Width | Thickness | L:W Ratio | Size[a] |
|---|---|---|---|---|---|
| (*n* = 16) | | | | | |
| mean | 59.2 | 17.9 | 5.7 | 3.4 | 60.6 |
| sd | 8.7 | 2.8 | 1.8 | 0.7 | 24.8 |
| min | 42.4 | 14.0 | 2.7 | 2.1 | 23.8 |
| max | 71.3 | 22.6 | 9.0 | 4.4 | 112.0 |

[a]Size = length × width × thickness/100.

Table 8.28. Dimensional Data (mm) for Retouched Tools from Wadi Aghar (J433)

| | Length | Width | Thickness | Nose thickness | Bit width | Size[a] | L : W Ratio |
|---|---|---|---|---|---|---|---|
| Endscrapers[b] | | | | | | | |
| Complete (*n* = 4) | | | | | | | |
| mean | 57.8 | 23.7 | 8.1 | 7.2 | NA | 125.0 | 2.8 |
| sd | 18.6 | 11.8 | .5 | 1.6 | | 75.5 | 0.9 |
| min | 30.3 | 7.8 | 7.5 | 5.6 | | 17.8 | 2.0 |
| max | 69.2 | 33.5 | 8.8 | 9.1 | | 191.3 | 3.9 |
| Incomplete (*n* = 9) | | | | | | | |
| mean | 63.8 | 30.4 | 10.7 | 10.1 | NA | 240.0 | 2.4 |
| sd | 9.5 | 11.9 | 5.4 | 5.4 | | 238.8 | 0.9 |
| min | 53.2 | 17.3 | 5.5 | 5.5 | | 91.0 | 1.4 |
| max | 75.7 | 50.7 | 23.6 | 23.0 | | 821.5 | 3.6 |
| Burins | | | | | | | |
| Complete (*n* = 3) | | | | | | | |
| mean | 50.5 | 25.0 | 8.9 | NA | 8.4 | 112.5 | 2.1 |
| sd | 7.9 | 2.7 | 1.1 | | 5.4 | 23.2 | 0.5 |
| min | 41.7 | 22.0 | 7.7 | | 3.2 | 87.9 | 1.5 |
| max | 56.9 | 27.2 | 9.8 | | 13.9 | 134.1 | 2.6 |
| Incomplete (*n* = 3) | | | | | | | |
| mean | 68.0 | 30.6 | 10.1 | NA | 9.4 | 225.0 | 2.4 |
| sd | 12.0 | 11.2 | 2.8 | | 1.0 | 154.4 | 0.8 |
| min | 55.2 | 24.2 | 8.1 | | 8.6 | 117.8 | 1.6 |
| max | 79.1 | 43.6 | 13.2 | | 10.6 | 401.9 | 3.3 |
| Other Tools | | | | | | | |
| Complete Blades (*n* = 6) | | | | | | | |
| mean | 72.7 | 28.6 | 9.1 | NA | NA | 187.7 | 2.7 |
| sd | 10.6 | 10.4 | 2.0 | | | 79.6 | 0.7 |
| min | 62.9 | 19.1 | 7.3 | | | 95.2 | 1.9 |
| max | 92.5 | 47.8 | 12.4 | | | 329.5 | 3.9 |

[a]Size = length × width × thickness/100.
[b]Includes multiple use tools.

Other than the single pointed blade that is retouched, the tool assemblage is typologically Upper Paleolithic. It does, however, lack the diversity of Upper Paleolithic tool forms, as well as retouched bladelets and the characteristic el-Wad points of Early Ahmarian assemblages in the southern Levant.

## Comments

Evaluation of this assemblage for the purpose of establishing industrial affiliation suggests that it shares limited technological and typological attributes with other Upper Paleolithic assemblages. The fact that it includes what appear to be residual Middle Paleolithic technological features implies that it is transitional between the Middle and Upper Paleolithic or very early Upper Paleolithic. Comparisons with the transitional assemblage from Boker Tachtit (Marks 1983b) and the early Upper Paleolithic assemblages from K'sar Akil (Ohnuma and Bergman 1990) provide a basis of comparison with which to place the assemblage temporally and technologically.

The uppermost level at Boker Tachtit (Level 4) culminates in a technological shift from a Levallois-based technology to a single platform blade-core technology (Marks 1983b). Some of the important features of this technological shift include an increase in unidirectional removals, a decrease in core preparation, points that are morphologically Levallois but technologically are not Levallois, and a mixture of faceted and unfaceted platforms (Marks 1983b:74–77).

Many of these are features that are documented in the cores and debitage at Wadi Aghar—features which are not typical of later Upper Paleolithic industries. The significant difference between the Boker Tachtit, Level 4 assemblage and Wadi Aghar pertains to platform features on the debitage. The Wadi Aghar debitage exhibits a dominance of unfaceted platforms—hallmarks of Upper Paleolithic industries. Metric comparisons of the Wadi Aghar pointed blades with blade measurements from Boker Tachtit show that the transitional blades from Level 4 (Marks 1983b:Table 2.3) are longer and larger than those in the Wadi Aghar assemblage (see Table 8.27). When the Wadi Aghar lithics were compared to Upper Paleolithic assemblages, the overall larger dimensions of the Wadi Aghar assemblage were significant, emphasizing the *relatively* smaller lithic elements that typify the Upper Paleolithic sites in this study area.

Early Upper Paleolithic levels from K'sar Akil in northern Lebanon provide additional early Upper Paleolithic assemblages for comparison. Two subphases have been identified as Transitional Upper Paleolithic at K'sar Akil: Levels XXV–XXIV, and Levels XXIII–XXI/XX (Azoury 1986:34; Ohnuma and Bergman 1990:133). Levels XXIII–XXII are similar in many respects to Level 4 at Boker Tachtit in marking a shift to a single platform blade-core technology. They describe the technology of Levels XXIII–XXI/XX as a non-Levallois technology in which there is a prevalence of non-Levallois *prismatic* single platform cores and a primary focus on the production of blades (Ohnuma and Bergman 1990:111–112,114). Large, multifaceted platforms characterize the majority of debitage throughout Levels XXIII–XXI/XX, and many of the blades are morphologically identical to Levallois points. Like Level 4 at Boker Tachtit, these levels represent a transition out of the Middle Paleolithic. Blade technology in the subsequent early Upper Paleolithic levels (XX/XIX–XVI) was focused on cores that continue to be prismatic but feature opposed striking platforms. The "Levallois" points that were common in the earlier levels decrease and are replaced by K'sar Akil points (Ohnuma and Bergman 1990:110–112).

Clearly the descriptions for the transitional and early Upper Paleolithic assemblages from K'sar Akil bear only limited similarities to the Wadi Aghar lithics. The later Upper Paleolithic assemblages from K'sar Akil (XX/XIX–XVI) are more similar with an emphasis on blade technology featuring plain platforms and pointed tools. The correspondence to K'sar Akil blade technology and that evidenced at Wadi Aghar, however, is only superficially similar. Plain platforms on K'sar Akil blades are generally quite small and feature platform abrasion, while those from Wadi Aghar are large and lack platform edge regularization. K'sar Akil points are distinguished by retouch used to shape the blades into points, and the majority of points from Level XVII are 40–55 mm in length and 13–14 mm in width (Bergman 1981:324). The elongated points from Wadi Aghar feature large, unabraded platforms, retouch on only one example, and larger dimensions than the K'sar Akil points (see Table 8.27). On the other hand, the Wadi Aghar assemblage is similar to transitional or early Upper Paleolithic assemblages from Boker Tachtit and K'sar Akil in that it is typologically Upper Paleolithic in its tool types, although less diverse.

The conclusion that may be drawn from these comparisons is that the Wadi Aghar assemblage represents a technological stage that succeeds the very early Upper Paleolithic at Boker Tachtit, Level 4 and K'sar Akil, Levels XXIII–XXI/XX. The Wadi Aghar lithics exhibit important typological and technological attributes that are distinctly early Upper Paleolithic but different from both Boker Tachtit or K'sar Akil. And yet the assemblage is dissimilar to the Early Ahmarian and Levantine Aurignacian assemblages. The Wadi Aghar assemblage appears to represent a technological stage between the local Levantine Mousterian and subsequent Upper Paleolithic assemblages in the south Jordan study area. Marks (1983b:83) has suggested that the assemblage from Tor Sabiha in south Jordan represents a temporal bridge between the Early Levantine Mousterian in the Negev and the later transitional sequence at Boker Tachtit. There are strong morphological similarities in the mix of elongated points with both faceted and unfaceted platforms from Tor Sabiha (J8), Tor Faraj (J430), and Wadi Aghar, especially those from Tor Sabiha featuring plain, unfaceted platforms. The technological similarities between the Wadi Aghar points and those from Tor Sabiha and Tor Faraj suggest the possibility of an *in situ* technological evolution and continuity between the local Levantine Mousterian and the Upper Paleolithic in southern Jordan. The technological stage represented at Wadi Aghar needs to be defined in greater detail using a larger sample, and to be placed within the context of the larger Levant.

## DISCUSSION

No radiometric dates are available for the Upper Paleolithic assemblages in the Jebel Qalkha study area that would establish a temporal order among sites. Technologically and typologically, the majority of the assemblages fit within recently-defined Upper Paleolithic traditions. The current model explaining assemblage variability in the Upper Paleolithic suggests two assemblage types—the Levantine Aurignacian and the Ahmarian—with the latter incorporating an early and a late stage (Gilead 1981, 1991; Marks 1981; Marks and Ferring 1988). Similarity of the South Jordan Upper Paleolithic assemblages to these defined assemblage types, however, is variable. Some variation may reflect regional differences since the nature of the Jordanian Upper Paleolithic is only now being defined.

Four of the six assemblages fall into two distinct groups on the basis of differences in technology and typology:

(1) Early Ahmarian—Tor Hamar (F–G) (J431) and Tor Aeid (J432),
(2) Levantine Aurignacian—Jebel Humeima (J412) and Tor Fawaz (J403).

Between the two groups, differences in overall assemblage composition associated primarily with typology are illustrated in Figure 8.30, in which differences in the relative frequencies of tool classes clearly define some of the most salient dissimilarities between Early Ahmarian (Figure 8.30a) and Levantine Aurignacian (Figure 8.30b) assemblages. Early Ahmarian assemblages are characterized by pointed tools that are absent in the Levantine Aurignacian.

The surface collection from Site J440 in the Wadi Aghar shares resemblances in technology and typology with both the Levantine Aurignacian and Early Ahmarian assemblages in the Jebel Qalkha study area. The blade technology is more sophisticated than that found in the Levantine Aurignacian, but the assemblage lacks typical features found in the Early Ahmarian assemblages. Since the assemblage consists solely of surface collections, a representative sample of technological and typological features may be lacking. Although the assemblage is distinctly Upper Paleolithic, a more specific affiliation is not possible on the basis of the surface collection.

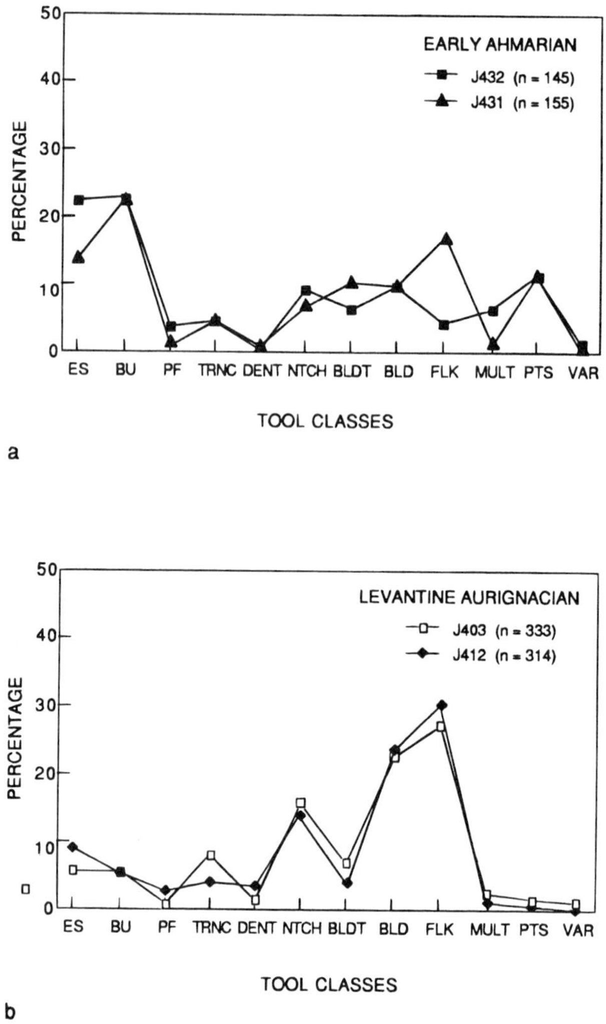


Figure 8.30. Percentage comparisons of tool classes between Early Ahmarian and Levantine Aurignacian assemblages.

The Wadi Aghar site of J433 is unique among the assemblages in this study, exhibiting both Middle and Upper Paleolithic features. Since it is also unlike the known Initial Upper Paleolithic assemblage from Boker Tachtit (Marks 1977, 1983a,b) and K'sar Akil (Ohnuma and Bergman 1990), it appears to document a somewhat more evolved stage in the transition from the local Mousterian to the Early Ahmarian in south Jordan.

## The Early Upper Paleolithic

### Wadi Aghar (J433)

The Wadi Aghar site of J433 is unusual among the assemblages in this study because its lithic artifacts exhibit a number of technological characteristics found in transitional and early Upper Paleolithic assemblages at Boker Tachtit (Level 4) in the Central Negev, K'sar Akil (Levels

XXIII–XII) in northern Lebanon, as well as features reminiscent of local Mousterian assemblages at Tor Faraj (J430) and Tor Sabiha (J8). And yet, the Wadi Aghar debitage and tools feature distinctly Upper Paleolithic attributes that separate them from the more transitional technologies. Although the tool assemblage from Wadi Aghar lacks the diversity of most Upper Paleolithic assemblages, the tools that are present would fit easily into any Upper Paleolithic assemblage. It does lack important tool types that would identify it with either the Early Ahmarian or the Levantine Aurignacian. On the basis of somewhat mixed technological features and the lack of Upper Paleolithic tool types, the Wadi Aghar appears to represent a technological stage that succeeds local Middle Paleolithic assemblages and other Levantine Middle/Upper Paleolithic transitional assemblages. Locally, it appears to represent an *in situ* technological evolution and reflects an intermediate stage between the Levantine Mousterian at Tor Sabiha and Tor Faraj and the Early Ahmarian at Tor Hamar and Tor Aeid.

## The Early Ahmarian

### Tor Hamar (J431) and Tor Aeid (J432)

Although assemblages from Tor Aeid and Layers F and G at Tor Hamar have no associated dates, the technological and typological characteristics of these assemblages resemble, for the most part, those designated Early Ahmarian from the Negev and Sinai. Following Ronen's (1976) initial recognition of non-Aurignacian assemblages in northern Israel, the Ahmarian has become more widely known on the basis of assemblages from southern regions of the Levant. Marks and Ferring (1988) distinguish an early Ahmarian blade and bladelet technology, such as that found at Boker A in the central Negev at ca. 37,000 B.P., from the late Ahmarian blade and bladelet technology represented at Ein Aqev East in the Negev and Ain el-Buhira (WHS 618) in the Wadi Hasa, Jordan, dated to 20,300 B.P. (Coinman 1990:49–50; Clark *et al.* 1988:240).

Early Ahmarian technology is characterized by serial blade production in which large blanks produced early in reduction were used primarily for endscrapers and burins, while smaller blades and bladelets were struck during the later stages of reduction (Marks and Ferring 1988:56). The smaller blades were typically made into el-Wad points—probably the most diagnostic feature of most Early Ahmarian assemblages. Similar early Upper Paleolithic assemblages have been identified at Lagaman sites in the Gebel Maghara, which are dominated by el-Wad points (e.g., Lagama VI, VII, XV, XVI) (Gilead 1983, 1991; Bar-Yosef and Belfer 1977). Other early Upper Paleolithic assemblages from Sinai are the Abu Noshra sites in the south (Phillips 1988) and the Qadesh Barnea sites in northeastern Sinai, where el-Wad points are less frequent (Gilead and Bar-Yosef 1993). These Early Ahmarian assemblages between ca. 30,000 and 38,000 B.P. are similar in technology, but they tend to be more variable in typological characteristics. Relative proportions and type varieties of retouched tools vary considerably with some endscrapers and burins made on large, thick blanks, while retouched blades and bladelets and el-Wad points all vary in size, retouch, and relative frequency among the known assemblages.

The technology of the Early Ahmarian assemblages from Layers F–G at Tor Hamar (J431) and Tor Aeid (J432) are best illustrated by the reconstructed core reduction strategies. Discarded cores are relatively small but exhibit final removal facets that tend to be elongated and are consistent with a blade technology and production directed at more elongated debitage. The majority of cores suggest an emphasis on single platforms with subpyramidal shapes, but double and multiple platforms are also evident. Core preparation is indicated by platform edge abrasion and regularization. Rejuvenation of core platforms was accomplished primarily through the core tablet technique, but partial removal of the core platform was also important. On-site primary processing of cores is evident in the production of primary elements. The size and shape of primary removals varies between the two sites. Primary elements from Tor Hamar are much smaller and more blade-like than those from Tor Aeid. Hard hammer percussion is most evident in the lack of platform lipping and the presence of *fractures a languette* on the bulbs of both primary and secondary elements. Lipped platforms are less frequent, suggesting that soft hammers were also used in primary as well as secondary reduction stages.

Blank production for tools can be divided into three discrete groups: Group 1, consisting of large, less elongated tools, such as endscrapers, burins, and other flake tools; Group 2, comprised of smaller, thinner blade tools corresponding to blade debitage; and Group 3, made up of very specialized el-Wad points made on elongated blades and bladelets. Selection of blanks indicates a preference for non-cortical interior products.

Typologically, the Early Ahmarian assemblages from Tor Hamar and Tor Aeid are distinguished by pointed tools in the form of el-Wad points. These vary between the two assemblages with those from Tor Hamar being larger and occurring more frequently on curved and twisted blanks. Those from Tor Aeid are generally smaller and made on blanks that are incurvate but rarely twisted. Retouch is similar on el-Wad points in both assemblages, in that it is used for shaping blanks into points. It varies from fine to semi-abrupt or abrupt backing, and it occurs invariably on obverse edges, in contrast to other el-Wad varieties that feature inverse retouch.

While the tool assemblages from these Early Ahmarian assemblages are compositionally quite similar, blade tools other than the more specialized el-Wad tools from Tor Hamar tend to be smaller and more elongated than those from Tor Aeid.

Endscrapers and burins are typical tools of the Early Ahmarian assemblages. Scrapers, especially endscrapers, are relatively more frequent at Tor Aeid. Many were made on flakes, on core tablets, or on large, thick blanks, with a number of carinated examples. Burins from Tor Aeid are among the largest recovered in the Upper Paleolithic assemblages, occurring most often on thick, core-like blanks. More typical burins in both assemblages are angle and dihedral varieties made on blades. The variability in endscrapers and burins documented in these Early Ahmarian assemblages suggests some typological variability is more extensive than indicated by the current descriptions for this assemblage type. Some of that variability in burins may be due to the fact that, typologically, this tool class is extremely heterogeneous, making the functional correlates of such a variety of tools far from clear.

The retouched bladelets in the Early Ahmarian assemblages do not appear to represent a significant portion of the tool assemblages and do not comprise a group of standardized tools. Some stratigraphic mixing in Layer F at Tor Hamar resulted in the presence of intrusive microlithic varieties, while those at Tor Aeid are unremarkable.

## Levantine Aurignacian

### Jebel Humeima (J412) and Tor Fawaz (J403)

Jebel Humeima (J412) and Tor Fawaz (J403) are among assemblages that were initially identified as Levantine Aurignacian (Henry 1982:16–17; 1986:427–430). They can best be described as "non-Ahmarian," fitting only loosely into the descriptions of the Levantine Aurignacian. The Levantine Aurignacian was *re-defined* by Gilead (1981:339) and Marks (1981:347) and characterized as a *flake* technology focused on flake debitage and tools and exhibiting low percentages of blades and bladelets. More recently, it has been described as a relatively inferior *blade* technology which produced large, thick blades but lacked a true bladelet technology (Marks and Ferring 1988:46). In this respect, Jebel Humeima and Tor Fawaz are characterized by a blade technology that produced large, thick, bulky debitage and tools. Discarded cores and debitage characteristics suggest reduction strategies which utilized single platform cores primarily. There is no evidence of cresting or *lames à crête*, and core preparation appears to be minimal. Platforms were rejuvenated using the core tablet technique at Jebel

Humeima and a variation of the tablet technique at Tor Fawaz. In both assemblages, platforms on cores were also maintained or the orientations changed by the removal of only portions of the platform. Both hard and soft hammer percussion was used in what appears to be interchangeable modes during primary and secondary stages of reduction.

Comparatively more on-site primary processing took place at Jebel Humeima and Tor Fawaz than at the other Upper Paleolithic sites. Primary removals were, on average, considerably shorter and more flake-like than subsequent interior products. After the smaller cortical pieces were removed, reduction shifted to producing large blanks, mostly blades, from which specific groups of tools were made. Production goals appear to have been directed at blade blanks since the majority of tools in both assemblages were made on blades. At Tor Fawaz the debitage components are equally divided between blade and flake debitage, while the Jebel Humeima assemblage is less *lamellar* and dominated by debitage with flake characteristics and dimensions. In both assemblages, tools made on cortical blanks represent an important percentage of the retouched pieces.

The dimensional parameters of endscrapers and burins from Tor Fawaz are more restricted than those from Jebel Humeima, but both assemblages are characterized by very large endscrapers and relatively smaller burins. Other tools were generally made on much smaller blade and flake blanks at both sites.

Typologically, the tool assemblages are dominated by less formal classes of tools, such as general retouched pieces and notches, followed by scrapers and burins, which are considerably less frequent at both sites. Endscrapers and burins occur about equally on flake and blade blanks. At Jebel Humeima, endscrapers tend to exhibit thick noses and reflect a certain amount of carination, while only one carinated scraper was recovered from Tor Fawaz. Burins are quite heterogeneous in both assemblages, but in general, do not reflect thick carinated blanks. Burins on snaps or old surfaces are most typical, but each assemblage has low frequencies of a variety of types.

Pointed tools are not characteristic of these two assemblages, but variously retouched blades occur in similar proportions. Retouched bladelets are infrequent at Jebel Humeima but relatively more frequent at Tor Fawaz. Retouch on blades and bladelets is generally not formal or standardized and occurs most often on the obverse sides. Inversely and alternating retouch is less typical, as is backing.

In summary, the assemblages from Tor Fawaz and Jebel Humeima vary technologically and typologically from the descriptions of the Levantine Aurignacian. The variation may in part be due to regional differences, but more likely, the variability from established assemblage criteria suggests the Levantine Aurignacian, as it is currently defined, does not accurately describe all non-Ahmarian assemblages in the southern Levant.

Appendix 8.1. Inventory of Lithic Assemblage From Tor Fawaz (J403)

| | Unit | | | | | | |
|---|---|---|---|---|---|---|---|
| | 1 | 2 | 3 | 4 | 5 | Total | % |
| Tools | | | | | | | |
| Scrapers | 1 | 8 | 4 | 2 | 5 | 20 | 6.0 |
| Burins | 5 | 4 | 3 | 2 | 4 | 18 | 5.4 |
| Perforators | | 1 | | | 2 | 3 | .9 |
| Truncations | 2 | 7 | 6 | 6 | 7 | 28 | 8.4 |
| Denticulates | 1 | 1 | 1 | | 3 | 6 | 1.8 |
| Notches | 3 | 7 | 8 | 11 | 24 | 53 | 15.9 |
| Retouched Bladelets | 3 | 6 | 5 | 3 | 7 | 24 | 7.2 |
| Retouched Blades | 11 | 21 | 14 | 14 | 17 | 77 | 23.1 |
| Retouched Flakes | 15 | 13 | 29 | 9 | 25 | 91 | 27.3 |
| Multiple Use Tools | | 1 | 4 | 1 | 1 | 7 | 2.1 |
| Points | 1 | 1 | | | 1 | 3 | 0.9 |
| Massive Pieces | | | | | 3 | 3 | 0.9 |
| Subtotal | 42 | 70 | 74 | 48 | 99 | 333 | 99.9 |
| Debitage | | | | | | | |
| Cores | 6 | 5 | 3 | — | 9 | 23 | 1.4 |
| Blades | 88 | 152 | 142 | 144 | 154 | 680 | 41.0 |
| Flakes | 96 | 198 | 215 | 116 | 123 | 748 | 45.1 |
| CTE | 4 | 6 | 5 | 5 | 15 | 35 | 2.1 |
| Primary Elements | 25 | 47 | 48 | 17 | 26 | 163 | 9.8 |
| Burin Spalls | — | — | — | — | 10 | 10 | .6 |
| Subtotal | 219 | 408 | 413 | 282 | 337 | 1659 | 100.0 |
| Debris | | | | | | | |
| Chips | 139 | 273 | 361 | 227 | 253 | 1253 | 62.9 |
| Chunks | 83 | 187 | 261 | 71 | 136 | 738 | 37.1 |
| Subtotal | 222 | 460 | 622 | 298 | 389 | 1991 | 100.0 |
| Debitage | 219 | 408 | 413 | 282 | 337 | 1659 | 41.6 |
| Debris | 222 | 460 | 622 | 298 | 389 | 1991 | 50.0 |
| Tools | 42 | 70 | 74 | 48 | 99 | 333 | 8.4 |
| Total | 483 | 938 | 1109 | 628 | 825 | 3983 | 100.0 |
| Ratios | | | | | | | |
| Tools: Debitage | 1: 5.5 | | | | | | |
| Cores: Tools | 1:13.1 | | | | | | |
| Cores: Debitage | 1:71.1 | | | | | | |
| Debitage: Chips | 1: 0.7 | | | | | | |
| Primary Elements: Debitage | 1: 9.2 | | | | | | |
| Cores: Primary Elements | 1: 7.1 | | | | | | |
| CTE: Cores | 1: 0.6 | | | | | | |
| Flakes: Blades | 1: 0.9 | | | | | | |

Appendix 8.2. Blank Characteristics of Debitage from Tor Fawaz (J403)

| | Primary Elements | | Debitage Blades | |
|---|---|---|---|---|
| | *n* | % | *n* | % |
| Cortical Coverage | 111 | | 100 | |
| Total | | 22.5 | | — |
| Partial | | 77.5 | | 16.0 |
| None | | — | | 84.0 |
| Lipping | 42 | | 72 | |
| Lipped | | 31.0 | | 11.1 |
| Unlipped | | 57.1 | | 77.8 |
| Unlipped with scar | | 9.5 | | — |
| Pointed | | 2.4 | | 11.1 |
| Facet Pattern | 80 | | 92 | |
| Unidirectional, Parallel | | 55.0 | | 78.3 |
| Bidirectional, Opposed | | 6.2 | | 6.5 |
| Truncated (Blades) | | NA | | 2.2 |
| Centripedal (Flakes) | | 32.5 | | NA |
| Converging | | 6.2 | | 13.0 |
| Platform Facets | 42 | | 85 | |
| Single | | 73.8 | | 67.1 |
| Dihedral | | 2.4 | | 3.5 |
| Multiple | | — | | 3.5 |
| Crushed | | 7.1 | | 23.5 |
| Cortex | | 16.7 | | 2.3 |
| Distal Profile | 71 | | 68 | |
| Feathered | | 74.6 | | 86.8 |
| Overpassed | | 2.8 | | 5.9 |
| Underpassed | | 16.9 | | 2.9 |
| Stepped | | 5.6 | | 4.4 |
| Lateral Profile | 110 | | 100 | |
| Curved | | 50.9 | | 8.0 |
| Flat | | 24.3 | | 54.0 |
| Twisted | | 9.9 | | 38.0 |
| Rippled | | 2.7 | | — |
| Convex | | 11.7 | | — |
| Cross Section (Blades) | NA | | 98 | |
| Single | | | | — |
| Double | | | | 48.9 |
| Triple | | | | 50.0 |
| Multiple | | | | 2.0 |

Appendix 8.3. Detailed Typology for Tor Fawaz (J403)

| Type | Unit 1 | 2 | 3 | 4 | 5 | Total | % |
|---|---|---|---|---|---|---|---|
| Scraper Class | | | | | | 20 | 06.0 |
| Endscraper on flake | | 2 | 2 | | | 4 | |
| Endscraper on retouched flake | | 1 | | | 1 | 2 | |
| Endscraper on blade | | | | 2 | | 2 | |
| Endscraper on retouched blade | 1 | 4 | 1 | | 2 | 8 | |
| Endscraper, thumbnail | | | 1 | | | 1 | |
| Endscraper, carinated | | 1 | | | | 1 | |
| Endscraper with lateral notch | | | | | 2 | 2 | |
| Burin Class | | | | | | 18 | 5.4 |
| Burin, dihedral | | | | | 2 | 2 | |
| Burin, angle on snap or old surface | 2 | 3 | | 2 | 2 | 9 | |
| Burin, angle on concave truncation | 1 | | | | | 1 | |
| Burin, angle on convex truncation | | | 1 | | | 1 | |
| Burin, multiple mixed | | 1 | | | | 1 | |
| Burin, plan | | | 1 | | | 1 | |
| Burin, carinated | 1 | | | | | 1 | |
| Burin on platform | | | 1 | | | 1 | |
| Burin, multiple angle on old surface | 1 | | | | | 1 | |
| Truncations | | | | | | 28 | 8.4 |
| Truncated blade/bladelet | 2 | 7 | 3 | 2 | 4 | 18 | |
| Truncated flake | | | 3 | 2 | 3 | 8 | |
| Bitruncated blade | | | | 2 | | 2 | |
| Retouched Bladelets | | | | | | 24 | 7.2 |
| Bladelet, unilateral obverse | | 2 | 1 | 1 | 1 | 5 | |
| Bladelet, unilateral inverse | | | 1 | | 1 | 2 | |
| Bladelet, bilateral obverse | 1 | | 1 | | | 2 | |
| Bladelet, bilateral inverse | | | 1 | | | 1 | |
| Bladelet, alternate retouch | 2 | 1 | | | | 3 | |
| Bladelet, partial retouch | | 2 | 1 | 1 | 3 | 7 | |
| Bladelet, pointed straight | | | | | 1 | 1 | |
| Bladelet, Ouchtata retouch | | | | | 1 | 1 | |
| Bladelet, partially backed | | 1 | | 1 | | 2 | |
| Retouched Blades | | | | | | 77 | 23.1 |
| Blade, unilateral obverse | 5 | 4 | | 4 | 1 | 14 | |
| Blade, unilateral inverse | 3 | 1 | 3 | | 4 | 11 | |
| Blade, bilateral obverse | 1 | 1 | 2 | 3 | | 7 | |
| Blade, bilateral inverse | 1 | | | | | 1 | |
| Blade, alternate retouch | 1 | 10 | 1 | 1 | 2 | 15 | |
| Blade, partially retouched | | 4 | 8 | 6 | 10 | 28 | |
| Blade, pointed backed | | 1 | | | | 1 | |
| Retouched Flakes | | | | | | 91 | 27.3 |
| Flake, unilateral obverse | 2 | 2 | 6 | | 1 | 11 | |
| Flake, unilateral inverse | 1 | 2 | 2 | | 5 | 10 | |
| Flake, bilateral obverse | 3 | 1 | | 1 | | 5 | |
| Flake, bilateral inverse | | | 1 | | | 1 | |
| Flake, alternate retouch | | | 2 | 1 | 3 | 6 | |
| Flake, partially retouched | | 5 | 2 | 2 | 9 | 18 | |

Appendix 8.3. (*Continued*)

| Type | Unit 1 | 2 | 3 | 4 | 5 | Total | | % |
|---|---|---|---|---|---|---|---|---|
| Retouched piece, fragment | 9 | 3 | 16 | 5 | 7 | 40 | | |
| Denticulate Class | | | | | | | 6 | 1.8 |
| Denticulated piece | 1 | 1 | | | 1 | 3 | | |
| Denticulated piece with retouch | | | 1 | | 1 | 3 | | |
| Notch Class | | | | | | | 53 | 15.9 |
| Single notch | 1 | 1 | 2 | 1 | 4 | 9 | | |
| Single notch with retouch | 2 | 5 | 5 | 3 | 12 | 27 | | |
| Multiple notch | | | | 6 | 5 | 11 | | |
| Multiple notch with retouch | | | | | 2 | 2 | | |
| Micro-notch | | 1 | 1 | 1 | 1 | 4 | | |
| Multiple Use Class | | | | | | | 7 | 2.1 |
| Endscraper–Burin | | | 1 | | | 1 | | |
| Sidescraper–Notch | | 1 | 1 | | 1 | 3 | | |
| Burin–Perforator | | | 1 | | | 1 | | |
| Burin Core–Notch | | | | 1 | | 1 | | |
| Notch–Truncation | | | 1 | | | 1 | | |
| Perforator | | | | | | | 3 | 0.9 |
| Perforator | | 1 | | | 2 | 3 | | |
| Massive Pieces | | | | | | | 3 | 0.9 |
| Endscraper on a massive piece | | | | | 1 | 1 | | |
| Notch on a massive piece | | | | | 2 | 2 | | |
| Points | | | | | | | 3 | 0.9 |
| Levallois point | | 1 | 1 | | | 1 | 3 | |
| Total | | 42 | 70 | 74 | 48 | 99 | 333 | 99.9 |

Appendix 8.4. Inventory of Lithic Assemblage from Jebel Humeima (J412)

| | Unit 1 | | Unit 2 | | Site total | |
|---|---|---|---|---|---|---|
| | *n* | % | *n* | % | *n* | % |
| Tools | | | | | | |
| Scrapers | 0019 | 007.7 | 0010 | 015.2 | 0029 | 009.2 |
| Burins | 13 | 5.2 | 5 | 7.6 | 18 | 5.7 |
| Perforators | 6 | 2.4 | 2 | 3.0 | 8 | 2.6 |
| Truncations | 12 | 4.8 | 1 | 1.5 | 13 | 4.1 |
| Denticulates | 8 | 3.2 | 3 | 4.5 | 11 | 3.5 |
| Notches | 34 | 13.7 | 11 | 16.7 | 45 | 14.3 |
| Retouched Bladelets | 8 | 3.2 | 5 | 7.6 | 13 | 4.1 |
| Retouched Blades | 63 | 25.4 | 12 | 18.2 | 75 | 23.9 |
| Retouched Flakes | 80 | 32.3 | 15 | 22.7 | 95 | 30.3 |
| Multiple Use Tools | 3 | 1.2 | 1 | 1.5 | 4 | 1.3 |
| Points | 1 | .4 | 1 | 1.5 | 2 | .6 |
| Varia | 1 | .4 | — | | 1 | .3 |
| Subtotal | 248 | 99.9 | 66 | 100.0 | 314 | 99.9 |

Appendix 8.4. (*Continued*)

| | Unit 1 | | Unit 2 | | Site total | |
|---|---|---|---|---|---|---|
| | *n* | % | *n* | % | *n* | % |
| Debitage | | | | | | |
| Cores | 24 | 2.3 | 11 | 3.0 | 35 | 2.5 |
| Blades | 221 | 21.5 | 76 | 21.1 | 297 | 21.4 |
| Flakes | 460 | 44.8 | 209 | 57.9 | 669 | 48.2 |
| CTE | 20 | 1.9 | 6 | 1.6 | .26 | 1.9 |
| Primary Elements | 300 | 29.2 | 59 | 16.3 | 359 | 25.9 |
| Burin Spalls | 1 | .1 | — | | 1 | .1 |
| Subtotal | 1026 | 99.8 | 361 | 99.9 | 1387 | 100.0 |
| Debris | | | | | | |
| Chips | 3748 | 63.8 | 887 | 69.4 | 4635 | 64.8 |
| Chunks | 2122 | 36.2 | 392 | 30.6 | 2514 | 35.2 |
| Subtotal | 5870 | 100.0 | 1279 | 100.0 | 7149 | 100.0 |
| Debitage | 1026 | 14.4 | 361 | 21.2 | 1387 | 15.7 |
| Debris | 5870 | 82.2 | 1279 | 74.9 | 7149 | 80.8 |
| Tools | 248 | 3.4 | 66 | 3.9 | 314 | 3.5 |
| Total | 7144 | 100.0 | 1706 | 100.0 | 8850 | 100.0 |
| Ratios | | | | | | |
| Tools: Debitage | 1: 4.1 | | 1: 5.5 | | 1: 4.4 | |
| Cores: Tools | 1:10.3 | | 1: 6.0 | | 1: 9.0 | |
| Cores: Debitage | 1:41.7 | | 1:31.8 | | 1:38.6 | |
| Debitage: Chips | 1: 3.6 | | 1: 2.5 | | 1: 3.3 | |
| Primary Elements: Debitage | 1: 2.4 | | 1: 5.1 | | 1: 2.9 | |
| Cores: Primary Elements | 1:12.5 | | 1: 5.4 | | 1:10.2 | |
| CTE: Cores | 1: 1.2 | | 1: 1.8 | | 1: 1.3 | |
| Flakes: Blades | 1: 0.5 | | 1: 0.4 | | 1: 0.4 | |

**Appendix 8.5. Blank Characteristics of Debitage from Jebel Humeima (J412)**

| | Primary Elements | | Debitage Blades | |
|---|---|---|---|---|
| | *n* | % | *n* | % |
| Cortical Coverage | 164 | | 100 | |
| Total | | 8.5 | | — |
| Partial | | 91.5 | | 10.0 |
| None | | — | | 90.0 |
| Lipping | 88 | | 73 | |
| Lipped | | 29.5 | | 5.5 |
| Unlipped | | 56.8 | | 80.8 |
| Unlipped with scar | | 12.5 | | — |
| Pointed | | 1.1 | | 13.7 |
| Facet Pattern | 143 | | 100 | |
| Unidirectional, Parallel | | 67.1 | | 87.0 |

Appendix 8.5. (*Continued*)

| | Primary Elements | | Debitage Blades | |
|---|---|---|---|---|
| | *n* | % | *n* | % |
| Bidirectional, Opposed | | 12.5 | | — |
| Truncated (Blades) | | — | | 1.0 |
| Centripedal (Flakes) | | 17.5 | | — |
| Converging | | 2.8 | | 11.0 |
| Platform Facets | 86 | | 84 | |
| Single | | 53.5 | | 67.9 |
| Dihedral | | 12.8 | | 8.3 |
| Multiple | | 3.5 | | 2.4 |
| Crushed | | 3.5 | | 21.4 |
| Cortex | | 26.7 | | — |
| Distal Profile | 112 | | 75 | |
| Feathered | | 64.3 | | 70.6 |
| Overpassed | | .9 | | 1.3 |
| Underpassed | | 30.3 | | 20.0 |
| Stepped | | 4.5 | | 8.0 |
| Lateral Profile | 164 | | 100 | |
| Curved | | 46.9 | | 16.0 |
| Flat | | 24.4 | | 47.0 |
| Twisted | | 7.9 | | 36.0 |
| Rippled | | 6.7 | | 1.0 |
| Convex | | 12.2 | | — |
| Cross Section (Blades) | NA | | 96 | |
| Single | | | | — |
| Double | | | | 63.5 |
| Triple | | | | 32.3 |
| Multiple | | | | 4.2 |

Appendix 8.6. Detailed Typology for Jebel Humeima (J412)

| | Unit | | | |
|---|---|---|---|---|
| Type | 1 | 2 | Total | % |
| Scraper Class | | | 29 | 9.2 |
| Endscraper on flake | 4 | 1 | | |
| Endscraper on retouched flake | 5 | | | |
| Endscraper on blade | 5 | 3 | | |
| Endscraper on retouched blade | 2 | | | |
| Endscraper, carinated | 1 | 5 | | |
| Endscraper, shouldered/nosed | 1 | | | |
| Endscraper, bit (fragment) | 1 | 1 | | |
| Burin Class | | | 18 | 5.7 |
| Burin, dihedral | 1 | | | |
| Burin, angle on snap or old surface | 6 | 4 | | |
| Burin, angle on straight truncation | 2 | | | |
| Burin, transverse on lateral retouch | 1 | | | |
| Burin, transverse on lateral notch | 1 | | | |

Appendix 8.6. (*Continued*)

| Type | Unit 1 | Unit 2 | Total | % |
|---|---|---|---|---|
| Burin, plan | 2 | | | |
| Burin, transverse on plain lateral surface | | 1 | | |
| Perforator | | | 8 | 2.6 |
| Perforator | 6 | 2 | | |
| Truncations | | | 13 | 4.1 |
| Truncated blade/bladelet | 10 | 1 | | |
| Truncated flake | 1 | | | |
| Bitruncated blade/bladelet | 1 | | | |
| Denticulate Class | | | 11 | 3.5 |
| Denticulated piece | 8 | 2 | | |
| Denticulate with retouch | | 1 | | |
| Notch Class | | | 45 | 14.3 |
| Single notch | 15 | 9 | | |
| Single notch with retouch | 9 | 2 | | |
| Multiple notch | 7 | | | |
| Multiple notch with retouch | 3 | | | |
| Retouched Bladelets | | | 13 | 4.1 |
| Bladelet, unilateral obverse | 1 | 2 | | |
| Bladelet, unilateral inverse | 3 | | | |
| Bladelet, bilateral obverse (Ouchtata) | | 2 | | |
| Bladelet, alternate retouch | | 1 | | |
| Bladelet, partially retouched | 4 | | | |
| Retouched Blades | | | 75 | 23.9 |
| Blade, unilateral obverse | 18 | 8 | | |
| Blade, unilateral inverse | 6 | | | |
| Blade, bilateral inverse | 1 | | | |
| Blade, alternate retouch | 4 | | | |
| Blade, partially retouched | 33 | 4 | | |
| Blade, partially backed | 1 | | | |
| Retouched Flakes | | | 95 | 30.3 |
| Flake, unilateral obverse | 26 | 3 | | |
| Flake, unilateral inverse | 6 | 1 | | |
| Flake, bilateral obverse | 2 | | | |
| Flake, bilateral inverse | 1 | | | |
| Flake, alternate retouch | 1 | | | |
| Flake, partially retouched | 37 | 9 | | |
| Retouched piece, fragment | 7 | 2 | | |
| Multiple Use Class | | | 4 | 1.3 |
| Endscraper–Burin | 1 | 1 | | |
| Burin–Notch | 1 | | | |
| Truncation–Perforator | 1 | | | |
| Points | | | 2 | .6 |
| Levallois point | 1 | 1 | | |
| Varia | | | 1 | .3 |
| Lame à mâchure | 1 | | | |
| Total | 248 | 66 | 314 | 99.9 |

**Appendix 8.7. Inventory of Lithic Assemblage from Tor Hamar, Layers F and G (J431)**

| | Layer F | | Layer G | | Total | |
|---|---|---|---|---|---|---|
| | *n* | % | *n* | % | *n* | % |
| Tools | | | | | | |
| Scrapers | 14 | 13.1 | 7 | 14.5 | 21 | 13.6 |
| Burins | 23 | 21.5 | 12 | 25.0 | 35 | 22.5 |
| Perforators | 1 | 0.9 | 1 | 2.1 | 2 | 1.2 |
| Truncations | 6 | 5.6 | 1 | 2.1 | 7 | 4.5 |
| Denticulates | | | 1 | 2.1 | 1 | 0.6 |
| Notches | 5 | 4.7 | 6 | 12.5 | 11 | 7.1 |
| Retouched Bladelets | 15 | 14.0 | 1 | 2.1 | 16 | 10.3 |
| Retouched Blades | 9 | 8.4 | 6 | 12.5 | 15 | 9.7 |
| Retouched Flakes | 15 | 14.0 | 11 | 22.9 | 26 | 16.8 |
| Points | 16 | 14.9 | 2 | 4.2 | 18 | 11.6 |
| Multiple Use Tools | 2 | 1.9 | | | 2 | 1.2 |
| Massive Pieces | 1 | 0.9 | | | 1 | 0.6 |
| Subtotal | 107 | 99.9 | 48 | 100.0 | 155 | 99.7 |
| Debitage | | | | | | |
| Cores | 11 | 1.1 | 4 | 0.7 | 15 | 0.9 |
| Blades | 595 | 59.2 | 369 | 61.4 | 964 | 60.0 |
| Flakes | 316 | 31.4 | 166 | 27.6 | 482 | 30.0 |
| CTE | 22 | 2.2 | 9 | 1.5 | 31 | 1.9 |
| Primary Elements | 56 | 5.6 | 48 | 8.0 | 104 | 6.5 |
| Microburins | 1 | 0.1 | — | — | 1 | 0.1 |
| Burin Spalls | 4 | 0.4 | 5 | 0.8 | 9 | 0.6 |
| Subtotal | 1005 | 100.0 | 601 | 100.0 | 1606 | 100.0 |
| Debris | | | | | | |
| Chips | 1251 | 97.6 | 620 | 96.4 | 1871 | 97.2 |
| Chunks | 31 | 2.4 | 23 | 3.6 | 54 | 2.8 |
| Subtotal | 1282 | 100.0 | 643 | 100.0 | 1925 | 100.0 |
| Debitage | 1005 | 42.0 | 601 | 46.5 | 1606 | 43.6 |
| Debris | 1282 | 53.5 | 643 | 49.8 | 1925 | 52.2 |
| Tools | 107 | 4.5 | 48 | 3.7 | 155 | 4.2 |
| Total | 2394 | 100.0 | 1292 | 100.0 | 3686 | 100.0 |
| Ratios | | | | | | |
| Tools: Debitage | 1: 9.4 | | 1: 12.5 | | 1: 10.4 | |
| Cores: Tools | 1: 9.7 | | 1: 12.0 | | 1: 10.7 | |
| Cores : Debitage | 1:90.4 | | 1:149.2 | | 1:106.1 | |
| Debitage: Chips | 1: 1.2 | | 1: 1.0 | | 1: 1.2 | |
| Primary Elements: Debitage | 1:16.9 | | 1: 11.5 | | 1: 14.4 | |
| Cores: Primary Elements | 1: 5.1 | | 1: 12.0 | | 1: 6.9 | |
| CTE: Cores | 1: 0.5 | | 1: 0.4 | | 1: 0.5 | |
| Flakes: Blades | 1: 1.9 | | 1: 2.2 | | 1: 2.0 | |

Appendix 8.8. Black Characteristics of Debitage at Tor Hamar, Layers F and G (J431)

| | Primary Elements | | Debitage Blades | |
|---|---|---|---|---|
| | *n* | % | *n* | % |
| Cortical Coverage | 80 | | 287 | |
| Total | | 18.5 | | — |
| Partial | | 81.5 | | 15.7 |
| None | | — | | 84.3 |
| Lipping | 37 | | 285 | |
| Lipped | | 37.8 | | 41.1 |
| Unlipped | | 48.6 | | 48.7 |
| Unlipped with scar | | 13.5 | | 3.4 |
| Pointed | | — | | 6.8 |
| Facet Pattern | 65 | | 285 | |
| Unidirectional, Parallel | | 60.0 | | 62.1 |
| Bidirectional, Opposed | | 18.5 | | 10.5 |
| Truncated (Blades) | | — | | 9.8 |
| Centripedal (Primary Elements) | | 20.0 | | — |
| Converging | | 1.5 | | 17.5 |
| Platform Facets | 37 | | 264 | |
| Single | | 75.7 | | 50.0 |
| Dihedral | | 5.4 | | 3.0 |
| Multiple | | — | | 3.0 |
| Crushed | | 10.8 | | 43.9 |
| Cortex | | 8.1 | | — |
| Distal Profile | 63 | | 208 | |
| Feathered | | 60.3 | | 81.7 |
| Overpassed | | 4.8 | | 2.4 |
| Underpassed | | 25.4 | | 13.0 |
| Stepped | | 9.5 | | 2.9 |
| Lateral Profile | 81 | | 287 | |
| Curved | | 46.9 | | 36.2 |
| Flat | | 19.7 | | 15.0 |
| Twisted | | 16.1 | | 43.5 |
| Rippled | | 6.2 | | 1.4 |
| Convex | | 11.1 | | 3.8 |
| Cross Section (Blades) | NA | | 287 | |
| Single | | | | 0.3 |
| Double | | | | 37.6 |
| Triple | | | | 50.2 |
| Multiple | | | | 11.8 |

**Appendix 8.9. Detailed Typology of Retouched Tools from Tor Hamar, Layers F and G (J431)**

| Type | Layer F | Layer G | Total | | % |
|---|---|---|---|---|---|
| Scraper Class | | | | 21 | 13.6 |
| Endscraper on flake | 04 | 01 | 05 | | |
| Endscraper on retouched flake | 2 | 1 | 3 | | |
| Endscraper on blade | 2 | 4 | 6 | | |
| Endscraper on retouched blade | 3 | | 3 | | |
| Endscraper, denticulated | 1 | | 1 | | |
| Endscraper, shouldered/nosed | 1 | | 1 | | |
| Endscraper on Aurignacian blade | 1 | | 1 | | |
| Sidescraper | | 1 | 1 | | |
| Burin Class | | | | 35 | 22.5 |
| Burin, dihedral | 5 | 3 | 8 | | |
| Burin, angle on snap or old surface | 9 | 6 | 15 | | |
| Burin, plan | 1 | 1 | 2 | | |
| Burin, carinated | 2 | 1 | 3 | | |
| Burin on platform | 5 | 1 | 6 | | |
| Multiple angle burins | 1 | | 1 | | |
| Perforator | | | | 2 | 1.2 |
| Perforator | 1 | 1 | 2 | | |
| Truncations | | | | 7 | 4.5 |
| Truncated blade/bladelet | 2 | | 2 | | |
| Truncated flake | 1 | 1 | 2 | | |
| Bitruncated blade | 3 | | 3 | | |
| Denticulate Class | | | | 1 | 0.6 |
| Denticulated piece | | 1 | 1 | | |
| Notch Class | | | | 11 | 7.1 |
| Single notch | 3 | 1 | 4 | | |
| Single notch with retouch | 1 | | 1 | | |
| Multiple notch | 2 | 4 | 6 | | |
| Retouched Bladelets | | | | 16 | 10.3 |
| Bladelet, unilateral obverse | 1 | | 1 | | |
| Bladelet, unilateral inverse | 1 | | 1 | | |
| Ouchtata bladelet | 1 | | 1 | | |
| Straight backed bladelet | 3 | 1 | 4 | | |
| Arched backed bladelet | 3 | | 3 | | |
| Partially backed piece | 1 | | 1 | | |
| Fragment of a backed piece | 5 | | 5 | | |
| Retouched Blades | | | | 15 | 9.7 |
| Blade, unilateral obverse | 1 | 2 | 3 | | |
| Blade, unilateral inverse | 1 | | 1 | | |
| Blade, bilateral obverse | 3 | 2 | 5 | | |
| Blade, alternately retouched | 2 | | 2 | | |
| Blade, partially retouched | | 1 | 1 | | |
| Blade, straight backed | 1 | | 1 | | |
| Blade, backed fragment | 1 | 1 | 2 | | |
| Retouched Flakes | | | | 26 | 16.8 |
| Flake, unilateral inverse | | 1 | 1 | | |

Appendix 8.9. (*Continued*)

| Type | Layer F | Layer G | Total | | % |
|---|---|---|---|---|---|
| Flake, bilateral obverse | | 1 | 1 | | |
| Flake, partially retouched | 1 | 3 | 4 | | |
| Flake, retouched fragment | 14 | 6 | 20 | | |
| Points | | | | 18 | 11.6 |
| El Wad point | 12 | 2 | 14 | | |
| La Mouillah | 2 | | 2 | | |
| Levallois point | 1 | | 1 | | |
| Lamelle Dufour | 1 | | 1 | | |
| Multiple Use Class | | | | 2 | 1.2 |
| Endscraper–Burin | 1 | | 1 | | |
| Burin–Notch | 1 | | 1 | | |
| Massive Piece Class | | | | 1 | 0.6 |
| Denticulate on massive piece | 1 | | 1 | | |
| Total | 107 | 48 | | 155 | 99.7 |

Appendix 8.10. Inventory of Lithic Assemblage from Tor Aeid (J432)

| | Unit 1 | | Unit 2 | | Site total | |
|---|---|---|---|---|---|---|
| | *n* | % | *n* | % | *n* | % |
| Tools | | | | | | |
| Scrapers | 16 | 20.2 | 16 | 24.2 | 32 | 22.1 |
| Burins | 22 | 27.8 | 10 | 15.2 | 32 | 22.1 |
| Perforators | 4 | 5.1 | 1 | 1.5 | 5 | 3.4 |
| Truncations | 2 | 2.5 | 4 | 6.0 | 6 | 4.1 |
| Denticulates | 1 | 1.3 | | | 1 | 0.7 |
| Notches | 9 | 11.4 | 4 | 6.0 | 13 | 9.0 |
| Retouched Bladelets | 6 | 7.6 | 3 | 4.5 | 9 | 6.2 |
| Retouched Blades | 4 | 5.1 | 11 | 16.7 | 15 | 10.3 |
| Retouched Flakes | 2 | 2.5 | 3 | 4.5 | 5 | 3.4 |
| Multiple Use Tools | 3 | 3.8 | 6 | 9.1 | 9 | 6.2 |
| Points | 10 | 12.7 | 6 | 9.1 | 16 | 11.0 |
| Varia | | | 2 | 3.0 | 2 | 1.4 |
| Subtotal | 79 | 100.0 | 66 | 99.8 | 145 | 99.9 |
| Debitage | | | | | | |
| Cores | 27 | 2.6 | 29 | 2.2 | 56 | 2.3 |
| Blades | 548 | 52.1 | 734 | 54.5 | 1282 | 53.5 |
| Flakes | 362 | 34.4 | 416 | 30.9 | 778 | 32.4 |
| CTE | 28 | 2.7 | 34 | 2.5 | 62 | 2.6 |
| Primary Elements | 80 | 7.6 | 124 | 9.2 | 204 | 8.5 |
| Burin Spalls | 7 | .6 | 9 | .7 | 16 | 0.7 |
| Subtotal | 1052 | 100.0 | 1346 | 100.0 | 2398 | 100.0 |

Appendix 8.10. (*Continued*)

| | Unit 1 | | Unit 2 | | Site total | |
|---|---|---|---|---|---|---|
| | *n* | % | *n* | % | *n* | % |
| Debris | | | | | | |
| Chips | 1277 | 96.5 | 1580 | 96.2 | 2857 | 96.3 |
| Chunks | 47 | 3.5 | 63 | 3.8 | 110 | 3.7 |
| Subtotal | 1324 | 100.0 | 1643 | 100.0 | 2967 | 100.0 |
| Debitage | 1052 | 42.9 | 1346 | 44.0 | 2398 | 43.5 |
| Debris | 1324 | 53.9 | 1643 | 53.8 | 2967 | 53.8 |
| Tools | 79 | 3.2 | 66 | 2.2 | 145 | 2.6 |
| Total | 2455 | 100.0 | 3055 | 100.0 | 5510 | 100.0 |
| Ratios | | | | | | |
| Tools:Debitage | 1:13.3 | | 1:20.4 | | 1:16.5 | |
| Cores:Tools | 1: 2.9 | | 1: 2.3 | | 1: 2.6 | |
| Cores:Debitage | 1:38.0 | | 1:45.4 | | 1:41.8 | |
| Debitage:Chips | 1: 1.2 | | 1: 1.2 | | 1: 1.2 | |
| Primary Elements:Debitage | 1: 12.2 | | 1: 9.8 | | 1:10.7 | |
| Cores:Primary Elements | 1: 3.0 | | 1: 4.3 | | 1: 3.6 | |
| CTE:Cores | 1: 9 | | 1: 8 | | 1: 9 | |
| Flakes:Blades | 1: 1.5 | | 1: 1.7 | | 1: 1.6 | |

**Appendix 8.11. Blank Characteristics of Debitage from Tor Aeid (J432)**

| | Primary Elements | | Debitage blades | |
|---|---|---|---|---|
| | *n* | % | *n* | % |
| Cortical Coverage | 144 | | 111 | |
| Total | | 13.2 | | — |
| Partial | | 86.8 | | 4.5 |
| None | | — | | 95.5 |
| Lipping | 84 | | 91 | |
| Lipped | | 33.3 | | 18.7 |
| Unlipped | | 41.7 | | 65.9 |
| Unlipped with scar | | 23.8 | | — |
| Pointed | | — | | 15.4 |
| Facet Pattern | 121 | | 107 | |
| Unidirectional, Parallel | | 58.7 | | 81.3 |
| Bidirectional, Opposed | | 19.0 | | 2.8 |
| Truncated (Blades) | | — | | .9 |
| Centripedal (Primary Elements) | | 17.3 | | — |
| Converging | | 4.9 | | 14.9 |
| Platform Facets | 84 | | 105 | |
| Single | | 54.8 | | 71.4 |
| Dihedral | | 4.7 | | 3.8 |

Appendix 8.11. (*Continued*)

| | Primary Elements | | Debitage blades | |
|---|---|---|---|---|
| | *n* | % | *n* | % |
| Multiple | | 4.7 | | — |
| Crushed | | 20.2 | | 24.8 |
| Cortex | | 15.5 | | — |
| Distal Profile | 97 | | 100 | |
| Feathered | | 78.3 | | 82.0 |
| Overpassed | | 3.2 | | 6.0 |
| Underpassed | | 15.4 | | 9.0 |
| Stepped | | 3.1 | | 3.0 |
| Lateral Profile | 135 | | 110 | |
| Curved | | 62.2 | | 27.1 |
| Flat | | 16.3 | | 31.5 |
| Twisted | | 11.1 | | 38.7 |
| Rippled | | 5.2 | | 1.8 |
| Convex | | 5.2 | | — |
| Cross Section (Blades) | NA | | 111 | |
| Single | | | | — |
| Double | | | | 32.4 |
| Triple | | | | 63.1 |
| Multiple | | | | 2.7 |

Appendix 8.12. Detailed Typology for Tor Aeid (J432)

| Type | Unit | | Total | % |
|---|---|---|---|---|
| | 1 | 2 | | |
| Scraper Class | | | 032 | 22.1 |
| Endscraper on flake | 03 | 02 | | |
| Endscraper on retouched flake | 3 | 2 | | |
| Endscraper on blade | 3 | 4 | | |
| Endscraper on retouched blade | 4 | 2 | | |
| Endscraper, carinated | 2 | | | |
| Endscraper, double opposed | 1 | | | |
| Endscraper with lateral notch | | 2 | | |
| Sidescraper | | 3 | | |
| Endscraper, bit | | 1 | | |
| Burin Class | | | 32 | 22.1 |
| Burin, dihedral | 4 | 2 | | |
| Burin, angle on snap or old surface | 7 | 3 | | |
| Burin, transverse on lateral retouch | 2 | 1 | | |
| Burin, multiple mixed | 2 | | | |
| Burin, plan | 1 | | | |
| Burin, carinated | 5 | 4 | | |
| Burin, transverse on lateral edge (unretouched) | 1 | | | |

Appendix 8.12. (*Continued*)

| Type | Unit 1 | Unit 2 | Total | % |
|---|---|---|---|---|
| Perforator | | | 5 | 3.4 |
| Perforator | 4 | 1 | | |
| Truncations | | | 6 | 4.1 |
| Truncated blade/bladelet | 2 | 4 | | |
| Denticulate Class | | | 1 | 0.7 |
| Denticulated piece | 1 | | | |
| Notch Class | | | 13 | 9.0 |
| Single notch | 2 | | | |
| Single notch with retouch | 4 | 4 | | |
| Multiple notch | 1 | | | |
| Multiple notch with retouch | 2 | | | |
| Retouched Bladelets | | | 9 | 6.2 |
| Bladelet, unilateral observe | 4 | | | |
| Bladelet, unilateral obverse (Ouchtata) | | 1 | | |
| Bladelet, unilateral inverse | | 1 | | |
| Bladelet, partially retouched | | 1 | | |
| Straight backed bladelet | 1 | | | |
| Partially backed piece | 1 | | | |
| Retouched Blades | | | 15 | 10.3 |
| Blade, unilateral obverse | 1 | 3 | | |
| Blade, unilateral inverse | 1 | 4 | | |
| Blade, bilateral obverse | | 1 | | |
| Blade, partially retouched | 1 | 2 | | |
| Backed piece, fragment | | 1 | | |
| Pointed blade, fragment | 1 | | | |
| Retouched Flakes | | | 5 | 3.4 |
| Flake, alternate retouch | | 1 | | |
| Flake, partially retouched | | 1 | | |
| Retouched piece, fragment | 2 | 1 | | |
| Multiple Use Class | | | 9 | 6.2 |
| Endscraper—Burin | | 1 | | |
| Endscraper—Burin—Notch | | 1 | | |
| Endscraper—Denticulate | 1 | | | |
| Burin—Notch | 1 | | | |
| Burin—Truncation—Scraper | 1 | | | |
| Perforator—Notch | | 1 | | |
| Denticulate—Notch | | 2 | | |
| Backed bladelet—Notch | | 1 | | |
| Points | | | 16 | 11.0 |
| El Wad point | 10 | 6 | | |
| Varia | | | 2 | 1.4 |
| Burin spall with retouch | | 2 | | |
| Total | 79 | 66 | 145 | 99.9 |

Appendix 8.13. Inventory of Lithic Assemblage from Site J440

| | Unit | | | | | | | |
|---|---|---|---|---|---|---|---|---|
| | 1 | 2 | 3 | 4 | 5 | 6 | Total | % |
| Tools | | | | | | | | |
| Scrapers | | 1 | 1 | 1 | | 2 | 5 | 9.4 |
| Burins | 1 | | 6 | 1 | 2 | 4 | 14 | 26.4 |
| Perforators | | | | | | | | |
| Truncations | | | 1 | 1 | 1 | 1 | 4 | 7.6 |
| Denticulates | | | | | | 1 | 1 | 1.9 |
| Notches | | | | 1 | | 2 | 3 | 5.7 |
| Retouched Bladelets | 1 | | 1 | 1 | 1 | 1 | 5 | 9.4 |
| Retouched Blades | 1 | | | 2 | 2 | | 5 | 9.4 |
| Retouched Flakes | | | | 1 | 4 | 8 | 13 | 24.5 |
| Multiple Use Tools | | | 1 | 1 | | 1 | 3 | 5.7 |
| Points | | | | | | | | |
| Subtotal | 3 | 1 | 10 | 9 | 10 | 20 | 53 | 100.0 |
| Debitage | | | | | | | | |
| Cores | 4 | 3 | 3 | 2 | | 5 | 17 | 11.4 |
| Blades | 14 | 14 | 17 | 5 | 8 | 4 | 62 | 51.7 |
| Flakes | 3 | 6 | 5 | 5 | 4 | 13 | 36 | 30.0 |
| CTE | 2 | 2 | 0 | | 3 | 3 | 10 | 8.3 |
| Primary Elements | 5 | 2 | 4 | | | 1 | 12 | 10.0 |
| Burin Spalls | | | | | | | | |
| Subtotal | 28 | 27 | 29 | 12 | 15 | 26 | 137 | 100.0 |
| Debris | | | | | | | | |
| Chips | | | | | 6 | | 6 | 50.0 |
| Chunks | | | | | 6 | | 6 | 50.0 |
| Subtotal | | | | | 12 | | 12 | 100.0 |
| Debitage | 28 | 27 | 29 | 12 | 15 | 26 | 137 | 67.8 |
| Debris | | | | | 12 | | 12 | 5.9 |
| Tools | 3 | 1 | 10 | 9 | 10 | 20 | 53 | 26.2 |
| Total | 31 | 28 | 39 | 21 | 37 | 46 | 202 | 99.9 |

Appendix 8.14. Detailed Typology for Site J440

| Type | Unit 1 | 2 | 3 | 4 | 5 | 6 | | Total | % |
|---|---|---|---|---|---|---|---|---|---|
| Scraper Class | | | | | | | | 5 | 9.4 |
| Endscraper on flake | | 1 | | | | 2 | 3 | | |
| Endscraper on retouched blade | | | 1 | | | | 1 | | |
| Sidescraper | | | | 1 | | | 1 | | |
| Burin Class | | | | | | | | 14 | 26.4 |
| Burin, dihedral | | | 1 | | 1 | 2 | 4 | | |
| Burin, angle on snap or old surface | 1 | | 2 | 1 | 1 | 1 | 6 | | |
| Burin, angle on oblique truncation | | | | | | 1 | 1 | | |
| Burin, angle on concave truncation | | | 1 | | | | 1 | | |
| Burin, plan | | | 1 | | | | 1 | | |
| Burin on platform | | | 1 | | | | 1 | | |
| Truncations | | | | | | | | 4 | 7.6 |
| Truncated blade/bladelet | | | | 1 | 1 | | 2 | | |
| Truncated flake | | | 1 | | | 1 | 2 | | |
| Denticulate Class | | | | | | | | 1 | 1.9 |
| Denticulated piece | | | | | | 1 | 1 | | |
| North Class | | | | | | | | 3 | 5.7 |
| Single notch | | | | | | 2 | 2 | | |
| Single notch with retouch | | | | 1 | | | 1 | | |
| Retouched Bladelets | | | | | | | | 5 | 9.4 |
| Bladelet, unilateral obverse | | | | | | 1 | 1 | | |
| Bladelet, unilateral inverse | | | | 1 | | | 1 | | |
| Bladelet, bilateral obverse | | | | | 1 | | 1 | | |
| Bladelet, bilateral inverse | 1 | | | | | | 1 | | |
| Bladelet, partial retouch | | | 1 | | | | 1 | | |
| Retouched Blades | | | | | | | | 5 | 9.4 |
| Blade, unilateral inverse | | | | 1 | | | 1 | | |
| Blade, bilateral obverse | | | | | 1 | | 1 | | |
| Blade, partially retouched | 1 | | | 1 | 1 | | 3 | | |
| Retouched Flakes | | | | | | | | 13 | 24.5 |
| Flake, unilateral obverse | | | | 1 | 1 | | 2 | | |
| Flake, unilateral inverse | | | | | 1 | 1 | 2 | | |
| Flake, bilateral obverse | | | | | | 1 | 1 | | |
| Flake, alternate retouch | | | | | | 3 | 3 | | |
| Flake, partially retouched | | | | | | 3 | 3 | | |
| Retouched piece, fragment | | | | | 2 | | 2 | | |
| Multiple Use Class | | | | | | | | 3 | 5.7 |
| Endscraper–Burin | | | 1 | | | | 1 | | |
| Burin–Notch–Scraper | | | | 1 | | | 1 | | |
| Perforator–Notch | | | | | | 1 | 1 | | |
| Total | 3 | 1 | 10 | 9 | 10 | 20 | | 53 | 100.0 |

Appendix 8.15. Inventory of Lithic Assemblage from Wadi Aghar (J433)

| | Units | | | Feature | Total | |
|---|---|---|---|---|---|---|
| | 1 | 2 | 3 | 1 | *n* | % |
| Tools | | | | | | |
| Scrapers | 3 | 8 | | | 11 | 40.7 |
| Burins | | 2 | | 1 | 3 | 11.1 |
| Notches | 2 | 1 | 5 | | 8 | 29.6 |
| Retouched blades | | 1 | 1 | | 2 | 7.4 |
| Multiple Use Tools | | 2 | 1 | | 3 | 11.1 |
| Subtotal | 5 | 14 | 7 | 1 | 27 | 99.9 |
| Debitage | | | | | | |
| Cores | 4 | 5 | 5 | 2 | 16 | 8.4 |
| Blades | 40 | 49 | 4 | 2 | 95 | 49.7 |
| Flakes | 18 | 29 | 6 | — | 53 | 27.7 |
| CTE | 2 | 4 | 1 | — | 7 | 3.7 |
| Primary Elements | 2 | 10 | 3 | 2 | 17 | 8.9 |
| Burin Spalls | 2 | — | 1 | — | 3 | 1.6 |
| Subtotal | 68 | 97 | 20 | 6 | 191 | 100.0 |
| Debris | | | | | | |
| Chips | 33 | 33 | 1 | — | 67 | 62.6 |
| Chunks | 18 | 19 | 3 | — | 40 | 37.4 |
| Subtotal | 51 | 52 | 4 | 0 | 107 | 100.0 |
| Debitage | 68 | 97 | 20 | 6 | 191 | 58.8 |
| Debris | 51 | 52 | 4 | — | 107 | 32.9 |
| Tools | 5 | 14 | 7 | 1 | 27 | 8.3 |
| Total | 124 | 163 | 31 | 7 | 325 | 100.0 |
| Ratios | | | | | | |
| Tools: Debitage | 1: 7.1 | | | | | |
| Cores: Tools | 1: 1.7 | | | | | |
| Cores: Debitage | 1:10.9 | | | | | |
| Debitage: Chips | 1: 0.3 | | | | | |
| Primary Elements: Debitage | 1:10.2 | | | | | |
| Cores: Primary Elements | 1: 1.1 | | | | | |
| CTE: Cores | 1: 2.3 | | | | | |
| Flakes: Blades | 1: 1.8 | | | | | |

Appendix 8.16. Blank Characteristics of Debitage from Wadi Aghar (J433)

| | Primary elements | | Debitage blades | |
|---|---|---|---|---|
| | *n* | % | *n* | % |
| Cortical Coverage | 12 | | 65 | |
| Total | | 8.3 | | — |
| Partial | | 91.7 | | 21.5 |
| None | | — | | 78.5 |
| Lipping | 11 | | 55 | |
| Lipped | | 36.4 | | 21.8 |
| Unlipped | | 36.4 | | 38.2 |
| Unlipped with scar | | 27.2 | | 34.5 |
| Pointed | | — | | 5.4 |
| Facet Pattern | 11 | | 65 | |
| Unidirectional, Parallel | | 54.5 | | 66.1 |
| Bidirectional, Opposed | | 27.3 | | 20.0 |
| Truncated | | — | | — |
| Centripedal | | 18.2 | | — |
| Unidirectional, Converging | | — | | 13.9 |
| Platform Facets (see Table 8.24) | | | | |
| Distal Profile | 11 | | 38 | |
| Feathered | | 55.6 | | 76.3 |
| Overpassed | | — | | — |
| Underpassed | | 11.1 | | 18.4 |
| Stepped | | 33.3 | | 5.3 |
| Lateral Profile | 12 | | 65 | |
| Curved | | 75.0 | | 50.8 |
| Flat | | — | | 20.0 |
| Twisted | | 8.3 | | 24.6 |
| Rippled | | — | | 1.5 |
| Convex | | 16.7 | | 3.1 |
| Cross Section (Blades) | NA | | 65 | |
| Single | | | | — |
| Double | | | | 30.8 |
| Triple | | | | 52.3 |
| Multiple | | | | 16.9 |

Appendix 8.17. Detailed Typology for Wadi Aghar (J433)

| | Unit | | | Feature | | | |
|---|---|---|---|---|---|---|---|
| Type | 1 | 2 | 3 | 1 | Total | | % |
| Scraper Class | | | | | | 11 | 40.7 |
| Endscraper on blade | 1 | 4 | | | 5 | | |
| Endscraper on retouched blade | 1 | 2 | | | 3 | | |
| Endscraper with lateral notch | 1 | 2 | | | 3 | | |
| Burin Class | | | | | | 3 | 11.1 |
| Burin, angle on snap or old surface | | 1 | | 1 | 2 | | |
| Burin, multiple mixed | | 1 | | | 1 | | |

Appendix 8.17. (*Continued*)

| Type | Unit 1 | Unit 2 | Unit 3 | Feature 1 | Total | | % |
|---|---|---|---|---|---|---|---|
| Retouched Blades | | | | | | 8 | 29.6 |
| Blade, unilateral inverse | 2 | | 2 | | 4 | | |
| Blade, alternating retouch | | | 1 | | 1 | | |
| Blade, partially retouched | | 1 | 2 | | | 3 | |
| Notch Class | | | | | | 2 | 7.4 |
| Single notch | | 1 | | | 1 | | |
| Single notch with retouch | | | 1 | | 1 | | |
| Multiple Use Class | | | | | | 3 | 11.1 |
| Endscraper—Burin | | 2 | 1 | | 3 | | |
| Total | 5 | 14 | 7 | 1 | | 27 | 99.9 |

## Chapter 9

# The Qalkhan Occupations

Qalkhan assemblages, found at four sites within the Jebel Qalkha study area (Figure 9.1), constitute a newly defined industry that appears to be the earliest within the Epipaleolithic succession of southern Jordan. Based upon stratigraphic evidence from two of these sites and technotypologic correlations with radiocarbon dated horizons from sites in the Azraq Basin of northeastern Jordan, the Qalkhan would appear to date to within an interval stretching from ca. 20,000–15,000 B.P.

Certain technotypologic characteristics, especially those linked to the microburin technique, occurred much earlier in the Qalkhan than within other complexes of the Levantine Epipaleolithic. Aside from the Qalkhan, we see the earliest evidence for an intensive use of the microburin technique within the Mushabian Complex beginning ca. 14,000 B.P. and extending from Sinai across the Negev into southern Jordan.

In regard to the cultural-historic framework of the Levantine Epipaleolithic, the Qalkhan appears to fill the void immediately preceding the Mushabian, at least in the region extending from southern through eastern Jordan. This would define an alternative developmental path within a portion of the arid zone that parallelled the cultural-historic sequence represented by the Kebaran–Geometric Kebaran–Natufian succession centered in the Mediterranean woodlands. From an evolutionary perspective, the Qalkhan then contributes to the evidence for a dichotomy between steppe-desert and woodland adaptations that stretched over much of the Epipaleolithic. In pointing to an origin for the Mushabian, the Qalkhan also indicates that this particular developmental track emerged within the Levant, most likely within the arid belt east of the Rift Valley, and not within northeast Africa as previously speculated (Henry 1974, 1989; Phillips and Mintz 1977; Bar-Yosef 1981; Bar-Yosef and Belfer-Cohen 1989).

### SITE SETTINGS AND RESOURCES

Three of the sites (J405, J406b, and J407) were discovered and tested during survey of the Wadi Humeima in 1980 (Henry 1982a; Jones 1983). Continued work was undertaken at the site of Wadi Humeima (J406b) during the winter of 1983 and a Qalkhan horizon was exposed in the deep deposit of the Tor Hamar (J431) rockshelter during the summer of 1988. Although the Qalkhan was initially defined on the basis of small assemblages recovered from shallow (30 cm) deposits at J405, J406b, and J407, the subsequent excavations of J406b and J431 furnished much larger assemblages from thicker stratified deposits. The deposits of these two sites not only provided an opportunity for better defining the Qalkhan, but also allowed for placing the Qalkhan in stratigraphic context relative to other Epipaleolithic complexes (i.e., the Kebaran and Mushabian).

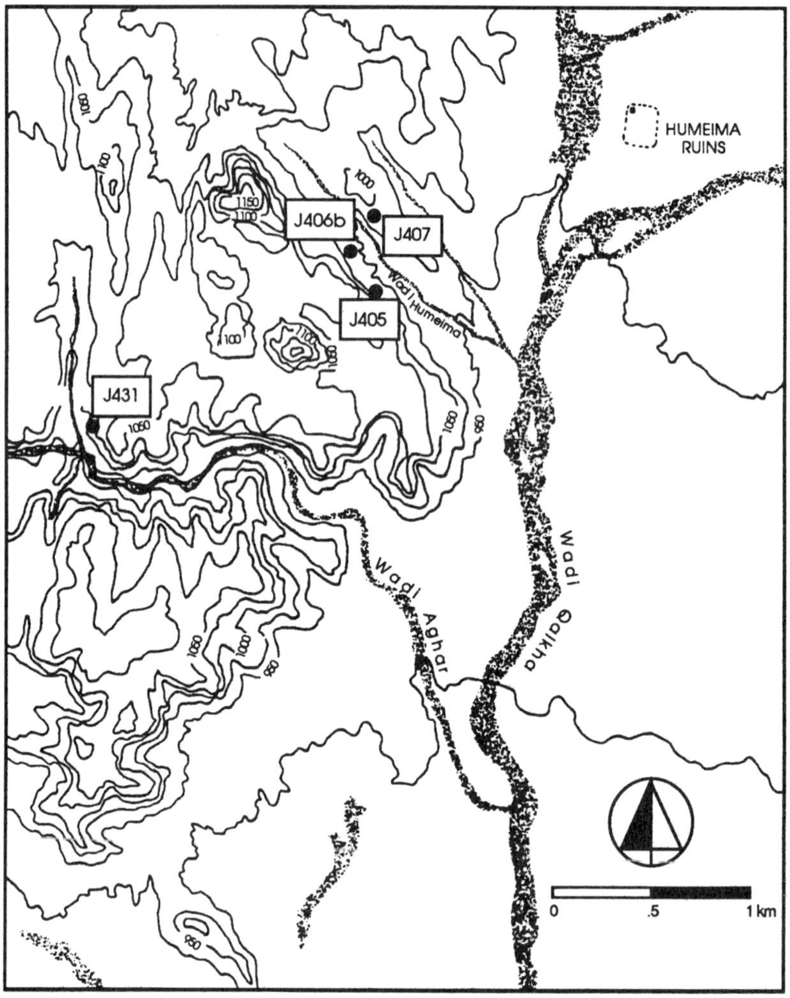


**Figure 9.1.** Map of Jebel Qalkha study area showing the locations of Qalkhan sites.

The three sites (J405, J406b, J407) located along the Wadi Humeima are in open settings positioned along the slopes that emerge from the northeastern escarpment of Jebel Humeima and cover areas ranging from 50–200 $m^2$. Although none of these sites are located closer than 50 m to the cliff-face, they can be said to have northeastern exposures in that the high jebel blocks both wind and sunlight from the south and west. In contrast, the occupation at Tor Hamar (J431) is within a rockshelter with a southwestern exposure. All of the sites fall within an elevational belt ranging from 950–1,020 masl.

Aside from the surface water that would have been available from the nearby wadi during the winter wet season, water is likely to have been found in the bedrock pools of the Wadi Aghar near Tor Hamar into the spring. Seeps and springs also occur along the base of the cliff overlooking the sites of the Wadi Humeima. Although none of these observed to flow during the ten years of study in the area, concentrations of vegetation, travertine, and a rock-cut cistern at one of the locations suggest at least some flow in the recent past (Farrand 1984).

## SITE PLANS AND STRATIGRAPHY

Following their discovery in 1980, the three sites along the Wadi Humeima were surface collected and test excavated (Jones 1983). All of the sites were small (ranging from 50–200 $m^2$) and apparently shallow with none of the soundings showing cultural materials beneath 40 cm. Because of time constraints, the largest and richest of these sites (J406b) received the least attention, however. So in 1983, additional soundings were made at Wadi Humeima (J406b) and these revealed a much deeper (+1.1m) deposit with a thick Qalkhan horizon overlain by 50 cm of Early Hamran artifacts. It was not until 1988 that another Qalkhan occupation was discovered. This 50–90 cm thick horizon was found in the excavation of Tor Hamar (J431) interstratified between Mushabian and Upper Paleolithic (Ahmarian) layers.

### Wadi Humeima (J406b)

The site consists of two concentrations of artifacts distributed on the surface of a gentle slope that falls away from the cliff-face of Jebel Humeima to the wadi (Figure 9.2). The higher and smaller of these concentrations (J406a) represents a Late Natufian occupation (discussed in Chapter 12). Excavation of two 1 $m^2$ units (1 and 3) near the lower edge of the Qalkhan concentration revealed bedrock at a depth of 25–30 cm (Figure 9.3). An extensive exposure of bedrock in the shallow drainage west of the concentration suggests that at least the lower portion of the site rests on a gently sloping bedrock bench, but the deposit thickens significantly upslope as evidenced by the +1.1m depth of units 2 and 4 without encountering bedrock. These units display a uniform red sand (2.5YR 4/5) throughout their depth. Two rock lenses (20 cm and 50 cm below surface) probably reflect scree deposits on old surfaces. Their near-level bedding suggests that these paleosurfaces followed more gentle slopes than that of the modern surface. The lower of these conforms to a cultural break separating Qalkhan from the overlying Early Hamran (Kebaran Complex) horizon.

### Tor Hamar (J431); Layer $E^2$

The Qalkhan occupation in the Tor Hamar rockshelter is defined only within the outside, downslope Block II of the excavation (Figure 8.5). Over most of Block II, Qalkhan artifacts were found within a powdery strata ($E^2$) of ash and sand that underlies the surface Layer A which contains Mushabian artifacts (Figure 9.4). Within the upslope units (8 and 10) of the block, a wedge of very compact, burnt sediment with a high density of fire-cracked rocks was recorded. This wedge, lacking diagnostic artifacts, may represent an extension of Layer $E^1$ (a prominent layer of burning within Block I) or merely a large hearth. In any case, it is clear that the thick Mushabian horizon of Block I (composing Layers A–D and perhaps E) pinches out considerably downslope. Layer F, consisting of a yellowish red silt, was exposed at the base of Block II, but failed to yield diagnostic artifacts.

## ARTIFACTS

The deposits of the four sites were partitioned to yield six artifact assemblages (Table 9.1). Artifacts recovered from the excavations at site J405 and J407 were lumped into single assemblages for each of the sites. The small samples of artifacts from these sites coupled with their recovery from shallow, nonstratified deposits justified this approach.

A single assemblage was also formed from the lower levels (50–110 cm) of units 2 and 4 at Wadi Humeima (J406b). Again, an absence of natural stratigraphy and an interest in obtaining adequate sample sizes for comparative purposes prompted this grouping. It should be noted that the decision to group the lowest 60 cm of the two units was made after level by level com-

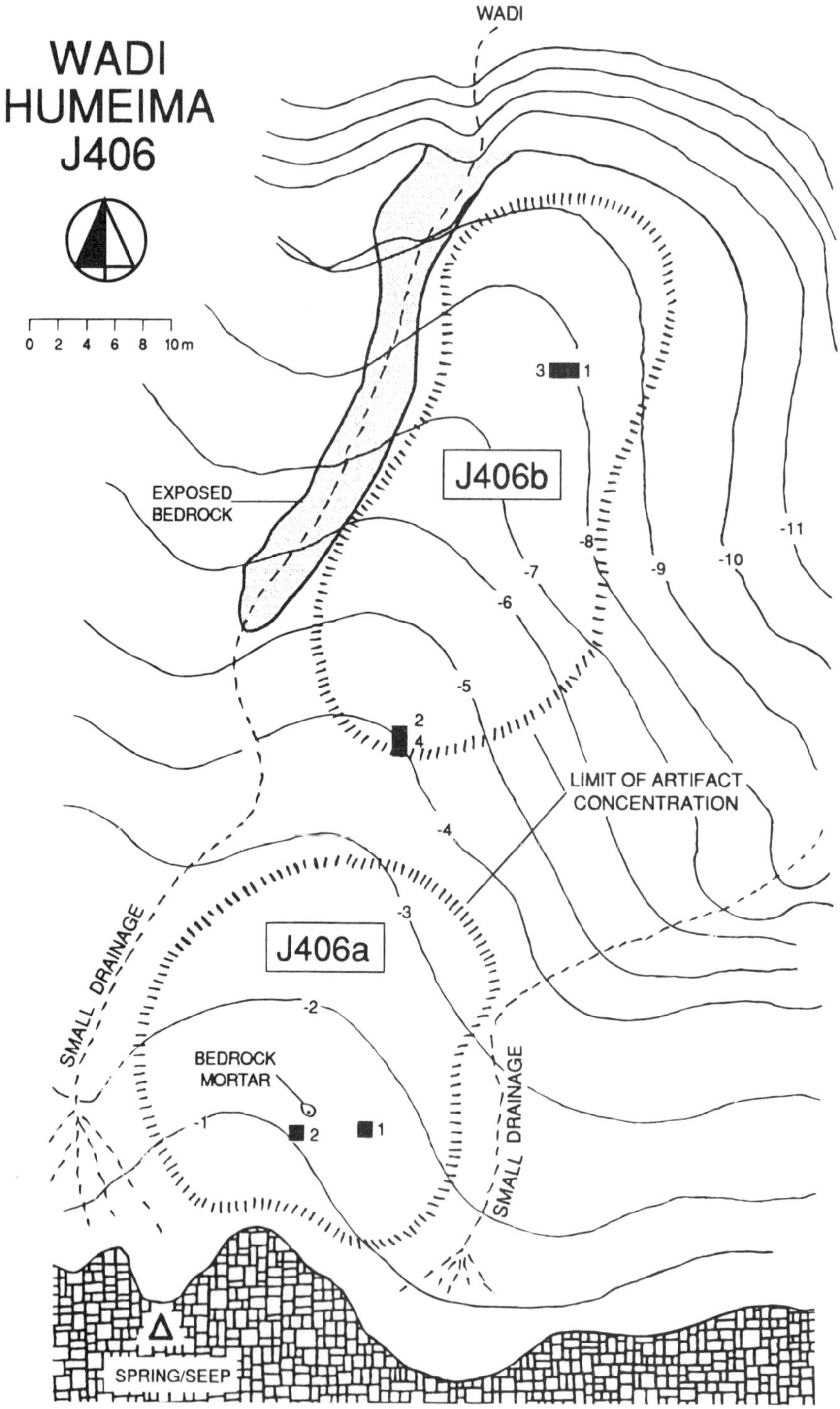


**Figure 9.2.** Contour map and site plan of the site of Wadi Humeima (J406). Note artifact concentrations (a and b) that define distinct occupations.

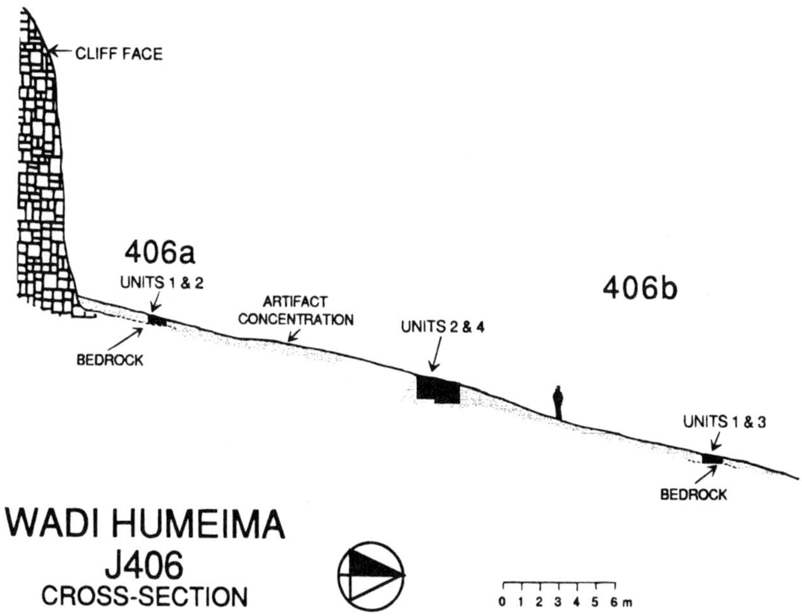


**Figure 9.3.** Cross-section of Wadi Humeima (J406) showing excavation units and deposits.

parisons defined a major typologic break between this and the upper 50 cm of the deposit. Qalkhan points and narrow arched backed pointed bladelets are restricted to the levels beneath 50 cm. The small samples of artifacts recovered from the shallow (30 cm) units 1 and 3 were not included in the technotypologic comparisons for fear that these apparently deflated deposits had experienced some mixture of the Qalkhan and Early Hamran horizons. The 22 Qalkhan points collected from these units are included in the attribute study of this unique and apparently culturally diagnostic type, however.

The Qalkan layer ($E^2$) of Tor Hamar (J431) was partitioned into three assemblages (i.e., upper, middle, lower) in order to examine stratigraphic trends. Each of these yielded adequate artifact samples.

## Typology

Qalkhan assemblages contain toolkits dominated by nongeometric microliths with narrow arched backed and pointed bladelets, straight backed bladelets with truncations opposite snaps, medial segments of straight backed bladelets, and La Mouillah points as the most prominent types (Appendix 9.1). Although occurring in lesser numbers, large triangular points (Qalkhan points) consistently are present within the assemblages and appear to be diagnostic of the industry.

### Scrapers

Endscrapers on flakes and blades dominate the class with the latter occurring most frequently (Figures 9.5, 9.6, 9.7). Most are well formed with relatively wide, symmetric working-bits that are often accompanied by lateral retouch. Although much less common, carinated, thumbnail, and circular types represent a small proportion of most assemblages.

Three of the assemblages (J406b-low; J431 $E^2$-up and mid) also exhibit very high proportions of scraper working-bits that have been removed either by breakage or by a chamfered blow (Figure 9.7:u). The removal of a dull bit by a lateral blow may well have served as a quick and

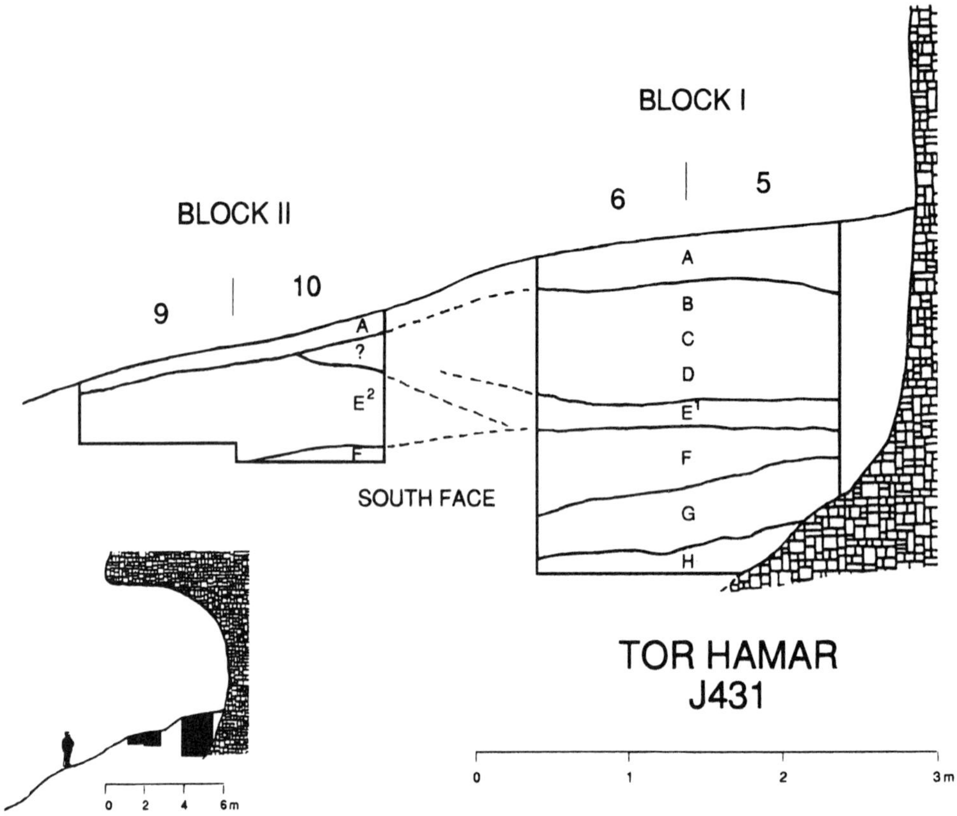


Figure 9.4. Cross-section and stratigraphy of Tor Hamar (J431).

efficient means of resharpening. As might be expected the bodies of such specimens are relatively rare (Figure 9.5:j). Their contextual association with the complete scrapers and bits does indicate, however, that a wide range of activities linked to the manufacture, use, rejuvenation, and discard of scrapers was undertaken at the two sites.

### Burins

This tool class, poorly represented in Qalkhan assemblages, is comprised mainly of angle burins struck from snaps and old surfaces and to a slightly lesser extent dihedral forms (Figure 9.5:i,j). The angle burins occur mainly on blades, whereas the dihedral burins appear mostly on flakes.

### Geometric Microliths

When present, geometrics occur only in the form of small trapeze/rectangles. In most of the assemblages they account for less than 1% of the tools and never exceed 3.6%.

### Nongeometric Microliths

Along with the points, this class contains types that are most diagnostic of the industry. Of particular importance are narrow arched backed and pointed bladelets, arched backed bladelets, straight backed and truncated bladelets opposite snaps, and medial segments of straight backed bladelets. Although included within the Point Class, La Mouillah and Qalkhan points also are

Table 9.1. Artifact inventories for Qalkhan assemblages

| Assemblage | J405[a] | J406b[b] | J407[a] | J431E[b] UP | J431E[a] Mid | J431E[b] Low |
|---|---|---|---|---|---|---|
| *Tools* | | | | | | |
| Scraper | 12.2 | 9.5 | 2.0 | 11.9 | 11.8 | 18.8 |
| Burin | 6.1 | 3.2 | 5.9 | 0.7 | 2.2 | 8.9 |
| Geometric | 0.0 | 0.3 | 0.0 | 0.7 | 2.2 | 3.6 |
| Nongeom | 2.0 | 21.8 | 11.8 | 58.9 | 57.0 | 21.4 |
| Abr_ret | 0.0 | 6.1 | 0.0 | 0.0 | 0.0 | 16.1 |
| Truncation | 0.0 | 4.8 | 0.0 | 3.0 | 0.0 | 3.6 |
| Retouched PC. | 49.0 | 34.0 | 54.9 | 10.0 | 9.7 | 14.3 |
| Denticulate | 4.1 | 4.0 | 0.0 | 1.5 | 0.0 | 0.0 |
| Notch | 20.4 | 11.1 | 5.9 | 3.3 | 5.4 | 3.6 |
| Multiple TL. | 0.0 | 0.3 | 0.0 | 0.0 | 0.0 | 0.0 |
| Perforator | 4.1 | 1.1 | 5.9 | 0.7 | 0.0 | 0.0 |
| Utilized PC. | 0.0 | 0.8 | 0.0 | 0.4 | 0.0 | 0.0 |
| Points | 2.0 | 3.2 | 13.7 | 8.9 | 11.8 | 9.8 |
| Varia | 0.0 | 0.0 | 0.0 | 0.0 | 0.0 | 0.0 |
| Total No. | 49 | 377 | 51 | 270 | 93 | 112 |
| *Debitage* | | | | | | |
| Core | 3.6 | 2.2 | 2.4 | 1.5 | 1.9 | 1.8 |
| Blade | 50.9 | 31.5 | 55.0 | 46.4 | 58.6 | 54.8 |
| Flake | 30.7 | 53.7 | 29.6 | 24.6 | 33.4 | 30.1 |
| Core Trom. El. | 1.2 | 1.9 | 4.8 | 2.1 | 2.0 | 1.5 |
| Primary El. | 13.0 | 8.9 | 6.5 | 6.8 | 8.3 | 9.6 |
| Microburin | 0.6 | 1.7 | 1.7 | 7.6 | 2.1 | 1.6 |
| Burin Spall | 0.0 | 0.0 | 0.0 | 0.2 | 0.4 | 0.6 |
| Total No. | 332 | 1396 | 291 | 2263 | 991 | 947 |
| *Debris* | | | | | | |
| Chips | | 96.4 | | 94.3 | 94.9 | 95.9 |
| Chunks | | 3.6 | | 5.7 | 5.1 | 4.1 |
| Total No. | 119 | 2019 | 94 | 2857 | 1828 | 1638 |
| *Total Artifacts* | 500 | 3792 | 436 | 5390 | 2912 | 2697 |

[a]Jones 1983
[b]Limited to Units 2&4

microlithic in scale and are in many ways related technologically and most likely even functionally (i.e., in serving as armatures) to the nongeometric microliths.

***Narrow arched backed and pointed bladelets***, the most common microlith, account for 15–31% of the class, including points (Table 9.1; Figure 9.8). Specimens are bi-pointed and spike-like with relatively narrow and thick cross-sections (Figure 9.7; Table 9.2). Most display some portion of a microburin scar that is reflective of the means by which specimens were truncated prior to their ends having been retouched into points.

***Arched backed bladelets***, present in moderate frequencies, are substantially wider than the narrow, pointed forms and typically display heavy, abrupt retouch all the way to their tips (Figure 9.7).

***Straight backed and truncated bladelets opposite snaps*** occur in moderate numbers. They are typically trapezoidal in form with the unretouched end having been produced by a snap as opposed to the microburin technique (Figure 9.7).

***Straight backed bladelet medial segments*** also appear in moderate numbers. Specimens take the shape of trapezes with the short side created by a straight backed edge and the ends

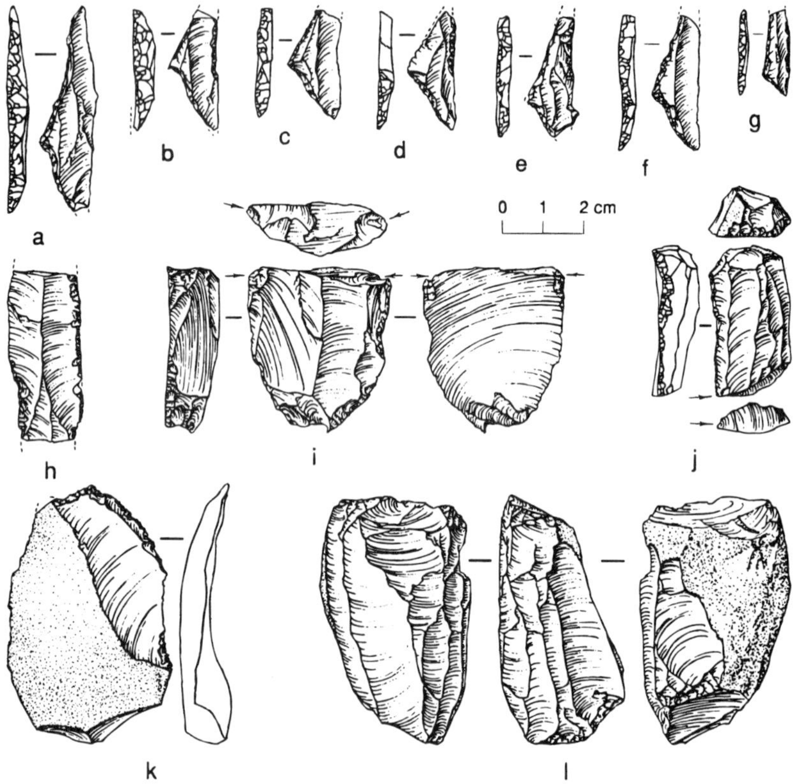


Figure 9.5. Illustrations of artifacts from Wadi Humeima (J406b): (a–g) Qalkhan points, (h) retouched blade, (i,j) burins opposite scrapers, (k) retouched primary flake, (l) opposed platform blade core.

formed by snaps (Figure 9.7). The microburin technique was rarely used in truncating specimens.

Dimensionally, the two straight backed and truncated types are near identical and restricted to relatively tight length, width, and thickness distributions (Table 9.2). This implies that rather than representing fragments of other types (e.g., straight-backed bladelets, trapezes), the "snaps" that determined their lengths were intentional and deliberate.

## Retouched Pieces

Although this is a prominent tool class in all of the assemblages, it is much more strongly represented in the assemblages from the sites along the Wadi Humeima. Specimens with obverse retouch are over twice as common as those inversely retouched. Bladelets are the most common blank form, followed by flakes and then blades.

## Points

The class is mainly represented by La Mouillah and Qalkhan points, although rare Microgravette points are also present.

***La Mouillah Points*** present an interpretative problem in that they may have served as finished tools, most likely armatures, and also as an intermediate stage in the production of backed

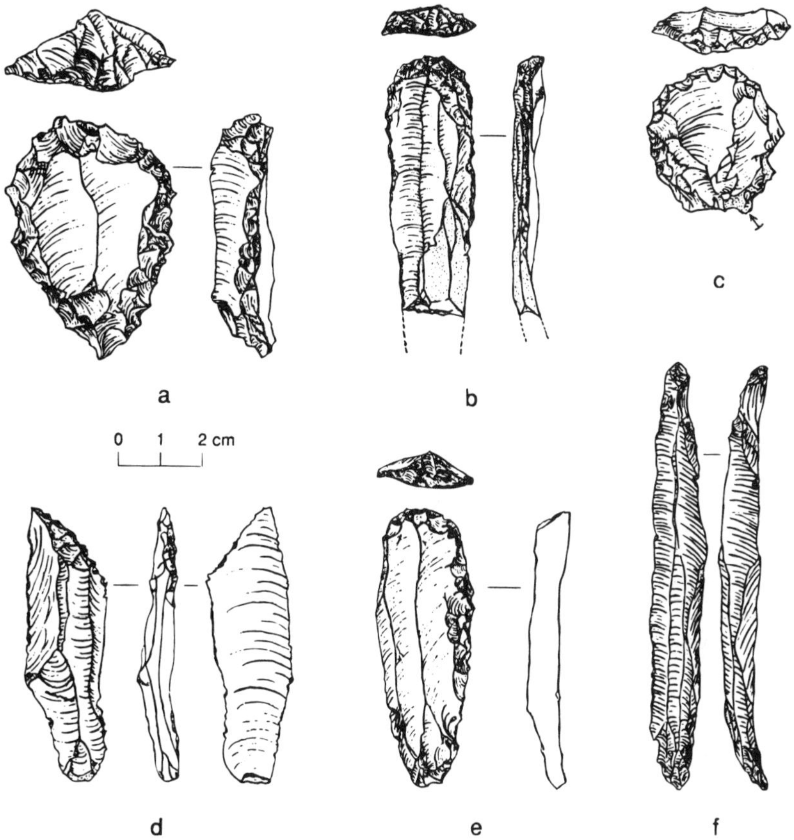


**Figure 9.6.** Illustrations of tools from Tor Hamar (J431) Layer $E^2$: (a,c) endscrapers on flakes with lateral retouch, (b,e) endscrapers on blades with lateral retouch, (d) blade with oblique truncation, (f) blade with retouched tips.

and truncated microliths such as arched backed bladelets (Figure 9.7). The production of La Mouillah points as a step toward fabrication of narrow arched backed pointed bladelets is clearly indicated by the remnants of microburin scars on the ends of specimens. A strong inverse relationship between the frequencies of La Mouillah points and narrow arched backed pointed bladelets is also consistent with the notion that La Mouillah points are reflective of a production stage. Metric data show that the mean widths and thicknesses of La Mouillah points are substantially larger than those of narrow arched backed pointed bladelets, however (Table 9.2). This implies that, at least, the largest of La Mouillah points did not result from an intermediate stage in production, but most likely served as finished tools.

***Qalkhan points*** are an apparent *fossile directeur* of the industry and, like La Mouillah points, are tied to the microburin technique in their production (Figures 9.5, 9.7). Dimensionally, they are significantly larger than other points and pointed nongeometric microliths with their length—width distributions falling near the upper range of microliths (Table 9.2).

Points were formed by heavy retouch or backing, always along the left margin of a blade or bladelet, which was terminated by a deep notch near the base of a specimen (Figure 9.9). The notch served to guide the removal of the base through the microburin technique, leaving

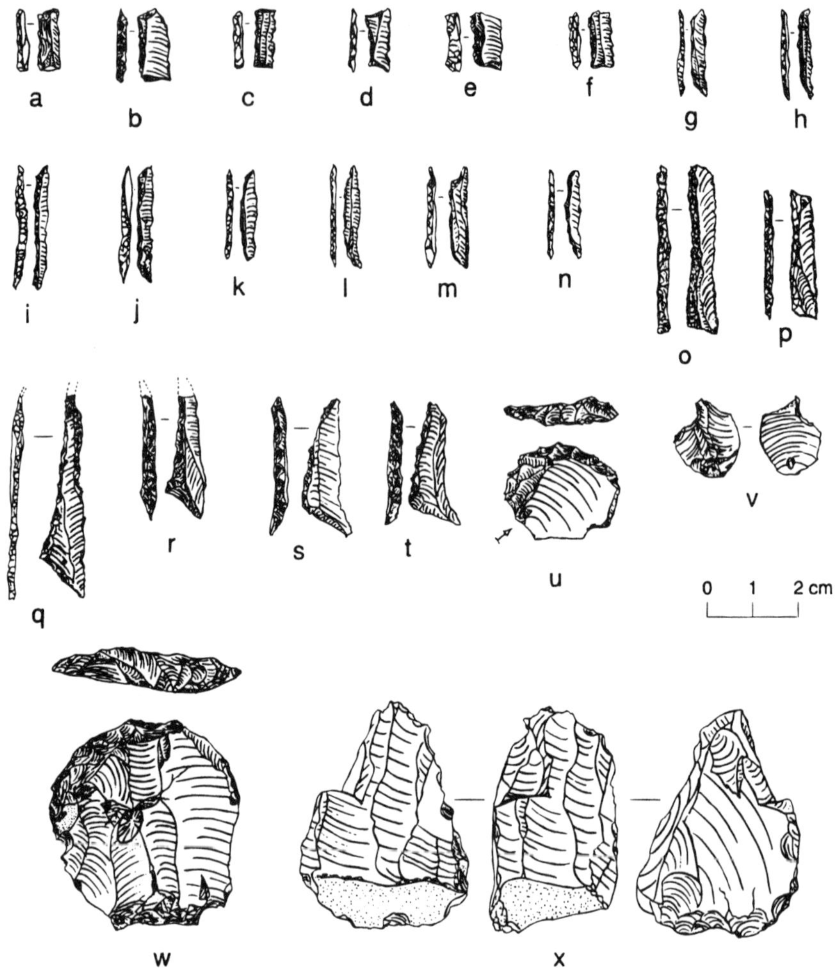


**Figure 9.7.** Illustrations of artifacts from Tor Hamar (J431) Layer E[2]: (a–d,o,p) straight backed bladelets with truncation opposite snap, (e,f) medial segments of straight backed bladelets, (g–n) narrow arch backed and pointed bladelets, (q–t) Qalkhan points, (u) bit of endscraper, (v) microburin, (w) endscraper on flake, (x) bladelet core.

a thin, pointed end and an overall form of a scalene triangle. The microburin scar was then retouched to varying degrees, most likely depending on how well the truncation fit the desired angle of the base. Some incomplete specimens were recovered in which the microburin truncation created too small an angle with the backed edge and thus prevented the formation of the basal tang (Figure 9.7:r–t). Aside from weakening a blade for its sectioning as part of the microburin technique, the lateral notch also left a spur on most specimens after final retouching and this may have facilitated the hafting of points.

## Other Tool Classes

Notches, denticulates, and perforators generally account for small proportions of Qalkhan toolkits. As with the other classes, blade and bladelet blanks are dominant for these tool forms.

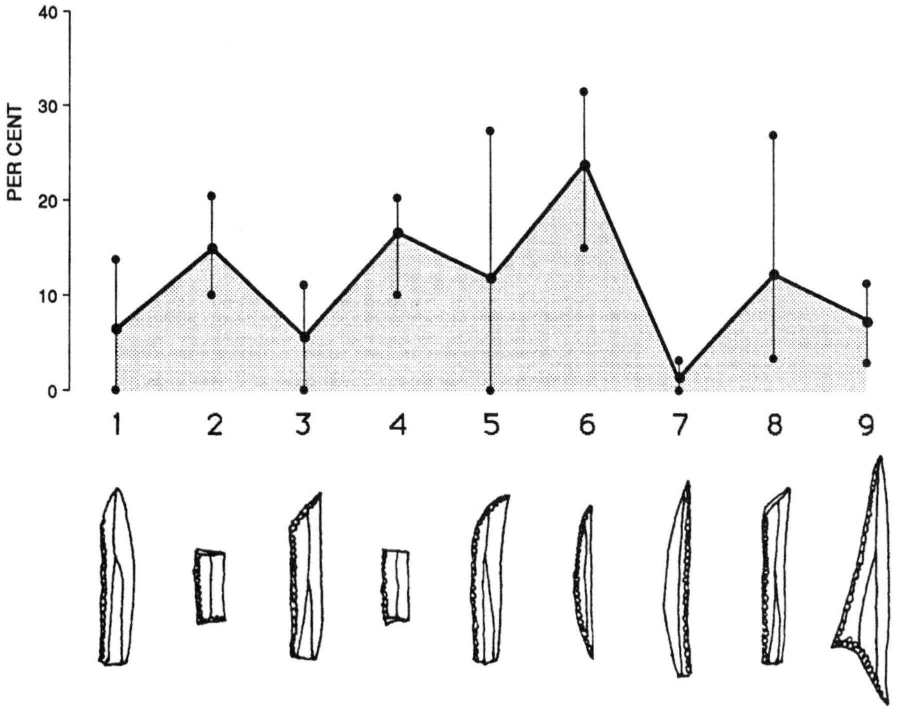


**Figure 9.8.** Histogram comparing the relative proportions of the major microliths for Qalkhan assemblages. Means are designated by the heavy line and ranges by the lighter, vertical line. Note that types 2,4,6, and 9 are well represented and tightly distributed about the means.

## Technology

The Qalkhan lithic technology focussed on the production of relatively wide bladelets from wedge-shaped (subpyramidal) single platform cores. The industry also is characterized by use of the microburin technique whereby bladelets were sectioned as a step toward their fabrication into points and nongeometric microliths.

### Raw Material

Raw material was obtained from *in situ* chert beds on the plateau and from small cobbles derived from these beds and found in the nearby Wadi Qalkha. Cortex when present on cores occurs slightly more often as an abraded, water-worn surface than as limestone, thus suggesting a slightly greater use of the nearby wadi cobbles than the more distant sources. Interestingly, the nearby Humeima chert nodules were seldom used and Rift Valley chert was not recorded in any of the assemblages. Although present in several assemblages, artifacts made from quartzite are rare.

### Cores

The most common approach to the shaping of cores appears to have followed a sequence initiated by the removal of one end of a small ovoid cobble or nodule for creation of a platform (Figure 9.10). An elongated primary element was then struck along the "edge" of the cobble

Table 9.2. Comparisons of the Dimensions of Dominant Microliths

| | J406b | J431E$^2$UP | J431E$^2$MD | J431E$^2$LOW |
|---|---|---|---|---|
| Straight-Backed Bladelet with Truncation opposite Snap | (15) | (13) | (7) | (5) |
| Mean Length | 15.7 | 13.4 | 14.8 | 14.9 |
| Mean Width | 6.2 | 6.2 | 6.7 | 5.1 |
| Mean Thickness | 2.4 | 2.4 | 2.4 | 2.2 |
| Medial Segment of Straight Backed Bladelet | (13) | (13) | (8) | (0) |
| Mean Length | 14.8 | 14.0 | 14.4 | 0.0 |
| Mean Width | 6.2 | 5.6 | 5.7 | 0.0 |
| Mean Thickness | 2.5 | 2.3 | 2.3 | 0.0 |
| Arched Backed Bladelet | (11) | (27) | (6) | (6) |
| Mean Length | 18.4 | 23.4 | 20.1 | 22.4 |
| Mean Width | 7.2 | 5.5 | 6.4 | 5.1 |
| Mean Thickness | 2.8 | 2.7 | 3.3 | 2.0 |
| Narrow Arched Backed and Pointed Bladelet | (23) | (28) | (13) | (5) |
| Mean Length | 17.8 | 21.6 | 18.9 | 19.9 |
| Mean Width | 4.1 | 3.6 | 3.5 | 3.8 |
| Mean Thickness | 1.9 | 1.9 | 2.0 | 1.7 |
| La Mouillah Point | (2) | (15) | (3) | (2) |
| Mean Length | 22.3 | 21.2 | 20.1 | 20.7 |
| Mean Width | 8.1 | 5.3 | 5.7 | 5.1 |
| Mean Thickness | 2.7 | 2.2 | 2.6 | 2.4 |
| Qalkhan Point | (25)[a] | (4) | (1) | (5) |
| Mean Length | 34.6 | 36.4 | 35.4 | 37.4 |
| Mean Width | 10.1 | 9.2 | 14.1 | 9.5 |
| Mean Thickness | 3.3 | 2.9 | 3.6 | 2.5 |

[a]Includes specimens from all units.

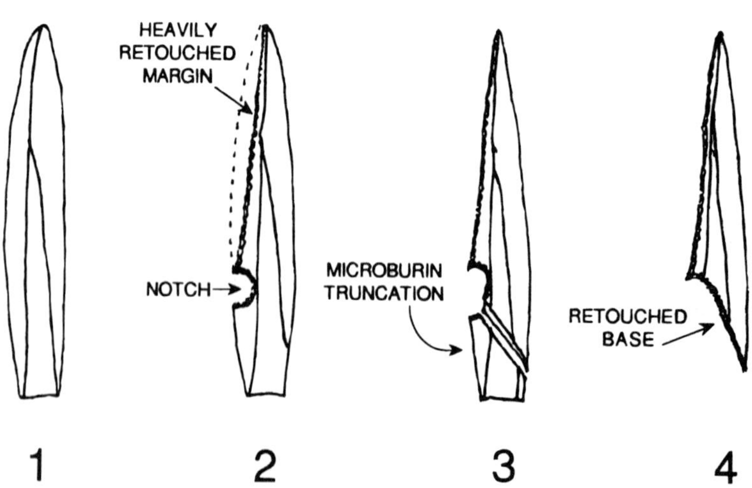


Figure 9.9. Illustration of the steps involved in the fabrication of a Qalkhan point.

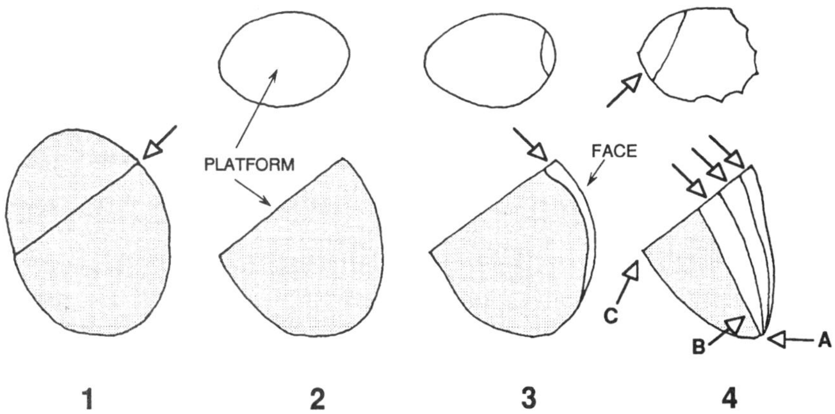


**Figure 9.10.** Illustration of the methods employed in core shaping and rejuvenation. 4A–C represent alternative means of platform renewal.

and this scar was then used to guide the removal of subsequent bladelets. In not employing the *lame à crête* technique to initially shape cores for bladelet production, the Qalkhan approach may well have been somewhat more efficient for reducing small cobbles and nodules. But in lacking an initial narrow, well-defined ridge to guide removals, Qalkhan cores were prone to yielding bladelets that were relatively wide for their lengths. After exhaustion most cores were abandoned without efforts to extend bladelet production through the creation of additional platforms (Table 9.3). When efforts were made to extend the lives of cores, Qalkhan knappers selected between three options (Figure 9.10) in a roughly balanced manner and generated "twisted," "opposed," and "ninety-degree" cores.

As Jones (1983) noted in her initial study, the Qalkhan assemblages from the Wadi Humeima sites show an emphasis on initial processing and this pattern is duplicated in the assemblages from Tor Hamar (J431). Relatively high frequencies of cores and primary elements (Table 9.1) certainly imply extensive core shaping and reduction within the occupations.

A dimensional comparison of the debitage indicates that most of the cores were exhausted at the time of abandonment, with the mean lengths of core facets measuring some 2–5 mm less than mean bladelet lengths. What is so interesting here is that few of the exhausted cores show evidence of rejuvenation such as the formation of new platforms. The great majority are associated with unfacetted, single platforms. Also, Jones (1983) noted in her study that 60–80% of the cores showed less than 50% of their platform circumferences had been utilized. The simplest explanation for why so few of the cores show signs of rejuvenation may be linked to the initial size of the raw material (i.e., chert nodules and stream cobbles) relative to the desired dimensions of the bladelets. It may well be that once production of bladelets from the initial platforms had ended, the overall dimensions of cores were too small to generate new platforms that would yield the desired blank dimensions. This notion is supported by the dimensional data for Qalkhan cores in which their facets and overall dimensions are quite similar.

## Bladelets

The bladelet attributes of Qalkhan assemblages are remarkably homogeneous and consistent with having been struck predominantly from cores with single, unfacetted platforms (Figure 9.11, Appendix 9.3).

Table 9.3. Attributes of Cores from Qalkhan Assemblages

| | Assemblages | | | | | | | |
|---|---|---|---|---|---|---|---|---|
| | J406b | | J431E$^2$UP | | J431E$^2$MID | | J431E$^2$LOW | |
| Core Attributes | Mean | Range | Mean | Range | Mean | Range | Mean | Range |
| Core Length | 38.5 | 18–62 | 39.8 | 18–72 | 32.9 | 24–48 | 35.8 | 19–51 |
| Core Width | 35.6 | 24–51 | 38.1 | 19–59 | 38.2 | 30–54 | 29.9 | 16–43 |
| Facet Length | 30.7 | 13–56 | 32.8 | 17–71 | 24.5 | 20–34 | 28.9 | 18–47 |
| Facet Width | 12.8 | 6–28 | 16.2 | 6–45 | 11.2 | 8–13 | 14.4 | 6–25 |
| Weight | 37.9 | 8–100 | 48.1 | 7–148 | 44.9 | 20–109 | 26.7 | 4–65 |
| N | 19 | 34 | 4 | 19 | | | | |
| *Platform Orientation* (%) | | | | | | | | |
| Unidirectional | 68.4 | | 76.5 | | 50.0 | | 68.5 | |
| Opposed | 15.8 | | 17.7 | | 25.0 | | 10.5 | |
| Twisted | 5.3 | | 2.9 | | 25.0 | | 10.5 | |
| Change/90 | 10.5 | | 2.9 | | 0 | | 10.5 | |
| *Platform Facetting* | | | | | | | | |
| Single | 42.1 | | 64.7 | | 75.0 | | 79.0 | |
| Faceted | 57.9 | | 35.3 | | 25.0 | | 15.7 | |
| Cortex | 0 | | 0 | | 0 | | 5.3 | |
| *Platform Regularization* | | | | | | | | |
| Present | 94.7 | | 97.1 | | 100.0 | | 68.4 | |
| Absent | 5.3 | | 2.9 | | 0 | | 31.6 | |
| *Cortical Cover* | | | | | | | | |
| Absent | 31.6 | | 58.8 | | 50.0 | | 79.0 | |
| Limestone | 31.6 | | 11.8 | | 25.0 | | 10.5 | |
| Water Worn | 36.8 | | 29.4 | | 25.0 | | 10.5 | |
| *Core Shape* | | | | | | | | |
| Pyramidal | 0 | | 14.7 | | 25.0 | | 26.3 | |
| Wedge (Sub-pyr.) | 84.2 | | 79.4 | | 75.0 | | 63.2 | |
| Cuboid | 10.5 | | 5.9 | | 0 | | 10.5 | |
| Irregular | 5.3 | | 0 | | 0 | | 0 | |

Bladelets are relatively wide to their lengths with length:width ratios averaging about 2.6:1. The large majority display curved lateral profiles and unfacetted platforms, an expression of their having been removed from wedge-shaped cores with single, unfacetted platforms. An unexpected high frequency of bladelets exhibit diffuse bulbs, an attribute linked to hard-hammer percussion. Pointed and lipped bulbs, normally associated with more precise forms of loading (i.e., punch and soft-hammer techniques), are found on only about 19% of the bladelets. This configuration may indicate a shift from hard-hammer to soft-hammer and punch technique during the course of bladelet production. A predominance of pointed distal shapes, converging lateral edges, and parallel facet patterns are all consistent with the previously discussed core characteristics. The very high proportion of bladelets with feathered terminations denotes the fine control that Qalkhan knappers maintained over removals, even when employing hard-hammer percussion.

## Microburin Technique

Although microburins are present in all Qalkhan assemblages, they yield variable indices, especially when only traditional IMbt's are compared (Table 9.4). Variability is significantly reduced when restricted (i.e., confined to geometrics, nongeometric microliths, and points) mi-

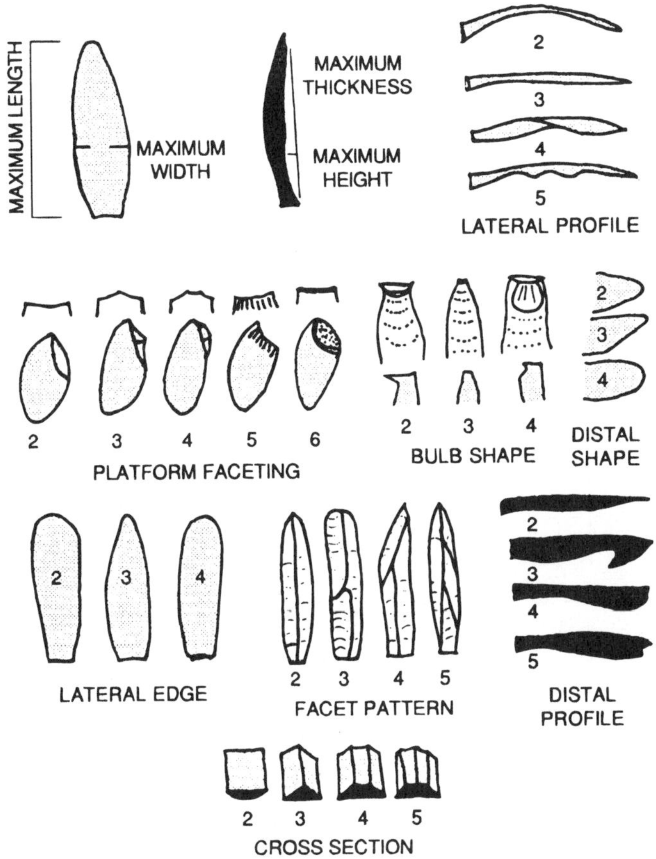


**Figure 9.11.** Schematic illustration of technological attributes observed for blades.

croburin indices are compared. The microburins are large relative to other Epipaleolithic industries associated with the technique. This perhaps stems in part from their production as part of sectioning large bladelets in the fabrication of Qalkhan points. The large majority of Qalkhan microburins represent the proximal end of a bladelet and originate from a notch formed on the left edge. While this pattern mirrors the procedure by which the bases of bladelets were removed in the production of Qalkhan points, many of the proximal, left-oriented microburins are too small to have been associated with point production and must have been tied to the fabrication of other microliths.

## COMPARISONS AND CONCLUSIONS

### Local Settlement and Procurement Patterns

Although the settings of the open-air sites along the Wadi Humeima differ significantly from that of the Tor Hamar rockshelter, they do not appear to have induced major differences in prehistoric activities, at least, as can be detected from artifact analyses. Artifact densities (Table 9.5), technological attributes, and typological profiles are remarkably similar across Qalkhan assemblages.

Table 9.4. Attributes of Microburins from Qalkhan Assemblages

| | J405 | J406b | J407 | J431E²UP | J431E²MD | J431E²LW |
|---|---|---|---|---|---|---|
| *Indices* | | | | | | |
| IMbt | 3.9 | 005.9 | 8.9 | 038.9 | 18.4 | 011.8 |
| rIMbt | 50.0 | 20.3 | 27.8 | 48.0 | 24.1 | 27.8 |
| N | 2 | 24 | 5 | 172 | 21 | 15 |
| Tools | 49 | 377 | 51 | 270 | 93 | 112 |
| rIMbt Tools | 2 | 94 | 13 | 186 | 66 | 39 |
| *Orientation* (%) | | | | | | |
| Proximal | — | 83.3 | — | 63.8 | 67.9 | — |
| Distal | — | 17.7 | — | 37.2 | 32.1 | — |
| Right | — | 25.0 | — | 23.4 | 17.9 | — |
| Left | — | 75.0 | — | 77.6 | 82.1 | — |
| *Dimensions* (mm) | | | | | | |
| Mean Length | — | 8.6 | — | 13.8 | 15.5 | — |
| Mean Width | — | 8.8 | — | 11.5 | 9.0 | — |

Table 9.5. Comparison of the Artifact Densities of Qalkhan Occupations

| | | $m^3$ | Artifacts | Density N Art/0.1 $m^3$ |
|---|---|---|---|---|
| J406b units 2 & 4 | | 1.2 | 3,786 | 316 |
| J431E² | Upper | 1.2 | 5,390 | 449 |
| | Middle | 0.9 | 2,912 | 324 |
| | Lower | 0.9 | 2,689 | 299 |

Tools do account for a higher proportion of the artifact inventories in the Wadi Humeima sites, but this may well be an expression of the differences in depositional environments between the sites. In having been recovered from unsheltered, alluvial fan deposits, the Wadi Humeima assemblages experienced greater sheet wash and natural abrasive forces than those artifacts collected from the protected deposit of Tor Hamar. Light, marginal retouch induced by natural abrasion is difficult to distinguish from cultural retouch. Therefore, artifacts buried within these clastic deposits are likely to have had greater numbers recorded in the "retouched piece class" and this would have inflated the overall tool inventories of the Wadi Humeima assemblages.

The Qalkhan occupations appear to represent relatively ephemeral encampments that covered small areas (ranging from 50–200 $m^2$). With the exception of Tor Hamar, the mostly shallow (30–60 cm) cultural deposits suggest that they were irregularly revisited. In that the *sondages*, representing from only 1–4% of the site areas, yielded assemblages with roughly similar technotypologic profiles, there is a high probability that the occupations as a whole were relatively homogeneous with little in the way of spatially segregated and specialized activities.

The prominence of primary elements and cores denotes an emphasis on initial processing activities (i.e, core shaping and blank production) within the occupations, but the presence of microburins, reworked scrapers, and basal fragments of points indicates that a full range of reduction activities was undertaken at the sites. Beyond generating blanks, primarily as bladelets, Qalkhan knappers engaged in tool production, use, and rejuvenation—all apparently within the same locus. The toolkits are dominated by types that are most likely to have served as armatures. Given their light weights, they probably tipped arrows. The next most prominent formally retouched tool classes include scrapers and notches, both of which are linked to a hunting toolkit. Scrapers would have served in the processing of hides, and notches, for the fabrication of shafts.

In the absence of organic evidence, clues to the environmental setting and seasonality of the occupations come from indirect sources, namely the exposures of the sites and the seasonal availability of water. With its southwestern exposure, Tor Hamar makes a poor warm season camp, especially in the late afternoon. Also, unless the climate had been considerably more moist, then the nearby Wadi Aghar and its bedrock pools would have lacked water from the late spring until the fall rains. A winter wet-season occupation of the shelter appears the most likely. With their positions along the northeastern flank of the Jebel Humeima, the open-air sites of J405, J406, and J407 are exposed to prevailing winter winds and lose sunlight in the early afternoon. These conditions, coupled with the sites' lack of protection from precipitation and their high elevations, make them unlikely places for winter camps during late Last Glacial times. During the warm season, water may not have been available in the wadis, but may have been obtained, at least during wet years, from the springs along the base of the jebel.

Although Qalkhan sites were not found in the survey of the Wadi Judayid study area, the substantial presence of plateau cherts (with limestone cortex) in the assemblages points to Qalkhan groups having spent some time in the uplands during the course of their annual round. Interestingly, not a single specimen of the black chert located northeast of Jebel Qalkha on the eastern flank of the Rift Valley was recorded among Qalkhan artifacts. Ornamental shells from the Red Sea (Chapter 17), 60 km south of the area, suggest some form of contact with the coast.

## Chronology and Regional Relationships

Stratigraphic comparisons of assemblages from the deposits of Wadi Humeima (J406b) and Tor Hamar (J431) fail to show technotypologic patterns that might be viewed as time-trends within the Qalkhan horizon. These comparisons do allow for placing the Qalkhan into a broader cultural-stratigraphic framework, however.

At Wadi Humeima (J406b), the 60 cm thick Qalkhan horizon abruptly gives way to an overlying 50 cm thick Early Hamran (Kebaran Complex) horizon. Artifacts diagnostic of the Qalkhan (i.e., narrow arched backed and pointed bladelets and Qalkhan points) are absent from the upper 50 cm of the deposit, replaced by high frequencies of typically Early Hamran microliths (i.e., straight-backed bladelets with truncations opposite snaps and medial segments). At Tor Hamar (J431), the Qalkhan horizon is confined to Layer $E^2$ where it overlies the early Ahmarian (Upper Paleolithic) horizon of Layers F and G and underlies a Madamaghan (Mushabian Complex) horizon found in Layers A–$E^1$.

The overlying Early Hamran and Madamaghan horizons provide a means of establishing "ceiling dates" for the Qalkhan. Unfortunately, only the upper part of the Madamaghan occupation (upper Layer C) has furnished radiocarbon dates (12,680 ± 320 and 12,320 ± 90 B.P.). Although not directly dated, seriation of artifacts found at the beginning of the occupation in Layer $E^2$ suggests an early placement in the Mushabian Complex sequence which is relatively well dated to between 14,170–11,700 B.P. (Henry 1989:115). The temporal placement of the Early Hamran is more problematic. This is not only because the industry lacks radiocarbon assays, but also because the cultural chronology of the Kebaran Complex is poorly understood (Bar-Yosef and Vogel 1987; Hovers and Marder 1991). Dates available for the complex do indicate, however, that it extended from ca. 19–15,000 B.P. Therefore, the radiocarbon dates of the complexes represented in the horizons overlying the Qalkhan suggest that the industry should date to before some 14,000–15,000 years B.P.

## Regional Comparisons

When looking beyond the study area (Figure 9.12), assemblages that show strong affinities with those of the Qalkhan have been reported from the nearby Petra area in southern Jordan (Schyle and Uerpmann 1988), the Azraq Basin of northeastern Jordan (Garrard *et al.* 1985, 1986, 1987, 1988; Garrard and Byrd 1992; Byrd 1990), the Yabrud rockshelter of western Syria (Rust 1950) and the El Kowm Oasis in northeastern Syria (Cauvin *et al.* 1979; Cauvin 1981; Cauvin and Coqueugniot 1990).

*Petra Area.* Two assemblages recovered from the surfaces of Sabra 3 and Adh Dhaman near Petra are sufficiently similar to those described for the Qalkhan that they should be included within the industry. Although their observations are not quantified, Schyle and Uerpmann (1988) note the similarities to the Qalkhan and illustrate numerous Qalkhan points and microburins. The presence of chamfered pieces denotes another parallel with the assemblages of the Jebel Qalkha area. While Schyle and Uerpmann (1988) also suggest affinities with Geometric Kebaran assemblages from the Mediterranean coast of Israel (i.e., Kiryath Aryeh I and Hofith),

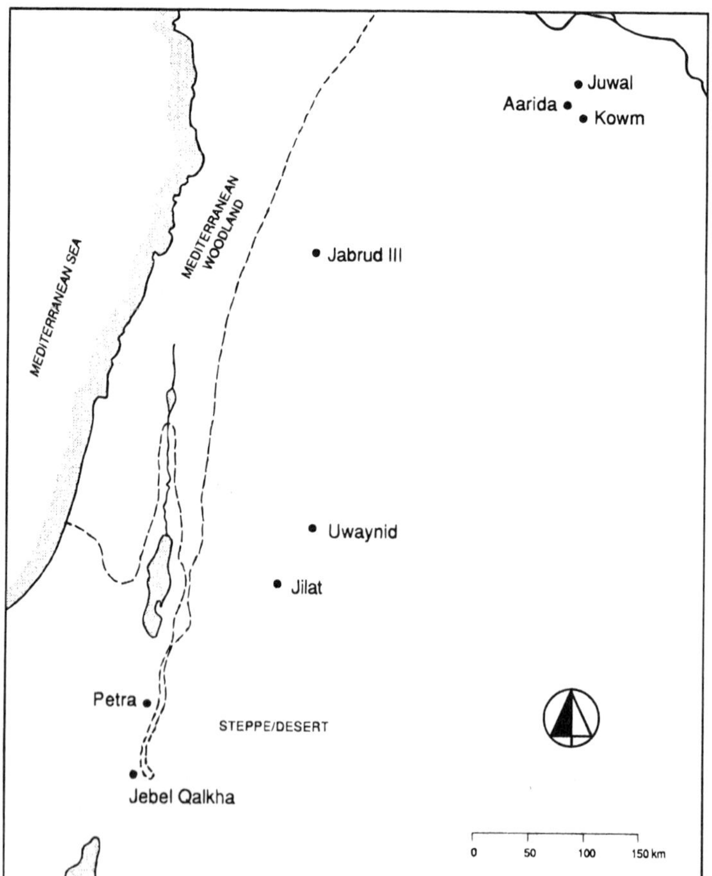


Figure 9.12. Map showing the distribution of Qalkhan assemblages relative to the major biotic zones of the Levant.

these assemblages lack Qalkhan and La Mouillah points as well as evidence for use of the microburin technique (Bar-Yosef 1970).

*Azraq Basin.* The sites of Uwaynid 14, Uwaynid 18, and Jilat 6 produced stratified assemblages that appear to represent an extension of the Qalkhan Industry to northeastern Jordan. Although Qalkhan points were found only in the middle phase of Jilat 6, other diagnostic elements of the Qalkhan such as narrow arched backed and pointed bladelets and La Mouillah points form substantial parts of the assemblages (Byrd 1990; Garrard and Byrd 1992; Table 9.6). Also, the microburin technique was consistently employed as evidenced by restricted indices ranging from 20.8–54.9 (Byrd 1990). Like the Jebel Qalkha assembalges, those of the Azraq Basin are essentially nongeometric with only the middle phase of Jilat 6 yielding a few trapeze/rectangles. Byrd (1988) notes the similarities between the Azraq Basin assemblages and a time-trend in which the narrow arched backed and pointed bladelets were replaced by larger microliths such as La Mouillah points and double truncated backed bladelets.

The Qalkhan occupations of the Azraq Basin are substantially larger (1,200–1,400 $m^2$) than those of the Jebel Qalkha area. Their cultural deposits exhibit a bimodal distribution with respect to their thicknesses, which Byrd (1990) interprets as an expression of short-term as opposed to more intensive, repeated occupations. Aside from the chipped stone artifacts, basalt grinding stones, bone points, lumps of ochre, and ornamental shells from the Mediterranean and Red Sea were also found.

As with the occupations of the Jebel Qalkha area, the contexts and artifact configurations of the Azraq Basin sites point to a subsistence strategy focussed on hunting. Byrd (1990) notes the emphasis upon the fabrication of microliths, presumably to tip projectiles, as expressed in high debitage:core and tool:core ratios and low debitage:tool ratios. While the artifact densities

**Table 9.6. Relative proportions of dominant microliths from Azraq Basin in Qalkhan assemblages from Azraq Basin**

| Assemblage Phase | Uwaynid 14[a] | | | | Uwaynid 18[a] | | Jilat 6[b] | | | |
|---|---|---|---|---|---|---|---|---|---|---|
| | Upper | | Middle | | Up 35 cm | | Middle | | Lower | |
| | N | % | N | % | N | % | N | % | N | % |
| Trapeze/Rectangle | — | — | — | | — | | 004 | 003.7 | 0 | 0 |
| Narrow Arched Backed and Pointed Bladelet | 003 | 002.8 | 042 | 025.0 | 166 | 049.3 | 4 | 3.7 | 55 | 067.8 |
| Straight Backed and Truncated Bladelet Opposite Snap[c] | 85 | 79.4 | 109 | 64.9 | 167 | 49.6 | 48 | 44.4 | 13 | 16.1 |
| La Mouillah Point | 19 | 17.8 | 17 | 10.1 | 4 | 1.1 | 49 | 45.4 | 13 | 16.1 |
| Qalkhan Point | — | — | — | — | — | — | 3 | 2.8 | — | 0 |
| N | 107 | 100.0 | 168 | 100.0 | 337 | 100.0 | 108 | 100.0 | 81 | 100.0 |

[a]Garrard et al. 1987.
[b]Garrard et al. 1986.
[c]Described as "backed fragments with truncations."

of the Azraq occupations are some 3–4 times higher on the average than those of the Jebel Qalkha area, the tool:debitage ratios of the assemblages from the two areas are quite similar. The main difference in the assemblages of the two areas rests in the much greater degree of initial processing in the Jebel Qalkha sites as evidenced by their higher proportions of cores and primary elements.

Based upon radiocarbon dates from three sites, the Azraq Basin occupations (grouped within the "Early Epipaleolithic") are thought to fall between ca. 20,000–15,500 B.P., most likely in the early half of this interval (Table 9.7; Byrd 1990; Garrard and Byrd 1992). The dates from Jilat 6 are problematic in that they come from both the Upper (post-Qalkhan) and the Middle Phase and do not agree with their stratigraphic order. This has caused the researchers to question their accuracy and suggest possible contamination (Garrard *et al.* 1988; Byrd 1990).

The Azraq occupations contain steppe fauna dominated by gazelle (*Gazella subgutturosa*), but also include the bones of lesser numbers of aurochs (*Bos primigenius*), hare (*Lepus capensis*), equid (*Equus* sp.), camel (*Camelus* sp.), wolf (*Canis lupus*), ovicaprid and tortoise (*Testudo graeca*). Ostrich eggshell fragments were retrieved, as well. In concert with the fauna, the generally arid character of the environment is indicated by occupations resting within aeolian silts. But the presence of paleosols suggests higher water tables, more vegetation, and more moist conditions than exist in the area today (Garrard *et al.* 1988).

*Yabrud III Rockshelter.* From this high elevation site in western Syria, Rust (1950) recovered assemblages from layers 4—7 (his Jüngeres Nebekian and Spätcapsien) that show several affinities to the Qalkhan. Within his artifact illustrations, Rust shows narrow arched backed and pointed bladelets, scalene triangles, probable Qalkhan points (plate 103, numbers 10 and 12), and La Mouillah points (plate 102, 9; plate 104, 20 and 21) along with microburins. He also noted a stratigraphic trend in which narrow obliquely truncated bladelets and points declined through time, while broad truncated bladelets increased. Byrd (1990) noted a similar trend within the occupations of the Azraq Basin.

*El Kowm.* Three sites in the vicinity of the El Kowm Oasis have yielded assemblages with technotypologic configurations resembling that of the Qalkhan (Cauvin et al. 1979, Cauvin 1981, Cauvin and Coqueugniot 1990). Qalkhan points, described as large triangles and "scalène tronques d'El Kowm," and La Mouillah points furnish important typologic similarities. Chamferred

**Table 9.7. Radiocarbon Dates Associated with the Azraq Basin Epipaleolithic Occupations**

| Site/Provenience | Date | Lab Number | Material |
|---|---|---|---|
| Uwaynid 14 | | | |
| Upper and Middle | 18,400 ± 250 | OXA 866 | Charcoal |
| Phases | 18,900 ± 250 | OXA 865 | Charcoal |
| Uwaynid 18 | | | |
| Trench I, Hearth | 19,500 ± 250 | OXA 868 | Charcoal |
| Trench II, Top | 19,800 ± 350 | OXA 864 | Charcoal |
| Trench II, Base, Hearth | 23,200 ± 400 | OXA 867 | Charcoal |
| Jilat 6 | | | |
| Upper and Middle | 11,450 ± 200 | OXA 523 | Charcoal |
| Phases (not in | 11,740 ± 200 | OXA 522 | Charcoal |
| stratigraphic | 15,520 ± 200 | OXA 524 | Charcoal |
| order) | 16,010 ± 200 | OXA 525 | Charcoal |

[a]All but Uwaynid, Trench II—Base Hearth and Jilat 6, Upper Phase are linked to Qalkhan assemblages (from Garrard *et al.*, 1988).

pieces also are reported at El Kowm I reminiscent of similar finds in the Petra and Jebel Qalkha areas. An emphasis on the microburin technique within the El Kowm assemblages denotes a significant technological affinity with the Qalkhan. In their comparisons, Cauvin and Coqueugniot (1990) note the similarities of the El Kowm assemblages to Jilat 6 and J407 as well as Yabrud III,3 and a Mushabian assemblage (Mushabi V).

## SUMMARY

The Qalkhan represents a newly defined microlithic industry found within the arid belt along the eastern margins of the Mediterranean woodlands stretching from southern Jordan to near the Euphrates of northeastern Syria. Based upon stratigraphic contexts and associated radiocarbon assays, it appears to date to between 20,000–15,500 B.P.

The principal typologic characteristics of the industry include Qalkhan and La Mouillah points, narrow arched backed and pointed bladelets, truncated bladelets, and low frequencies of trapeze/rectangles. Well formed endscrapers on blades and flakes along with chamferred blades form an important part of the industry's large tool component. Technologically, the microburin technique was consistently used in controlling the sectioning of bladelets. These were mainly produced from single platform, unfacetted wedge shaped cores.

The sizes of the sites vary significantly with a bimodal distribution of large (1,200–1,400 m$^2$) and small (<200$^2$) occupations with relatively thin cultural deposits. The sites appear to have enjoyed a steppic environmental setting with more moisture available than today as evidenced by geomorphic and faunal data. Many of the sites are located near active or ephemeral springs.

In that Qalkhan toolkits are dominated by microliths that were most likely fabricated for armatures, subsistence activities appear to have focussed on hunting. Groundstone items are quite rare. This emphasis on hunting with an attendant highly mobile settlement pattern may well have represented an arid land adaptation to the steppe zone bordering the eastern fringes of the Mediterranean woodlands. The presence of Red Sea shells from occupations in the Azraq Basin (Reese 1991) denotes some form of north-south interaction within the steppe-desert region east of the Rift Valley. The Qalkhan therefore appears to provide evidence for arid land adaptation during late Last Glacial times and a different, although parallel, line of development with the Kebaran Complex of the Mediterranean woodlands.

**Appendix 9.1 Type List for Epipaleolithic Assemblages**

A. Scraper Class

1. End scraper on flake
2. End scraper on flake with lateral retouch
3. End scraper on blade
4. End scraper on blade with lateral retouch
5. Circular scraper
6. Thumbnail scraper
7. Denticulated end scraper
8. Carinated end scraper
9. Shouldered or nosed scraper
10. Doubled opposed end scraper
11. End scraper with lateral notch
12. Core scraper
13. End scraper on Aurignacian blade

Appendix 9.1 (*Continued*)

14. End scraper working bit
15. Side scraper

B. Burin Class

16. Dihedral burin
17. Angle burin on snap or old surface
18. Angle burin on straight truncation
19. Angle burin on oblique truncation
20. Angle burin on concave truncation
21. Angle burin on convex truncation
22. Transverse burin on lateral retouch
23. Transverse burin on lateral notch
24. Multiple burin on truncation
25. Multiple dihedral burin
26. Multiple mixed burin

C. Microlithic Geometric

27. Lunate
28. Lunate with bifacial retouch (Helwan)
29. Lunate atypical or incomplete
30. Isosceles triangle
31. Isosceles triangle atypical or imcomplete
32. Trapeze/rectangle
33. Trapeze/rectangle with bifacial retouch
34. Trapeze/retangle atypical or incomplete
35. Scalene triangle
36. Fragment of geometric

D. Nongeometric Microliths

37. Straight backed bladelet
38. Straight backed bladelet with basal truncation
39. Straight backed bladelet with distal truncation
40. Straight backed bladelet with truncation opposite snap
41. Straight backed with bitruncation
42. Bilaterally backed bladelet
43. Straight backed bladelet with medial segment
44. Arched backed bladelet
45. Arch backed with truncation
46. Bifacially retouched (Helwan) bladelet
47. Irregular or concave
48. Pointed straight backed bladelet
49. Pointed straight backed bladelet with basal truncation
50. Narrow arched backed and pointed bladelet
51. Arched back with micro burin
52. Partially backed piece
53. Scalene bladelet
54. Ouchata bladelet
55. Fragment of backed piece
56. Various backed piece

E. Abruptly Retouched Blades and Flakes

57. Arched backed blade

### Appendix 9.1 (*Continued*)

58. Straight backed blade
59. Bilaterally backed blade
60. Irregularly backed blade
61. Partially backed blade (<1/3 edge)
62. Pointed backed blade
63. Fragment of backed blade
64. *Piece a cran*
65. Backed flake

F. Truncations

66. Truncated blade/bladelet
67. Truncated flake
68. Bitruncated blade/bladelet

G. Retouched Pieces

69. Bladelet retouched unilateral-obverse
70. Bladelet retouched unilateral-inverse
71. Bladelet retouched bilateral-obverse
72. Bladelet retouched bilateral-inverse
73. Bladelet alternately retouched
74. Bladelet partially retouched (<1/3 edge)
75. Pointed straight retouched bladelet
76. Pointed arched retouched bladelet
77. Blade retouched unilateral-obverse
78. Blade retouched unilateral-inverse
79. Blade retouched bilateral-obverse
80. Blade retouched bilateral-inverse
81. Blade alternately retouched
82. Blade partially retouched
83. Flake retouched unilateral-obverse
84. Flake retouched unilateral-inverse
85. Flake retouched bilateral-obverse
86. Flake retouched bilateral-inverse
87. Flake alternately retouched
88. Flake partially retouched
89. Fragment of retouched piece

H. Denticulate Class

90. Denticulated piece
91. Denticulated piece with retouch

I. Notch Class

92. Single notch
93. Single notch with retouch
94. Multiple notch
95. Multiple notch with retouch
96. Micro-notch

J. Multiple Use Class

K. Perforator

L. Massive Piece Class

97. End scraper on massive piece

**Appendix 9.1 (*Continued*)**

98. Denticulate on massive piece
99. Notch on massive piece
100. Bifacially retouched massive piece

M. Utilized Piece Class
101. Sickle Blade
102. *Lame a machure*
103. Hammerstone

N. Points
104. Microgravette
105. El Wad Point
106. Arrowheads (describe individually)
107. *La Mouillah*
108. *Piquant-Triedre*
109. Qalkhan Point

O. Varia
110. Describe individually

**Appendix 9.2. Type List for Qalkhan Assemblages**

| | J405 | J406b | J407 | J431 | J431 | J431 |
|---|---|---|---|---|---|---|
| Assemblages | | Lower | | $E^2$_UP | $E^2$_MD | $E^2$_LW |
| Number | 49 | 377 | 50 | 270 | 93 | 112 |
| A. Scraper Class | 12.0 | 9.5 | 1.9 | 11.0 | 11.8 | 18.7 |
| 1. | 2.0 | 3.7 | | 1.4 | 2.1 | 5.3 |
| 2. | | | | 0.7 | 2.6 | |
| 3. | | 0.3 | | 2.5 | 8.6 | 4.4 |
| 4. | | 1.3 | | 3.3 | | |
| 5. | 2.0 | | | 0.3 | | |
| 6. | | 1.6 | 1.9 | | 1.0 | 1.7 |
| 7. | | | | | | 0.8 |
| 8. | 6.0 | | | | | |
| 9. | | | | | | |
| 10. | | 0.3 | | 0.3 | | |
| 11. | | | | | | |
| 12. | 2.0 | 1.1 | | | | 3.5 |
| 13. | | | | | | |
| 14. | | 1.3 | | 2.5 | | |
| 15. | | | | | | |
| B. Burin Class | 6.0 | 3.2 | 5.7 | 0.7 | 2.1 | 8.9 |
| 16. | | 0.8 | | | 1.0 | 2.6 |
| 17. | | 1.3 | | | | 6.2 |
| 18. | | 0.3 | | | | |
| 19. | | | | | | |
| 20. | | 0.3 | 0.3 | | | |

**Appendix 9.2. (*Continued*)**

| | J405 | J406b | J407 | J431 | J431 | J431 |
|---|---|---|---|---|---|---|
| Assemblages | | Lower | | $E^{2}$_UP | $E^{2}$_MD | $E^{2}$_LW |
| 21. | | | | | | |
| 22. | 4.0 | 0.3 | 5.7 | 0.3 | | |
| 23. | | | | | | |
| 24. | | | | | | |
| 25. | 2.0 | | | | | |
| 26. | | 0.3 | | | 1.0 | |
| C. Microlithic Geometric | 0.3 | | 0.7 | 2.1 | 3.5 | |
| 27. | | | | | | |
| 28. | | | | | | |
| 29. | | | | | | |
| 30. | | 0.3 | | | | |
| 31. | | | | | | |
| 32. | | | | 0.7 | 2.1 | 3.5 |
| 33.–36. | | | | | | |
| D. Nongeometric Microliths | 2.0 | 21.8 | 11.3 | 58.3 | 56.7 | 21.4 |
| 37. | | 0.8 | | 7.0 | 4.3 | |
| 38. | | | | | 5.3 | |
| 39. | | 0.3 | | | | 2.6 |
| 40. | | 4.0 | 1.9 | 4.8 | 7.5 | 4.4 |
| 41. | | | | | | |
| 42. | | 0.8 | 1.9 | | | |
| 43. | | 3.4 | | 4.8 | 8.6 | 6.2 |
| 44. | | 2.1 | | 7.7 | | 2.6 |
| 45. | | | | 5.5 | | |
| 46. | | | | | | |
| 47. | | 1.6 | | 6.6 | 4.3 | |
| 48. | | | | | | |
| 49. | | | | | | |
| 50. | | 6.1 | | 10.3 | 12.9 | 4.4 |
| 51. | | | | | | |
| 52. | 2.0 | | 5.7 | 6.2 | 9.6 | |
| 53. | | | 1.9 | 2.2 | | |
| 54. | | 1.6 | | 0.3 | 2.1 | 0.8 |
| 55. | | | | 2.9 | | |
| 56. | | 1.1 | | | 2.1 | |
| E. Abruptly Retouched Blades and Flakes | 6.1 | | | | 16.0 | |
| 57.–59. | | | | | | |
| 60. | | 0.3 | | | | 5.3 |
| 61. | | | | | | |
| 62. | | | | | | |
| 63. | | 5.6 | | | | 10.7 |
| 64. | | | | | | |
| 65. | | 0.3 | | | | |

Appendix 9.2. (*Continued*)

| Assemblages | J405 | J406b | J407 | J431 | J431 | J431 |
|---|---|---|---|---|---|---|
| | | Lower | | $E^2$_UP | $E^2$_MD | $E^2$_LW |
| F. Truncations | | 4.8 | | 2.9 | | 3.5 |
| 66. | | 4.8 | | 2.2 | | 3.5 |
| 67. | | | | 0.7 | | |
| 68. | | | | | | |
| G. Retouched Pieces | 48.0 | 34.0 | 52.8 | 9.6 | 9.6 | 14.2 |
| 69. | 34.0 | 11.7 | 39.6 | | | 3.5 |
| 70. | 10.0 | 4.0 | 7.6 | 0.3 | 1.7 | |
| 71. | | | | 0.3 | | |
| 72. | | | | | 1.0 | |
| 73. | 4.0 | 1.9 | 5.7 | 1.8 | | |
| 74.–76. | | | | | | |
| 77. | | 0.3 | | 1.8 | 3.2 | |
| 78. | | 0.3 | | | | |
| 79. | | | | | | 2.6 |
| 80. | | | | | | |
| 81. | | 0.3 | | 0.3 | | |
| 82. | | | | | | |
| 83. | | 6.1 | | 2.9 | 5.3 | |
| 84. | | 2.1 | | | | |
| 85. | | 2.9 | | | | 6.2 |
| 86. | | | | | | |
| 87. | | 0.8 | | | | |
| 88. | | 3.7 | | 2.2 | | |
| 89. | | | | | | |
| H. Denticulate Class | 4.0 | 4.0 | | 1.4 | | |
| 90. | 2.0 | 2.1 | | 1.1 | | |
| 91. | 2.0 | 1.9 | | 0.3 | | |
| I. Notch Class | 20.0 | 11.1 | 5.7 | 3.3 | 5.3 | 3.5 |
| 92. | 12.0 | 1.9 | 3.8 | | 3.2 | 3.5 |
| 93. | | | | | | |
| 94. | 8.0 | 2.1 | 1.9 | | | |
| 95. | | 1.9 | | | | |
| 96. | | 5.3 | | 3.3 | 2.1 | |
| J. Multiple Use Class | 2.0 | 6.3 | 1.9 | | | |
| K. Perforator | 4.0 | 1.1 | 5.7 | 0.7 | | |
| L. Massive Piece Class | | | | | | |
| 97.–100. | | | | | | |
| M. Utilized Piece Class | | 0.8 | | 0.3 | | |
| 101. | | | | | | |
| 102. | | 0.5 | | | | |
| 103. | | 0.3 | | 0.3 | | |

Appendix 9.2. (*Continued*)

| | J405 | J406b | J407 | J431 | J431 | J431 |
|---|---|---|---|---|---|---|
| Assemblages | | Lower | | $E^2$_UP | $E^2$_MD | $E^2$_LW |
| N. Points | 2.0 | 3.2 | 13.2 | 8.8 | 11.6 | 9.8 |
| 104. | | 0.3 | | 1.4 | 1.0 | |
| 105. | | | | | 2.1 | |
| 106. | | | | | | |
| 107. | | 0.5 | | 5.9 | 3.2 | 8.0 |
| 108. | | 0.3 | | | 1.0 | |
| 109. | 2.0 | 2.1 | 13.2 | 1.4 | 4.3 | 1.7 |
| O. Varia | | | | | | |
| 110. | | | | | | |

Appendix 9.3 Bladelet Attributes of Qalkhan Assemblages

| Assemblages | J405[a] | J406b LOWER | J431 $E^2$_UP | J431 $E^2$_MD | J431 $E^2$_LW |
|---|---|---|---|---|---|
| Number | 66 | 94 | 177 | 61 | 75 |
| Maximum Length | | | | | |
| Range | —[b] | 20–54 | 20–81 | 21–71 | 20–75 |
| Mean | 28.3 | 31.4 | 37.1 | 35.4 | 34.1 |
| Standard Deviation | 7.3 | 9.0 | 13.4 | 11.4 | 11.2 |
| Maximum Width | | | | | |
| Range | — | 5–24 | 6–28 | 5–23 | 5–32 |
| Mean | 11.2 | 12.3 | 12.8 | 12.5 | 12.5 |
| Standard Deviation | 2.3 | 4.2 | 4.4 | 4.8 | 5.6 |
| Maximum Thickness | | | | | |
| Range | — | 2–11 | 1–10 | 1–13 | 1–12 |
| Mean | 2.6 | 4.4 | 3.7 | 3.9 | 4.0 |
| Standard Deviation | 1.6 | 2.1 | 1.7 | 2.4 | 2.5 |
| Lateral Profile | | | | | |
| 2. Curved | — | 19.2 | 31.6 | 41.0 | 8.0 |
| 3. Flat | — | 64.9 | 44.6 | 39.3 | 56.0 |
| 4. Twisted | — | 8.5 | 22.6 | 18.0 | 33.3 |
| 5. Rippled | — | 7.5 | 1.1 | 1.6 | 2.7 |
| Platform Facets | | | | | |
| 2. Single | — | 61.7 | 53.1 | 50.8 | 57.3 |
| 3. Dihedral | — | 5.3 | 10.2 | 13.1 | 17.3 |
| 4. Multiple | — | 4.3 | 2.3 | 4.9 | 4.0 |
| 5. Crushed | — | 26.6 | 22.0 | 23.0 | 17.3 |
| 6. Cortex | — | 2.1 | 2.3 | 1.6 | 1.3 |
| Shape of Bulb | | | | | |
| 2. Lipped | — | 8.5 | 29.9 | 42.6 | 9.3 |
| 3. Pointed | — | 10.6 | 21.7 | 9.8 | 5.3 |
| 4. Unlipped | — | 79.8 | 48.6 | 47.5 | 85.3 |

Appendix 9.3 *(Continued)*

| Assemblages | J405[a] | J406b LOWER | J431 E²_UP | J431 E²_MD | J431 E²_LW |
|---|---|---|---|---|---|
| Distal Shape | | | | | |
| 2. Pointed | — | 54.3 | 18.6 | 33.0 | 30.7 |
| 3. Oblique | — | 4.3 | 31.1 | 9.8 | 20.0 |
| 4. Blunt | — | 41.5 | 50.3 | 57.4 | 49.3 |
| Alignment of Lateral Edges | | | | | |
| 2. Expanding | — | 18.1 | 18.6 | 24.6 | 13.3 |
| 3. Converging | — | 46.8 | 47.5 | 44.3 | 52.0 |
| 4. Parallel | — | 35.1 | 33.9 | 31.2 | 34.7 |
| Facet Pattern | | | | | |
| 2. Parallel | — | 60.6 | 42.4 | 50.8 | 64.0 |
| 3. Parallel Opposed | — | 11.7 | 23.7 | 23.0 | 2.7 |
| 4. Parallel Truncated | — | 4.3 | 11.3 | 6.6 | 10.7 |
| 5. Parallel Converging | — | 18.1 | 20.3 | 19.7 | 18.7 |
| Distal Profile | | | | | |
| 2. Feathered | — | 81.9 | 77.4 | 72.1 | 78.7 |
| 3. Overpassed | — | 0 | 10.2 | 18.0 | 13.3 |
| 4. Underpassed | — | 12.8 | 7.9 | 6.6 | 4.0 |
| 5. Stepped | — | 5.3 | 2.8 | 3.3 | 4.0 |
| Cross Section | | | | | |
| 2. Single | — | 0 | 0.6 | 0 | 0 |
| 3. Double | — | 29.8 | 41.8 | 44.3 | 21.3 |
| 4. Triple | — | 54.3 | 41.8 | 47.5 | 49.3 |
| 5. Multiple | — | 15.9 | 15.3 | 8.2 | 29.3 |

[a]Jones 1983.
[b]Data not available.

# Chapter 10

# The Hamran Sites

The Hamran is the best represented of the Epipaleolithic taxa with occupations having been discovered in each of the four study areas (Figures 10.1, 10.2, 10.3). The sites in the Jebel Hamra area are of particular importance, however, for they reveal a stratigraphic succession that allows for sequencing the assemblages into four phases. Early Hamran assemblages, dominated by nongeometric microliths, are replaced by assemblages with trapeze/rectangles during the Middle Hamran. Lunates along with an intentional use of the microburin technique emerge in the Late Hamran and become dominant characteristics of toolkits in Final Hamran times.

The technotypologic evolution of the Hamran follows a very similar path to that recognized within the Epipaleolithic of the core Mediterranean zone west of the Rift Valley. There the succession of Kebaran, Geometric Kebaran, and Natufian complexes is reflected in nongeometric microlithic assemblages giving way to those dominated by trapeze/rectangles, which in turn are replaced by those characterized by lunates and the microburin technique. Although the Hamran succession largely follows that of the Mediterranean zone, at a more specific industrial level, the technotypologic characteristics of the Hamran assemblages fall outside the parameters conventionally used in defining the Kebaran and Geometric Kebaran (Bar-Yosef 1970, 1981). In addition to these technotypologic differences, the southern Jordan assemblages clearly trace a local evolutionary sequence from stratified contexts. For these reasons, new "industry level" labels (i.e., Early Hamran, Middle Hamran, etc.) are employed. In order to convey the affinities of these Hamran industries to those of the "Kebaran" and "Geometric Kebaran," I have elevated these labels to taxa of a higher order, that of "complex" (Chapter 3; Henry 1989a, 1990).

From a behavioral perspective, the Hamran succession shows a parallel development with the Epipaleolithic of the Mediterranean zone. With an emphasis upon straight-backed and truncated bladelets, the Early Hamran most closely resembles Bar-Yosef's (1981) "Cluster D" or Late Kebaran assemblages (Bar-Yosef and Belfer-Cohen 1989). With a geometric component formed solely of trapeze/rectangles, the Middle Hamran is most similar to the "Group II" assemblages of the Geometric Kebaran (Bar-Yosef 1981). The Late\Final Hamran assemblages, with their significant lunate and microburin representation, are unique, lacking parallels within the sequence of the core Mediterranean zone. The lunate and microburin component of Late/Final Hamran assemblages indicate, however, a very late position within the Geometric Kebaran sequence, most likely transitional to the Early Natufian.

The artifact inventories, settings, and distributions of Hamran sites suggest a settlement-procurement pattern in which groups followed an annual cycle of transhumance. They appear to have aggregated into long-term camps in low elevation rockshelters during the winter as reflected in those sites found on the floor of the Wadi Hisma. During the warm season,

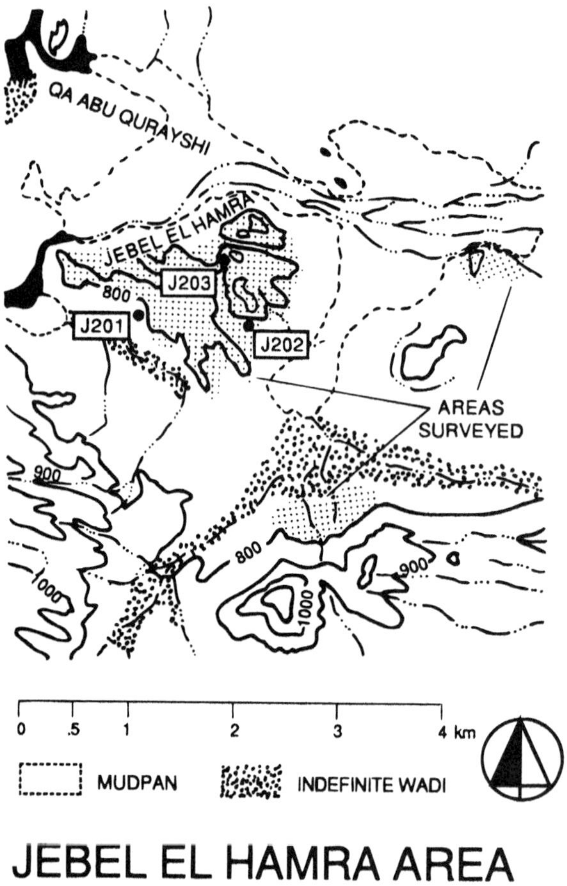


**Figure 10.1.** Map of Jebel Hamra study area showing locations of Hamran sites.

groups apparently dispersed into smaller, more mobile units and occupied the higher elevations of the piedmont and uplands. Although this pattern is most completely and clearly reflected in the Early and Middle Hamran, its remnants are found throughout the Hamran sequence.

## EARLY HAMRAN

Early Hamran assemblages were recovered from seven sites and each of the study areas (Figures 10.1, 10.2, 10.3). Based upon elevation, physiographic setting, and exposure, they can be broken into two groups. These include the lowland sites found in rockshelters at the foot of inselbergs on the floor of the Wadi Hisma: Jebel Hamra (J201), Jebel Mishraq (J504), and Jebel Muheimi (J520). The other group consists of open-air sites located in the higher elevations of the piedmont including the sites of Wadi Humeima (J406b), J21, and J22.

The rockshelter sites also contain evidence of later Middle Hamran occupations, either as overlying layers (e.g., at J201) or on the surface (at J504 and J520). Of the open-air sites, only Wadi Humeima (J406b) contains another component: an underlying Qalkhan horizon as previously described (Chapter 9).

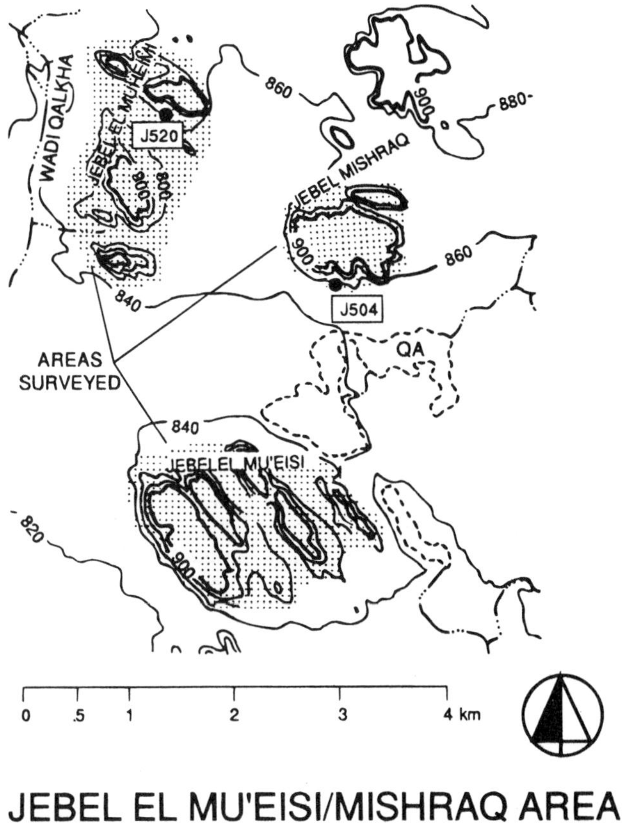


**Figure 10.2.** Map of Jebel Mishraq/Mu'eisi/Muheimi study area showing locations of Hamran sites.

## Settings, Excavation Plans, and Stratigraphy

The most extensive excavation was undertaken at Jebel Hamra (J201) and this consisted of only 8 m$^2$. The other sites were investigated by *sondages* of from 2–5 m$^2$ or large surface (and immediate subsurface) collections of deflated occupations (J21 and J22).

### Jebel Hamra (J201) Layer C

The site (830 masl) is associated with a shallow rockshelter found at the base of a sheer sandstone cliff that forms the southern face of Jebel Hamra (Figure 10.4). The site overlooks an extension of the Qa Oumm Salab, a dry lake bed, which is located some 23 m beneath the floor of the shelter and about 0.8 km away. A dense surface scatter of artifacts extend from the back wall of the shelter downslope, covering an area of about 320 m$^2$. Bedrock terraces stretch from the cliff face in the eastern portion of the site and extensive rockfall marks the western limits of the site. A large panel of petroglyphs covers much of the wall in this area.

Subsequent to identifying *in situ* materials in two test units (1 and 2) in 1979, a 6 m long stratigraphic trench was excavated into the slope during the 1980 season. This revealed a 90–100 cm thick deposit composed of three layers (A, B, and C) of sandy silt that vary in ash content, color, and compaction (Figure 10.5). Artifacts were found throughout the deposit to weathered sandstone bedrock which was exposed in the upper end of the trench. Layer A (5–10 cm thick)

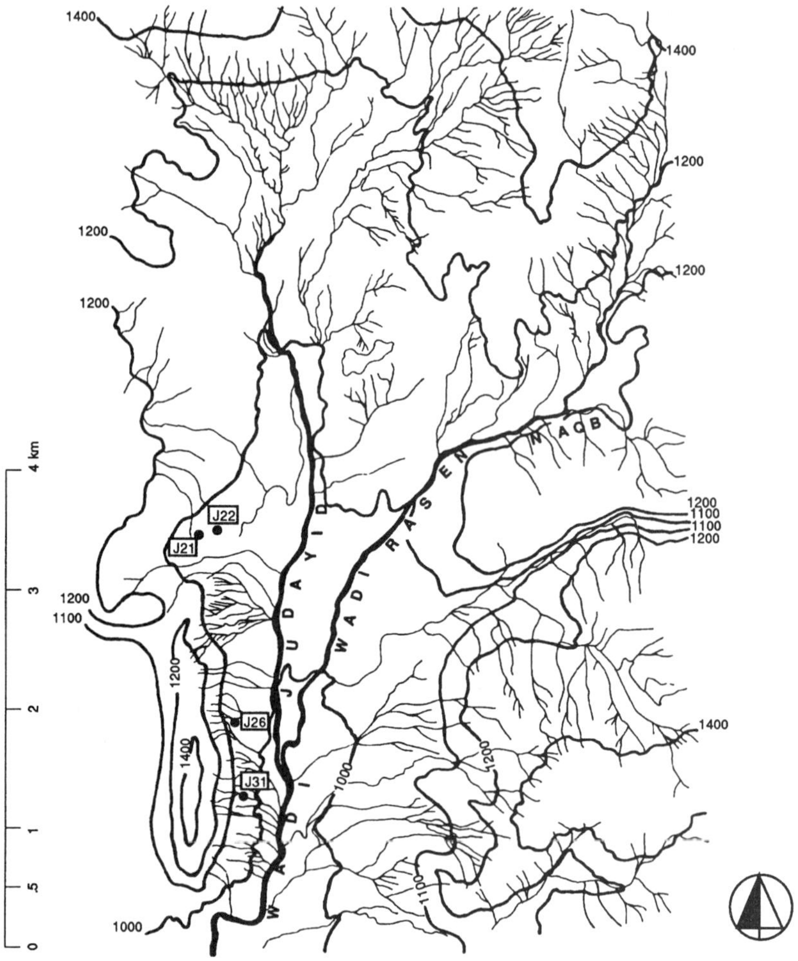


**Figure 10.3.** Map of Judayid Basin study area showing locations of Hamran sites.

consists of a red (2.5YR 5/6) sandy-silt with small angular sandstone rubble. Layer B (60–80 cm thick) is composed of a more compact red (2.5 YR 4/8) sandy-silt with little rubble. Layer C (+30 cm thick) is formed of a very compact, light red (2.5 YR 1/6) silt. The Layer B/C contact is well defined and conforms to both cultural and paleobotanic breaks in the stratigraphic record. Early Hamran artifacts were confined to Layer C, whereas Middle Hamran materials were recovered from Layers A and B. Pollen indicative of a rather arid setting was found in Layers A and B, but Layer C was sterile of pollen.

## Jebel Mishraq (J504)

The site (860 masl) is situated at the base of a steep, 100 m high, vertical wall that creates the south face of Jebel Mishraq (Figures 10.6 and 10.7). A very dense surface artifact scatter, covering an area of about 600 m$^2$, extends from a shallow overhang and falls away down a gentle slope. To the south the site overlooks a small unnamed Qa which rests some 20 m below the floor of the shelter and some 0.6 km away.

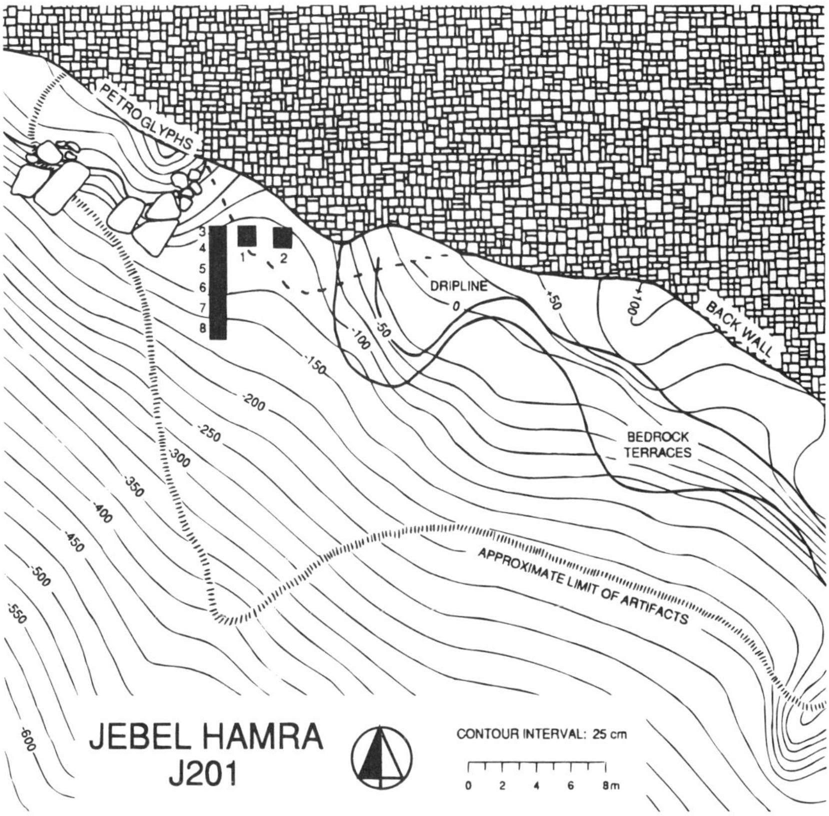


**Figure 10.4.** Topographic map of Jebel Hamra (J201) showing the artifact distribution, excavation plan, and location of petroglyphs.

A cluster of small bedrock mortars ("cupmarks") has been formed in a low bedrock shelf near the west end of the shelter and to the west of these a panel of petroglyphs of ibexes were inscribed. In addition to making surface collections from two 1 $m^2$ units (3 and 4), sondages (units 1 and 2) were excavated in the floor of the shelter. These exposed a red silty sand to a depth of 60 cm with two strata (layers A and B). Cultural material extends throughout the deposit, but occurs in the highest density within Layer A.

Layer A, forming the upper 40 cm of the deposit, consists of a reddish-brown silty sand with a high density of platy sandstone detritus. Charcoal from a hearth (25 cm below surface in Unit 1) yielded a $C^{14}$ date of 11,985 ± 109 B.P. (SMU 905), a determination much too young for a nongeometric industry. The charcoal may well have been contaminated by extraneous carbon from a nearby overlying layer of modern goat dung. Layer B is composed of a horizontally bedded, reddish-brown, silty-sand. It is more compact than the overlying layer and contains scatters of ash or carbonate flecks.

## Jebel Muheimi (J520)

The site (870 masl) is located a little over 2 km northwest of Jebel Misraq in the northernmost inselberg of the Jebel el Muheimi group. Like the sites of Jebel Hamra (J201) and Jebel Misraq (J504), Jebel Muheimi (J520) is associated with a shallow, south-facing rockshelter

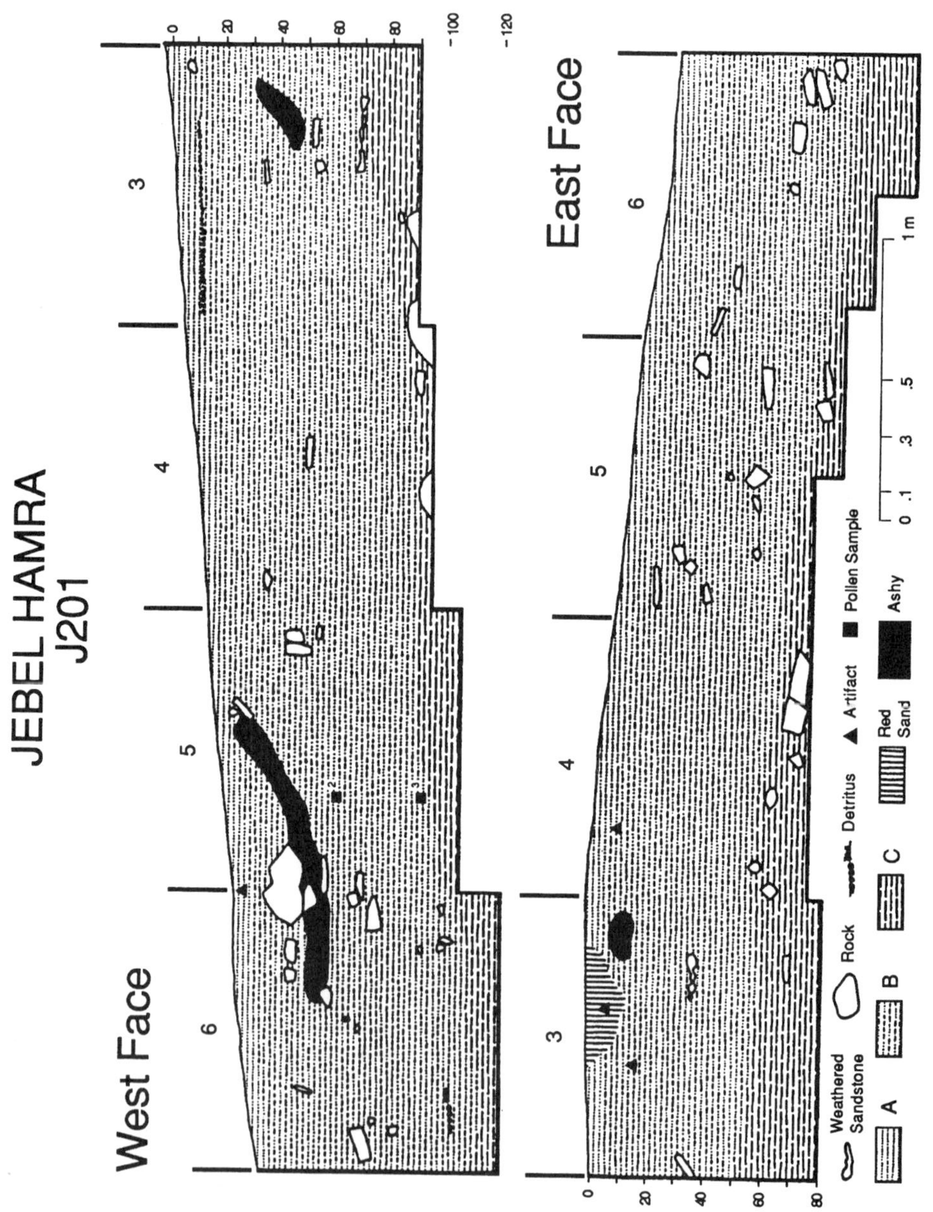


Figure 10.5. Stratigraphic section drawing of Jebel Hamra (J201) deposit.

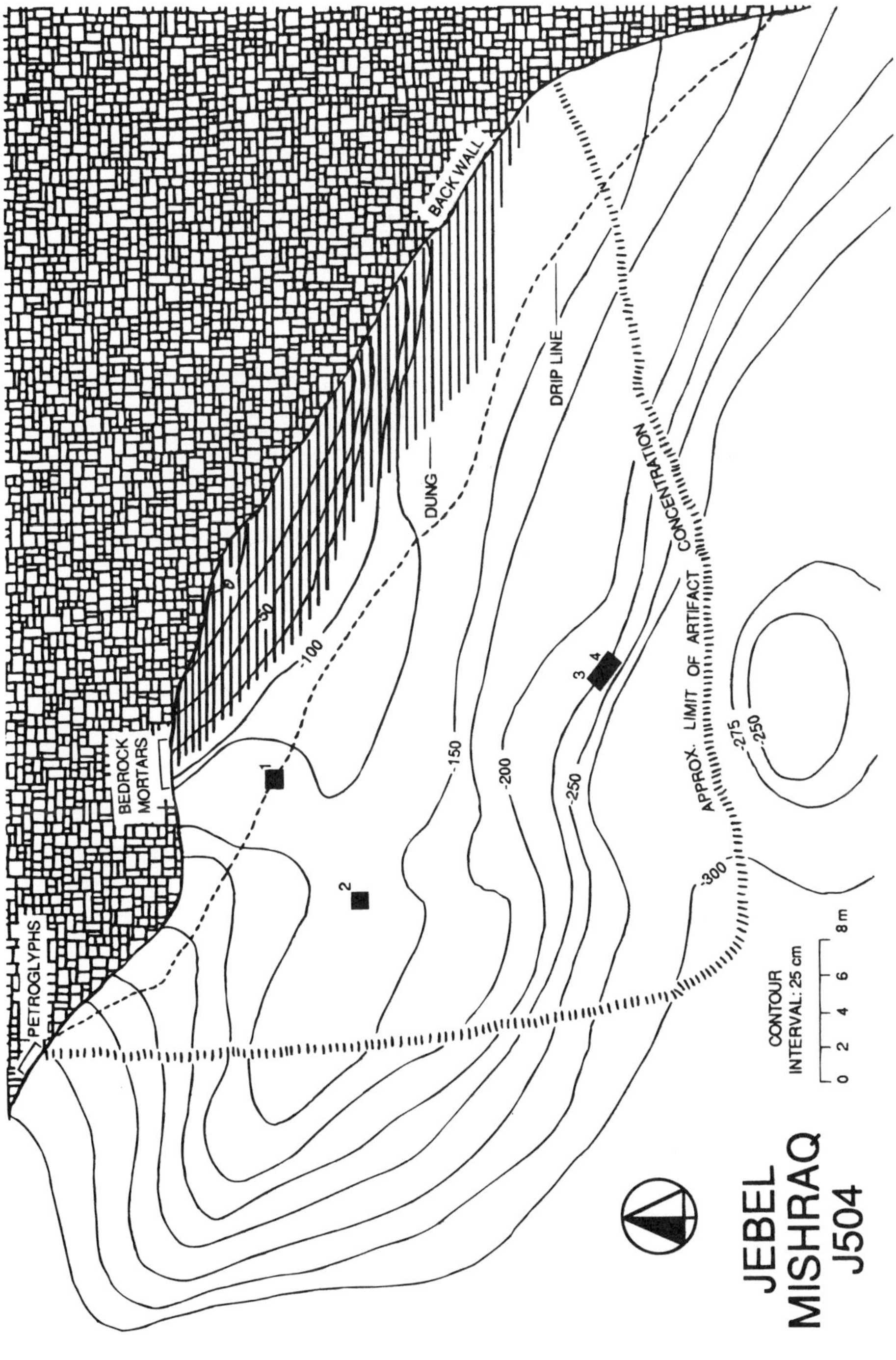


Figure 10.6. Topographic map of Jebel Mishraq (J504) showing artifact distribution, excavation units, and location of petroglyphs.

Figure 10.7. Photograph of Jebel Mishraq (J504) looking from the southwest. Note vehicle located near the eastern limit of the site for scale.

formed at the base of a sheer walled inselberg (Figure 10.8). Artifacts are scattered over an area of about 500 $m^2$ within a deep embayment of the jebel. Petroglyphs have been inscribed on the back wall of the shelter at its eastern end. A large area of drift sand is encroaching upon the site from the south and drainages from the jebel also are cutting it as they form erosional gullies.

Five 1 $m^2$ units were surface collected and two of these (units 1 and 2) were excavated to 40 cm. Most of the artifacts were found in the upper 20 cm of a red sandy silt.

### Wadi Humeima (J406b) Upper

As previously described (Chapter 9), the site (1,020 masl) is located along the Wadi Humeima in an open setting at the foot of the slope that falls away from the northeastern escarpment of Jebel Humeima. Although an open-air site, the high jebel throws the site into its shadow by mid-afternoon. Early Hamran artifacts were collected from the surface and a 50 cm thick, red sand deposit overlying a Qalkhan horizon. Other than a lens of rubble, thought to denote the scree of a paleosurface, there is no apparent lithological difference between the Qalkhan and Early Hamran layers.

### J21 and J22

These two open sites (1,080 masl) were found about a kilometer west of the modern channel of the Wadi Judayid near the mouth of a deep embayment in the western rim of the basin. The sites, separated by less than 500 m, rest on near level surfaces eroded from a coalesced alluvial fan (Member II) that filled the Judayid Basin during the late Last Glacial (Chapter 2; Henry *et al.* 1983). Both sites display low densities of white patinated artifacts scattered over large areas (500–1,000 $m^2$), but the areas in which artifacts are concentrated cover only some 100–120 $m^2$. Artifacts were found to be mainly confined to the surface, although some specimens were found to be shallowly (5 cm) buried. The predominant surface setting, patination, and large dispersal of artifacts imply some degree of natural disturbance of the occupations. This appears to have been minimal, however, as specimens exhibit very fresh edges. Also, artifact specimens (microflakes or chips) that are normally removed from sites that have experienced significant deflation or sheet erosion are present in large numbers.

Each of the sites was systematically collected by establishing grids (J21—53 $m^2$, J22—28 $m^2$) within the areas of artifact concentrations. The surfaces were scraped and screened through 1 mm mesh in order to recover minute specimens, thus maintaining the standard recovery procedures.

## Flint Artifacts

In that vertical and horizontal comparisons of artifacts from excavation (collection) units at each of the sites failed to identify significant technotypologic differences, the artifacts from each site were analyzed as a single assemblage (Table 10.1).

### Typology

Early Hamran assemblage are characterized by nongeometric microliths dominated by straight-backed bladelets with truncations opposite snaps and medial segments of straight backed bladelets (Appendix 10.1). In a few assemblages, retouched pieces constitute the largest class, but they show little patterning with respect to orientation or location of retouch. Of the more formally retouched large tool classes, scrapers, notches, and burins occur in moderate frequencies.

***Scrapers*** primarily consist of well formed endscrapers on blades followed by ones on flakes (Figure 10.9). Many of these display lateral or bilateral retouch.

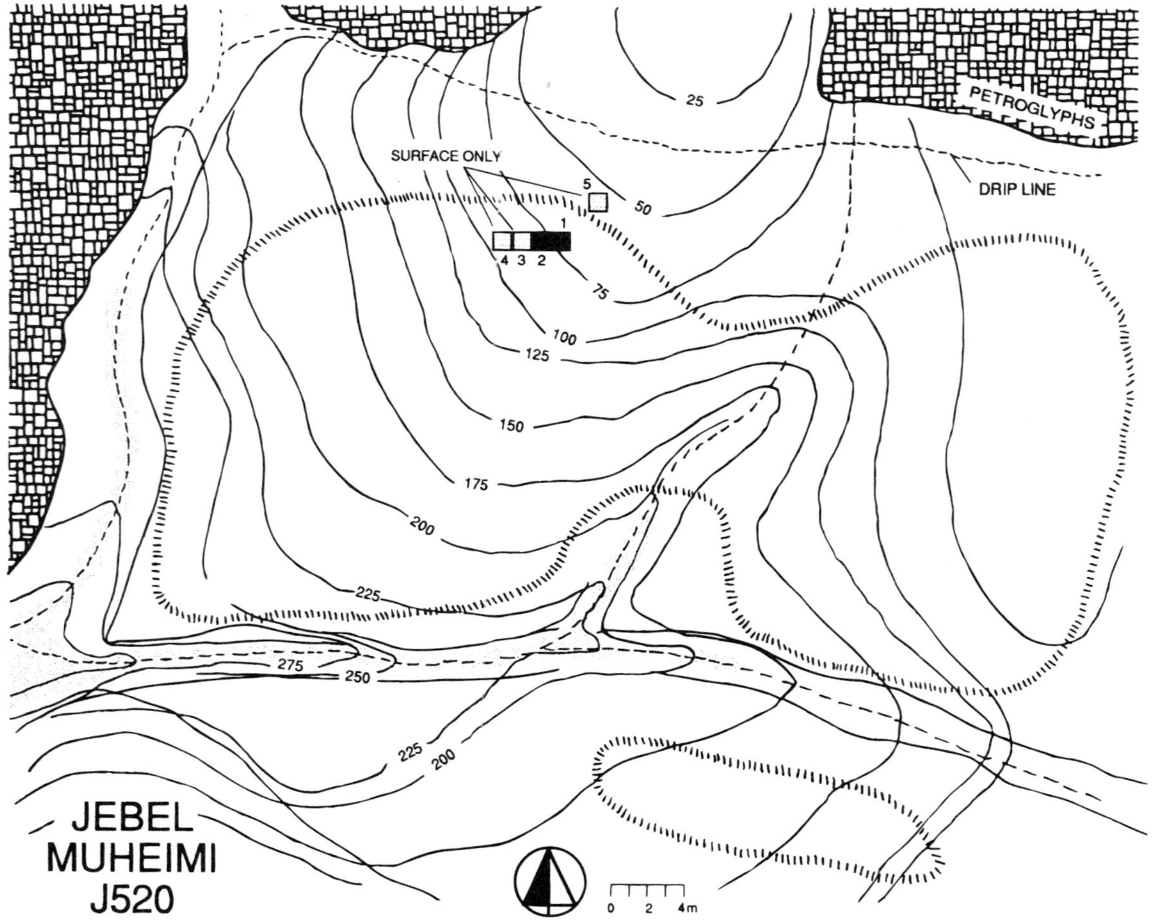


Figure 10.8. Topographic map of Jebel Muheimi (J520) showing artifact distribution, excavation units, and location of petroglyphs.

Table 10.1. Artifact Inventories for Early Hamran Assemblages

| Assemblage | J201 | J504 | J520 | J406B | J21 | J22 |
|---|---|---|---|---|---|---|
| *Tools* | | | | | | |
| Scraper | 5.2 | 12.8 | 11.8 | 10.0 | 7.1 | 13.0 |
| Burin | 3.2 | 2.7 | 3.6 | 2.4 | 3.6 | 8.7 |
| Geometric | 0.0 | 0.3 | 0.7 | 0.3 | 1.8 | 0.0 |
| Nongeometric | 64.3 | 28.3 | 26.1 | 40.7 | 19.6 | 13.0 |
| Abrupt Ret. | 0.0 | 1.0 | 1.8 | 0.3 | 0.0 | 0.0 |
| Truncation | 2.6 | 4.4 | 2.9 | 1.1 | 3.6 | 0.0 |
| Retouched Pc. | 16.2 | 31.3 | 32.9 | 35.5 | 42.9 | 52.2 |
| Denticulate | 0.0 | 3.7 | 1.8 | 1.4 | 5.4 | 0.0 |
| Notch | 3.2 | 8.4 | 14.3 | 6.2 | 5.4 | 13.0 |
| Multiple Tl. | 0.0 | 1.3 | 0.4 | 0.0 | 0.0 | 0.0 |
| Perforator | 3.2 | 3.4 | 3.2 | 0.0 | 1.8 | 0.0 |
| Utilized Pc. | 0.0 | 0.0 | 0.0 | 0.0 | 0.0 | 0.0 |
| Points | 0.0 | 0.0 | 0.0 | 0.8 | 0.0 | 0.0 |
| Varia | 1.9 | 2.4 | 0.7 | 1.4 | 8.9 | 0.0 |
| Total No. | 154 | 297 | 280 | 369 | 56 | 23 |
| *Debitage* | | | | | | |
| Core | 0.8 | 1.8 | 1.7 | 1.8 | 0.5 | 1.9 |
| Blade | 76.2 | 60.2 | 18.9 | 39.6 | 57.4 | 47.2 |
| Flake | 17.9 | 30.6 | 68.3 | 46.4 | 29.8 | 41.5 |
| Core Trim. El. | 2.9 | 2.3 | 0.7 | 1.7 | 2.0 | 0.6 |
| Primary El. | 2.2 | 4.8 | 8.1 | 8.4 | 9.3 | 8.8 |
| Microburin | 0.0 | 0.3 | 2.4 | 2.0 | 0.0 | 0.0 |
| Burin Spall | 0.0 | 0.0 | 0.0 | 0.1 | 0.0 | 0.0 |
| Total No. | 736 | 954 | 1030 | 1221 | 551 | 159 |
| *Debris* | | | | | | |
| Chips | | | 77.8 | 97.4 | | |
| Chunks | | | 22.2 | 2.6 | | |
| Total No. | 555 | 813 | 2272 | 2011 | 405 | 175 |
| *Total Artifacts* | 1445 | 2064 | 3302 | 3601 | 1012 | 357 |

***Burins*** are dominated by angle burins on truncations with lesser numbers struck from snaps or old surfaces (Figure 10.10). Dihedral burins occur less commonly, but are present in most assemblages (Figure 10.9).

***Nongeometric Microliths*** account for 26–64% of the toolkits of the excavated assemblages, but only 13–20% of the toolkits from the predominantly surface sites of J21 and J22. Although the lower proportions of microliths at the two surface sites might be viewed as a reflection of their loss through erosion, this is not likely, given that chips (flakes <20 mm) are present at J21 and J22 in roughly the same proportions as in the excavated sites.

Within the class, four variants of straight-backed bladelets are dominant (Figure 10.11; Table 10.2). Of these, medial segments and straight-backed bladelets with truncations opposite snaps stand out (Figure 10.10). Complete and pointed forms occur at lower frequencies. As Jones (1983:221) noted in her study, backing on the specimens is roughly balanced between normal abrupt and fine abrupt retouch.

The dimensions of the snapped and truncated forms (types 40 and 43) are near identical (Figure 10.11), thus suggesting that these forms were created through the intentional sectioning of bladelets rather than through their accidental breakage. The sectioning was accomplished by a blow to the obverse or inverse surface, rather than by weakening the bladelet with a lateral notch as is done with the microburin technique. In examining Early Hamran microliths

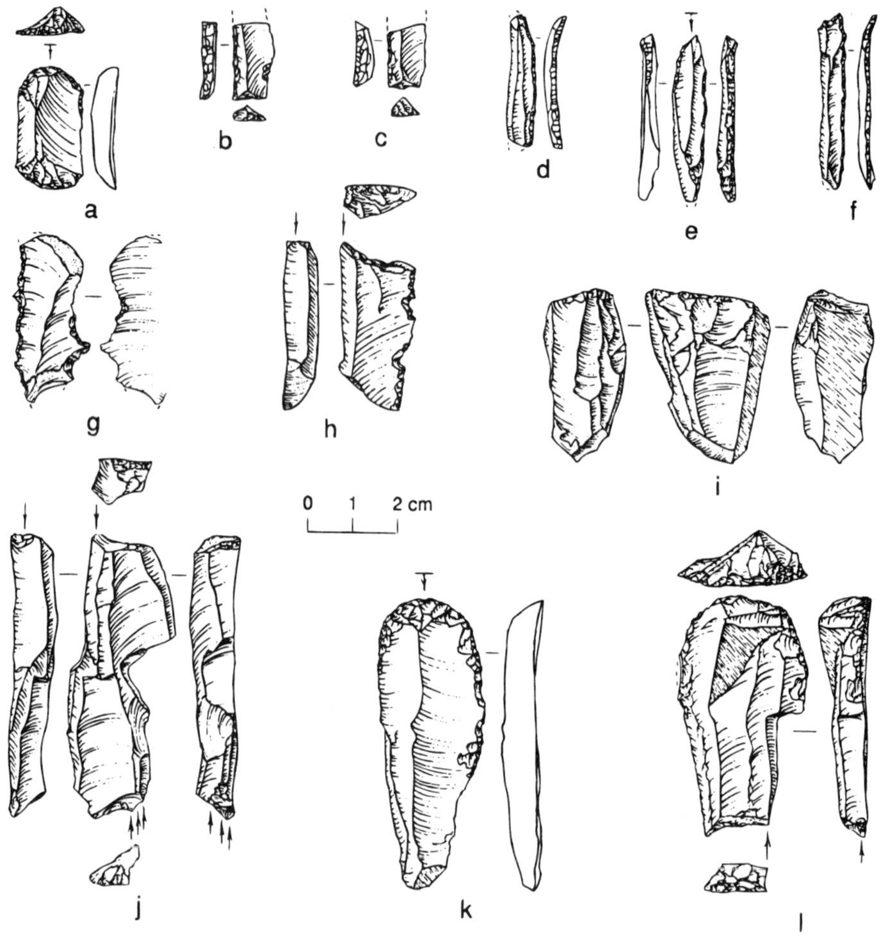


**Figure 10.9.** Illustrations of Early Hamran artifacts from Jebel Mishraq (J504): (a) double opposed endscraper, (b,c) straight-backed bladelets with truncation opposite snap, (d–f) backed bladelets, (g) denticulate, (h) angle burin on truncation, (i) bladelet core, (j) multiple burin, (k) endscraper on blade, (l0) endscraper opposite burin on truncation.

under low magnification, Jones (1983:222) found ridge or hinge scars to be visible on their snapped ends when viewed in profile. Following Speth's (1974:13) observations, she concluded that a snap would likely flare or hinge opposite the face which received the blow. The rough balance between specimens with hinges on the inverse (57%) and obverse (43%) portions of the snap (Jones 1983:224) implies that little preference was given to either surface in striking the bladelets for sectioning.

Jones (1983) also found 34–50% of snapped and truncated forms to exhibit wear in the form of continuous removals on the edge opposite the backing, thereby indicating that they had been used. Her findings are consistent with those of other researchers who have noted edge damage (Fujimoto 1979) and mastic (Bar-Yosef and Goring-Morris 1977) on trapeze/rectangles, suggesting that they had been used as transverse arrowheads.

*Notches* make up a moderate portion of most toolkits and mainly appear as single notches on blades.

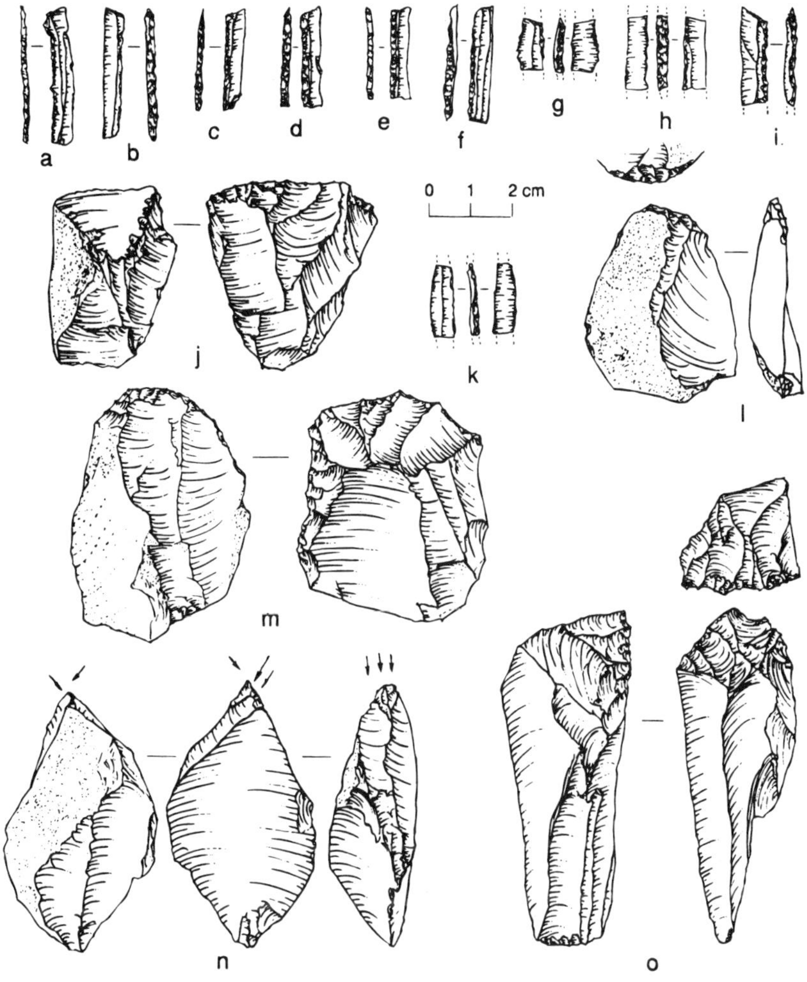


**Figure 10.10.** Illustrations of Early Hamran artifacts from Wadi Humeima (J406b): (a–f,i) straight-backed bladelets with truncation opposite snap, (g–h,k) medial segments of straight-backed bladelets, (j,m) bladelet cores, (l) endscraper on flake, (n) dihedral burin, (o) core scraper.

***Retouched pieces*** are composed largely of specimens with unilateral obverse retouch with blades and flakes being about equally represented.

***Multiple tools*** are rare, but almost exclusively represented by endscrapers on blades opposite angle burins (Figure 10.8).

***Perforators***, present in moderate frequencies in the lowland assemblages, are characteristically formed by contiguous retouched concavities that create a thick, pointed beak.

## Technology

Early Hamran technology emphasized the production of relatively wide bladelets from multiple platform pyramidal and subpyramidal cores. These were rejuvenated primarily through the creation of opposed or change of orientation platforms. Following production, bladelets were subsequently abruptly retouched into various types of nongeometric microliths. When the

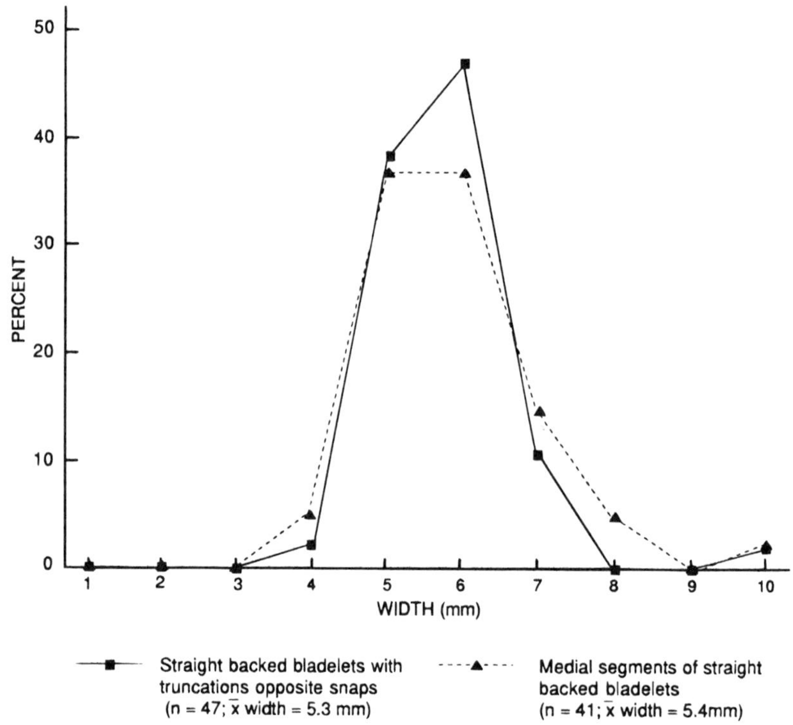


Figure 10.11. Comparison of the widths of straight-backed and truncated bladelets opposite snaps and medial segments of straight-backed bladelets.

**Table 10.2. Relative Frequencies of Dominant Microliths Within Early Hamran Assemblages**

| Assemblage | J201 Layer C | J504 | J520 | J40b Upper | J21 | J22 |
|---|---|---|---|---|---|---|
| *Microlith* | | | | | | |
| Straight-Backed Bladelet | 0 | 5.3 | 25.8 | 8.8 | 0 | 0 |
| Straight-Backed Bladelet with Truncation Opposite Snap | 17.2 | 47.4 | 32.3 | 41.2 | 33.3 | 66.7 |
| Medial Segment of Straight-Backed Bladelet | 69.0 | 47.4 | 38.7 | 36.0 | 66.7 | 33.3 |
| Pointed Straight-Backed Bladelet | 13.8 | 0 | 3.2 | 14.0 | 0 | 0 |
| N | 58 | 57 | 31 | 114 | 3 | 3 |

manufacturing process called for the controlled sectioning of a specimen, this was accomplished through a blow to either surface which created a snap at one or (with a second blow) both ends of a bladelet. In some cases the snaps were retouched (e.g., straight backed and truncated) and in others left unmodified (e.g., medial segments).

***Raw Material*** was obtained mainly from *in situ* chert beds on the plateau and from small cobbles derived from these beds and found in the Wadi Qalkha. Small quartzite cobbles furnished a much lesser used local material. Cobble to pebble size nodules of quartzite skirt the bases of most of the inselbergs in the Wadi Hisma having eroded from the sandstone. Although none of the Early Hamran cores were fashioned from quartzite, other classes of debitage and micolithic tools were made from the nearly translucent material.

***Cores***, for the most part, were so extensively exhausted that it is difficult to reconstruct with any detail the technique by which they were shaped. It is clear from the core trimming elements, however, that the *lame à crête* technique was not used in the initial shaping of cores. Subsequent reduction and rejuvenation did not employ the removal of core tablets. Consistent with the strong degree of exhaustion, cortex is uncommon on cores and rarely extensive. The productive lives of cores were extended through the creation of opposed and reoriented platforms; single platform cores are uncommon (Table 10.3).

**Table 10.3. Attributes of Cores from Early Assemblages**

| | Assemblages J406b | |
|---|---|---|
| Core Attributes | Mean | Range |
| Core Length | 37.0 | 22–60 |
| Core Width | 36.9 | 20–63 |
| Facet Length | 25.6 | 13–45 |
| Facet Width | 13.3 | 5–23 |
| Weight | 44.5 | 9–130 |
| N | 26 | |
| *Platform Orientation (%)* | | |
| Unidirectional | 69.2 | |
| Opposed | 19.2 | |
| Twisted | 0.0 | |
| Change/90 | 11.5 | |
| *Platform Faceting* | | |
| Single | 38.5 | |
| Faceted | 57.7 | |
| Cortex | 3.9 | |
| *Platform Regularization* | | |
| Present | 80.8 | |
| Absent | 19.2 | |
| *Cortical Cover* | | |
| Absent | 50.0 | |
| Limestone | 26.9 | |
| Water Worn | 23.1 | |
| *Core Shape* | | |
| Pyramidal | 3.9 | |
| Wedge (Sub–pyr.) | 76.9 | |
| Cuboid | 15.4 | |
| Irregular | 3.9 | |

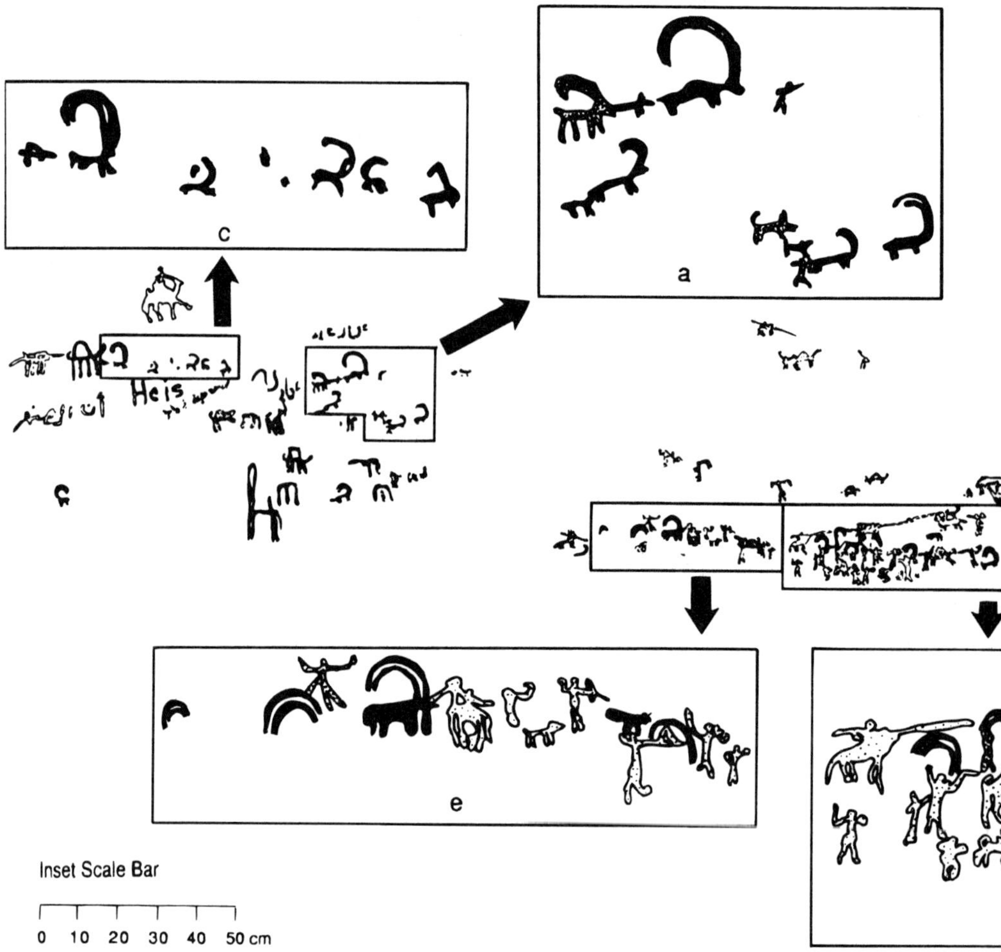


Figure 10.12. Main petroglyph panel of the Jebel Hamra (J201) rockshelter. The petroglyphs shown in black

*Bladelets* display length:width ratios averaging about 2.5:1. High proportions of crushed platforms and unlipped, diffuse bulbs indicate a substantial use of hard-hammer percussion (Appendix 10.2). Lipped and pointed bulbs, attributes of more precise forms of loading (e.g., punch and soft-hammer techniques), occur on only about 18% of the bladelets. Even with an emphasis upon hard-hammer percussion, however, Early Hamran knappers were able to control bladelet production with good precision as evidenced by a high percentage (84.3) of feathered terminations.

## Other Artifacts and Petroglyphs

Aside from the chipped stone assemblages, Early Hamran artifact inventories are confined to ornamental marine shells (Chapter 17) and a few examples of ground stone. The ground stone consists of a cluster of shallow mortars, "cup marks" ground into a low bedrock shelf near the back wall of the Jebel Mishraq (J504) rockshelter. A portable mortar fashioned from a slab of sandstone also was found on the surface of the site near the center of the artifact concentra-

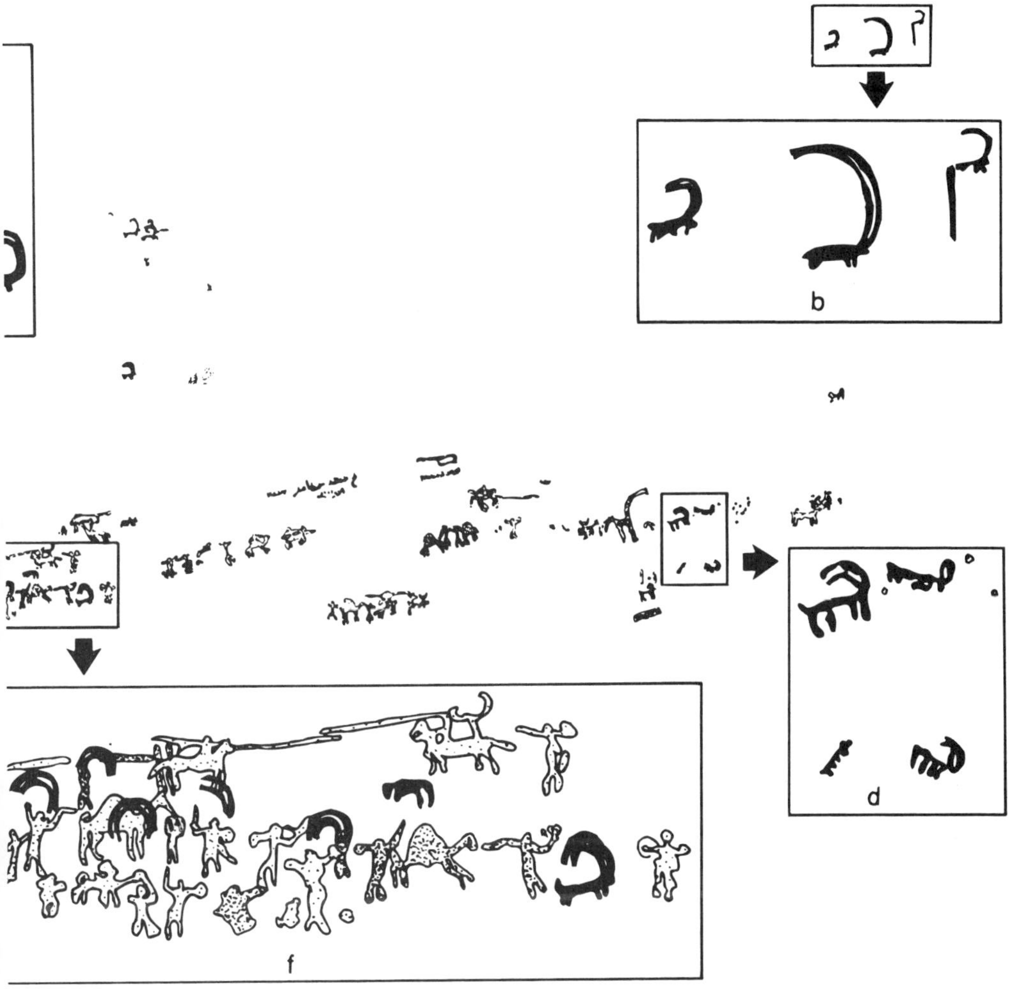


have a similar patina to the natural rock face. Selected petroglyphs are shown at an enlarged scale.

tion. It is possible, of course, that the ground stone items were associated with an occupation other than the Early Hamran, but other artifacts of such an occupation are not evident.

## Petroglyphs

Panels of petroglyphs were found engraved on the rock faces at the backs of the Jebel Hamra (J201), Jebel Mishraq (J504), and Jebel Muheimi (J520) rockshelters. At each of the sites, figures of wild goats (*Capra ibex*) appear as individuals or in herds. They follow a common style in which a "stick figure" body is dwarfed by exaggerated, overarching horns (Figures 10.12, 10.13).

Although there is a strong spatial correlation between the ibex petroglyphs and Early Hamran occupations, as in all such rock art, it is difficult to establish the precise age of the engravings. Several lines of evidence suggest the petroglyphs are of considerable antiquity, however. The rock faces of the inselbergs have developed a dark grayish brown varnish from weathering and in most cases the varnish of the ibex petroglyphs is indistinguishable from that of the surrounding natural surface. Also, the bodies of the figures often have been weathered away leaving only the more deeply incised arched horns as remnants. In the panel of Jebel Hamra (J201), petroglyphs with three lighter shades of varnish are present and in some cases these overlay the ibex figures. These oldest of the post-ibex petroglyphs portray battle scenes with armed horsemen and camel riders, and they are accompanied by ancient Thalmudic script. The script, generally dated to ca. 2,500–1,300 B.P. (Harding, 1959), thus provides a "ceiling" age for the earliest petroglyphs. Similar petroglyphs have been described at the Kilwa Oasis, some 150 km southeast of the study area, also in association with Paleolithic materials.

### Summary

The technotyplogic configurations of Early Hamran assemblages show strong affinities to the Kebaran, especially Bar-Yosef's (1981) "Cluster D" which is characterized by straight-backed and obliquely truncated bladelets (Kebara points). Such assemblages generally have been thought to fall late within the Kebaran sequence that stretched from ca. 20–14,500 B.P. (Bar-Yosef and Vogel 1987; Bar-Yosef and Belfer–Cohen 1989). While recognizing that backed and obliquely truncated bladelets do come to dominate the backed microliths in assemblages from stratified contexts, Hovers and Marder (1991) point to radiocarbon dates which suggest a much greater time-depth for such assemblages.

At a more specific scale, the Early Hamran exhibits some subtle technotypological differences with other Kebaran industries. Early Hamran bladelets appear wider (averaging about 12.5 mm) than those previously published for the Kebaran which generally fall under 12 mm (Bar-Yosef and Belfer-Cohen 1989). Whereas the cores of the Kebaran are described as predominantly single-platform varieties, Early Hamran cores display mostly multiple platforms (i.e., opposed and change of orientation). Also, the approach to core shaping as defined by Bar-Yosef (1991) for Kebaran assemblages that are distributed mainly east of the Rift is not present in the Early Hamran. In this, *lames à crête* and core tablets are produced in shaping narrow single platform cores (Bar-Yosef 1991). From a typological perspective, the micropoints (curved and pointed backed bladelets) common to Late Kebaran assemblages are absent from the Early Hamran. Although backed and obliquely truncated bladelets dominate Early Hamran microliths, these are mainly in the form of truncations opposite snaps, not complete straight-backed and truncated specimens. Finally, it should be noted that "Falita points" (a Gravette-like blade) also reported from Kebaran sites east of the Rift are not found in the Early Hamran.

In summary, it would appear that the Early Hamran assemblages represent a regional variant of those defined within Bar-Yosef's "Cluster D" of the Kebaran Complex and most likely fall late (ca. 16,000–14,500 B.P.) within the sequence.

## MIDDLE HAMRAN

Middle Hamran assemblages were recovered from four sites including the lowland rockshelters of Jebel Hamra (J201) and El Quweira (J203) and the open-air, piedmont sites of J26 and J31 (Figures 10.1 and 10.3). The rockshelter sites of Jebel Mishraq (J520) and Jebel Muheimi (J520) also appear to have had Middle Hamran occupations as evidenced by trapeze/rectangles in the artifacts deflated on the surfaces of these sites. Another Middle Hamran site (J437) was recorded, but never excavated, as an open-air occurrence in the Jebel Qalkha area.

### Excavation Plans and Stratigraphy

Middle Hamran horizons were exposed in the upper layers (A and B) of Jebel Hamra (J201) and in the lower part (Layers B and C) of the El Quweira (J203) deposit. The occupations of the Judayid Basin sites are all open-air, single component occurrences. They were found eroding from relatively shallow deposits which form the basin's late Last Glacial fill (Member II).

## Jebel Hamra (J201): Layers A and B

Middle Hamran artifacts were recovered from Layer A, a relatively thin (5–10 cm thick) red sandy-silt and sandstone rubble stratum, and Layer B, a 60–80 cm thick, compact red sandy-silt stratum. Ash lenses are common to both layers.

## El Quweira (J203): Layers B and C

A light scatter of artifacts was found on the north side of a narrow canyon which separates the main body of Jebel el Hamra from a northeastern extension of the inselberg. The canyon, providing a corridor through the jebel, is currently being invaded by drift sand which has formed high dunes. Such a dune is encroaching on the site and creates the southwestern boundary of the surface artifact distribution (Figure 10.14). A high, near vertical face of the jebel and a shallow overhang form the northeastern limits of the site. To the southeast the shelter overlooks the head of a narrow, deeply incised channel which drains south into the Qa Oumm Salab and to the northwest the site offers a commanding view of the nearby Qa Abu Qurayshi and the distant town of El Quweira.

Five 1 $m^2$ test units were excavated within the surface concentration of artifacts with the deepest of these reaching 1.1m beneath the surface. Artifacts were found to a depth of 80 cm in units 3, 4, and 5, whereas the deposits of units 1 and 2 appear to have been reworked with artifacts occurring only within the upper 20 cm. The soundings revealed a stratigraphy consisting of a red (2.5 YR 5/8), moderate grained sand that extends from the surface to about 60–70 cm depth (Figure 10.15). This layer (A) grades into a 10 cm thick rubble layer (B) composed of strongly oxidized, tabular fragments of indurated sandstone. Similar rubble veneers the modern surface skirting the Jebel Hamra and extends out from the inselberg some 25–100 m. The underlying Layer C is composed of a compact, red sand.

Middle Hamran artifacts were recovered from Layer B and C, whereas Late Hamran specimens were found in Layer B1.

## J26

Artifacts were found eroding from the edge of an alluvial fan which drops away from the foot of Jebel Queisa on the western edge of the Judayid Basin. The site, at an elevation of about 1,080 masl, is situated roughly midway between the edge of the jebel and the Wadi Judayid, close to its confluence with the Wadi Ras en Naqb. Although an open-air site, Site J26 essentially has northern and eastern exposures due to its proximity to the high Jebel Queisa which blocks sunlight from the south and west by mid-afternoon.

Artifacts were found scattered over an area of about 200 $m^2$ along the southern edge of an alluvial fan (Figure 10.16). Test excavations placed along the upslope edge of the artifact scatter exposed a concentration of artifacts accompanied by rocks that casually delimited a firepit within a 10 cm occupation horizon (Figure 10.17). The deposit of the site consists of an undifferentiated moderate grained, red sand which forms the Member II fill of the basin (Henry *et al.* 1983).

## J31

Site J31 is located about 0.6 km south of Site J26 with a quite similar setting. Artifacts were found eroding from an alluvial fan which forms part of the Member II fill of the basin. They were concentrated in an area of ca. 200 $m^2$ bounded by drainages that define the edges of the fan (Figure 10.18).

Figure 10.13. Photograph of petroglyphs on the back wall of the Jebel Muheimi (J520) rockshelter.

Eleven 1 $m^2$ test units were excavated near the center of the concentration to a maximum depth of 1.8 m below surface. These revealed two layers (A and B) and artifacts to a depth of 70 cm. Layer A, 40–50 cm thick, is composed of a silty-sand with thin, steeply bedded lenses that follow the modern slope. It appears to represent reworked Layer B sediments from upslope deposits. Layer B, composed of a moderate grained red sand, displays a much higher artifact density accompanying near level bedded ash lenses (Figure 10.19).

## Flint Artifacts

Assemblages were formed within the Jebel Hamra (J201, Layers A and B) and El Quweira (J203 B and C) rockshelters according to natural layers, whereas artifacts from the entire excavated deposits of the open sites of J26 and J31 were grouped into single assemblages (Table 10.4).

### Typology

In showing a continuity from the Early Hamran, Middle Hamran assemblages are characterized by nongeometric microliths dominated by straight-backed bladelets with truncations opposite snaps and medial segments of straight-backed bladelets (Appendix 10.3). A significant proportion of the assemblages contain geometric microliths, however. These occur exclusively as trapeze rectangles. Formally retouched large tool classes are not well represented.

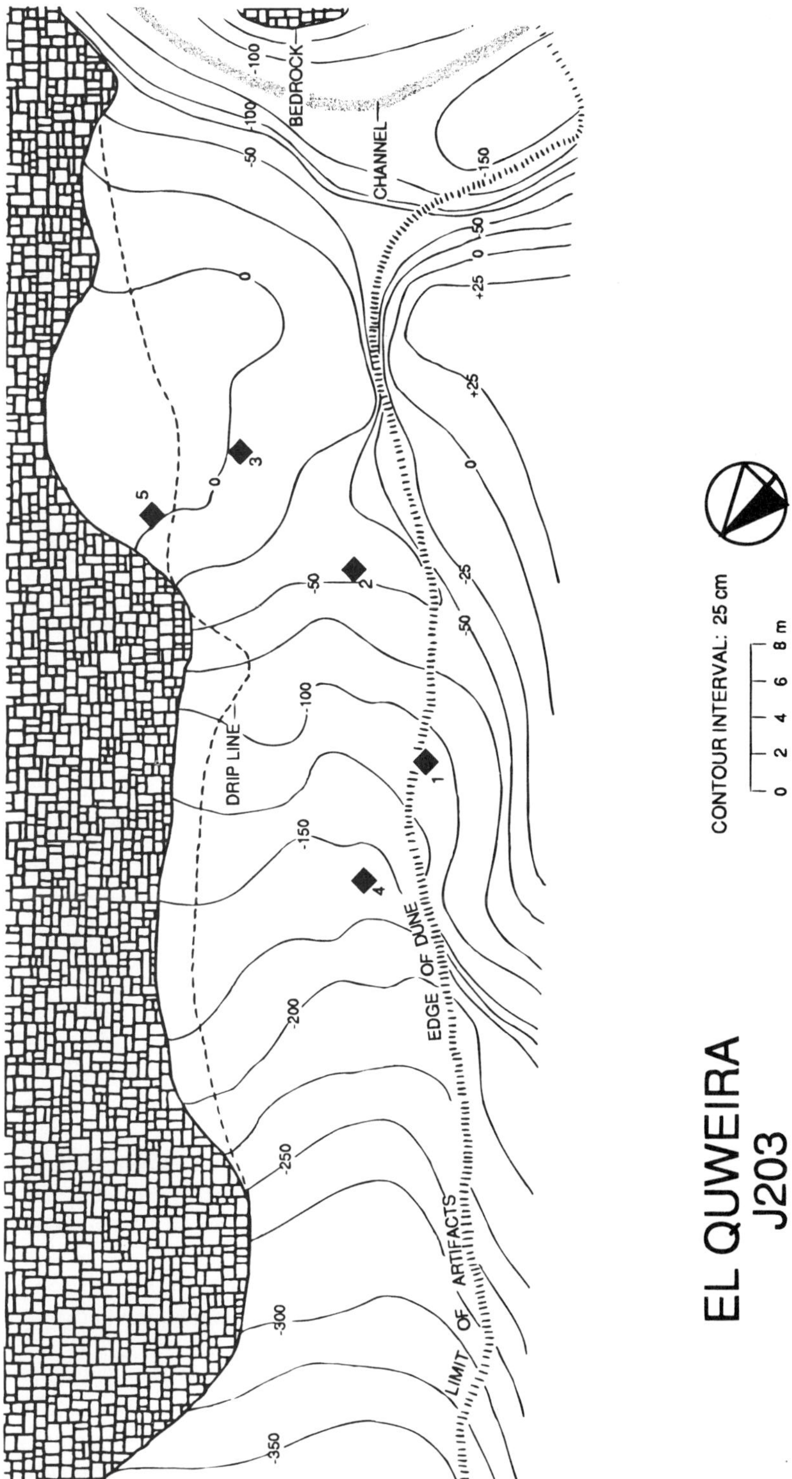


Figure 10.14. Topographic map of El Quweira (J203) showing artifact distribution and excavation plan.

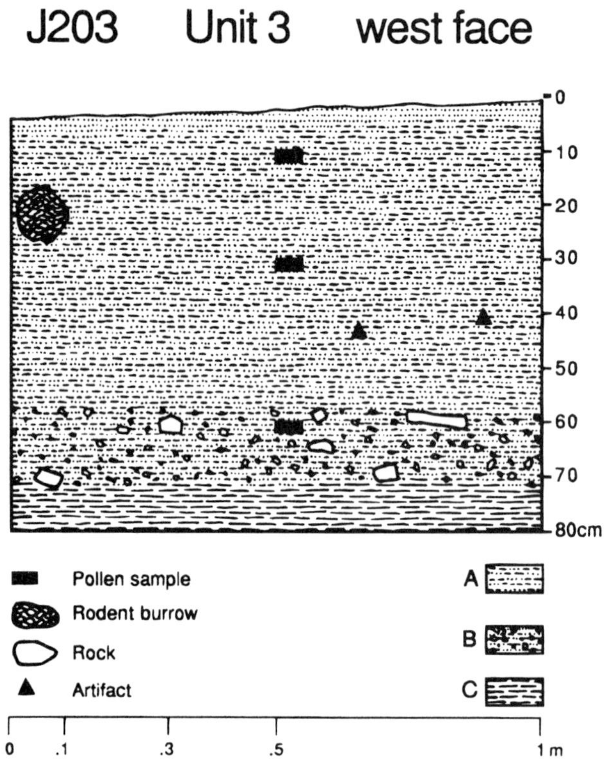


**Figure 10.15.** Section drawing of deposit at El Quweira (J203).

*Scrapers*, present in low frequencies, occur most often as simple end scrapers on blades followed by ones on flakes (Figure 10.20).

*Geometric microliths* in the form of trapeze/rectangles represent moderate proportions of the toolkits (Figure 10.20). The large majority were fashioned with normal abrupt retouch followed by fine abrupt retouch; inverse and *sur enclume* backing are quite rare (Jones 1983:223). The ends of the trapeze/rectangles are predominantly straight-oblique with concave and convex forms accounting for less than 4%. The ends have been truncated by snaps as opposed to the microburin technique.

*Nongeometric microliths* account for 58–73% of the toolkits with straight-backed bladelets with truncations opposite snaps and medial segments of straight-backed bladelets standing out (Figure 10.20). As in the Early Hamran assemblages, the dimensions of the two types are near identical, thus suggesting that these forms were created through the intentional sectioning of bladelets rather than their accidental breakage. As in the Early Hamran this sectioning was accomplished by a direct blow to the surface of a specimen rather than by use of the microburin technique.

*Truncations* make up a moderate portion of most Middle Hamran toolkits and occur most often as straight-oblique truncations on bladelets. They may well represent an early step in the production of trapeze/rectangles. *Retouched pieces* are composed largely of specimens with unilateral obverse retouch with blades and flakes being about equally represented. *Notches and denticulates*, present in low frequencies, mainly occur on bladelets.

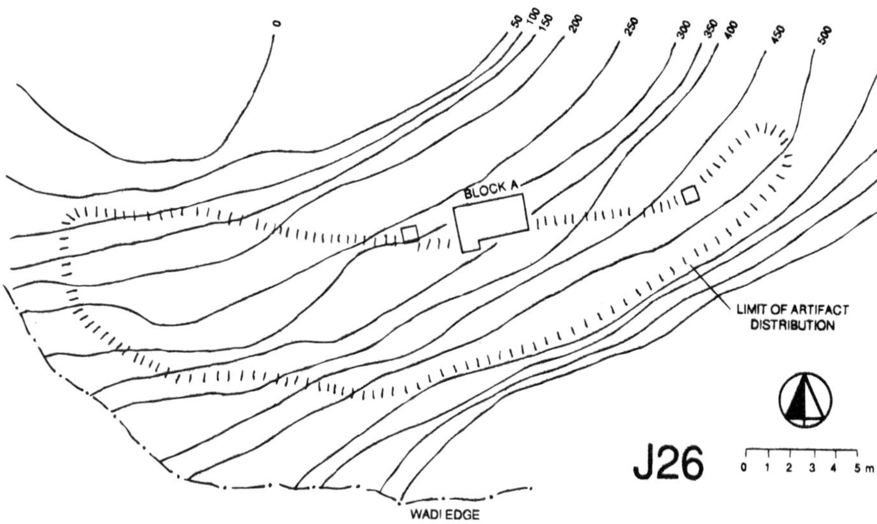


**Figure 10.16.** Topographic map of Site J26 showing artifact distribution and excavation plan.

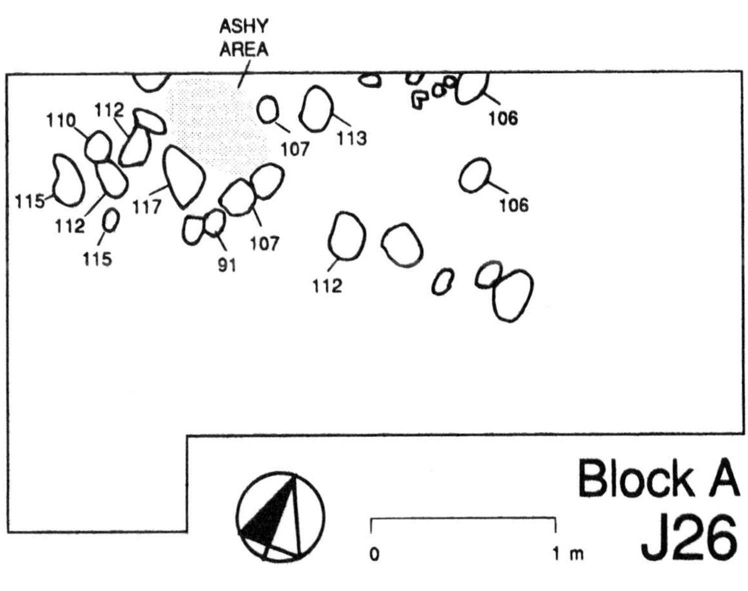


**Figure 10.17.** Plan of firepit at Site J26.

## Technology

Middle Hamran technology follows the pattern of the Early Hamran with an emphasis on the production of relatively wide bladelets from mostly multiple platform pyramidal and sub-pyramidal cores. Cores were rejuvenated primarily through the reorientation of platforms as opposed to removal of core tablets.

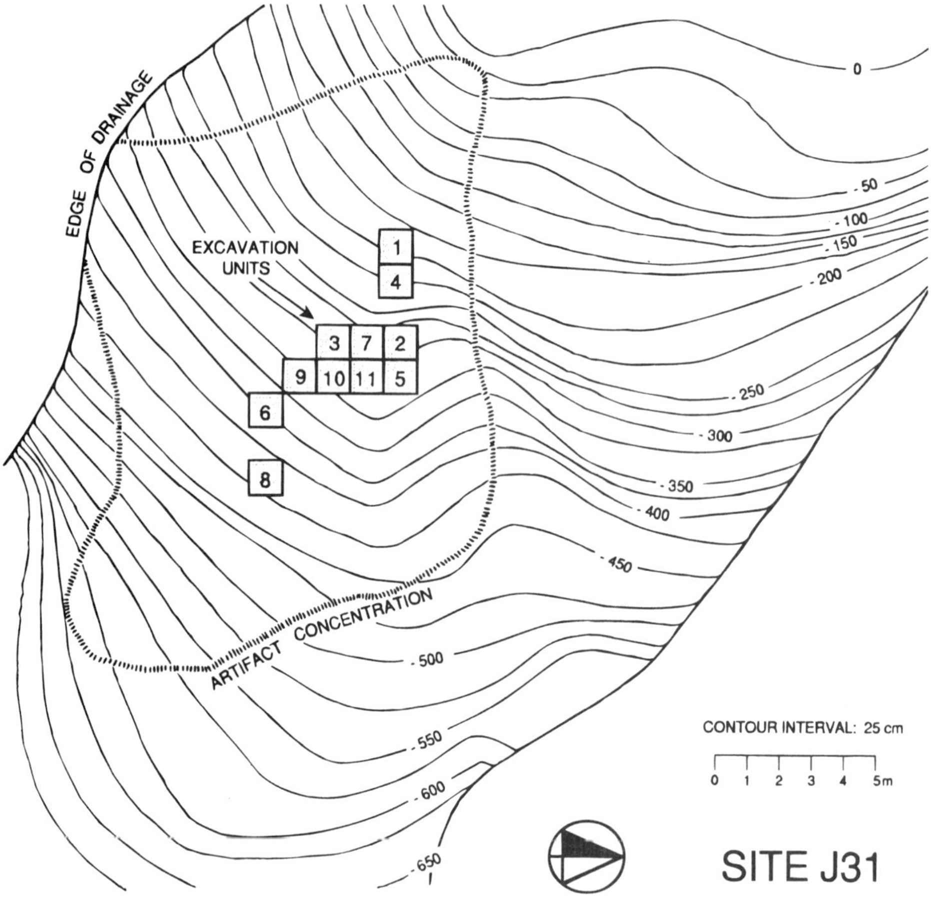


**Figure 10.18.** Topographic map of Site J31 showing artifact distribution and excavation plan.

***Raw material*** exploitation also followed the earlier pattern. Chert was obtained from both *in situ* sources on the plateau and from small cobbles derived from these beds and found in the Wadi Qalkha. Quartzite cobbles furnished a much lesser used local material for those occupations on the floor of the Wadi Hisma.

***Cores***, not well represented in the assemblages, when present occur mostly as multiple platform, pyramidal (subpyramidal) varieties. The small sample, coupled with rather extensive exhaustion, hinders the reconstruction of core shaping methods. Core trimming elements show, however, that the *lame à crête* technique was not used in the initial shaping of cores and subsequent rejuvenation did not employ the removal of core tablets. Cortex is uncommon and rarely extensive. The productive lives of cores appear to have been extended through the creation of opposed and reoriented platforms (Table 10.5).

***Bladelets*** are relatively wide for their lengths with length:width ratios averaging about 2.8:1. High proportions of crushed platforms and unlipped, diffuse bulbs indicate a substantial use of hard-hammer percussion (Jones 1983:178). Lipped and pointed bulbs, attributes of more precise forms of loading (e.g., punch and soft-hammer techniques), occur much less often. Even with an emphasis upon hard-hammer percussion, however, Middle Hamran knappers were able

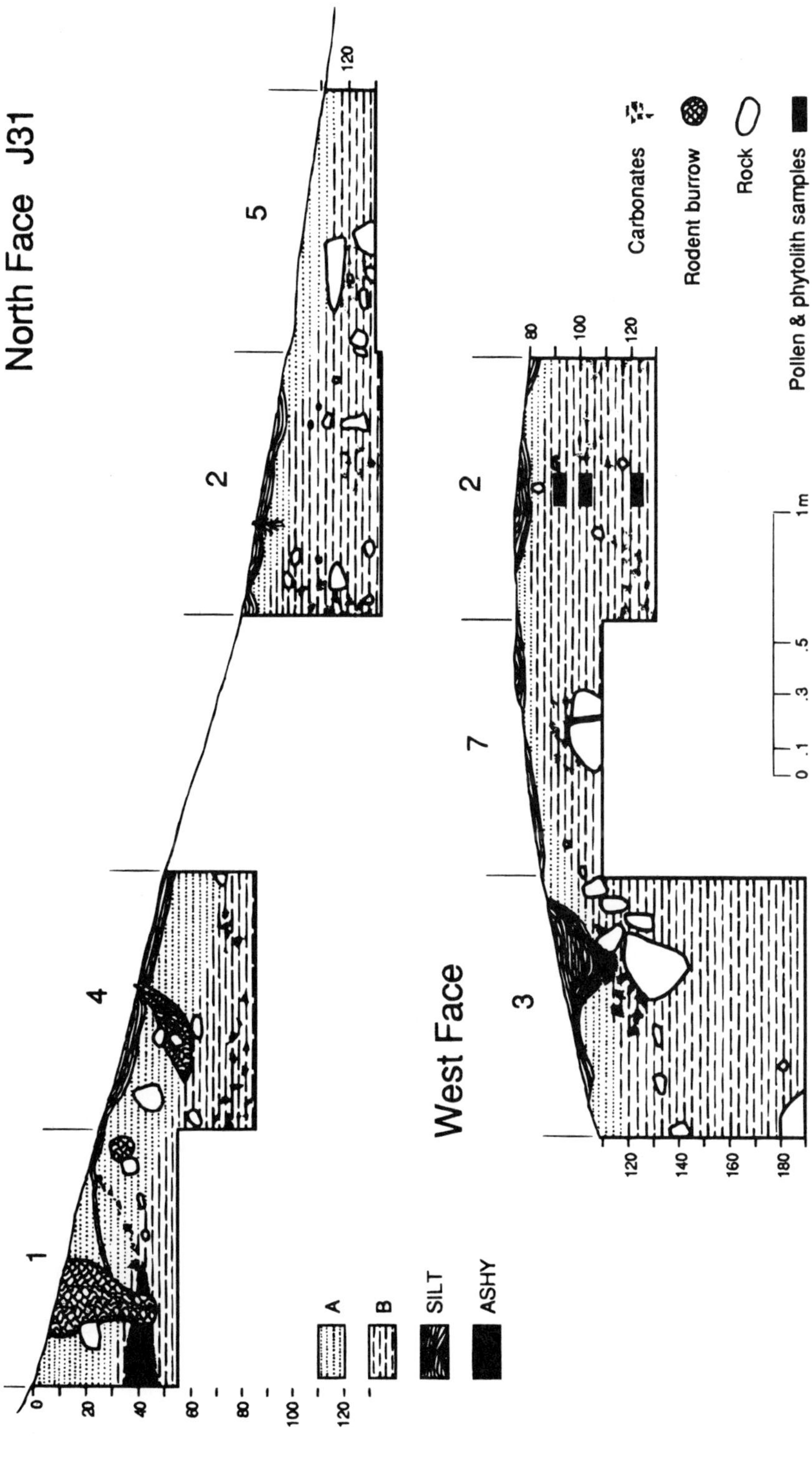


Figure 10.19. Section drawing of the deposit at Site J31.

Table 10.4. Artifact Inventories for Middle Hamran Assemblages

| Assemblage | J201 Layer A | J201 Layer B | J203 Lower | J26 | J31 |
|---|---|---|---|---|---|
| *Tools* | | | | | |
| Scraper | 3.9 | 4.4 | 1.3 | 2.1 | 2.0 |
| Burin | 0.0 | 0.6 | 1.3 | 0.7 | 0.0 |
| Geometric | 19.6 | 15.0 | 6.7 | 9.2 | 10.1 |
| Nongeometric | 57.5 | 59.2 | 73.3 | 71.4 | 63.1 |
| Abrupt Ret. | 0.6 | 0.0 | 0.0 | 0.0 | 0.0 |
| Truncation | 2.2 | 2.2 | 5.3 | 2.8 | 11.4 |
| Retouched Pc. | 11.7 | 10.9 | 10.7 | 10.0 | 7.4 |
| Denticulate | 0.6 | 1.6 | 0.0 | 0.0 | 2.0 |
| Notch | 2.8 | 5.9 | 1.3 | 2.8 | 1.3 |
| Multiple Tl. | 0.0 | 0.0 | 0.0 | 0.0 | 2.0 |
| Perforator | 1.1 | 0.0 | 0.0 | 0.0 | 0.0 |
| Utilized Pc. | 0.0 | 0.0 | 0.0 | 0.0 | 0.0 |
| Points | 0.0 | 0.0 | 0.0 | 0.7 | 0.0 |
| Varia | 0.0 | 0.3 | 0.0 | 0.0 | 0.7 |
| Total No. | 179 | 321 | 75 | 140 | 149 |
| *Debitage* | | | | | |
| Core | 2.0 | 0.7 | 2.0 | 3.1 | 1.7 |
| Blade | 75.7 | 70.5 | 77.7 | 38.6 | 70.1 |
| Flake | 19.4 | 22.5 | 15.9 | 52.0 | 23.5 |
| Core Trim. El. | 0.3 | 1.9 | 2.4 | 0.5 | 0.2 |
| Primary El. | 2.6 | 4.4 | 2.0 | 5.5 | 4.4 |
| Microburin | 0.0 | 0.0 | 0.0 | 0.3 | 0.0 |
| Burin Spall | 0.0 | 0.0 | 0.0 | 0.0 | 0.0 |
| Total No. | 304 | 1052 | 251 | 383 | 519 |
| *Debris* | | | | | |
| Chips | | | | 85.0 | |
| Chunks | | | | 15.0 | |
| Total No. | 550 | 927 | 119 | 492 | 651 |
| *Total Artifacts* | 1,033 | 2,300 | 445 | 1,015 | 1,319 |

to control bladelet production with good precision as evidenced by the low proportions of hinged and overpassed terminations.

## Other Artifacts

Beyond the chipped stone artifacts, Middle Hamran assemblages yielded a single ground stone specimen, a small quartzite pebble hammer stone, and a substantial number of ornamental marine shells. A hand stone fashioned from granodiorite was recovered from Layer B of Jebel Hamra (J201). It appears to have been pecked and then ground into form with the underside showing substantial polish, most likely through use (Figure 10.21). Granodiorite is common to the nearby Precambrian mountains located to the south and west of the site. A small, oval quartzite pebble was recovered from the hearth area of J26. It is unmodified with the exception of "pecking scars" on both ends, thus suggesting that it was used as a knapping stone. The most common sources of the quartzite pebbles are from the sandstone formations of the Wadi Hisma.

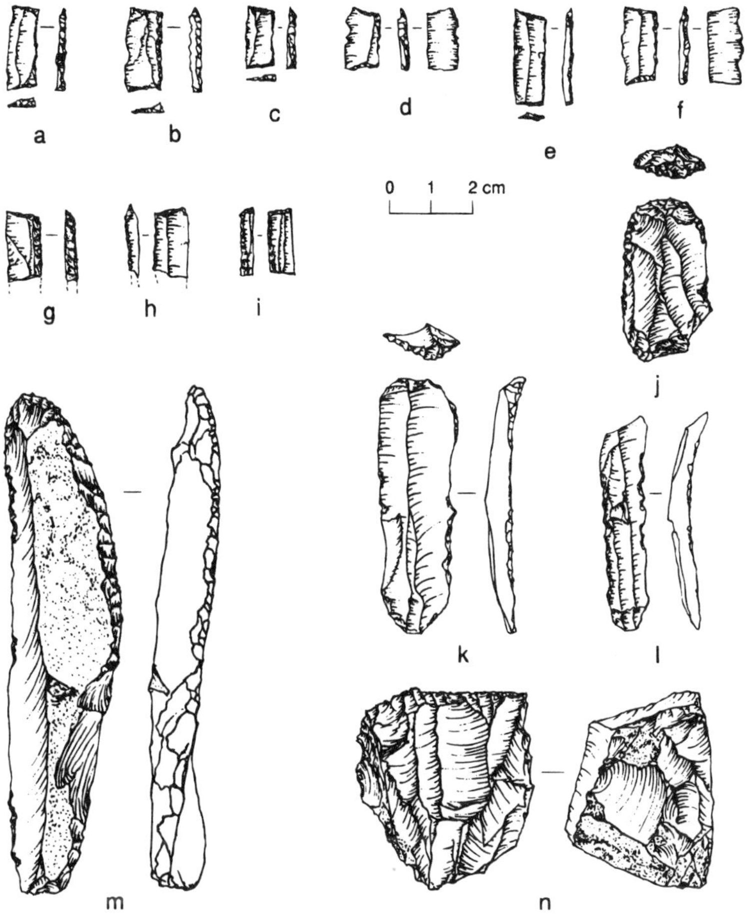


**Figure 10.20.** Illustrations of Middle Hamran artifacts from Jebel Hamra (J201): (a–f) trapeze-rectangles, (g–i) straight-backed bladelets with truncation opposite snap, (j–m) endscrapers on blades, (n) bladelet core.

## Summary

The technotypological configurations of the Middle Hamran assemblages strongly resemble those of the Geometric Kebaran Complex, especially the Group II Industry (Henry 1989:153–55; Bar-Yosef 1981:297–298). Group II assemblages are distinguished from those of other Geometric Kebaran industries by geometric microliths which occur almost exclusively as trapeze/rectangles. While in the other industries of the Geomeric Kebaran, trapeze/rectangles are almost always dominant, other microliths, especially triangles and lunates, comprise significant portions of the toolkit. The trapeze/rectangles of Group II assemblages are also substantially wider (averaging about 7 mm) than those of the other Geometric Kebaran industries.

In fact, Bar-Yosef and Belfer-Cohen (1989) suggest that differences in the widths of trapeze/rectangles may serve as a "chronological yardstick" for the complex. They point to a likely continuity between the narrow trapeze/rectangles of Group I assemblages and the narrow

Table 10.5. Artifact Inventories for Late and Final Hamran Assemblages

| Assemblage | Final J202 Layer A–B2 | Late J202 Layer C | J203 Layer A Up | J203 Layer A Low |
|---|---|---|---|---|
| *Tools* | | | | |
| Scraper | 3.7 | 4.5 | 3.3 | 3.4 |
| Burin | 1.1 | 0.8 | 0.0 | 0.1 |
| Geometric | 19.7 | 13.7 | 13.5 | 10.1 |
| Nongeometric | 34.7 | 47.6 | 69.2 | 75.6 |
| Abrupt Ret. | 0.1 | 0.0 | 0.2 | 0.0 |
| Truncation | 4.1 | 6.1 | 3.1 | 2.7 |
| Retouched Pc. | 17.1 | 15.3 | 5.4 | 4.7 |
| Denticulate | 2.1 | 1.6 | 1.0 | 0.4 |
| Notch | 13.5 | 8.2 | 4.0 | 1.6 |
| Multiple Tl. | 0.6 | 0.0 | 0.0 | 0.0 |
| Perforator | 2.1 | 1.6 | 1.0 | 0.4 |
| Utilized Pc. | 0.0 | 0.0 | 0.0 | 0.0 |
| Points | 0.0 | 0.0 | 0.0 | 0.0 |
| Varia | 1.1 | 0.8 | 0.0 | 0.9 |
| Total No. | 1070 | 380 | 481 | 675 |
| *Debitage* | | | | |
| Core | 2.6 | 2.2 | 0.7 | 0.9 |
| Blade | 58.0 | 60.4 | 76.7 | 82.7 |
| Flake | 28.6 | 29.4 | 19.5 | 13.9 |
| Core Trim. El. | 1.2 | 1.4 | 0.8 | 0.3 |
| Primary El. | 5.3 | 4.6 | 2.1 | 2.2 |
| Microburin | 4.2 | 1.8 | 0.1 | 0.0 |
| Burin Spall | 0.1 | 0.3 | 0.0 | 0.0 |
| Total No. | 3878 | 1531 | 2149 | 2143 |
| *Debris* | | | | |
| Total No. | 3265 | 1443 | 1431 | 1357 |
| *Total Artifacts* | 8213 | 3354 | 4061 | 4175 |

microliths of the Kebaran assemblages of northern Israel and Lebanon and thus conclude that Group II assemblages should have fallen later in time than those of Group I. Unfortunately, firmly fixed radiometric dates are available only for Group II assemblages and these fall between about 14.5–13/12.8K B.P. (Bar-Yosef and Vogel 1987; Henry 1989:11). The Group II assemblages also show a distribution within the arid zone which stretches from northeastern Sinai across the Negev and into southern and perhaps even eastern Jordan.

## LATE AND FINAL HAMRAN

Assemblages representing the Late and Final Hamran were found at only two sites, Qa Salab (J202) and El Quweira (J203). Both are located in the Jebel el Hamra and occur as south, facing rockshelters which overlook nearby dry lake beds. The technotypologic successions of the two deposits define a Late Hamran horizon at El Quweira (Layer A) and Late (Layer C) and Final (Layers A-B2) Hamran horizons at Qa Salab (J202).

The assemblages clearly show a technotypologic time-trend in which trapeze/rectangles are replaced by lunates accompanied by a rise in the microburin technique. This smooth, gradual trend remains uninterrupted across stratigraphic boundaries within the two deposits. The

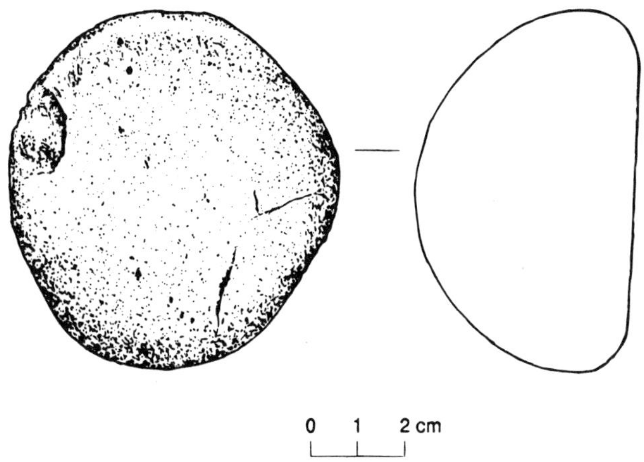


**Figure 10.21.** Illustration of shaped pestle recovered from the Middle Hamran horizon of Jebel Hamra (J201).

distinction, then, between the Late and Final Hamran is purely arbitrary. The Late Hamran is defined merely by a dominance of trapeze/rectangles, whereas the Final Hamran is identified on the basis of a dominance of lunates.

## Settings, Excavation Plans, and Stratigraphy

The rockshelter of Qa Salab (J202) is located about 0.5 km south of the site of El Quweira (described in the Middle Hamran section) on the southern flank of the Jebel el Hamra. The sites rest at opposite ends of a narrow canyon that cuts the inselberg and provides access between the playas lying to the north and south of the jebel.

### Qa Salab (J202)

Artifacts were found covering an area of ca. 388 m$^2$ along the base of the jebel associated with a shallow overhang (Figure 10.22). The shelter has been occupied recently by Bedouin as evidenced by hearths, discarded items, and a rock tomb on the slope to the southeast.

Over three field seasons (1979, 1980, 1983), twelve 1 m$^2$ units were excavated within the surface concentration of artifacts. While the eastern block of units (7–10) failed to yield buried material, the eight units dug in the western end of the shelter revealed a 60–70 cm thick deposit containing rock-lined hearths, animal bones, and a high density of artifacts (Figure 10.23).

Four stratigraphic layers (A, B1, B2, and C) were identified in the deposit. Layer A is composed of a thin (1–10 cm thick) ashy sand stratum containing goat dung, twigs, charcoal, and Bedouin items intermixed with prehistoric artifacts. Layers B1 and B2 are very ashy with the lower of these (B2) being composed of a darker gray ash with a greater amount of charcoal fines. Also, five small rock-lined hearths were exposed in the bottom 10 cm of Layer B2 (Figure 10.24). Layer C consists of a reddish-brown silty sand.

## Flint Artifacts

Artifacts within the Qa Salab (J202) deposit are partitioned into "Upper" (Layers A, B1, and B2) and "Lower" (Layer C) assemblages. At El Quweira (J203), the artifacts from the 60 cm thick Layer A also are divided into "Upper" (0–40 cm) and "Lower" (40–60 cm) assemblages (Table 10.6).

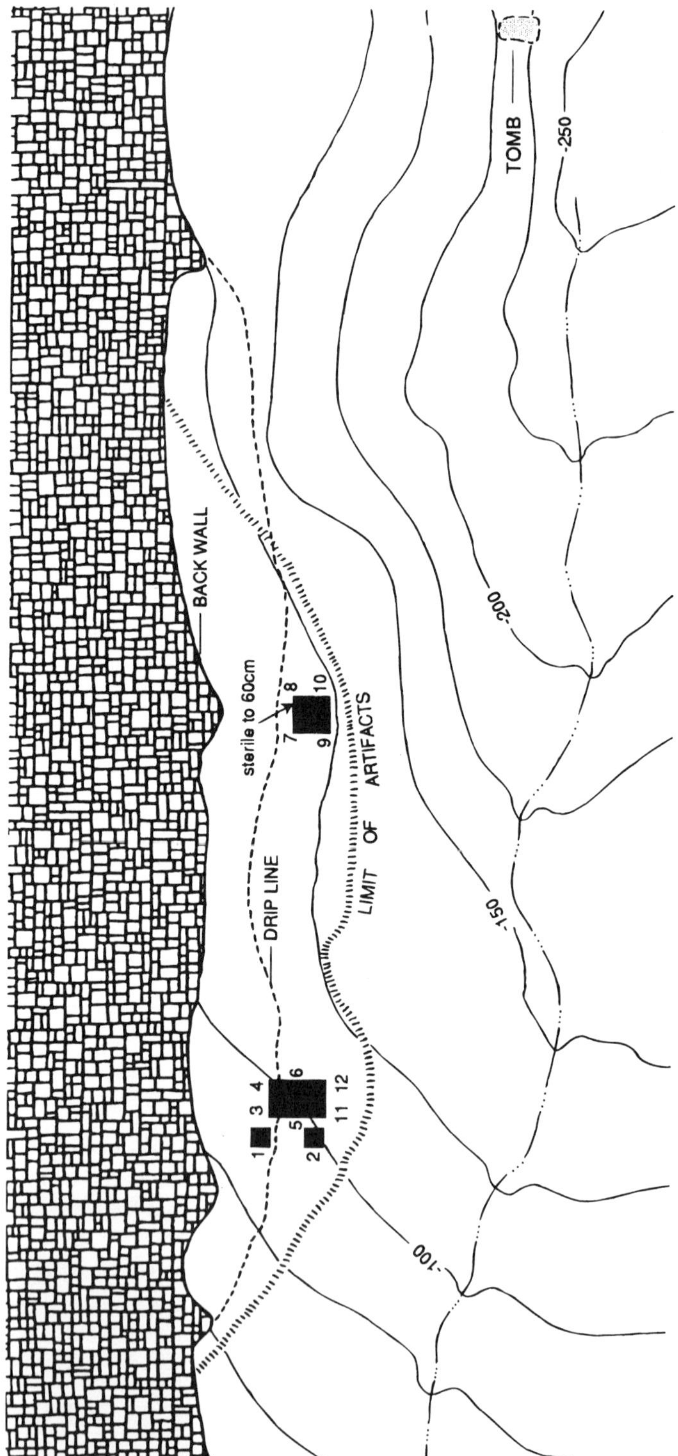


Figure 10.22. Topographic map of Qa Salab (J202) showing the artifact distribution and excavation plan.

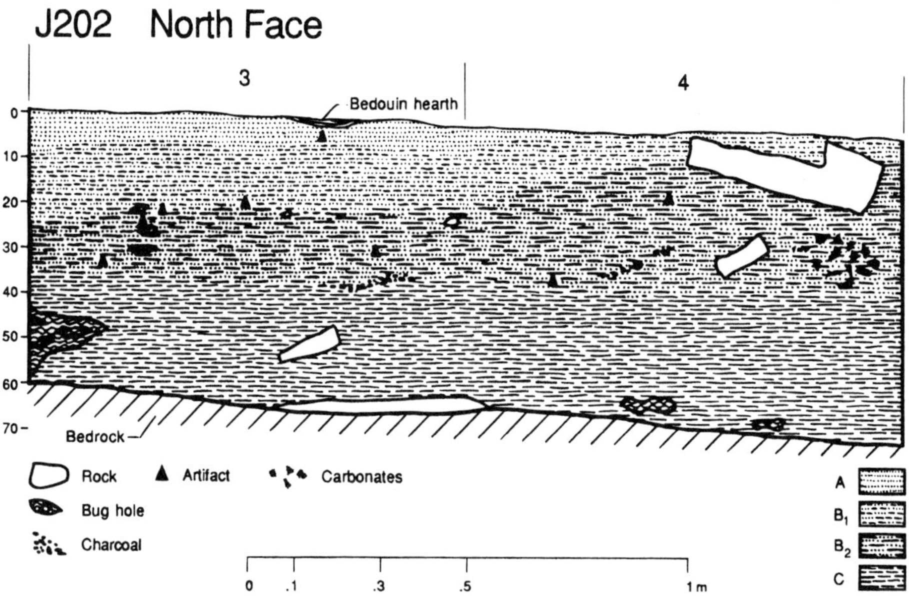


Figure 10.23. Section drawing of deposit at Qa Salab (J202).

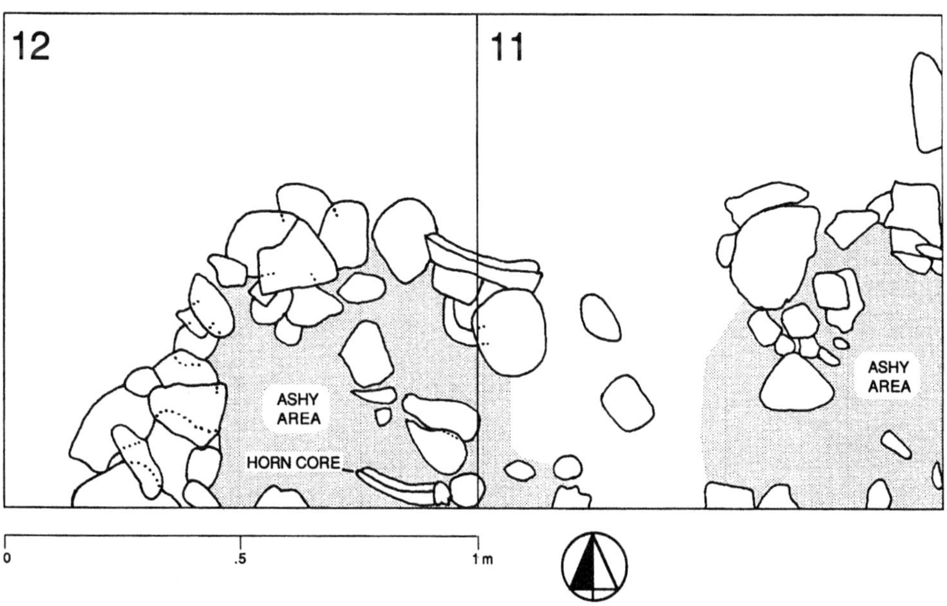


Figure 10.24. Plan of hearth at Qa Salab (J202).

Table 10.6. A Comparison of Hamran Sites Located in the Piedmont and the Lowland Zones with Regard to the Size, Thickness of Cultural Deposit, Direction of Exposure, and Artifact Density

| | Piedmont Sites (1,100–1,200 masl) | | | | | | Lowland Sites (800–840 masl) | | | | | |
|---|---|---|---|---|---|---|---|---|---|---|---|---|
| Site | 21 | 22 | 26 | 31 | 406b | mean | 504 | 520 | 201 | 202 | 203 | mean |
| Site size ($m^2$) | 100 | 120 | 170 | 180 | 320 | 178 | 600 | 260 | 320 | 390 | 650 | 444 |
| Cultural deposit (cm) | 5 | 5 | 20 | 30 | 50 | 22 | 60 | 30 | 80 | 60 | 80 | 62 |
| Exposure o-open | 0 | 0 | E | E | E | — | SW | S | S | S | SW | — |
| Artifact density ($N/m^3$) | 36 | 13 | 17 | 19 | 350 | 89 | 172 | 412 | 114 | 240 | 294 | 246 |
| Total No. of Artifacts | 1012 | 357 | 1035 | 1319 | 3601 | 1465 | 2064 | 3302 | 4778 | 11567 | 8236 | 5989 |

## Typology

The assemblages from Qa Salab (J202) and El Quweira (J203) show a continuity with the Middle Hamran in displaying high frequencies of trapeze/rectangles accompanied by straight-backed and truncated bladelets opposite snaps and medial segments of straight-backed bladelets. When placed in stratigraphic succession, however, the assemblages also show the appearance and progressive rise in lunates which ultimately come to replace trapeze/rectangles as the main geometric. Although the toolkits are dominated by microliths, large, formally retouched tools are well represented across all of the major tool classes, but they occur in low numbers with few classes accounting for more than 5% of the toolkit (Appendix 10.4).

From a technological perspective, a habitual utilization of the microburin technique emerged within the Late Hamran and became a dominant part of lunate production by Final Hamran times. The adoption of the technique appears to have been tied to a shift from trapeze/rectangle to lunate production.

*Scrapers* occur in low frequencies most often as simple end scrapers on blades, but with significant numbers also occurring on flakes (Figure 10.25). A striking number of endscrapers opposite burins are present in the Final Hamran assemblage.

*Geometric microliths* roughly double in their proportionate representation (ca. 10 to 20%) from Late to Final Hamran times. This rise in geometrics is mainly a consequence of a striking increase in lunates. Whereas trapeze/rectangles decline slightly from ca. 10 to 6%, lunates increase from less than 1 to almost 20% of the toolkit by the Final Hamran. Lunates are formed mainly through normal abrupt retouch, but a significant proportion exhibit an alternate semisteep retouch in which the tips of a specimen are retouched on one face and the central portion is retouched on the opposite face (Figure 10.25). This style of retouch may well represent a precursor to true bifacial or Helwan retouch. Only a few specimens with Helwan retouch are present and these are confined to the Final Hamran. The dimensions of the lunates are remarkably uniform with their lengths averaging ca. 18.3 mm. Their forms are largely slender and symmetric with length:width ratios averaging ca. 3.3:1.

An especially interesting difference between the trapeze/rectangles and the lunates relates to their modes of production. While bladelets destined for the production of trapeze/rectangles appear to have been truncated by "snaps," those used in the fabrication of lunates were segmented by use of the microburin technique. This technological shift may well have been linked to the greater control offered by the microburin technique in the creation of the sharply oblique truncations that ultimately formed the tips of a lunate. Remnants of such microburin scars are commonly displayed on the lunates.

*Nongeometric microliths* dominate the toolkits with straight-backed bladelets with truncations opposite snaps and medial segments of straight-backed bladelets accounting for the large majority of the class. Stratigraphically, geometric-microliths show a continual, marked decline (i.e., 76–35%) through the Late/Final Hamran sequence. As in Early and Middle Hamran assemblages, the dimensions of the two types are near identical thus suggesting that these forms were created through the intentional sectioning of bladelets rather than their accidental breakage. As in the earlier phases of the Hamran this sectioning was accomplished by a direct blow to the surface of a specimen rather than by use of the microburin technique.

*Notches* show a clear time-trend with their progressive rise during the course of the Late/Final Hamran sequence. The frequency of notches parallels that of lunates (and microburins) and is a likely expression of the initial step of notching bladelets as part of the microburin technique.

## Technology

In the main, Late/Final Hamran technology continues those patterns established during Early and Middle Hamran times. Relatively wide bladelets were struck from mostly multiple platform pyramidal and subpyramidal cores. Cores, in turn, were rejuvenated primarily through

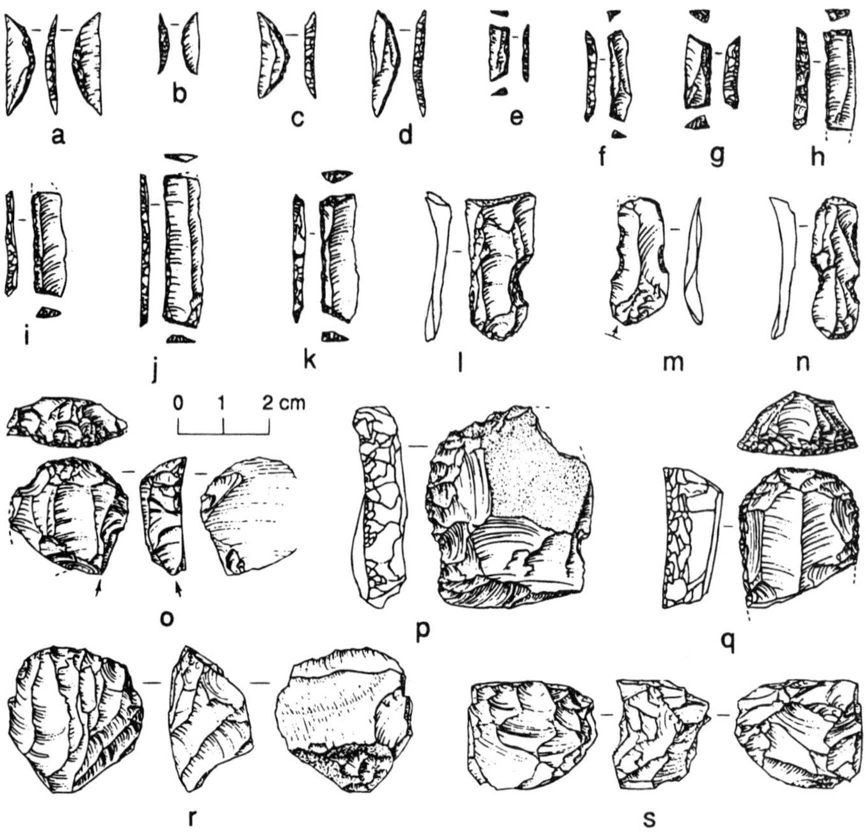


**Figure 10.25.** Illustrations of Late/Final Hamran artifacts from Qa Salab (J202): (a) Helwan lunate, (b–d) abrupt backed lunates, (e–g,j,k) trapeze-rectangles, (h,i) straight-backed bladelets with truncation opposite snap, (l–m) notches, (o) burin opposite scraper, (p,q) endscrapers, (r,s) bladelet cores.

the reorientation of platforms as opposed to removal of core tablets. While these procedures of blank production characterize the Hamran sequence as a whole, the adoption of the microburin technique within the Late and Final Hamran denotes an important technological departure from the earlier Hamran industries.

***Raw Material*** exploitation appears to have followed earlier patterns as well, at least as can be detected for the lowland sites. Chert was obtained from both *in situ* sources on the plateau and from small cobbles derived from these beds and found in the Wadi Qalkha. Quartzite cobbles furnished a much lesser used local material for those occupations on the floor of the Wadi Hisma such as Qa Salab (J202) and El Quweira (J203).

***Cores*** occur mostly as multiple platform, pyramidal (subpyramidal) varieties. As might be expected with their considerable distances to chert sources, the cores from the Qa Salab (J202) and El Quweira (J203) assemblages are mostly exhausted. Efforts at rejuvenation largely relied upon the creation of opposed and reoriented platforms (Table 10.7).

***Bladelets*** are relatively wide for their lengths in Late and Final Hamran assemblages with average length:width ratios ranging from about 2.7–2.9:1 (Appendix 10.5). As in earlier phases of the Hamran, the high proportions of crushed platforms and unlipped, diffuse bulbs reflect the emphasis that was placed upon hard-hammer percussion (Jones 1983:178). Lipped and

Table 10.7 A Comparison of Assemblage Groups by Tool Classes, Production Ratios, and the Percentage of Tools to all Blanks

| | Piedmont | | | | Group A Lowland | | | | Group B Lowland | | | |
|---|---|---|---|---|---|---|---|---|---|---|---|---|
| Tool Classes | Absolute Range | Range | X | SD | Absolute Range | Range | X | SD | Absolute Range | Range | X | SD |
| Scraper | 2–14.9 | 12.9 | 8.6 | 5.5 | 1.4–12.2 | 10.8 | 4.9 | 4.2 | 3.3– 5.1 | 1.8 | 4.0 | 0.9 |
| Burin | 0– 9.0 | 9.0 | 3.7 | 3.3 | 0– 2.8 | 2.8 | 1.6 | 1.0 | 0– 3.2 | 3.2 | 1.0 | 1.5 |
| Denticulate | 0– 7.8 | 7.8 | 2.4 | 3.2 | 0– 4.9 | 4.9 | 2.1 | 1.8 | 0– 1.9 | 1.9 | 1.9 | 0.8 |
| Notch | 1.3–13.6 | 12.3 | 6.5 | 4.5 | 1.4–13.4 | 12.0 | 6.9 | 4.8 | 1.8– 5.9 | 4.1 | 3.7 | 1.7 |
| Retouched Pc. | 7.4–63.8 | 56.4 | 46.8 | 23.9 | 5.1–32.5 | 27.4 | 14.5 | 11.3 | 4.8– 16.0 | 11.2 | 9.3 | 5.3 |
| Perforator | 0 | 0 | 0 | 0 | 0– 3.5 | 3.5 | 3.8 | 4.6 | 0– 3.2 | 3.2 | 1.0 | 1.5 |
| Truncation | 0–11.4 | 11.4 | 3.1 | 5.0 | 2.2– 6.0 | 3.8 | 4.5 | 1.5 | 2.6– 3.1 | 0.5 | 2.8 | 0.2 |
| Bk. Microlith | 9–73.2 | 64.2 | 23.8 | 28.2 | 30.8–84.5 | 53.7 | 64.0 | 21.7 | 65.4– 86.5 | 21.1 | 77.1 | 9.5 |
| Other | 0– 2.1 | 2.1 | 0.4 | 0.9 | 0– 0.5 | 0.5 | 2.0 | 0.3 | 0– 1.3 | 1.3 | 0.4 | 0.6 |
| Production Ratios | | | | | | | | | | | | |
| Blanks: | | | | | | | | | | | | |
| 1 core | 50–60 | 10 | 53.8 | 4.1 | 37–56 | 19 | 47.4 | 7.0 | 112–221 | 109 | 156 | 46.2 |
| Tools: | | | | | | | | | | | | |
| 1 core | 7–19 | 12 | 13.8 | 5.8 | 11–30 | 19 | 17.0 | 7.8 | 24–26 | 20 | 34.8 | 8.4 |
| % of Tools to all blanks | 20.4% | | | | 26.4% | | | | 18.2% | | | |

pointed bulbs, attributes of more precise forms of loading (e.g., punch and soft-hammer techniques), occur in much lower frequencies. But the reliance upon hard-hammer percussion does not appear to have adversely affected the ability of knappers to deliver well formed bladelets as evidenced by the very high proportions (i.e., 85–91%) of feathered terminations.

### Other Artifacts

Beyond the chipped stone specimens, other artifacts consist of numerous ornamental marine shells and small oval quartzite hammer stones.

### Summary

Certain technotypological features of Late and Final Hamran assemblages are unparalleled by other assemblages of the Geometric Kebaran Complex. Specifically these relate to the presence and even dominance of lunates within the geometric microliths coupled with an intensive utilization of the microburin technique. When viewed in the broader context of the late Levantine Epipaleolithic, however, these technotypologic configurations suggest a placement near the end of the Geometric Kebaran and most likely transitional to the Early Natufian.

Although this succession seems most likely, some important typological differences exist between Late/Final Hamran and Early Natufian assemblages. Perhaps the most obvious of these is the rarity of Helwan retouch within the Late/Final Hamran, whereas it represents the stylistic *fossile directeur* of Early Natufian and synchronous assemblages. While rare, the mere presence of Helwan retouch indicates a very late temporal position within the Geometric Kebaran for Late/Final Hamran assemblages. The significant proportions of alternate bifacially retouched lunates in the Late/Final Hamran may well point to the stylistic precursor of Helwan retouch.

## CONCLUSIONS

The assemblages that form the Hamran succession share several technotypologic characteristics while at the same time showing clear stratigraphic trends. Whereas these technotypologic time-trends parallel those identified within the Levantine Epipaleolithic as a whole, a few specific features and combinations are unique to the Hamran.

From a behavioral perspective, a settlement-procurement strategy following a pattern of transhumance appears to have persisted throughout the Hamran sequence. This local ecologic pattern also seems to have broad similarities in the Epipaleolithic record of the Levant. Low elevation sites are typically linked to larger occupations of longer duration than their upland counterparts (Bar-Yosef and Belfer-Cohen 1989; Kaufman 1992). At a regional scale, the Hamran would appear to represent a local variant of the Kebaran–Geometric Kebaran succession, culminating with a transition to the Natufian.

### Chronology and Time Trends

Given that the radiocarbon determinations from Early (J504) and Late/Final (J202) Hamran deposits are obviously too young, an absolute chronology must rely upon cross-reference dates from Kebaran and Geometric Kebaran occurrences.

#### Radiocarbon Chronology

In most closely resembling Bar-Yosef's "Cluster D" assemblages of the Kebaran Complex, Early Hamran assemblages would be expected to fall late (ca. 16–14,500 B.P.) within the sequence. Middle Hamran assemblages fit well within the technotypologic parameters of Group

II of the Geometric Kebaran and are thus likely to date from about 14,500–13/12,800 B.P. (Bar-Yosef and Vogel 1987; Henry 1989:11). In being known only from southern Jordan, the Late/Final Hamran assemblages lack reference dates, but their technotypologic characteristics indicate a temporal placement at the end of the Geometric Kebaran immediately preceding the Early Natufian. In that radiocarbon determinations for Early Natufian deposits in southern Jordan (Beidha and Wadi Judayid) cluster ca. 12,500 B.P., Final Hamran occupations are most likely to date ca. 13,000–12,500 B.P.

### Artifact Trends

Time trends seen in Hamran artifact inventories are confined largely to five types of microliths (Figure 10.26). All of these were probably employed as armatures for arrows as indicated by the presence of mastic and edge-wear on such forms within the Geometric Kebaran (Jones 1983; Fujimoto 1979; Bar-Yosef and Goring-Morris 1977). Straight-backed bladelets and the medial segments of straight-backed bladelets decline sharply from Early Hamran to Middle Hamran times accompanied by rises in trapeze/rectangles and backed and truncated specimens. From the Middle to Late Hamran, the most striking trends relate to the virtual disappearance of straight-backed bladelets, a slight decline in trapeze/rectangles, and the appearance of lunates. Lunates increase greatly by the Late Hamran in conjunction with noticeable drops in the medial segmented and truncated straight-backed bladelets. In that the microburin technique was tied to the production of lunates, an expected strong correlation is seen between microburin indices and lunate frequencies. The few lunates found in the Early Hamran are largely of the Krukowski variety and were thus most likely generated accidentally.

A technical examination of the trends manifested by the five types of microliths shows a progressive change toward a greater amount of edge modification. This line of evolution carries from the straight-backed bladelets and medial segments, which lack end-modification, through to specimens with retouch on one end, then both ends, and finally the entire back. Although stylistic preference may have influenced this trend to some degree, the fundamental changes in form and degree of retouch most likely resulted from functional requirements. The widespread distribution of the trend in both time and space (much of Africa and the Near East) suggests that it was tied to some technical aspect of the design of microliths, probably having to do with joining specimens to arrowshafts.

Several archaeologic and ethnographic examples of microliths as armatures show them to have been mounted by gum or resin to shafts as transverse arrowheads as well as barbs (Clark 1970:155; Clark *et al.* 1974). In either case the retouched portion of a specimen's edge represents that part which is bonded to a shaft, leaving the unmodified portion as the cutting edge. It would appear then that the increasingly greater amount of edge modification of Hamran microliths would have progressively improved their hafting qualities. From an evolutionary perspective, this may simply mean that once microliths had come to be mounted as transverse arrowheads or points, there were relatively few options to increasing the amount of modified edge beyond the design changes seen in the trend.

## Hamran Transhumance

The distribution, settings, and artifact content of Hamran sites point to a transhumant pattern of settlement and procurement in which groups coalesced in the lowlands during the winter and dispersed into the uplands during the warm season (Figure 10.27). Evidence for this pattern comes from a persistent, time-transgressive dichotomy between sites in the lowlands and those of the piedmont as expressed in several characteristics. These include clear differences in the sizes, occupational thicknesses, exposures, artifact densities, tool-debitage-core production ratios, and toolkit configurations between sites of the two elevational belts.

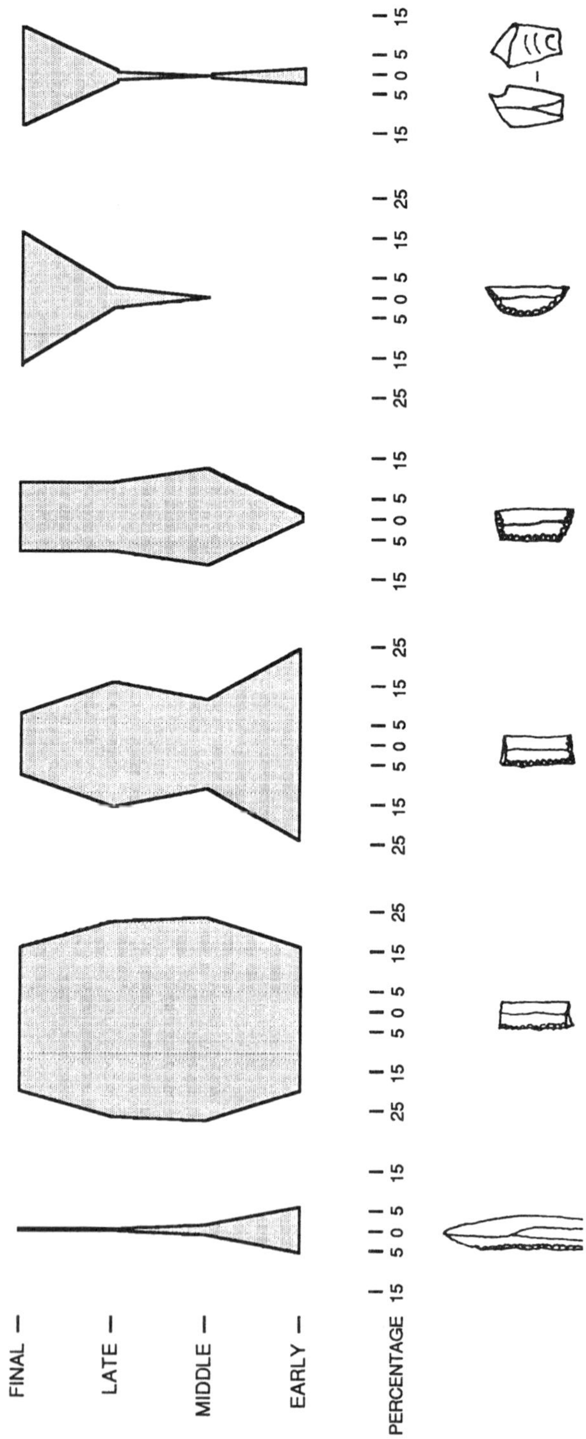


Figure 10.26. Time-trends observed for the principal microliths of Hamran assemblages.

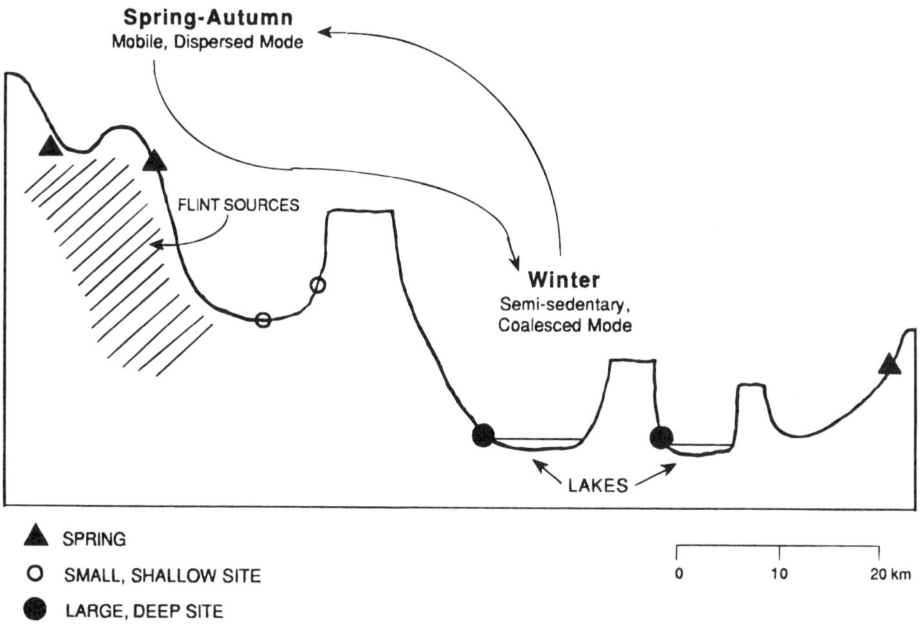


**Figure 10.27.** Transect of the area showing the distributions of resources, the varieties of Hamran sites, and the proposed pattern of transhumance.

## Settings, Site Characteristics, and Seasonality

Sites in the piedmont are all small with thin cultural deposits. All are open sites, but many are situated close enough to the faces of high jebels that they have limits on their exposures and in all cases these are to the east. In contrast to the sites of the piedmont, lowland sites are substantially larger with much thicker cultural deposits. They occur only in sheltered settings and are oriented consistently to the south and southwest.

These differences in settings and site characteristics are thought to reflect differences in the season and the intensity in which the Hamran camps of the two zones were inhabited. The small, thin cultural deposits of the piedmont are viewed as the remnants of ephemeral camps occupied by small groups, and rarely revisited. Their open settings, often with eastern exposures, would have made unlikely camps during the winter wet season at such high elevations, especially during the Late Glacial with its depressed temperatures.

The large, thick cultural deposits of the lowlands imply longer term residences by larger groups who regularly revisited the sites. Their settings in rockshelters with exposures to the south and southwest are most consistent with winter occupations. These settings would have afforded the occupants of the sites protection from the elements as well as exposure to afternoon sunlight. In extending winter sunlight the inhabitants of the shelters would have benefited from greater warmth during the day and also at night. Investigation of sites with such settings during the winter of 1983–84 revealed an important nighttime advantage of their prolonged exposure to sunlight. The dark red sandstone that forms the backwalls of the shelters is heated by sunlight during the afternoon. This heat is then radiated at night when temperature falls to a diurnal low and heat is most needed. In January of 1984 with nighttime temperatures hovering just above 0 °C, warmth from the backwalls of Jebel Muheimi, (J520) and Jebel Mishraq (J504) could be detected at a distance of 3–5 m. These lowland shelters would not only have offered protection from wind and precipitation, they also would have furnished evening warmth from natural solar furnaces.

## Artifact Densities and Production Ratios

Artifact densities are markedly higher for the lowland sites with only Wadi Humeima (J406b) showing such high densities for a site in the piedmont. This may well be because of its setting adjacent to a spring, a feature that would have increased the periodicity and duration of occupations. Beyond higher artifact densities, other indicators of the greater occupational intensity of the lowland sites come from their much greater number of hearths and other fixed features (e.g., bedrock mortars and perhaps even petroglyphs).

An examination of the debitage frequencies, strategies in core shaping, and blank dimensions for Hamran assemblages shows a remarkable continuity through the sequence. Some major differences in production ratios are present, but these are associated with differences in elevation, not time-frames. The piedmont sites average about 54 blanks (blades/bladelets) per core, while the lowland occupations fall into two groups. One of these resembles those of piedmont assemblages with low blank to core ratios, but the other averages about three times that number.

Most directly, these patterns might be interpreted as expressions of economizing behaviors on the part of the inhabitants of the lowland camps, given that most of their chert sources were restricted to piedmont and upland settings some 25–30 km away. Among their options for improving the efficiency of raw material consumption, they could have imported large raw material elements and cores to the lowlands. They also could have extended the productive lives of cores. Comparisons of the dimensions of primary elements, core facets, and blanks confirm that neither of these strategies was employed, however (Henry 1987, 1989b). Instead, the inhabitants of the lowland shelters apparently chose to import bladelets on a large scale, accounting for as much as two-thirds of those in an assemblage (Henry 1989b).

The importation of bladelets probably represented only one of a number of adjustments in procurement that accompanied movement from the piedmont to the lowlands. During their occupation of the piedmont, small, mobile Hamran groups are likely to have "mapped onto" resources, but in the context of their long-term lowland camps groups were likely to have adopted a logistically oriented strategy. The importation of bladelets would have allowed for the reduction of raw material to its most portable dimension, while at the same time, leaving open the opportunity for the production of a wide range of tool forms. In the Hamran, bladelets account for some 75–90% of the toolkits for the lowland sites. Thus, by importing bladelets over cores or unmodified chert, little of the potential variety of the toolkit was sacrificed, but efficiency in procurement of chert from the distant uplands was greatly enhanced.

## Toolkit Configurations

Comparison of the Hamran assemblages by tool class reveals statistical differences between the piedmont and lowland sites in only three of the nine classes (Table 10.9). Retouched pieces occur in much higher frequencies in the piedmont sites, while backed microliths (geometrics and nongeometric combined) and perforators are more common in the lowland sites. The specific functional explanation for these contrasts is unknown, but the dominance of expediently manufactured "retouched pieces" is consistent with the proposed ephemeral nature of the piedmont camps. The higher frequencies of "backed microliths" and "perforators," tools requiring more detail in design and greater effort in fabrication, are in accord with the proposed longer duration camps of the lowlands.

Although significant differences in the toolkits from the two zones are confined to only three classes, in all but one class (perforators) the assemblages of the piedmont zone show greater interassemblage variability than those of the lowland zone. This is expressed in their greater ranges and standard deviations. The greater variability between toolkits of the piedmont zone is consistent with proposed differences in occupational intensities of the two elevational belts. When inhabiting an ephemeral camp in the piedmont, a Hamran group is likely to have exploited a relatively narrow range of resources given both the temporal and spatial limits placed upon the catchment of the camp. Camps in the piedmont, therefore, would have varied significantly from one camp to another in regard to the resources available for procurement and processing. With such a settlement pattern a high degree of variability would be expected between the toolkits of different camps. The longer-term occupations of the lowland camps would have commanded larger catchments with a greater chance for diverse resources to come into and go out of phase. The greater temporal and spatial scopes of their catchments would have acted to reduce the degree of variability between the resources available to these lowland sites. Their activities would have been less specialized and toolkits would have varied less between sites.

## Regional Comparisons

In comparing the Hamran succession to neighboring regions, the Early Hamran appears to represent an extension of late Kebaran groups from the Mediterranean woodland core area into what was then a marginal woodland and steppe zone along the southern edge of the Jordanian Plateau. The Phase B occupation of Kharaneh IV denotes a similar extension along the eastern periphery of the woodland zone in northeastern Jordan (Muheisen 1985). West of the Rift Valley, the Kebaran was apparently restricted to the Mediterranean woodland core area, while the arid zone was inhabited by very late Upper Paleolithic groups such as at the Highland Negev sites of Ein Aqev (Marks 1976b) and Ein Aqev East (Ferring 1977). Similar very late Upper Paleolithic occurrences recently have been noted east of the Rift Valley on the Jordanian Plateau in the re-excavation of the Madamagh rockshelter (Schyle and Uerpmann 1988) and at the site of Ain el-Buhira (WHS 618) in the Wadi Hasa (Coinman 1990:49–50; Clark *et al.* 1988:240). Ein el-Buhira is dated to ca. 20.3K B.P.

In the Middle Hamran the strong artifactual affinities with Group II assemblages of the Geometric Kebaran point to an ecological shift from the earlier connections with the Mediterranean woodland zone to exploitation of the steppe-desert zone. Group II sites are distributed in the arid zone along an arc stretching from northeastern Sinai across the Negev and into southern (and perhaps eastern) Jordan.

Given their site settings, Late and Final Hamran groups apparently continued to exploit the arid zone along the margins of the woodlands. But in the absence of similar assemblages outside of southern Jordan, broader regional distribution patterns remain unknown.

**Appendix 10.1. Type List for Early Hamran Assemblages**

| Assemblages | J201 Layer C | J504 | J520 | J406b | J21 | J22 |
|---|---|---|---|---|---|---|
| Number | 154 | 297 | 280 | 368 | 56 | 22 |
| A. Scraper Class | | | | | | |
| 1. | 1.3 | 4.4 | 3.9 | 2.2 | 1.8 | 9.1 |
| 2. | | 1.0 | 1.1 | 1.4 | | |
| 3. | 1.3 | 4.7 | 0.7 | 2.2 | | |
| 4. | | 1.3 | 3.9 | 1.4 | | |
| 5. | | | | | | |
| 6. | 1.3 | | | | | |
| 7. | | 1.0 | 0.4 | 0.3 | 1.8 | |

Appendix 10.1. (*Continued*)

| Assemblages | J201 Layer C | J504 | J520 | J406b | J21 | J22 |
|---|---|---|---|---|---|---|
| Number | 154 | 297 | 280 | 368 | 56 | 22 |
| 8. | 1.3 | | | 1.9 | | |
| 9. | | | | 0.5 | 3.6 | |
| 10. | | | | | | |
| 11. | | | | 0.3 | | |
| 12. | | | 0.4 | | | |
| 13. | | | | | | |
| 14. | | | 1.4 | | | |
| 15. | | 0.3 | | | | |
| B. Burin Class | | | | | | |
| 16. | 1.9 | | 0.4 | 0.8 | 3.6 | 4.5 |
| 17. | 0.6 | 0.7 | 1.1 | 0.8 | | |
| 18. | 0.6 | 1.3 | 0.7 | 0.3 | | 4.5 |
| 19. | | | 1.1 | | | |
| 20. | | | 0.4 | 0.3 | | |
| 21. | | | | | | |
| 22. | | 0.3 | | | | |
| 23. | | | | 0.3 | | |
| 24. | | | | | | |
| 25. | | | | | | |
| 26. | | 0.3 | | | | |
| C. Microlithic Geometric | | | | | | |
| 27.–29. | | | | | | |
| 30. | | | | 1.8 | | |
| 31. | | | | | | |
| 32. | | 0.3 | 0.4 | 0.3 | | |
| 33. | | | | | | |
| 34. | | | 0.4 | | | |
| 35. | | | | | | |
| 36. | | | | | | |
| D. Nongeometric Microliths | | | | | | |
| 37. | | 1.0 | 2.9 | 2.7 | | |
| 38. | 1.3 | | 0.4 | | | |
| 39. | | | | 0.8 | | |
| 40. | 6.5 | 9.1 | 3.6 | 12.8 | 1.8 | 9.1 |
| 41. | | | | 0.8 | | |
| 42. | 0.6 | 0.7 | 0.4 | | | |
| 43. | 26.0 | 9.1 | 4.3 | 11.1 | 3.6 | 4.5 |
| 44. | 6.5 | | 1.4 | 1.6 | | |
| 45. | | | | 0.3 | | |
| 46. | | | | | | |
| 47. | 0.6 | | 0.4 | | | |
| 48. | 5.2 | | 0.4 | 4.3 | | |
| 49.–51. | | | | | | |
| 52. | 3.2 | 0.7 | | 2.4 | | |
| 53. | | | | | | |

Appendix 10.1. (*Continued*)

| Assemblages | J201 Layer C | J504 | J520 | J406b | J21 | J22 |
|---|---|---|---|---|---|---|
| Number | 154 | 297 | 280 | 368 | 56 | 22 |
| 54. | 1.9 | | | 0.8 | | |
| 55. | 11.7 | 6.7 | 11.1 | 1.9 | 14.3 | |
| 56. | 0.6 | 1.0 | 1.4 | 1.1 | | |
| E. Abruptly Retouched Blades and Flakes | | | | | | |
| 57. | | | | | | |
| 58 | | | 1.1 | | | |
| 59.–64. | | | | | | |
| 65. | | 1.0 | 0.7 | | | |
| F. Truncations | | | | | | |
| 66. | 2.6 | 4.4 | 2.5 | 0.53.6 | | |
| 67. | | | 0.4 | 0.5 | | |
| 68. | | | | | | |
| G. Retouched Pieces | | | | | | |
| 69. | | | 5.4 | 6.3 | | |
| 70. | | | 0.4 | 2.2 | | |
| 71. | | | 0.4 | 3.3 | | |
| 72. | | | | 0.3 | | |
| 73. | | | | 0.5 | | |
| 74. | | | | 2.2 | | |
| 75. | | | | | | |
| 76. | | | | | | |
| 77. | 9.7 | 17.2 | 3.9 | 2.2 | 28.6 | 31.8 |
| 78. | | 0.7 | | 2.2 | 3.6 | |
| 79. | | | 1.8 | 0.5 | | |
| 80. | | | | | | |
| 81. | 5.2 | 3.4 | | 1.1 | | |
| 82. | | | | 1.9 | | |
| 83. | 1.3 | 5.7 | 12.9 | 6.0 | 8.9 | 13.6 |
| 84. | | 1.7 | 1.4 | 0.3 | 1.8 | 9.1 |
| 85. | | | 2.5 | 0.8 | | |
| 86. | | | | 0.3 | | |
| 87. | | 2.7 | 1.8 | 1.4 | | |
| 88. | | | 0.7 | 4.3 | | |
| 89. | | | 1.8 | | | |
| H. Denticulate Class | | | | | | |
| 90. | | 3.7 | 1.8 | 1.4 | 5.4 | |
| 91. | | | | | | |
| I. Notch Class | | | | | | |
| 92. | 3.2 | 6.1 | 8.9 | 2.2 | 5.4 | 13.6 |
| 93. | | | 4.6 | 2.4 | | |
| 94. | | 2.4 | 0.7 | 0.5 | | |
| 95. | | | | 0.5 | | |
| 96. | | | | 0.5 | | |
| J. Multiple Use Class | | 1.3 | 0.4 | | | |
| K. Perforator | 3.2 | 3.4 | 3.2 | 1.8 | | |
| L. Massive Piece Class | | | | | | |
| 97.–100. | | | | | | |

Appendix 10.1. (*Continued*)

| Assemblages | J201 Layer C | J504 | J520 | J406b | J21 | J22 |
|---|---|---|---|---|---|---|
| Number | 154 | 297 | 280 | 368 | 56 | 22 |
| M. Utilized Piece Class | | | | | | |
| 101.–103. | | | | | | |
| N. Points | | | | | | |
| 104.–109. | | | | | | |
| O. Varia | | | | | | |
| 110. | 1.9 | 2.4 | 0.7 | 1.4 | 8.9 | |

Appendix 10.2. Bladelet Attributes of Early Hamran Assemblages

| Assemblages | J201[a] Layer C | J504[a] Upper | J406b[a] | J21 & 22[a] |
|---|---|---|---|---|
| Number | 15 | 26 | 154 | 34 |
| Maximum Length | | | | |
| Range | — | — | 20–79 | — |
| Mean | 29.7 | 28.6 | 36.3 | 27.2 |
| Standard Deviation | — | — | 13.2 | — |
| Maximum Width | | | | |
| Range | — | 5–24 | 5–38 | 5–23 |
| Mean | 11.2 | 12.3 | 13.0 | 12.5 |
| Standard Deviation | 2.3 | 4.2 | 5.0 | 4.8 |
| Maximum Thickness | | | | |
| Range | — | 2–11 | 1–19 | 1–13 |
| Mean | 2.6 | 4.4 | 4.6 | 3.9 |
| Standard Deviation | 1.6 | 2.1 | 2.7 | 2.4 |
| Lateral Profile | | | | |
| 2. Curved | | | 17.7 | |
| 3. Flat | | | 67.7 | |
| 4. Twisted | | | 10.1 | |
| 5. Rippled | | | 4.0 | |
| Platform Facets | | | | |
| 2. Single | | | 62.1 | |
| 3. Dihedral | | | 10.5 | |
| 4. Multiple | | | 4.4 | |
| 5. Crushed | | | 21.4 | |
| 6. Cortex | | | 1.6 | |

Appendix 10.2. (*Continued*)

| Assemblages | J201[a] Layer C | J504[a] Upper | J406b[a] | J21 & 22[a] |
|---|---|---|---|---|
| Number | 15 | 26 | 154 | 34 |
| Shape of Bulb | | | | |
| 2. Lipped | | | 9.7 | |
| 3. Pointed | | | 8.1 | |
| 4. Unlipped | | | 80.7 | |
| Distal Shape | | | | |
| 2. Pointed | | | 43.2 | |
| 3. Oblique | | | 7.3 | |
| 4. Blunt | | | 49.6 | |
| Alignment of Lateral Edges | | | | |
| 2. Expanding | | | 14.9 | |
| 3. Converging | | | 42.7 | |
| 4. Parallel | | | 42.3 | |
| Facet Pattern | | | | |
| 2. Parallel | | | 64.9 | |
| 3. Parallel Opposed | | | 10.1 | |
| 4. Parallel Truncated | | | 3.6 | |
| 5. Parallel Converging | | | 17.7 | |
| Distal Profile | | | | |
| 2. Feathered | | | 84.3 | |
| 3. Overpassed | | | 0.8 | |
| 4. Underpassed | | | 10.5 | |
| 5. Stepped | | | 4.4 | |
| Cross Section | | | | |
| 2. Single | | | 0.0 | |
| 3. Double | | | 22.2 | |
| 4. Triple | | | 54.8 | |
| 5. Multiple | | | 22.6 | |

[a]Jones 1983.

Appendix 10.3. Type List for Middle Hamran Assemblages

| Assemblages | J26 | J31 | J201 Lay A | J201 Lay B | J203 Lay B&C |
|---|---|---|---|---|---|
| Number | 140 | 297 | 280 | 368 | 56 |
| A. Scraper Class | | | | | |
| 1. | 2.1 | 0.7 | 0.6 | 0.3 | |
| 2. | | | | 0.3 | |
| 3. | | 1.3 | 1.1 | 1.2 | |
| 4. | | | | 0.3 | |
| 5. | | | 0.6 | | |

Appendix 10.3. (*Continued*)

| Assemblages | J26 | J31 | J201 Lay A | J201 Lay B | J203 Lay B&C |
|---|---|---|---|---|---|
| Number | 140 | 297 | 280 | 368 | 56 |
| 6. | | | | 0.6 | 1.3 |
| 7. | | | 0.6 | 0.3 | |
| 8. | | | 0.6 | 0.3 | |
| 9. | | | | | |
| 10. | | | | 0.6 | |
| 11.–13. | | | | | |
| 14. | | | 0.6 | 0.3 | |
| 15. | | | | | |
| B. Burin Class | | | | | |
| 16. | 0.7 | | 0.3 | | |
| 17. | | | | 1.3 | |
| 18.–24. | | | | | |
| 25. | | | 0.3 | | |
| 26. | | | | | |
| C. Microlithic Geometric | | | | | |
| 27.–31. | | | | | |
| 32. | 9.2 | 10.1 | 19.6 | 15.0 | 6.7 |
| 33.–36. | | | | | |
| D. Nongeometric Microliths | | | | | |
| 37. | 9.5 | 0.7 | | 0.3 | |
| 38. | 0.6 | 2.0 | 1.1 | | |
| 39. | 1.2 | 4.0 | | | |
| 40. | 20.9 | 20.1 | 36.3 | 30 2 | 38.7 |
| 41. | | | | | |
| 42. | 0.6 | | | 0.3 | |
| 43. | 5.7 | 12.8 | 12.8 | 15.6 | 26.7 |
| 44. | 0.6 | | | 0.3 | |
| 45. | | | | | |
| 46. | | | | | |
| 47. | | | 0.6 | 0.3 | |
| 48. | 1.2 | 3.4 | | 0.6 | 1.3 |
| 49.–51. | | | | | |
| 52. | | 1.3 | | | |
| 53. | | | | | |
| 54. | | | | 0.3 | |
| 55. | 21.5 | 18.1 | 6.7 | 10.0 | 2.7 |
| 56. | 1.2 | 0.7 | | 1.2 | 4.0 |
| E. Abruptly Retouched Blades and Flakes | | | | | |
| 57. | | | 0.6 | | |
| 58.–65. | | | | | |
| F. Truncations | | | | | |
| 66. | 2.8 | 8.7 | 1.7 | 2.2 | 5.3 |
| 67. | | 2.7 | | | |
| 68. | | | 0.6 | | |

**Appendix 10.3. (*Continued*)**

| Assemblages | J26 | J31 | J201 Lay A | J201 Lay B | J203 Lay B&C |
|---|---|---|---|---|---|
| Number | 140 | 297 | 280 | 368 | 56 |
| G. Retouched Pieces | | | | | |
| 69. | 10.0 | 4.7 | 2.2 | 4.7 | 4.0 |
| 70. | | | | 0.9 | 4.0 |
| 71. | | | 1.7 | | |
| 72. | | | | | |
| 73. | | 1.3 | | 0.6 | |
| 74.–82. | | | | | |
| 83. | | 0.7 | 3.9 | 3.7 | 1.3 |
| 84. | 0.7 | | 0.6 | 1.3 | |
| 85. | | | | | |
| 86. | | | | | |
| 87. | | | 1.1 | 0.3 | |
| 88. | | | 2.8 | | |
| 89. | | | | | |
| H. Denticulate Class | | | | | |
| 90. | | 2.0 | 0.6 | 1.6 | |
| 91. | | | | | |
| I. Notch Class | | | | | |
| 92. | 2.8 | | 2.8 | 2.5 | |
| 93. | | 0.7 | | | |
| 94. | | 0.7 | | 3.4 | 1.3 |
| 95. | | | | | |
| 96. | | | | | |
| J. Multiple Use Class | 2.0 | | | | |
| K. Perforator | 1.1 | | | | |
| L. Massive Piece Class | | | | | |
| 97.–100. | | | | | |
| M. Utilized Piece Class | | | | | |
| 101.–103. | | | | | |
| N. Points | | | | | |
| 104.–106. | | | | | |
| 107. | 0.7 | | | | |
| 108. | | | | | |
| 109. | | | | | |
| O. Varia | | | | | |
| 110. | 0.7 | 0.3 | | | |

Appendix 10.4. Late/Final Hamran Type List

| Assemblages | J202 Upper | J202 Lower | J203 Upper | J203 Lower |
|---|---|---|---|---|
| Number | 1065 | 382 | 483 | 669 |
| A. Scraper Class | | | | |
| 1. | 1.1 | 0.8 | | 0.9 |
| 2. | 0.2 | 0.5 | | 0.1 |
| 3. | 0.6 | 1.0 | 1.0 | 1.2 |
| 4. | 0.4 | 0.3 | | 0.7 |
| 5. | | | 0.2 | |
| 6. | | 0.3 | 0.4 | 0.1 |
| 7. | 0.4 | | | |
| 8. | 0.4 | 0.8 | | |
| 9. | 0.1 | 0.5 | | |
| 10. | 0.1 | 0.3 | | |
| 11. | | | | |
| 12. | 0.3 | | 0.6 | |
| 13. | | | | |
| 14. | 0.3 | | 1.0 | 0.3 |
| 15. | | | | |
| B. Burin Class | | | | |
| 16. | 0.3 | | | |
| 17. | 0.6 | | | |
| 18. | | 0.5 | | |
| 19.–24. | | | | |
| 25. | 0.3 | 0.3 | | |
| 26. | | | | |
| C. Microlithic Geometric | | | | |
| 27. | 12.9 | 3.4 | 1.7 | 0.3 |
| 28. | 0.1 | | | |
| 29. | | | | |
| 30. | | | | |
| 31. | 0.2 | | | |
| 32. | 6.7 | 10.2 | 11.6 | 9.9 |
| 33. | | | 0.2 | |
| 34.–36. | | | | |
| D. Nongeometric Microliths | | | | |
| 37. | 0.3 | | | |
| 38. | 0.1 | 0.3 | 0.2 | |
| 39. | 0.4 | 2.6 | 0.2 | 0.4 |
| 40. | 16.3 | 19.6 | 35.2 | 43.0 |
| 41. | | | | |
| 42. | | | | |
| 43. | 6.5 | 13.4 | 23.6 | 26.0 |
| 44. | 0.4 | 0.3 | | |
| 45.–47. | | | | |
| 48. | 1.1 | 0.5 | | |
| 49.–51. | | | | |
| 52. | 0.8 | | | |
| 53. | | | | |

Appendix 10.4. (*Continued*)

| Assemblages | J202 Upper | J202 Lower | J203 Upper | J203 Lower |
|---|---|---|---|---|
| Number | 1065 | 382 | 483 | 669 |
| 54. | | | 0.4 | 0.6 |
| 55. | 7.9 | 9.4 | 7.5 | 5.7 |
| 56. | 1.1 | 1.3 | 1.9 | 0.4 |
| E. Abruptly Retouched Blades and Flakes | | | | |
| 57.–64. | | | | |
| 65. | 0.1 | | 0.2 | |
| F. Truncations | | | | |
| 66. | 3.7 | 5.8 | 3.1 | 2.7 |
| 67. | 0.4 | 0.3 | | |
| 68. | 0.1 | | | |
| G. Retouched Pieces | | | | |
| 69. | 7.0 | 4.7 | 3.7 | 3.6 |
| 70. | 1.0 | 1.6 | 0.2 | 0.6 |
| 71. | | | | |
| 72. | | | | |
| 73. | 2.3 | 1.6 | 0.2 | 0.1 |
| 74.–82. | | | | |
| 83. | 6.0 | 5.5 | 0.6 | 0.4 |
| 84. | 0.8 | 1.0 | 0.6 | |
| 85. | | | | |
| 86. | | | | |
| 87. | 0.1 | 0.8 | | |
| 88. | | | | |
| 89. | | | | |
| H. Denticulate Class | | | | |
| 90. | 2.2 | 1.6 | 1.0 | 0.4 |
| 91. | | | | |
| I. Notch Class | | | | |
| 92. | 10.1 | 6.5 | 3.9 | |
| 93. | | | | |
| 94. | 3.4 | 1.6 | | 1.6 |
| 95. | | | | |
| 96. | | | | |
| J. Multiple Use Class | 0.6 | | | |
| K. Perforator | 2.2 | 1.6 | 0.2 | 0.4 |
| L. Massive Piece Class | | | | |
| 97.–100. | | | | |
| M. Utilized Piece Class | | | | |
| 101.–103. | | | | |
| N. Points | | | | |
| 104.–109. | | | | |
| O. Varia | | | | |
| 110. | 0.7 | 1.3 | 0.4 | |

Appendix 10.5. **Bladelet Attributes for Late and Final Hamran Assemblages**

| Assemblages | J202 Upper | J202 Lower |
|---|---|---|
| Number | | |
| Maximum Length | | |
| Range | 20–47 | 20–61 |
| Mean | 27.9 | 28.6 |
| Standard Deviation | 5.1 | 8.3 |
| Maximum Width | | |
| Range | 6–18 | 5–30 |
| Mean | 10.6 | 10.8 |
| Standard Deviation | 2.6 | 3.7 |
| Maximum Thickness | | |
| Range | 0.4–9.5 | 1.5–7.8 |
| Mean | 2.6 | 3.3 |
| Standard Deviation | 1.2 | 1.4 |
| Lateral Profile | | |
| 1. Indeterminate | | |
| 2. Curved | 19.4 | 39.7 |
| 3. Flat | 70.4 | 41.0 |
| 4. Twisted | 9.3 | 14.1 |
| 5. Rippled | 1.0 | 5.1 |
| Platform Facets | | |
| 1. Indeterminate | | |
| 2. Single | 67.6 | 70.5 |
| 3. Dihedral | 8.3 | 5.1 |
| 4. Multiple | 1.0 | 2.6 |
| 5. Crushed | 22.2 | 19.2 |
| 6. Cortex | 1.0 | 2.6 |
| Shape of Bulb | | |
| 1. Indeterminate | | |
| 2. Lipped | 2.8 | 18.0 |
| 3. Pointed | 7.4 | 14.1 |
| 4. Unlipped | 76.9 | 61.5 |
| Distal Shape | | |
| 1. Indeterminate | | |
| 2. Pointed | 36.1 | 52.6 |
| 3. Oblique | 7.4 | 12.8 |
| 4. Blunt | 55.6 | 34.6 |
| Alignment of Lateral Edges | | |
| 1. Indeterminate | | |
| 2. Expanding | 22.2 | 23.1 |
| 3. Converging | 21.3 | 30.8 |
| 4. Parallel | 56.5 | 46.2 |
| Facet Pattern | | |
| 1. Indeterminate | 6.5 | |
| 2. Parallel | 52.8 | 55.1 |
| 3. Parallel Opposed | 2.8 | 12.8 |
| 4. Parallel Truncated | 8.3 | 3.9 |
| 5. Parallel Converging | 22.2 | 24.4 |

Appendix 10.5. (*Continued*)

| Assemblages<br><br>Number | J202<br>Upper | J202<br>Lower |
|---|---|---|
| Distal Profile | | |
| 1. Indeterminate | | |
| 2. Feathered | 85.2 | 91.0 |
| 3. Overpassed | 1.0 | |
| 4. Underpassed | 8.3 | 2.6 |
| 5. Stepped | 5.6 | 6.4 |
| Cross Section | | |
| 1. Indeterminate | | |
| 2. Single | 1.0 | |
| 3. Double | 36.1 | 33.3 |
| 4. Triple | 54.6 | 59.0 |
| 5. Multiple | 8.3 | 7.7 |

# Chapter 11

# The Madamaghan Sites

DONALD O. HENRY AND CHEN SHEN

Since its discovery in the basin of Wadi Mushabi of northeastern Sinai in the early 1970s (Phillips and Mintz 1977), evidence of the late Epipaleolithic, Mushabian Complex has been found over much of the arid zone of the southern Levant (Bar-Yosef and Belfer-Cohen 1989; Henry 1989). But it has only been in the last decade that Mushabian occurrences have been identified east of the Rift Valley (Henry 1983, 1988; Henry and Garrard 1988; Shen 1992). This recognition of an eastern extension of the complex, the Madamaghan Industry, comes from the discovery of Mushabian occupational horizons in the deposits of the Tor Hamar (J431, Layers A–E$^{1}$) and Jebel Fatma (J436) rockshelters, both located in the Jebel Qalkha study area (Figure 11.1). Discovery and analyses of assemblages from these deposits also prompted a reconsideration of the taxonomic affiliation of the most recent horizon of the nearby Wadi Madamagh rockshelter located near Petra. Although Kirkbride (1958) referred to this assemblage from Wadi Madamagh as "Micro-Kebaran" (almost 20 years before the Mushabian was first reported in detail), it is clear from her description of microburins, La Mouillah points, and arched-backed bladelets that it is not a Kebaran assemblage at all, but one that fits comfortably within the Mushabian Complex as a whole (Henry 1988, 1989a). When the technotypologic characteristics of these assemblages are compared to those of Sinai and the highland Negev, a clear pattern of regional facies emerges with the Madamaghan Industry representing that of southern Jordan (Henry 1983, 1988, 1989a).

## ADAPTATION TO THE ARID ZONE

A fascinating aspect of the Mushabian is the clarity by which it provides definition of the differing adaptations of southern Levantine populations to steppe-desert and Mediterranean woodland settings. As an expression of a mobile foraging population adapted to exploitation of the arid zone, the Mushabian persisted on the fringes of the Mediterranean woodlands from ca. 14–12,000 B.P. In the lusher setting of the Mediterranean woodlands, a parallel, partly synchronous but different succession took place. Mobile foraging groups of the Geometric Kebaran came to intensively exploit wild cereal and nut resources, adopt a more sedentary existence, and assume all of the consequences of complex foragering as manifested in the Early Natufian (Henry 1989a). By inhabiting an environment outside of the natural distribution of wild cereals and nuts, the triggering resources of complex foraging, Mushabian populations persisted as simple foragers, never developing sedentary lifeways.

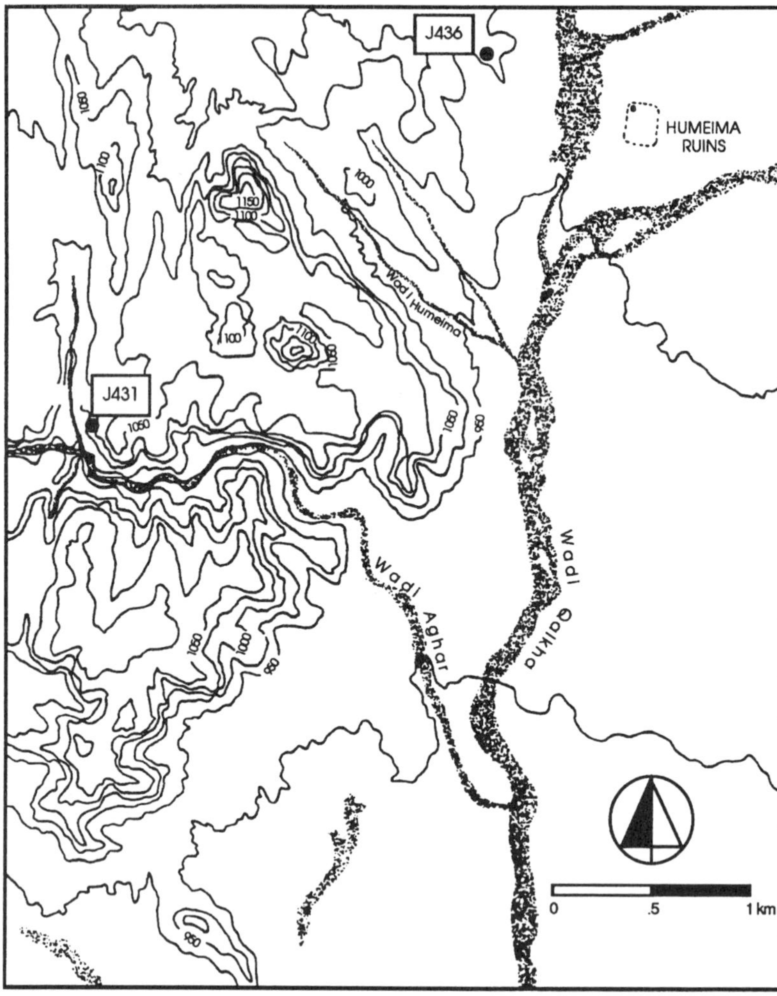


**Figure 11.1.** The locations of the Madamaghan sites of Tor Hamar (J431) and Jebel Fatma (J436) in the Jebel Qakha study area.

## REGIONAL FACIES AND ORIGIN

The artifact inventories of Mushabian sites are confined mainly to chipped stone specimens that form microlithic assemblages. Ornamental shells are common to sites, while ground stone specimens and bone tools represent rare finds. Ochre appears to have been extensively used, having been reported as worked lumps, as a stain on ground stone, and as part of the mastic on chipped stone points (Phillips and Mintz 1977; Henry and Garrard 1988).

The technotypologic configurations of Mushabian lithic assemblages show three distinctive regional patterns centered in northeastern Sinai, the highland Negev, and southern Jordan (Henry 1989a). These regional facies of the Mushabian Complex have created something of a terminological quagmire and this is especially true of those Mushabian assemblages of the Negev. No less than six taxonomic labels have been applied to assemblages included within this facies

(Henry 1983, 1989a; Goring-Morris 1987). This ambiguity comes primarily from the fact that, as with Kirkbride's (1958) classification of the assemblage from Madamagh, the Mushabian assemblages of the Negev were discovered (Yizraeli 1967; Bar-Yosef 1970, 1975, 1977; Marks *et al.* 1971; Marks and Simmons 1977) *before* they had been defined from stratified and radiometrically dated contexts in northeastern Sinai. Thus, in being obligated to use Kebaran and Geometric Kebaran assemblages as their principal referents, researchers proposed labels such as "Geometric Kebaran B" (Bar-Yosef 1970, 1975, 1977), and "Negev Kebaran" (Marks and Simmons 1977; Marks 1977b). There is now general agreement that these assemblages of the highland Negev were not related to the Kebaran, but were in some way tied to the Mushabian (Bar-Yosef and Belfer-Cohen 1987; Goring-Morris 1987; Henry 1983, 1989a). Conflicting terminology persists, however. While Henry (1983, 1989a) introduced the label "Negev Mushabian," Goring-Morris (1987) has proposed the term "Ramonian" to define those assemblages of the highland Negev that he believes were an outgrowth of the Mushabian.

Beyond confirming another regional facies of the Mushabian Complex, the assemblages of southern Jordan extend the range of the complex east of the Rift Valley and indicate continuity with the earlier Qalkhan Industry. This, in turn, challenges our traditional notions of an origin for the Mushabian in northeast Africa and suggests that we should look within the arid zone east of the Rift for a source.

## SITE SETTINGS AND RESOURCES

Tor Hamar (J431) and Jebel Fatma (J436), found at 950–1,000 masl, occupy southwest facing rockshelters located in canyons opening on the Wadi Qalkha. As previously described (Chapter 8), the Tor Hamar (J431) shelter is situated deep within the narrow Wadi Aghar canyon system some 3 km from its mouth. The Jebel Fatma (J436) shelter, on the other hand, is found near the mouth of a shallow, unnamed canyon close to its confluence with the Wadi Qalkha and enjoys a commanding view of the plain forming the western end of the Wadi Hisma. Both occupations would have had access to water from run-off in the nearby wadi beds during the winter wet season and from pools well into the spring. Also, seeps along the base of Jebel Humeima may have even provided water during the dry season for the inhabitants of Jebel Fatma (J436). Chert sources within the catchments of the sites are restricted mainly to the bed of the Wadi Qalkha and the limestone hills located just east of Jebel Fatma (J436).

## SITE PLANS AND STRATIGRAPHY

Both sites, discovered during survey of the Jebel Qalkha study area in 1983/84, display concentrations of flint artifacts and bones within the ashy deposits of shallow rockshelters. A deep conical mortar fashioned in a large sandstone block was also found within a Bedouin windbreak at Tor Hamar.

### Tor Hamar (J431)

The concentration of artifacts and habitation debris cover an area of ca. 300 m$^2$ extending from the backwall of the shelter downslope toward the wadi.

#### Site Plan

During the first season at the site in the winter of 1983/84, a *sondage* (units 1 and 2) was excavated to a depth of 1.3 m, exposing a Mushabian Complex (Madamaghan Industry) horizon (Henry and Garrard 1988). Eight additional units, formed by two 2 m × 2 m blocks, were excavated to a depth of 2.2 m the following season.

### Stratigraphy

The excavation exposed eight relatively level-bedded strata (A–H) within 22 arbitrary levels (Figures 8.6 and 9.4). The upper 1.6 m of the deposit contained Madamaghan artifacts (Layers A–$E^1$) overlaying Qalkhan (Layer $E^2$) and Upper Paleolithic-Ahmarian (Layers F–G) horizons. Limited exposure of Layer H yielded only a few artifacts.

***Layer A*** consists of a fine, powdery light grey ash. The upper (A1) 5–10 cm contains dung, twigs, grass seed, and other organic material of recent origin. The lower part of the layer (A2) is free of such material.

***Layers B, C, and D*** are differentiated by varying proportions of ash and sand. Ash lenses point to shallow hearths underlain by fire-reddened sediments that denote a relatively level-bedded stratigraphy. While these layers could be distinguished during excavation, they were virtually impossible to identify in the sections after the sediments had dried. Spraying the sections with a fine water mist also failed to reveal their contacts. Layer B differs from layer A in having a higher yellow-red sand content and in being more compact and less powdery. Layer B grades into Layer C at ca. 110–120 cm BD (below datum) with the latter being composed of a predominantly yellow-red sand. Layer C, in turn, grades into Layer D (more compact and ashier) at about 140 cm BD.

Two radiocarbon dates were obtained from charcoal recovered from upper Layer C: 12,680 ± 320 B.P. (SMU 1399); 12,320 ± 90 B.P. (ETH 806).

***Layer E¹*** is composed of a 20 cm thick sandstone rubble zone associated with dark grey ash and charcoal. The rubble, having fallen from the ceiling of the shelter, appears to have been associated with a large fire which is confined to the easternmost excavation block. The layer displays sharp contacts with contiguous strata and in units 4 and 6 is capped by a large 50 cm thick block of roof-fall.

### Feature

A feature representing a large hearth and possibly a portion of the stone foundation of a wind-break or shelter was exposed near the top of Layer C between 90–120 cm below datum (Figure 11.2). The fire-pit appears to have been used in at least two burnings as indicated by two different ash strata. These are surrounded by fire-reddened sand. The eastern portion of the pit shows two concentric rock-lines joining a large, inclined bedrock slab on the southern edge of the hearth. The innermost ring is formed of small stones bordering a whitish-grey ashy sand. This is separated from the outer, much larger line of rocks by fire-reddened sand. This outer rock-line may represent the foundation of a windbreak or shelter. Burnt bones were found concentrated near the center of the hearth and a sandstone muller was recorded between the hearth and foundation stones.

## Jebel Fatma (J436)

Artifacts and habitation debris cover an area of ca. 214 m² extending from a bedrock shelf near the backwall of the shelter down the slope. Two 1 m² units, excavated behind the drip-line in the west end of the shelter, exposed a 70 cm thick deposit composed of mottled, grey-yellow ashy sand overlying bedrock (Figure 11.3).

## ARTIFACTS

The inventories of the two sites are dominated by flaked stone artifacts (Figures 11.4, 11.5, 11.6), with Tor Hamar (J431) also yielding ground stone and worked bone specimens. Ornamental shell was recovered from both sites. While the deep Tor Hamar (J431) deposit al-

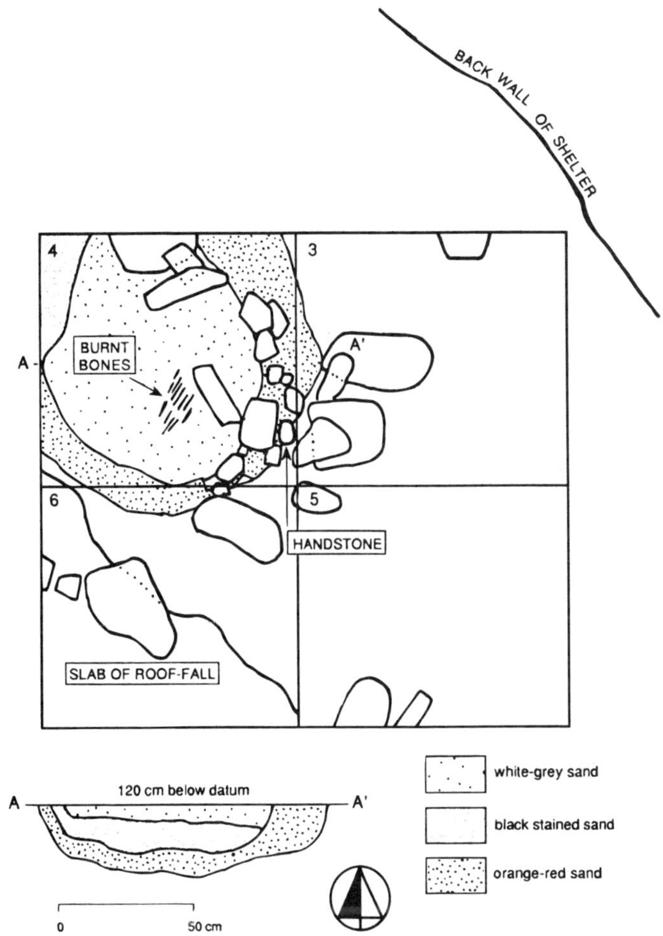


**Figure 11.2.** Plan and profile of hearth and possible foundation of windbreak or shelter exposed in Layer C of Tor Hamar (J431).

lowed for a comparison of five stratified assemblages from layers A–E[1], Jebel Fatma (J436) is represented by a single assemblage (Table 11.1).

## Lithic Typology

In accounting for 36–65% of the toolkits, nongeometric microliths dominate the assemblages, followed by smaller proportions of scrapers, retouched pieces, notches, and points (Appendix 11.1).

### Scrapers

Simple endscrapers on flakes and blades dominate the class with the latter occurring most frequently (Figure 11.4). Most are well formed with relatively wide, symmetric working-bits that are often accompanied by lateral retouch. Carinated, shouldered, denticulated, and notched types are common, but represent only a small proportion of most assemblages. The working bits of endscrapers also occur consistently, although in low numbers.

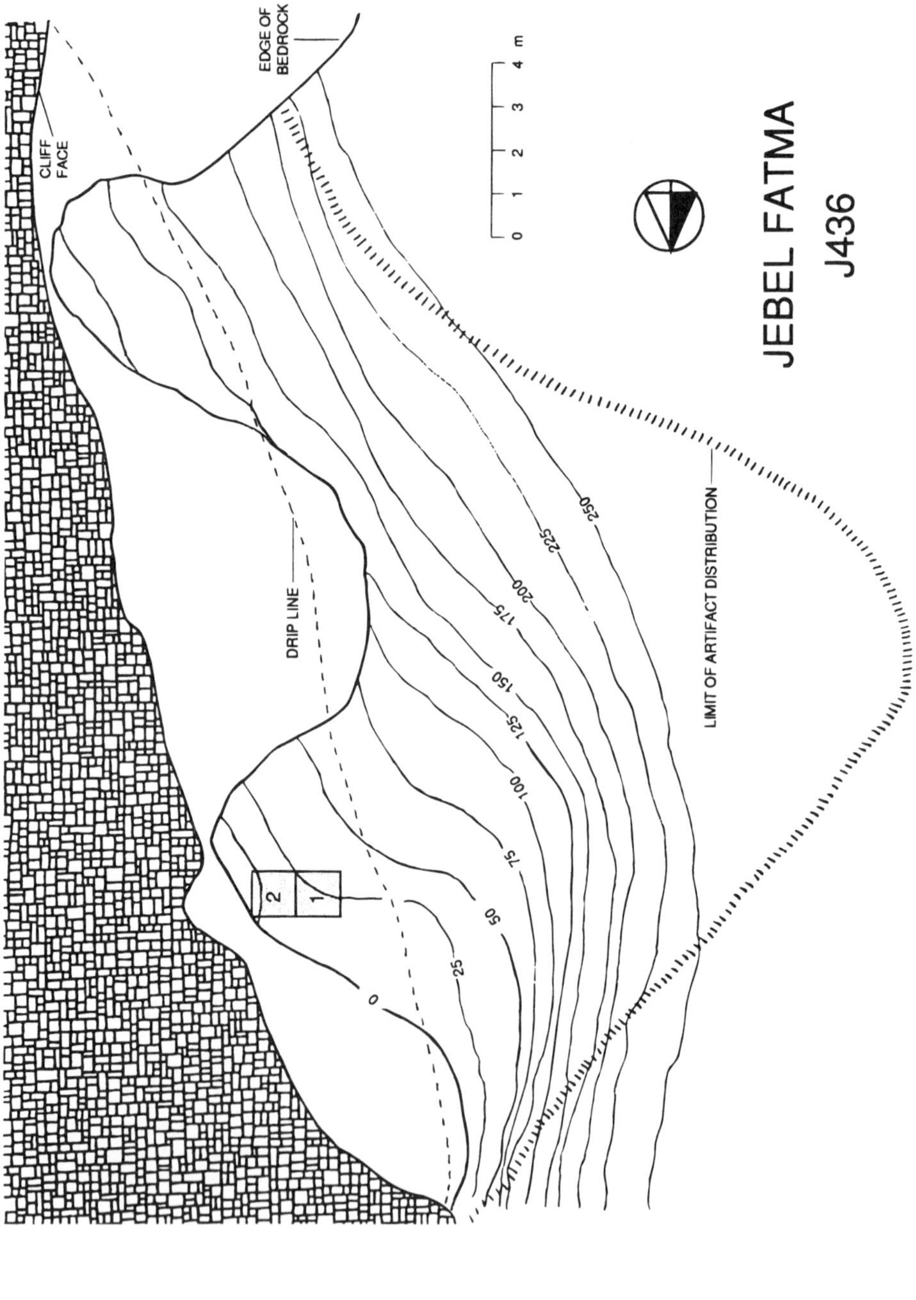


Figure 11.3. Topographic map of Jebel Fatma (J436) showing artifact distribution and excavation plan.

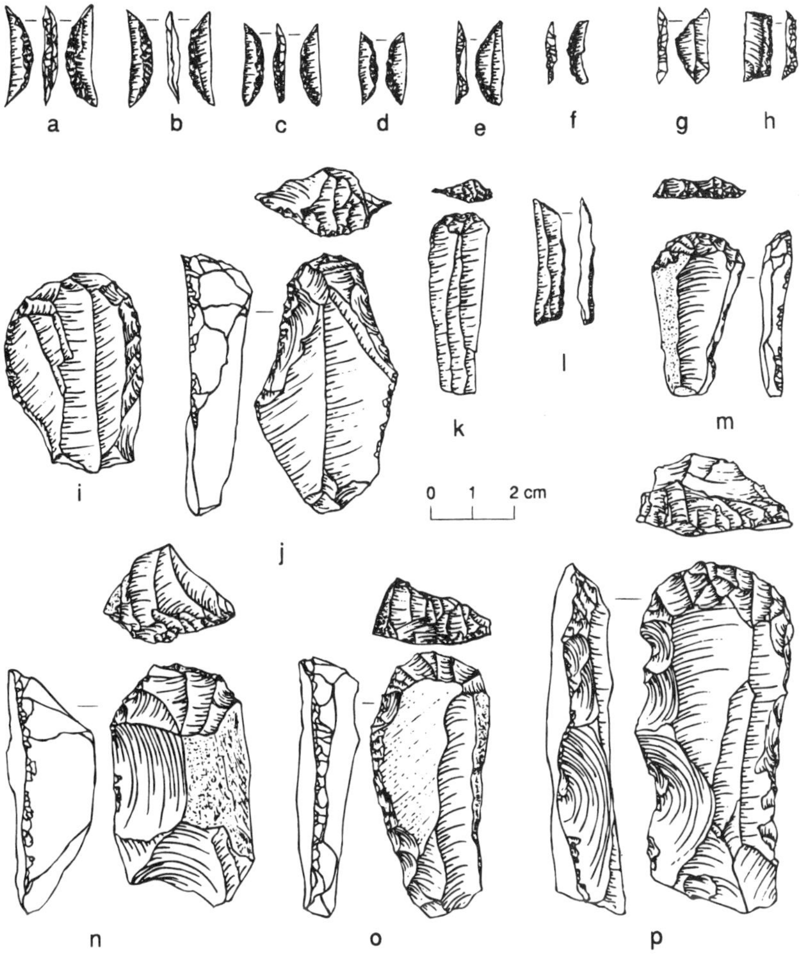


**Figure 11.4.** Illustrations of tools from the Madamaghan horizon of Tor Hamar (J431): (a–f) Helwan lunates, (g) triangle, (h) trapeze-rectangle, (i) endscraper on flake, (j,k,m–p) endscrapers on blades.

## Burins

This class, poorly represented in Madamaghan assemblages, is comprised mainly of angle burins struck from snaps and old surfaces (Figure 11.5).

## Geometric Microliths

Although the class never exceeds 2.6% of the Madamaghan toolkits, with the exception of the Jebel Fatma (J436) assemblage, geometrics are consistently represented in the form of Helwan lunates (Figure 11.4). Trapeze/rectangles and triangles are rare.

## Nongeometric Microliths

Straight-backed and arched-backed bladelets dominate the class with the latter often showing remnant microburin scars on the arch (Figure 11.5). Scalene backed bladelets, although less common, form a significant proportion of the nongeometric microliths, especially in the lower

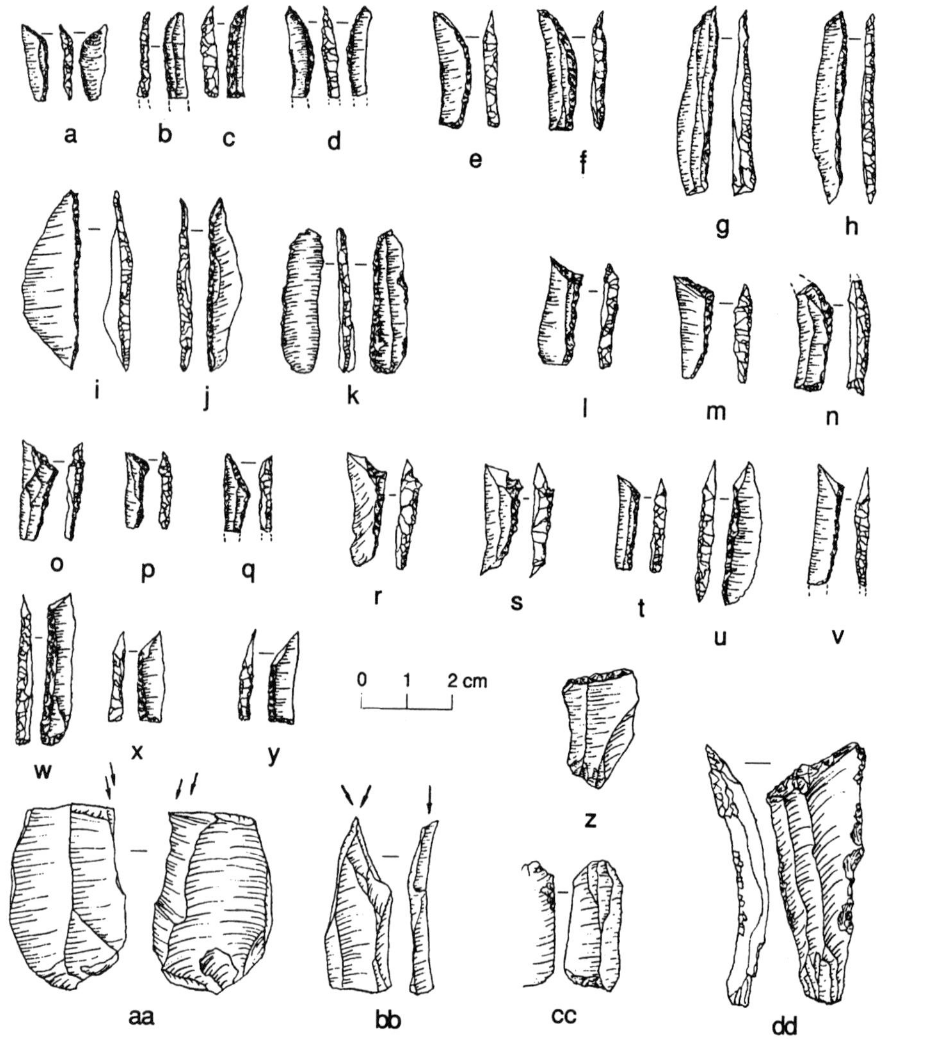


**Figure 11.5.** Illustrations of tools from the Madamaghan horizon of Tor Hamar (J431): (a–f) arched-backed bladelets, (g–j) microgravette points, (k) straight-backed bladelet, (l–q) straight-backed and obliquely truncated (scalene) bladelets, (r–y) *La Mouillah* points, (z,cc) truncations, (aa,bb) burins, (dd) truncation and denticulate.

layers. Like the arched-backed forms, many of these show a remnant microburin scar on the truncation. Goring-Morris (1987:204) refers to such scalene bladelets produced by the microburin technique as "Ramon points," a guide fossil to the Mushabian assemblages centered in the highland Negev.

## Retouched Pieces, Notches, and Denticulates

Blades and bladelets with unilateral, obverse retouch constitute the largest proportion of these classes (Chen 1992:94,99). Notches and denticulates form larger proportions of the toolkit of the upper layers at Tor Hamar (J431).

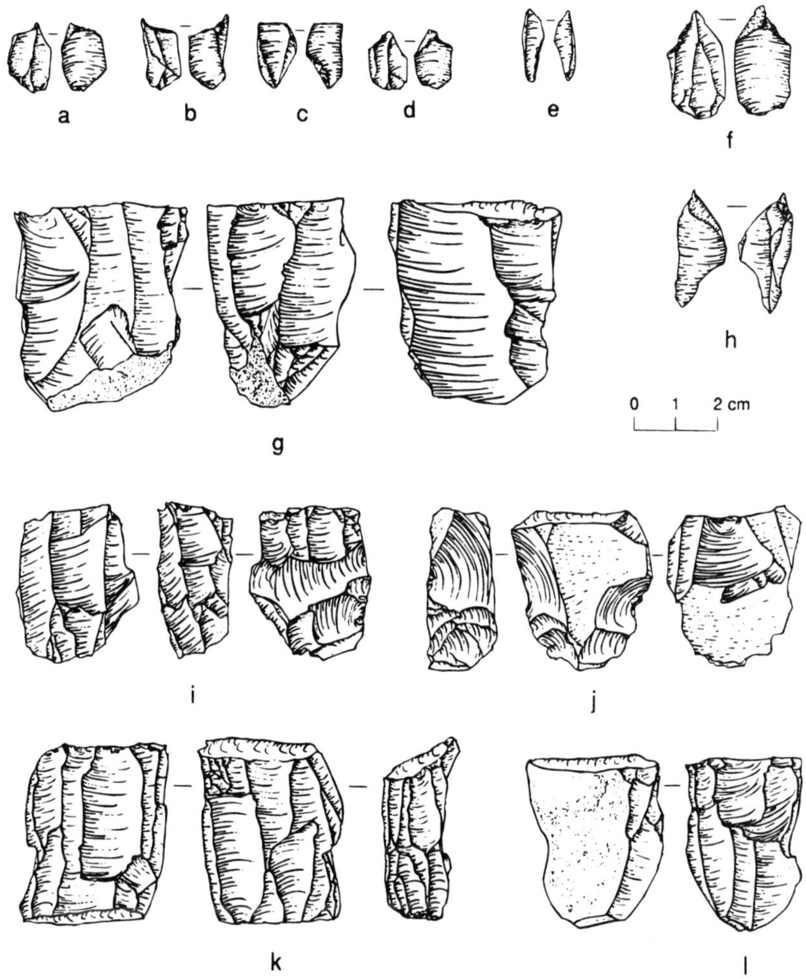


**Figure 11.6.** Illustrations of artifacts from the Madamaghan horizon of Tor Hamar (J431): (a–f,h) microburins, (g–l) bladelet cores.

## Points

Apart from rare microgravettes, the class is composed solely of La Mouillah points. As previously discussed (Chapter 9), La Mouillah points present somewhat of an interpretative problem in that they may have served as finished tools, most likely armatures, and also as an intermediate stage in the production of microliths such as arched-backed and scalene bladelets. Remnants of microburin scars on both of these types confirm that La Mouillah points were generated as an earlier step in their production. Comparison of the dimensions of La Mouillah points and backed bladelets, however, shows that the mean length of points are substantially larger than those of backed bladelets in Layer A of Tor Hamar (J431) (Shen 1992:103). In cases where La Mouillah points were fabricated into arched-backed and scalene bladelets, one would expect the finished specimens to display lengths on slightly less than those of the points. The

Table 11.1. Artifact Inventories for Madamaghan Assemblages (data for J431, Layers A-D & J436 from Shen 1992)

| Assemblage Layer | J431 A | J431 B | J431 C | J431 D | J431 E[1] | J436 |
|---|---|---|---|---|---|---|
| *Tools* | | | | | | |
| Scraper | 15.6 | 13.3 | 6.9 | 5.3 | 27.8 | 6.3 |
| Burin | 0.3 | 0.0 | 0.4 | 0.5 | 2.5 | 0.0 |
| Geometric | 2.6 | 2.6 | 1.3 | 1.1 | 1.2 | 8.8 |
| Nongeom | 36.4 | 44.2 | 64.6 | 62.3 | 49.4 | 45.6 |
| ABR_RET | 0.0 | 1.2 | 2.4 | 5.0 | 0.0 | 1.3 |
| Truncation | 2.0 | 2.6 | 0.7 | 0.0 | 1.2 | 0.0 |
| Retouched Pc. | 22.8 | 14.6 | 8.2 | 9.5 | 8.8 | 5.6 |
| Denticulate | 4.0 | 2.6 | 1.3 | 4.0 | 0.0 | 0.0 |
| Notch | 8.6 | 6.8 | 2.8 | 0.3 | 1.2 | 4.9 |
| Multiple Tl. | 0.3 | 0.0 | 0.0 | 0.0 | 0.2 | 0.0 |
| Perforator | 1.7 | 0.0 | 0.4 | 0.7 | 0.2 | 0.0 |
| Utilized Pc. | 0.0 | 0.0 | 0.0 | 0.4 | 0.0 | 0.0 |
| Points | 5.3 | 11.6 | 10.8 | 8.9 | 5.0 | 27.5 |
| Varia | 0.3 | 0.0 | 0.0 | 0.0 | 2.5 | 0.0 |
| Total No. | 302 | 233 | 452 | 377 | 79 | 160 |
| *Debitage* | | | | | | |
| Core | 2.1 | 2.1 | 2.0 | 1.5 | 2.1 | 2.6 |
| Blade | 30.2 | 44.0 | 59.1 | 46.4 | 36.9 | 49.7 |
| Flake | 46.4 | 33.4 | 16.3 | 24.6 | 49.2 | 35.4 |
| Core Trim. El. | 2.2 | 1.9 | 1.0 | 2.1 | .5 | 5.7 |
| Primary El. | 6.3 | 4.8 | 8.6 | 6.8 | 5.4 | 2.8 |
| Microburin | 11.9 | 13.8 | 12.8 | 7.6 | 5.8 | 3.8 |
| Burin Spall | 0.1 | 0.0 | 0.1 | 0.2 | 0.1 | 0.0 |
| Total No. | 1581 | 1992 | 2479 | 991 | 781 | 683 |
| *Debris* | | | | | | |
| Chips | | | | | 1333 | |
| Chunks | | | | | 258 | |
| Total No. | 1676 | 2004 | 3402 | 3877 | 2372 | 731 |
| *Total Artifacts* | 3559 | 4229 | 6333 | 5390 | 3232 | 2697 |

dimensional data thus implys that some La Mouillah points served as armatures without subsequent modification.

## Lithic Technology

Madamaghan assemblages are characterized by the production of relatively wide bladelets from wedge-shaped (sub-pyramidal) single platform and ninety-degree cores. Removal of the bladelets appears to have involved both hard-hammer percussion as well as soft-hammer and punch techniques. The microburin technique was the principal means by which bladelets were sectioned as a step toward their fabrication into points and nongeometric microliths.

### Raw Material

Raw material was obtained from *in situ* chert beds on the plateau and from small cobbles derived from these beds and found in the nearby Wadi Qalkha. Although a wide variety of cherts occurs in the assemblages, a honey-grey chalcedony was the one most commonly used.

Black chert from the Rift Valley and quartzite from the floor of the Wadi Hisma are curiously absent.

### Cores

Most of the cores are subpyramidal in shape with single platforms, but ninety-degree platform orientations are common. Whereas blade/bladelet cores compose the large majority of the cores, flake cores typically represent a little more than a third of the samples and yield their highest frequencies in association with ninety-degree platforms. This relationship is most likely an expression of failed attempts to produce bladelets from newly created platforms as cores neared exhaustion. Evidence for the exhaustion of most of the cores is noted by Shen (1992:52) in the high (>80°) platform-face angles and the mean bladelet lengths which exceed mean core facet lengths by over 12 mm.

### Bladelets

Bladelets are wide relative to their lengths, with length:width ratios averaging about 2.7:1. Their lateral profiles are roughly balanced between curved and flat varieties with unfacetted platforms. Although most assemblages yield substantial proportions of lipped bulbs, an attribute linked to soft-hammer percussion, the dominance of diffuse bulbs and crushed platforms underscores the extensive use of a hard hammer in detaching bladelets. Use of a hard hammer for delivery of microliths, in conjunction with the exhausted nature of the cores, appears to have reduced the efficiency of bladelet production. Only some 72–80% of bladelets show feathered terminations.

### Microburin Technique

The Microburin Technique was extensively employed in the controlled sectioning of bladelets for production of backed and truncated microliths. Restricted indices range from 70–79 for Tor Hamar (J431), but only 27 for Jebel Fatma (J436).

In an effort to understand the specific way in which the technique was employed, Shen (1992:76–80) analyzed a large sample (495) of microburins from Tor Hamar focussing on the position and orientation of microburin scars following the method of Tixier (1963). He found a consistent dominance of proximal, right scars in the Tor Hamar (J431) assemblages and the Jebel Fatma (J436) assemblage showed a similar pattern (Table 11.2). When this pattern of microburin scars is compared to the corresponding distribution of scars on arched-backed bladelets, La Mouillah points, and scalene bladelets, an obvious correspondence is indicated (Figure 11.7; Table 11.3). This shows that most bladelets were sectioned to remove their butts (proximal, right microburins), leaving their distal extremities and microburin scars for retouch on the left edge (when viewed from the obverse face).

## Time Trends within Lithic Assemblages

The thick stratified deposit of Tor Hamar with its high densities of artifacts provides evidence of a number of technotypologic trends which Shen (1992:123–126) has partitioned into Phases I, II, and III.

### Phase I

This, the earliest of the phases, includes Layers C and D (and probably $E^1$) of Tor Hamar (J431), the Jebel Fatma (J436) assemblage, and most likely the assemblage from Wadi Madamagh, as reported by Kirkbride (1958). The $C^{14}$ dates from the top of Layer C place Phase I sometime *earlier than* ca. 12,500 B.P.

Table 11.2. Position of Microburin Scars by Layer at Tor Hamar (after Shen 1992)

| | Layer A | | Layer B | | Layer C | | Layer D | |
|---|---|---|---|---|---|---|---|---|
| | N | % | N | % | N | % | N | % |
| Proximal, right | 61 | 41.8 | 54 | 50.5 | 68 | 48.9 | 52 | 50.5 |
| Proximal, left | 33 | 22.6 | 19 | 17.8 | 34 | 24.5 | 27 | 26.2 |
| Distal, right | 26 | 17.8 | 17 | 15.9 | 13 | 9.4 | 7 | 6.8 |
| Distal, left | 12 | 8.2 | 11 | 10.3 | 12 | 8.6 | 8 | 7.8 |
| Double | 1 | 0.7 | 0 | 0.0 | 0 | 0.0 | 0 | 0.0 |
| Fragment | 7 | 4.8 | 3 | 2.8 | 10 | 7.2 | 4 | 3.9 |
| Krukowski | 6 | 4.1 | 3 | 2.8 | 2 | 1.4 | 5 | 4.9 |
| Total | 146 | 100 | 107 | 100.1 | 139 | 100 | 103 | 100 |

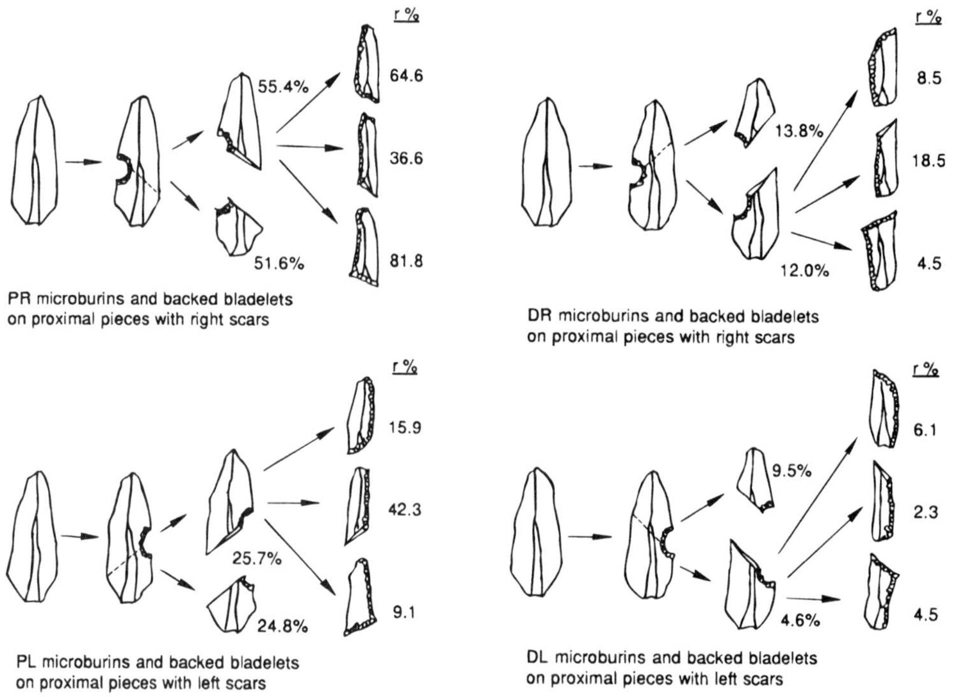


Figure 11.7. Schematic illustration showing the proportionate distributions for the different orientations of microburin scars on microburins and backed bladelets from Tor Hamar (J431). The r% refers to the restricted % of each type of backed bladelet.

Although the basic technology remains unchanged through the three phases, Phase I shows some minor differences in a greater utilization of opposed platform cores and hard-hammer percussion. Blanks also tend to be larger as expressed in longer and wider blades and bladelets. Microliths and blade/flake tools are also the largest in Phase I.

Typologically, microliths (i.e., geometric, nongeometrics, and points) account for 56–81% of the Phase I toolkits, the highest proportion of the three phases. Geometrics are primarily in the form of trapeze-rectangles with Helwan lunates being absent or very rare. Scalene bladelets (Ramon points) occur in relatively high frequencies.

Table 11.3. Comparison of the Positions of Microburin Scars on Backed Bladelet Types and on Microburins (after Shen 1992)

| | Backed Bladelets (restricted % of the class) | | | | | | | | | |
|---|---|---|---|---|---|---|---|---|---|---|
| | Point | | Arched | | Scalene | | Total | | Microb. | |
| | N | % | N | % | N | % | N | % | N | % |
| PR[a] | 13 | 18.3 | 7 | 8.5 | 1 | 4.5 | 21 | 12.5 | 235 | 51.6 |
| PL | 2 | 2.8 | 5 | 6.1 | 1 | 4.5 | 8 | 4.6 | 113 | 24.8 |
| DR | 26 | 36.6 | 53 | 64.6 | 18 | 81.8 | 97 | 55.4 | 63 | 13.8 |
| DL | 30 | 42.3 | 13 | 15.9 | 2 | 9.1 | 45 | 25.7 | 43 | 9.5 |
| D | 0 | 0.0 | 4 | 4.9 | 0 | 0.0 | 4 | 2.3 | 1 | 0.2 |
| Total | 71 | 100 | 81 | 100 | 22 | 99.9 | 175 | 100 | 455 | 100 |

[a]PR=Proximal on right, PL=Proximal on left, DR=Distal on right, DL=Distal on left, D=Double

## Phase II

Represented by material from Layer B of Tor Hamar, this appears to be a transitional phase with microliths accounting for only 54–58% of the toolkits. This is accompanied by a noticeable rise in the proportions of large tool classes (i.e., scrapers, retouched pieces, denticulates, and notches). Within geometrics, Helwan lunates (accounting for over 50%) replace trapeze/rectangles as the dominant type and among the non-geometrics scalene bladelets show a sharp decline.

## Phase III

Within this latest phase, from Layer A of Tor Hamar (J431), microliths show their progressive decline and constitute only 44% of the toolkit. Helwan lunates, now accounting for over 62% of the geometrics, are joined by normal, abrupt retouched lunates with trapeze/rectangles disappearing from the inventory.

### Armatures, Mastic, and Wear Patterns

Backed bladelets, La Mouillah points, and geometrics often display red or yellow ochre on their retouched edges. The ochre was probably mixed with some form of mastic, such as resin or gum from acacia, to facilitate fixing the specimens to shafts. This was a common practice during early Dynastic times in Egypt and among recent foragers in southern Africa (Clark *et al.* 1974). The ochre occurs almost exclusively on points and backed bladelets; implements most likely to have served as armatures. This use is, in fact, confirmed on many specimens by the presence of impact spalls on their tips and distal extremities.

A preliminary microscopic study revealed heavy polish along dorsal ridges and on retouched margins, especially near the bases of specimens (Figure 11.8). This polish is a likely result of the abrasion generated by specimens working in their hafts. The topography, brightness, and texture of the polish are consistent with those attributes associated with working wood as described by Keeley (1980:35–36). Topographic highs, having been leveled through abrasion and deposition, are linked up and produce a net-like pattern. Under magnification this results in a relatively smooth, bright surface. Although less common than polishes along backed edges, very light bifacial polish was sometimes observed along unretouched edges. When striae were observed they were oriented parallel to the edges of specimens.

The impact spalls, location of ochre, position of polish, and orientation of striae suggest that La Mouillah points and backed bladelets were mounted longitudinally on a shaft with the

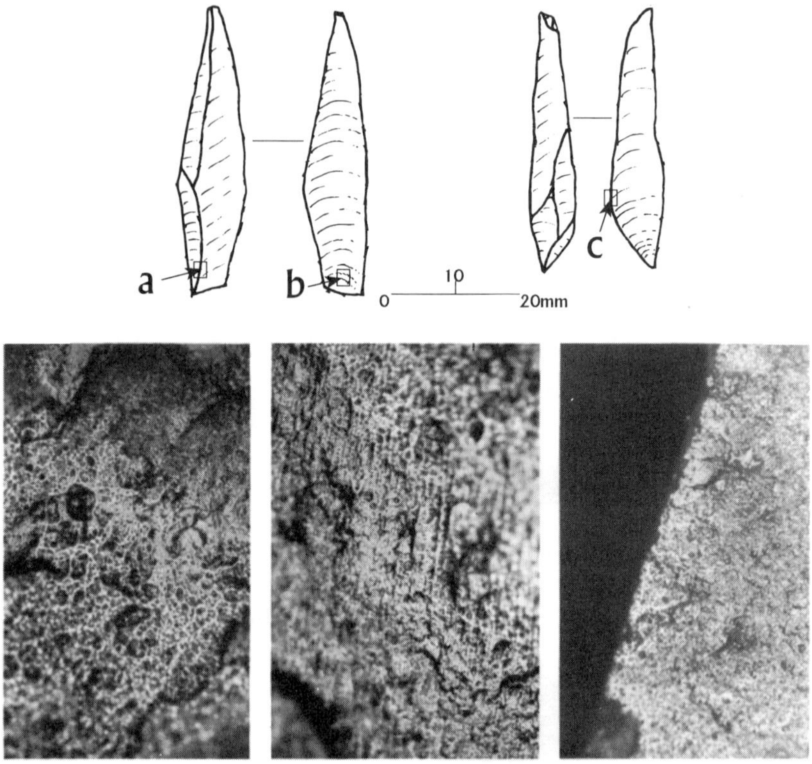


Figure 11.8. Photomicrograph (200X) of polish appearing on two point specimens from Tor Hamar (J431): (a) dorsal ridge near the base of a microgravette, (b) ventral surface near the base, (c) ventral surface near the base of a microgravette (after Henry and Garrard 1988).

backed edge resting in the haft, secured by mastic (Figure 11.9). Given the small sizes and light weights (mean = 0.5 g) of the specimens, they were probably projected by bow (Fenenga 1953; Thomas 1978; Shott 1993).

## Bone Tools

Five broken bone points were recovered from the Tor Hamar (J431) deposit with all but Layer B yielding a specimen (Figure 11.10). Most of the specimens appear to have been fashioned from split longbones by polishing the tips. One point also shows evidence of basal modification (Figure 11.10:c).

## Ground Stone Specimens

A large block containing a conical mortar was found as part of the Bedouin windbreak on the floor of the shelter and four other fragments were recovered from Layers A, C, and D of the deposit. Three of these would appear to be from mortars (or perhaps a mortar), whereas one is most likely from a bowl. All of the ground stone was fashioned from a granulite probably obtained from the mountains located a few kilometers south of the site.

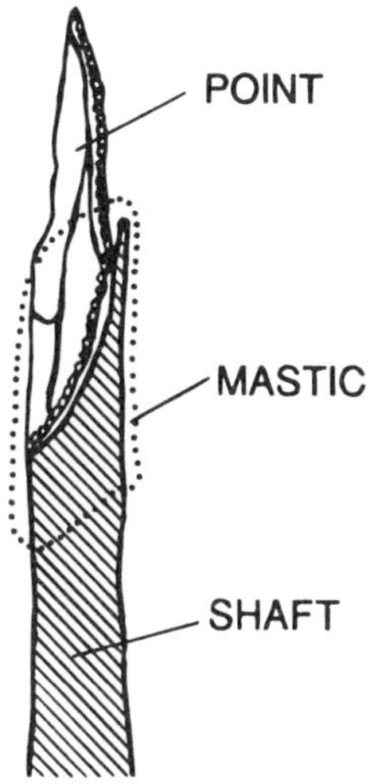


Figure 11.9. Illustration of the method by which points were likely to have been secured to shafts (after Henry and Garrard 1988).

## Ornamental Shell

The deposits of both sites yielded large numbers of shell including *Dentalium*, *Ancilla*, *Cerithium*, *Conus*, *Gibbula*, *Columbela*, *Glycymeris*, *Ostrea*, *Arcularia*, cowrie, scallops, and cockles (Chapter 17). Most come from the Red Sea, although a few specimens are of Mediterranean origin.

# FAUNAL REMAINS

## Economy and Environment

Tor Hamar (J431) produced a large faunal assemblage heavily dominated by the remains of gazelles and caprines (Henry and Garrard 1988, Chapter 20). The bones of hare, fox, jackal, ass/horse, aurochs, bird, and tortoise were also identified. The results of an analysis of the skeletal part representation of gazelles and caprines prompt Klein (Chapter 20) to suggest that the entire carcasses of these small species were returned to the shelter. Based upon the faunal assemblage, his environmental reconstruction for that of Madamaghan times differs little from the modern setting.

## Cementum Increments of Teeth: Seasonality

Lieberman's study (Chapter 18) of cementum increments of the teeth of *G. gazella*, *Capra sp.*, and undetermined *Bovidae* found all but one to display opaque outermost bands indicative of the winter, wet season as the time of death. The single translucent band was thin, implying that the animal had been killed in early spring. These data suggest that the occupations of the Tor Hamar (J431) shelter were seasonally specific and almost always occurred during the winter.

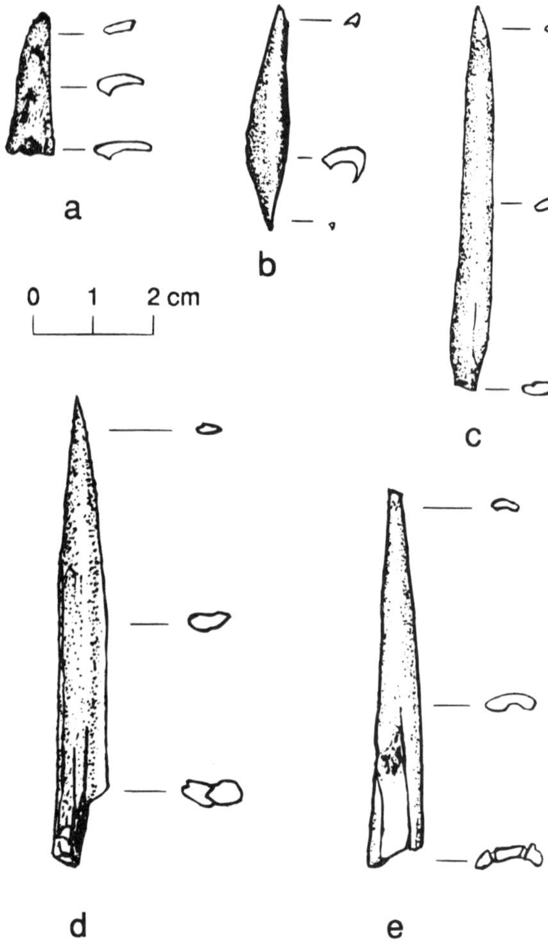


**Figure 11.10.** Illustrations of bone points from Tor Hamar (J431).

## POLLEN AND PHYTOLITHS

### Palynology

Emery-Barbier's (Chapter 16) palynological investigation of the Tor Hamar (J431) deposit identified adequately sized samples only in Layers A, B, and E. The spectra of Layers A and B, composed largely of artemisia, are indicative of relatively dry conditions. In contrast, the Layer E spectrum is composed almost exclusively of grasses. While this dominance is almost certainly an expression of grasses being introduced to the rockshelter for bedding or other purposes, their mere presence implies a more moist setting than that of the area today.

### Phytolith Analysis

Rosen (Chapter 19) found only Layer E to offer good preservation of phytoliths. She notes the large majority of phytoliths came from grasses, sedges, and/or rushes. Woody plants are also well represented and were likely used as fuel for fires. Silica skeletons resembling those of the genera *Cyperus* and *Juncus* are also present in moderate frequencies. Rosen observes that both of these genera commonly grow near desert springs and that their stems are used for baskets and mats. Also, she interprets the heavy domination of phytoliths of stems and leaves over those of floral parts or seeds to signal an occupation of the shelter when grasses were not flowering. That would mean sometime from late summer through winter. The high proportions of C4 grasses point to warm and relatively dry conditions.

## SUMMARY AND CONCLUSIONS

### Local Settlement-Procurement Strategies

All of the seasonal indicators associated with the occupations of Tor Hamar (J431) and Jebel Fatma (J436) point to the sites having been inhabited during the winter. Occupants of the shelters would have especially enjoyed their protected settings with southwestern exposures during the cold, wet season. It is at this time that water would have been immediately available from pools in the beds of nearby wadis. Beyond these indirect clues to the season(s) of occupation, Lieberman's (Chapter 18) cementum increment study and Rosen's (Chapter 19) analysis of phytoliths are in agreement in indicating a winter occupation of Tor Hamar (J431).

If these sites are representative of the exploitation of the lower edge of the piedmont by Madamaghan groups during the winter, wet season, what areas would they have been most likely to have inhabited during the warm season? While they may have migrated to the higher regions of the piedmont and uplands during the summer, this would have placed them in competition with contemporary foraging populations, initially (Phase I and II) represented by Middle-Final Hamran groups and later (Phase III) by Early Natufian groups. In that Madamaghan sites have not been found in the upland survey areas, where significant numbers of Hamran and Natufian sites have been discovered, it seems unlikely that the uplands were ever settled by Madamaghan groups. The most likely warm season shift in settlement for Madamaghan groups was to the lowlands of the Rift Valley and Red Sea coast. Here during most of the warm, dry season water would have been limited to springs, to which Madamaghan groups would have been tethered. During the spring and early summer the maturation of cereals and tubers in conjunction with aquatic and marine resources may have made such oasis settings along the Rift Valley to the Red Sea coast attractive. Many such springs exist today along the southern Araba to Aqaba near the confluences of major wadis (e.g., Wadi Sik, Wadi Rahma, and Aqaba).

The seasonal evidence from the Madamaghan sites allows for comparison and perhaps refinement of settlement-procurement models that have been proposed for the Mushabian (Goring-Morris 1987; Henry 1989a). In reconstructing settlement patterns for Mushabian (Ramonian) populations of the highland Negev, Goring-Morris (1987:255) argues for a warm season coalescence of groups in the uplands (i.e., ca. 900 masl) followed by settlement of lowland areas, such as coastal dunes during the winter. He bases his argument on the large sizes of the upland sites coupled with the observation that their high elevations would have made for uncomfortable winter camps. In contrast, Henry's (1989a:144–145) model favors an annual cycle that entailed a late winter (February–March) through midsummer (June–July) occupation of the uplands followed by a late summer (August–September) through mid-winter (December–January) settlement of the lowlands. Henry's reasoning is that while much of the winter may have been uncomfortably cold for occupation of the uplands, it was only during or immediately following the winter rainy season that a widespread occupation of the uplands would have been possible away from the more permanent water sources (e.g., springs and seasonal lakes). He draws upon limited seasonal indicators (wind direction and water availability) from Mushabi Basin sites in support of his model. In pointing to winter encampments, the evidence from the Tor Hamar (J431) and Jebel Fatma (J436) rockshelters supports at least a portion of Henry's model. The proposed upland residence of Madamaghan groups into midsummer, however, is not in accord with either the cementum increment data or the phytolith evidence. If the Tor Hamar (J431) occupations are representative of the settlement cycle, the upland occupation would appear to have been confined to the winter.

## Regional Relationships

The distribution of Mushabian Complex sites within the modern steppe-desert zone signals an arid land adaptation by the population (Bar-Yosef 1981; Bar-Yosef and Belfer-Cohen 1989; Henry 1989a). This finds expression in the regional facies of the complex as manifested in the Mushabian of Sinai, the Ramonian of the Negev, and the Madamaghan of southern Jordan. Small site areas, ephemeral features, and typically thin cultural horizons all point to a relatively mobile settlement system in which small foraging groups exploited arid settings up to the fringes of the Mediterranean woodlands. What is so interesting here is the evidence for cultural interaction with neighboring populations centered in the Mediterranean woodlands. Although such interaction between these arid zone and woodland foraging populations appears to have persisted for some 1,500–2,000 years (i.e., ca. 14–12/12,500 B.P.), these populations retained their distinctive cultural signatures.

### Site Distributions and Social Interaction

Within the Levant even subtle climatic perturbations induce striking shifts in the boundary separating the arid zone from the Mediterranean woodlands. This is especially so for those regions of marked relief, such as the Negev and southern Jordan, where woodland, steppe, and desert are tightly packed into narrow elevational belts. The close proximity and episodic shifts in distribution of these zones would have acted to inter-finger the habitats of steppe-desert and woodland adapted communities.

Such interaction is indicated by the diffusion of certain technotypologic traits between these neighboring foraging populations. During the early Mushabian we see the adoption of trapeze/rectangles and scalene bladelets from adjacent Geometric Kebaran groups who in turn appear progressively to have acquired the microburin technique from their arid land neighbors. With the emergence of the Natufian in the Mediterranean woodlands, cultural interplay between the boundary populations continued. This finds clear expression in the somewhat unique form of bifacial (Helwan) retouch that characterizes Early Natufian (12,500–11,000 B.P.) lunates as well as late Mushabian assemblages (Bar-Yosef 1981; Henry 1983, 1989; Gorring-Morris 1987, Bar-Yosef and Belfer-Cohen 1989).

### Transhumance, Drainages, and Territories

Synchronous and overlapping Mushabian and Geometric Kebaran site distributions were recognized in Sinai and the Negev sometime ago (Bar-Yosef and Phillips 1977), but their respective settlement patterns have not been isolated. Perhaps as a result of the striking elevational and environmental diversity in southern Jordan, there is greater clarity in the settlement differences of these synchronous populations. Hamran (Geometric Kebaran Complex) groups followed an annual cycle of coalescing in rockshelters near lakes on the floor of the Wadi Hisma (ca. 800 masl) during the winter, followed by a migration to the uplands accompanied by dispersal of the population into more ephemerally occupied camps during the warm season. In contrast, Madamaghan (Mushabian Complex) groups appear to have occupied sheltered settings at the lower edge of the piedmont (ca. 950–1,000 masl) during the winter, most likely followed by movement to the Rift Valley and coast during the warm, dry season.

Both Hamran and Madamaghan groups appear to have occupied portions of the lowland-lower piedmont belt of the broad Hisma Valley in the winter, but their site settings relative to the drainage patterns of the region differed significantly. Whereas Hamran sites are situated around lakes at the *lower end* of major drainages that are directly tied to the uplands, Mushabian sites are positioned near the *upper ends* of drainages that fall directly to the Rift Valley some 1,000 m below.

This site distribution relative to drainages may simply be an expression of natural routes for movement of groups between the different seasonal segments of their transhumant cycles. The Wadi Judayid provides direct access to the Judayid Basin and the surrounding uplands of the plateau from the lakes dotted across the Hisma valley. Not only would the Wadi Judayid drainage system have provided a natural avenue connecting the winter and summer ranges of Hamran groups, but it would also have provided a natural means by which such groups could monitor and anticipate available resources outside of the range that they were inhabiting. For example, from their winter camps on the lower end of the Wadi Judayid drainage they would have been able to establish the amount of precipitation that had fallen in the basin and uplands over the wet season. This, in turn, would have provided them with expectations on the abundance of surface water, plant resources, and game to be found there over the dry season. Perhaps more importantly, these expectations would have formed the basis for the scheduled movements of groups to the uplands. A similar linkage would have held for their warm season occupation of the piedmont and uplands. Near the end of the dry season, Hamran groups would have been able to judge when there had been sufficient precipitation in the uplands to charge the pot-holes and lakes downstream, thus making it possible to occupy these lower elevations. For the same reasons, Madamaghan groups would have benefitted from settling along those drainages falling from the lower edge of the piedmont to the Rift Valley.

## Subsistence Activities and Adaptive Distinction

Prior to the excavation of Tor Hamar (J431) we knew relatively little of the subsistence practices of Mushabian Complex groups (Bar-Yosef and Belfer-Cohen 1989). The animal remains (Chapter 20) and phytoliths (Chapter 19) from the deposit indicate a subsistence strategy differing little from the other simple foraging populations of the Levant. An emphasis upon the hunting of gazelles and sheep/goats along with the exploitation of seasonal wild plant resources parallels that of Geometric Kebaran groups. An absence of phytoliths of seeds, flowerets, and cereal husks suggests that the mortars and stone bowls of Tor Hamar (J431) were not used in the processing of wild cereals. They were probably used in preparing other plant foods or perhaps pigment.

The question then emerges as to what, if any, unique settlement-procurement strategies enabled Mushabian groups to inhabit the arid zone and how did these differ from their Geometric Kebaran contemporaries centered in the Mediterranean woodlands? Traditional technotypologic comparisons of chipped stone toolkits and ground stone specimens show no real functional differences between the complexes. As mentioned, the limited organic evidence from a Mushabian (Madamaghan) context fails to differ appreciably from that of Geometric Kebaran deposits. Aside from special knowledge of certain biotic resources restricted to the arid zone and a greater emphasis on hunting, it may well have been that the fundamental adaptive characteristic of Mushabian groups rested in their higher levels of settlement mobility.

## Origin of the Mushabian in the Qalkhan

The Mushabian traditionally has been viewed as representing a population expansion into the Levant from northeast Africa (Henry 1974, 1983, 1989a; Phillips and Mintz 1977; Bar-Yosef 1981; Bar-Yosef and Belfer-Cohen 1989). This interpretation was based upon four interrelated observations: (1) Mushabian Complex sites were restricted to those lands (the Sinai and Negev) of the Levant bordering northeast Africa; (2) certain technotypologic traits (microburin technique, La Mouillah points, arched-backed bladelets) of Mushabian Complex assemblages have affinities with those of North Africa (Tixier 1963; Camps 1975), the Nile Valley (Phillips 1973), and Nubia (Close 1978); (3) radiocarbon determinations for the Mushabian and similar northeast African complexes are near synchronous (Henry 1983; Bar-Yosef and Belfer-Cohen 1989); and (4) the Mushabian appeared to lack roots within the late Upper Paleolithic/early Epipaleolithic of the Levant (Goring-Morris 1987).

With new evidence for a Mushabian (Madamaghan Industry) presence east of the Rift Valley coupled with a technotypologic continuity with the earlier Qalkhan Industry (15,000–18,000 B.P.), also found in the arid-zone east of the Rift, the Mushabian seems more likely to have developed in the arid region along the eastern periphery of the Levant than in northeast Africa.

Aside from the Madamaghan assemblages from the mountains of southern Jordan, the assemblage reported from Wadi el-Jilat 8 (Site E2 of Waechter *et al.* 1938; Garrard *et al.* 1985b, 1986, 1988) near the Azraq Basin of eastern Jordan shares all of the technotypologic characteristics of the Mushabian. The assemblage is characterized by "la Mouillah points, curved, pointed arched-backed bladelets and occasional trapeze/rectangles" along with shell beads and grinding slabs (Garrard *et al.* 1988:321). Of the two radiocarbon dates from the deposit, the one determined from wood charcoal (13,310 ± 120 B.P., OxA 521) fits comfortably within the earlier (pre-Helwan lunate) Mushabian radiocarbon chronology. The other date (10,540 ± 160 B.P.), obtained from burnt bone, is much too young for the technotypologic configuration of the assemblage.

When comparing Qalkhan and Madamaghan assemblages, we see continuity expressed in the use of the microburin technique along with a dominance of arched-backed and pointed bladelets, straight-backed bladelets with truncations opposite snaps, medial segments of straight-backed bladelets, and La Mouillah points. While Qalkhan points and very narrow arched-backed pointed bladelets characterize the Qalkhan, these are replaced in the Madamaghan by wider arched backed bladelets, scalene bladelets, trapeze/rectangles, and ultimately Helwan lunates.

The Qalkhan—Madamaghan succession traces the development of a foraging population on the eastern fringes of the Mediterranean woodlands from ca. 20,000–12,500 B.P. The population appears to have extended its range west of the Rift Valley into Sinai and the Negev ca. 14,000 B.P. This development represents a parallel track with that of the Kebaran–Geometric Kebaran–Early Natufian succession centered in the Mediterranean woodlands.

**Appendix 11.1. Type List for Madamaghan Assemblages**

| Assemblages | | | J431 | | | J436 |
|---|---|---|---|---|---|---|
| Layer | A | B | C | D | E[1] | |
| Number | 302 | 233 | 452 | 377 | 79 | 132 |
| A. Scraper Class | | | | | | |
| 1. | 1.0 | 0.9 | 0.4 | 0.5 | 2.5 | 3.0 |
| 2. | 5.0 | 2.1 | 0.9 | 0.8 | | |
| 3. | 0.7 | 1.3 | 0.7 | | 3.7 | 2.2 |
| 4. | 4.6 | 4.3 | 2.1 | 1.3 | 3.7 | |
| 5. | | | | | | 0.7 |
| 6. | | | | | | |
| 7. | 0.7 | 0.4 | 0.2 | 0.5 | | 0.7 |

Appendix 11.1. (*Continued*)

| Assemblages | | | J431 | | | J436 |
|---|---|---|---|---|---|---|
| Layer | A | B | C | D | E[1] | |
| Number | 302 | 233 | 452 | 377 | 79 | 132 |
| 8. | 0.7 | 1.7 | 0.7 | 0.5 | 3.7 | |
| 9. | 0.3 | 0.9 | 0.7 | | | |
| 10. | 0.3 | 0.4 | | 0.3 | | |
| 11. | 0.7 | 0.4 | 0.4 | 0.3 | | |
| 12. | | | | | 5.0 | |
| 13. | 1.3 | 0.9 | | 0.8 | 1.2 | |
| 14. | | | | | 7.5 | |
| 15. | 0.3 | | 0.9 | 0.3 | | |
| B. Burin Class | | | | | | |
| 16. | | | 0.2 | | 2.5 | |
| 17. | | | 0.2 | 0.5 | | |
| 18.–25. | | | | | | |
| 26. | 0.3 | | | | | |
| C. Microlithic Geometric | | | | | | |
| 27. | 1.0 | | 0.2 | | | |
| 28. | 1.7 | 1.3 | 0.9 | 0.3 | | |
| 29. | | 0.4 | | | | |
| 30. | | | | | | |
| 31. | | 0.2 | 0.5 | | | |
| 32. | | 0.9 | | 0.3 | 1.2 | 7.5 |
| 33.–36. | | | | | | |
| D. Nongeometric Microliths | | | | | | |
| 37. | 4.3 | 4.7 | 9.3 | 6.6 | 6.3 | 2.2 |
| 38. | 1.7 | 0.4 | 1.8 | 2.1 | 2.5 | |
| 39. | 1.0 | 0.9 | 2.1 | 1.1 | | |
| 40. | | | | | 2.5 | 9.8 |
| 41. | 0.7 | 0.4 | | 0.8 | | |
| 42. | | | | | | 0.7 |
| 43. | 1.3 | 2.6 | 3.5 | 5.0 | | 12.8 |
| 44. | 5.3 | 9.0 | 11.4 | 9.3 | 15.1 | 15.1 |
| 45. | 1.0 | 4.7 | 2.5 | 4.8 | 11.3 | 0.7 |
| 46. | | | | | | |
| 47. | | | | | | |
| 48. | 2.3 | 4.7 | 4.0 | 5.6 | 1.2 | |
| 49. | 0.7 | 1.7 | 1.1 | 1.3 | | |
| 50. | | | | | | |
| 51. | 0.6 | 0.9 | 1.1 | 0.5 | | |
| 52. | 0.3 | 0.4 | 0.2 | 0.3 | | 0.7 |
| 53. | 1.0 | 0.9 | 1.9 | 3.7 | | 0.7 |
| 54. | 0.3 | 0.9 | 0.7 | 1.1 | | |
| 55. | 8.6 | 2.6 | 13.7 | 10.1 | 8.8 | 1.5 |
| 56. | 4.7 | 5.2 | 6.4 | 5.3 | 1.2 | 0.7 |

Appendix 11.1. (*Continued*)

| Assemblages | | | J431 | | | J436 |
|---|---|---|---|---|---|---|
| Layer | A | B | C | D | E[1] | |
| Number | 302 | 233 | 452 | 377 | 79 | 132 |
| E. Abruptly Retouched Blades and Flakes | | | | | | |
| 57. | | 0.4 | 0.9 | 3.2 | | |
| 58. | | 0.9 | 0.7 | 0.8 | | |
| 59. | 0.7 | 0.9 | 1.3 | 1.1 | | |
| 60. | 2.0 | 3.4 | 2.3 | 3.4 | | |
| 61. | | | 0.4 | 0.5 | | |
| 62. | | | | 0.5 | | |
| 63. | | | 0.2 | | | |
| 64. | | | | | | |
| 65. | | | 0.2 | | | 0.7 |
| F. Truncations | | | | | | |
| 66. | 1.3 | 1.7 | 0.2 | | | |
| 67. | 0.6 | 0.9 | 0.4 | | 1.2 | |
| 68. | | | | | | |
| G. Retouched Pieces | | | | | | |
| 69. | 3.6 | 0.9 | 0.9 | 0.8 | | 3.0 |
| 70.–76. | | | | | | |
| 77. | 11.9 | 8.2 | 4.0 | 5.0 | | |
| 78. | | | | | 1.2 | |
| 79. | | | | | 1.2 | 1.5 |
| 80. | | | | | | |
| 81. | | | | | | 1.5 |
| 82. | | | | | 3.7 | 0.7 |
| 83. | 3.0 | 4.3 | 2.5 | 2.7 | | 0.7 |
| 84. | | | | | | |
| 85. | | | | | 2.5 | |
| 86.–88. | | | | | | |
| 89. | 4.3 | 1.3 | 0.9 | 1.1 | | |
| H. Denticulate Class | | | | | | |
| 90. | 4.0 | 2.6 | 1.3 | 0.8 | | 1.5 |
| 91. | | | | | | |
| I. Notch Class | | | | | | |
| 92. | 6.0 | 3.8 | 2.4 | 2.7 | | 2.2 |
| 93. | | | | | | 1.5 |
| 94. | 2.6 | 3.0 | 0.4 | 0.5 | | |
| 95. | | | | | 1.2 | |
| 96. | | | | | | |
| J. Multiple Use Class | 0.3 | | | 0.3 | | |
| K. Perforator | 1.6 | | 0.4 | 0.3 | | |

Appendix 11.1. (*Continued*)

| Assemblages | | | J431 | | | J436 |
|---|---|---|---|---|---|---|
| Layer | A | B | C | D | E[1] | |
| Number | 302 | 233 | 452 | 377 | 79 | 132 |
| L. Massive Piece Class | | | | | | |
| 97. | | | | | 1.2 | |
| 98. | | | | | 1.2 | |
| 99. | | | | | | |
| 100. | | | | | | |
| M. Utilized Piece Class | | | | | | |
| 101. | | | | | | |
| 102. | | | | | | |
| 103. | | | | | | 1.5 |
| N. Points | | | | | | |
| 104. | 0.3 | | | 0.3 | | 11.3 |
| 105. | | | | | | |
| 106. | | | | | | |
| 107. | 4.9 | 11.6 | 10.8 | 11.1 | 3.7 | 12.8 |
| 108. | | 0.4 | 1.3 | | 1.2 | 0.7 |
| 109. | | | | | | |
| O. Varia | | | | | | |
| 110. | 0.3 | | | 0.3 | | |

## Chapter 12

# The Natufian Sites and the Emergence of Complex Foraging

This interval is linked to fundamental changes in the adaptive evolution of prehistoric foragers in the Levant (Henry 1981, 1989a; Bar-Yosef and Belfer-Cohen 1989; Bar-Yosef and Valla 1991). With the Natufian we see a transition from a foraging strategy based upon high residential mobility and movement to resources (i.e., simple foraging) to one of reduced mobility where intensification and the logistical acquisition of resources (i.e., complex foraging) were common (Henry 1989a; Tchernov 1991). This transition appears to have been triggered by the intensive exploitation of plant foods, especially cereals, nuts, and legumes. The demographic and social consequences of the economic change were striking. Population growth at regional and community levels appears to have been associated with increased social complexity as indicated by social ranking along with enhanced artistic and ceremonial activities (Henry 1989a, 1991; Belfer-Cohen 1991; Noy 1991).

Evidence of the Natufian occupation of the study areas is limited to two sites, Wadi Judayid (J2) and Wadi Humeima (J406a), both of which are located in the piedmont zone (Figures 12.1 and 12.2). The technotypologic configurations of the assemblages fall neatly within the parameters of those defined for the Natufian Complex (Henry *et al.* 1985; Henry 1988a; Sellars 1989). At a more detailed level, these artifact patterns place the Wadi Judayid (J2) settlement within the Early Natufian, whereas the occupation of Wadi Humeima (J406a) appears to have fallen during Late Natufian times. The temporal placement of Wadi Judayid (J2) gains additional support from radiocarbon dates which cluster ca. 12,500 B.P.

## SITE SETTINGS AND RESOURCES

Wadi Judayid (J2) rests at an elevation of ca. 1,100 masl at the foot of the plateau near the eastern margin of the Judayid Basin. It is an open-site located at the edge of a low terrace separating Wadi Judayid from Wadi Ras en Naqb about 2 km above the confluence of the two drainages. Water is available seasonally from the nearby large wadis. Also, a spring is reported to have flowed until recently just east of the site, emerging from the bank of the Wadi Ras en Naqb. Chert sources, found throughout the site's catchment, are concentrated in the wadi beds and on the slopes of the escarpment east of the site. Wadi Judayid's (J2) setting is typical of Natufian hamlets in resting in an ecotone between the Mediterranean woodlands of the uplands and grassland of the piedmont and basin floor (Henry 1989a).

Wadi Humeima (J406a), situated at an elevation of ca. 1,000 masl, rests at the lower edge of the piedmont. Although an open site, its position at the base of Jebel Humeima produces an

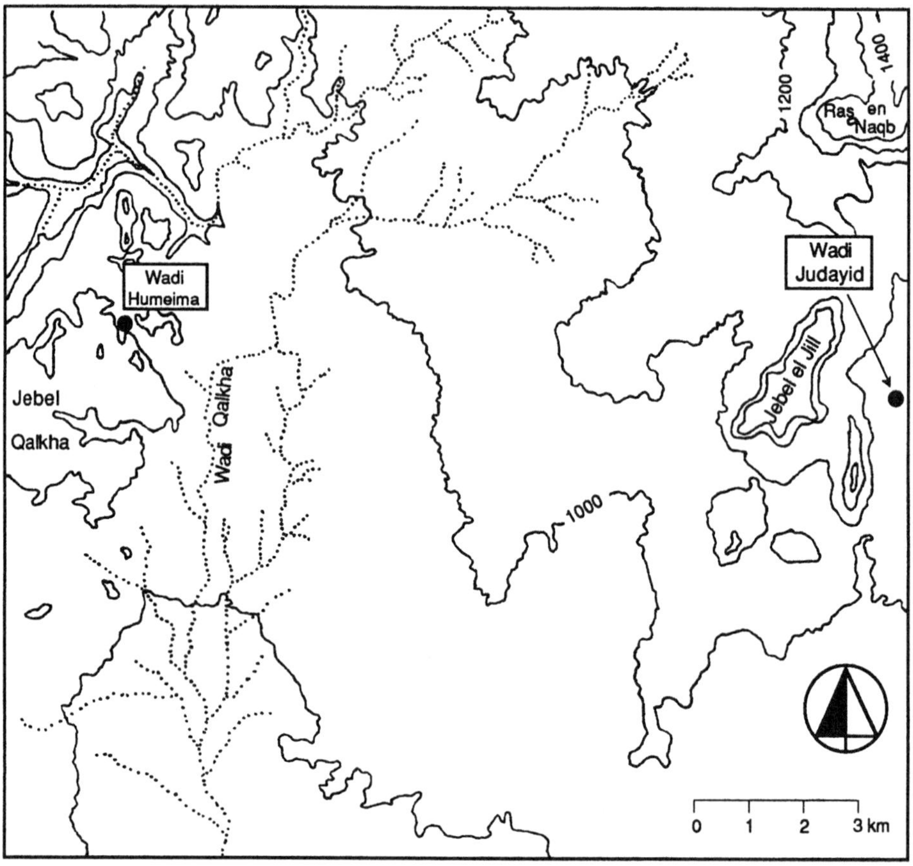


**Figure 12.1.** Map showing the locations of Wadi Judayid (J2) and Wadi Humeima (J406a).

eastern exposure. Beyond water being available as winter runoff from the nearby wadi, a spring at the base of the cliff is likely to have furnished a more dependable supply.

## SITE PLANS AND STRATIGRAPHY

### Wadi Judayid (J2)

The site of Wadi Judayid (J2) is defined by habitation debris (flint artifacts, bone, burnt rock, and ash) distributed over an area of ca. 400 $m^2$ (Figure 12.3). This concentration is eroding from a deposit set against the base of a 5 m high sandy terrace that represents the Member II alluvial fill of the basin. Early (J21 and J22) and Middle (J26 and J31) Hamran assemblages were found within this fill. Much of the original occupation area of the site appears to be buried within this deposit, thus the area of the site as determined by the surface scatter of artifacts is probably underestimated.

Excavation into the foot of the terrace revealed four stratigraphic layers and artifacts extending some 5m into the deposit (Figure 12.4). Layers A–C, composed of medium to fine grained sand, appear to have been formed by terrace colluvium interfingered with drift sand.

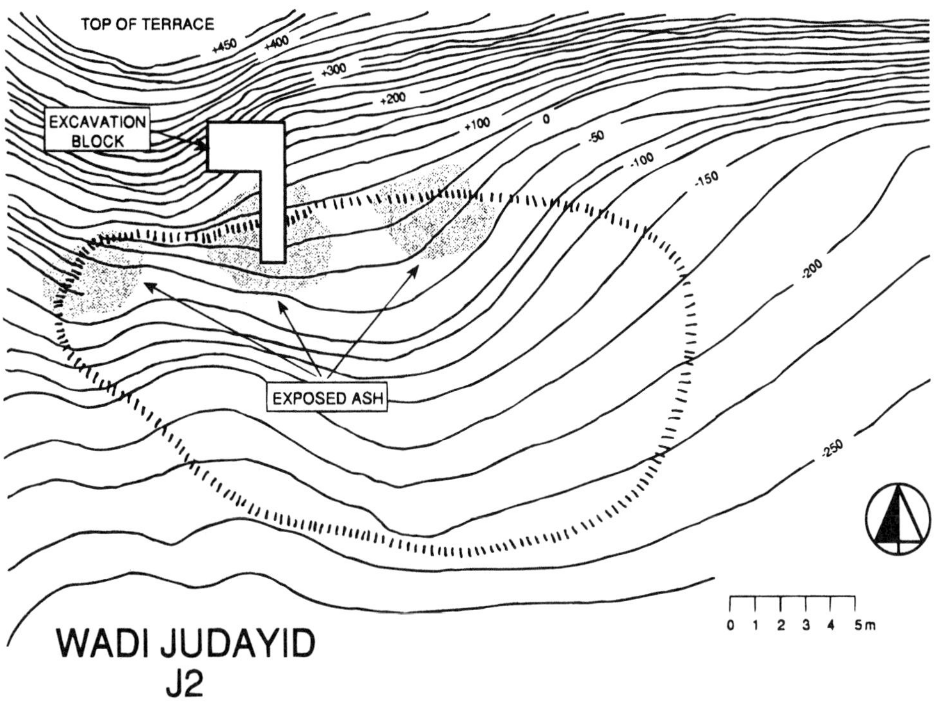


**Figure 12.2.** Topographic map showing the artifact distribution and excavation plan of Wadi Judayid (J2).

These rest on the silty sand of Layer D which represents an early Last Glacial alluvial fill (Member I) of the basin (Hassan, Chapter 2).

The occupation of the site was most intense during the deposition of the 50–60 cm thick Layer C as evidenced by the high densities of artifacts, burnt rock, ash, and bone within the stratum (Sellars 1989). Charcoal recovered from Layer C yielded radiocarbon dates of: 12,090 ± 800 B.P. (SMU-805), 12,750 ± 1,000 B.P. (SMU-806), 12,784 ± 650 B.P. (SMU-803). A hard lens of silty sand, high in carbonates, cements masses of habitation debris and defines the Layer C/D contact. This may well represent the compaction of a "living floor" by human traffic while the surface was wet. In any event, the lens must be anthropogenic as it is absent from off-site exposures of the Layer C/D contact.

### Wadi Humeima (J406a)

The site, defined by a concentration of artifacts distributed over an area of ca. 400 $m^2$, covers the upper portions of a slope extending from the base of Jebel Humeima (Figures 9.2, 9.3). The artifact distribution would appear to exaggerate the original occupation area of the site, given that many are heavily patinated and come from a shallow, steeply inclined cultural deposit. Two 1 $m^2$ units excavated within the artifact concentration revealed a shallow (20–30 cm) deposit of silty sand and sandstone rubble overlying bedrock with artifacts throughout.

## ARTIFACTS

Chipped stone specimens constitute most of the artifacts of the two sites with lesser numbers of ornamental marine shells and ground stone items being recorded (Table 12.1).

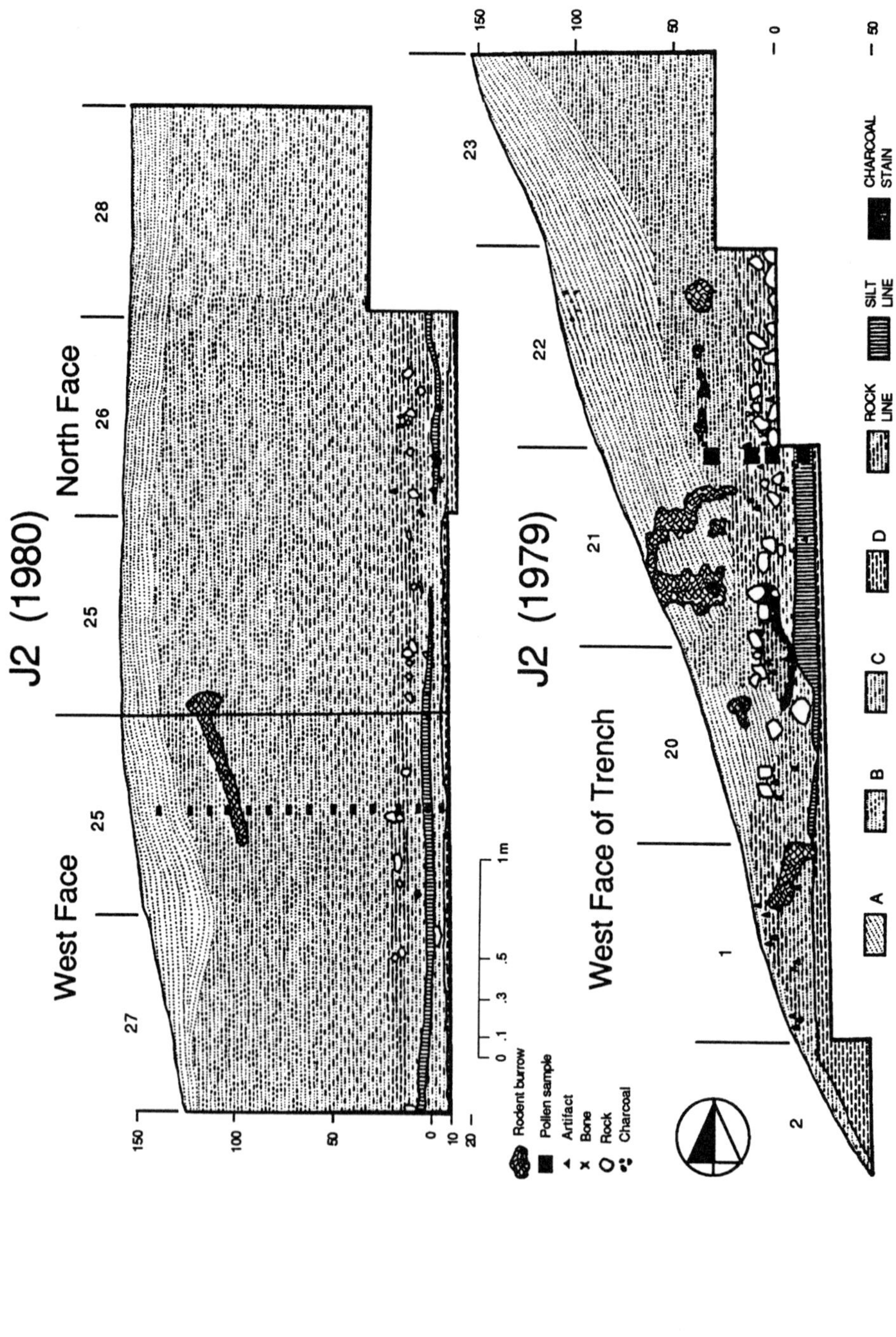


Figure 12.3. Stratigraphic profile of deposit at Wadi Judayid (J2).

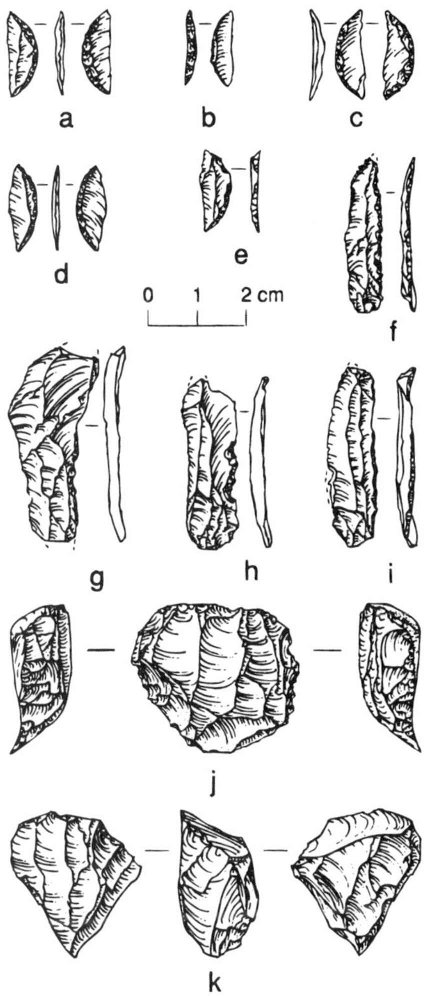


**Figure 12.4.** Illustrations of artifacts from the Early Natufian site of Wadi Judayid (J2): (a–d) Helwan lunates, (e) normal, abrupt-backed lunate, (f,i) straight-backed bladelets, (g) retouched bladelet, (j) core scraper, (k) bladelet core.

## Typology

Geometric microliths in the form of lunates dominate both assemblages. At the class level, the assemblages are remarkably similar, the only significant differences occurring with lunates and notches. Although the Wadi Judayid (J2) assemblage would appear to have greater diversity (i.e., more tool classes represented), this may be merely an expression of its much greater sample size (Figures 12.4 and 12.5).

### Time-Trends

At the type and attribute levels, the assemblages show some fundamental differences that are clearly linked to the different ages of the occupations (Appendix 12.1). Whereas Helwan (bifacially retouched) lunates and their variants (i.e., alternate and semi-steep forms) account

Table 12.1. Artifact inventories for the Natufian assemblages of Wadi Judayid and Wadi Humeima

| Assemblage | J2 | J406a |
|---|---|---|
| *Tools* | | |
| Scraper | 1.9 | 2.2 |
| Burin | 3.1 | 0.0 |
| Geometric | 60.5 | 31.9 |
| Nongeometric | 2.9 | 12.0 |
| Abrupt Ret. | 0.5 | 0.0 |
| Truncation | 5.2 | 3.3 |
| Retouched PC. | 14.0 | 18.7 |
| Denticulate | 0.2 | 4.4 |
| Notch | 9.4 | 24.2 |
| Multiple TL. | 0.3 | 0.0 |
| Perforator | 1.8 | 3.3 |
| Utilized PC. | 0.0 | 0.0 |
| Points | 0.2 | 0.0 |
| Varia | 0.2 | 0.0 |
| Total No. | 651 | 91 |
| *Debitage* | | |
| Core | 1.7 | 3.6 |
| Blade | 30.5 | 33.8 |
| Flake | 56.2 | 42.5 |
| Core Trim El. | 2.1 | 0.5 |
| Primary El. | 6.9 | 17.5 |
| Microburin | 2.6 | 2.1 |
| Burin Spall | 0.0 | 0.0 |
| Total No. | 12,316 | 560 |
| *Debris* | | |
| Chips | 81.2 | 89.2 |
| Chunks | 18.8 | 10.8 |
| Total No. | 22,181 | 1,391 |
| *Total Artifacts* | 35,148 | 2,042 |

for over 66% of those from Wadi Judayid (J2), none recorded in the Wadi Humeima (J406a) assemblage which is composed of normal and bipolar varieties (Table 12.2). Moreover, the dimensions of the lunates are dramatically different with those of Wadi Judayid (J2) being substantially larger. The large lunate sample drawn from a stratified context at Wadi Judayid (J2) also shows a progressive replacement of Helwan lunates by normal and bipolar forms through time (Sellars 1989). Also, this succession is accompanied by a sustained diminution of specimens.

It has long been recognized that assemblages of the Early Natufian exhibit greater proportions of Helwan retouch and that their lunates tend to be longer than those of the Late Natufian (Neuville 1951). This observation prompted seriation schemes based upon variation in retouch (Henry 1973a, 1977) and the lengths of lunates (Bar-Yosef and Valla 1979; Olszewski 1986). The general reduction in average lengths of lunates through time is in part an indirect shift in the frequency of Helwan to abrupt and bipolar retouched varieties, for, as initially observed by Neuville (1951:130), Helwan forms are generally longer than abrupt and bipolar forms. When

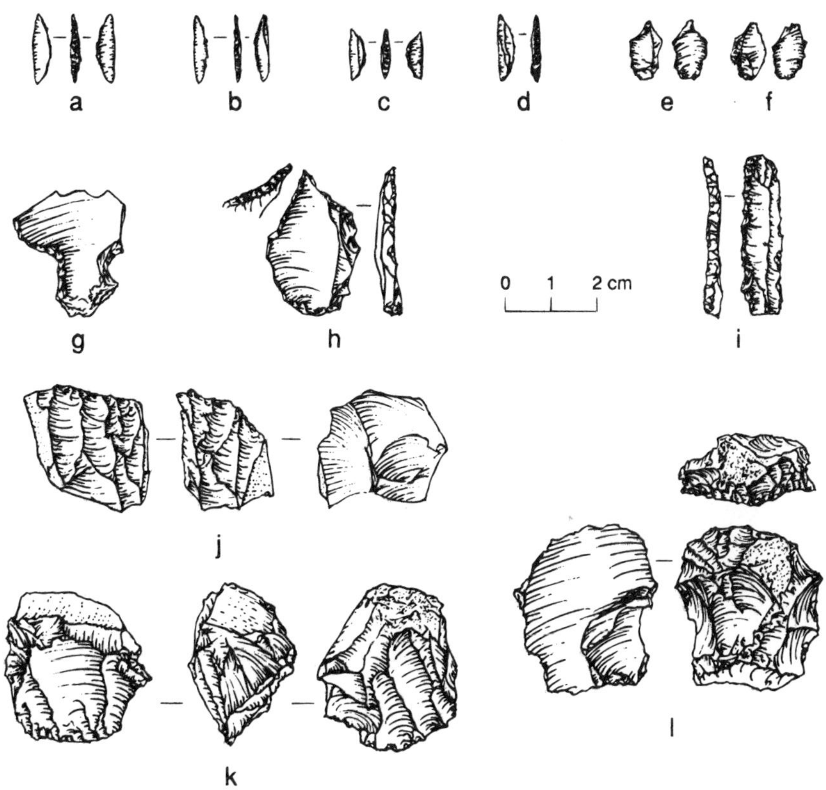


**Figure 12.5.** Illustrations of artifacts from the Late Natufian occupation of Wadi Humeima (J406a): (a–d) small abrupt-backed lunates, (e,f) microburins, (g) notch, (h) perforator, (i) straight-backed bladelet, (j,k) bladelet cores, (l) scraper.

**Table 12.2. Comparison of the varieties of retouch employed in the fabrication of lunates from Wadi Judayid (J2) and Wadi Humeima (J406a)**

| Retouch Variety | Wadi Judayid[a] | Wadi Humeima |
|---|---|---|
| Helwan | 37.3 | – |
| Normal Abrupt | 27.9 | 55.5 |
| Inverse Semi-steep | 10.7 | – |
| Obverse Semi-steep | 12.9 | – |
| Alternate | 3.3 | – |
| Bipolar | 2.0 | 54.5 |
| Incomplete/Atypical | 5.8 | – |
| No. | 394 | 29 |

[a] After Sellars

Helwan and abrupt/bipolar forms are partitioned a trend of diminution still generally holds, but there are some clear inconsistencies that have caused seriation schemes based upon varying lengths of lunates to be questioned (Henry 1985; Byrd 1987; Sellars 1991).

Radiocarbon dates from stratified deposits point to abrupt/bipolar retouch having come to be replaced by Helwan retouch as the dominant form by ca. 11,000 B.P.—an arbitrary divide between the Early and Late Natufian. The proportions of Helwan retouch and the radiocarbon determinations associated with the Wadi Judayid (J2) assemblage are consistent with the occupation's Early Natufian designation. As Sellars (1989, 1991) has noted, however, the dimensions of the lunate samples from Wadi Judayid would suggest a Late Natufian placement by dimensional criteria alone. The failure of the dimensional seriation schemes to provide reliable dating may rest in the importance of temporally independent functional forces in determining the dimensions of specimens (Marks and Larson 1977; Sellars 1991). Edge-wear studies show lunates to have been used as armatures (Anderson-Gerfaud 1983; Valla 1984), but it has not been determined whether they served as tips or barbs or both. These and other specific functions may have strongly influenced lunate dimensions.

Beyond fixing the Wadi Humeima (J406a) assemblage to the Late Natufian (i.e., 11,000–10,000 B.P.), the exceedingly small and exclusively abrupt/bipolar retouched lunates may suggest a placement near the very end of the sequence.

## Technology

Both assemblages display general technological configurations that agree closely with those defined for the Natufian. The assemblages reflect an emphasis upon the production of relatively short, wide bladelets struck from small globular and subpyramidal cores (Table 12.3). The cores primarily occur as single platform varieties, roughly divided between facetted and unfacetted forms. The dominance of crushed platforms and diffuse bulbs of percussion on bladelets suggests that they were removed by hardhammer.

### Raw Material

The inhabitants of Wadi Judayid (J2) had access to chert cobbles within the wadi bed some 50 m from the site in addition to abundant and ubiquitous chert from the nearby slopes of the plateau. Sellars (1989) reports evidence for heat-treating on many of the cores. Also noted in the assemblage are a few specimens of quartzite that point to limited utilization of material from the lowland. In exploiting the nearest source of chert, the occupants of Wadi Humeima (J406a) used the small cobbles from the bed of the Wadi Qalkha located some 1.5 km from site.

**Table 12.3. Comparison of the dimensions of lunates from Wadi Judayid (J2) and Wadi Humeima (J406a)**

| | J2[a] | | J406a |
|---|---|---|---|
| | Helwan | Normal | Normal |
| Length (mm) | | | |
| x | 15.8 | 13.8 | 12.9 |
| SD | 0.9 | 2.2 | 2.6 |
| Width (mm) | | | |
| x | 4.8 | 5.0 | 3.3 |
| SD | 0.5 | 0.8 | 0.7 |
| No. | 127 | 92 | 22 |

[a] After Sellars, 1989

### Reduction Strategies

Although full reduction streams were undertaken at both sites, the greater proportions of primary elements and cores at Wadi Humeima (J406a) appear to suggest a greater emphasis upon initial processing activities (Table 12.1). This may be merely an expression of the use of small wadi cobbles and the concomitant limited productive lives of the cores at Wadi Humeima (J406a). In fact, the core:debitage ratios of Wadi Humeima (26.4) are less than half those recorded for Wadi Judayid (56.3). This lower rate of blank production would act to inflate the proportionate representation of cores and primary elements in the assemblage. The smaller core dimensions and the higher proportions of multiple platforms reflect attempts by the Wadi Humeima (J406a) occupants to economize on raw material usage by extending the productive lives of cores.

After production, bladelets were truncated by use of the microburin technique for subsequent fabrication into lunates at both sites. The specific steps involved in this process may have varied considerably, but many of the lunates show clearly that the bladelets from which they were formed had been partially retouched prior to sectioning by the microburin technique. This specific application appears unique to the Natufian (Henry 1974).

### Other Artifacts

Ornamental marine shells, dominated by dentalia, were recovered from both sites (Reese, Chapter 17). These principally came from the Red Sea, but a few specimens of Mediterranean origin were recorded. Both sites also yielded a few groundstone specimens fashioned from sandstone. The surface artifact scatter at Wadi Judayid (J2) contained fragments of handstones and shaped mortars, while a deep holed conical mortar within a sandstone slab was found at Wadi Humeima (J406a). Interestingly, a cluster of similar mortars appear in a bedrock outcrop at the edge of a pothole ca. 1.5 km south of the site in the Wadi Aghar.

## PALEOENVIRONMENT AND SEASONALITY

Pollen (Emery-Barbier 1988; Chapter 16) and faunal remains (Henry *et al.* 1985; Chapter 18) were recovered only from the deposit of Wadi Judayid (J2). When integrated, these data point to steppic conditions that were becoming progressively moist during the occupation of the site. Moreover, Lieberman's study (Chapter 18) of the cementum layers of gazelle teeth shows Wadi Judayid (J2) to have been inhabited during more than one season.

### Faunal Remains

Prey species were dominated by wild goat (*Capra sp.*), undifferentiated sheep/goat, wild sheep (*Ovis orientalis*), and gazelle (*Gazella sp.*). Wild ass (*Equus sp.*), aurochs (*Bos primigenius*), Cape hare (*Lepus capensis*), leopard (*Panthera pardus*) and indeterminate bony fish and small bird remains were present in much lower numbers (Henry *et al.* 1985).

The range of animals exploited by the inhabitants of Wadi Judayid (J2) resemble the faunal assemblages of most other Natufian sites, differing only in the high representation of wild goats and sheep. The prominence of wild goat bones in the assemblage reflects the site's resource catchment which includes a prime goat habitat in the steep cliffs and uplands along the escarpment of the Jordanian Plateau. Wild goats also are well represented in other high elevation sites in southern Jordan (Beidha; Hecker 1989) and the Negev (Abu Salem; Davis *et al.* 1982; Butler *et al.* 1977). The presence of wild sheep in the Wadi Judayid (J2) assemblage provides the earliest evidence for the hunting of sheep in the Levant and significantly extends the zoogeographic range of the genus from the Zagros-Taurus arc (Uerpman 1981). Wild sheep have also been identified from the Late Natufian site of Rosh Horesha and the Harifian site of Abu Salem which are located in the highland Negev and dated to ca. 11,000–10,000 B.P. (Davis *et al.* 1982).

The presence of aurochs, ass, and leopard remains are indicative of a more moist setting than that of the area today. At least seasonally available surface water (needed daily by cattle and asses) and sufficient grasses for grazing would have been required to have supported the sheep, cattle, asses, and (indirectly) leopards. Beyond serving as environmental indicators, the animal remains confirm an exploitation of the site's diverse catchment, including the escarpment and uplands along with the gazelle habitat of the basin floor. The presence of fish vertebrae is also of interest. They indicate the exploitation of a perishable food-stuff from the Red Sea coast some 60 km away. Similar finds in the Natufian deposit at Hatula were attributed to fish having been introduced to the site after having been dried (Lernau 1985; Ronen and Lechavallier 1985).

### Pollen

The lower portion of the pollen diagram (Layer C) reflects dry and relatively cold conditions that become increasingly warm and moist after ca. 12,000 B.P. (Chapter 16). In addition to a rise in arboreal pollen, the presence of nearby aquatic vegetation points to increased humidity during Layer B times.

### Cementum Increments of Teeth

Lieberman's study (Chapter 18) of cementum increments of the teeth of *G. gazella*, *Capra sp.*, and undetermined *Bovidae* found four to lack cementum, another four to have opaque outermost bands, and seven to have translucent bands. Given that translucent bands are laid down in spring/summer and that opaque bands are formed in the fall/winter, this distribution is indicative of a multiseason occupation of Wadi Judayid.

## SUMMARY AND CONCLUSIONS

The occupation of Wadi Judayid (J2) demarks a fundamental shift in settlement-procurement strategy followed by the inhabitants of the area when compared to earlier Hamran patterns. In contrast to this earlier pattern of coalescence in long-term lowland encampments during the winter followed by dispersion of groups into the piedmont and uplands during the warm season, Early Natufian groups appear to have maintained long-term (if not year-round) camps at high elevations in the piedmont. This situation evidently was not unique to the study area, as substantial Early Natufian sites have been recorded in other high elevation settings in southern Jordan—Beidha (Byrd 1989a), Tabaqa (Byrd and Colledge 1991), and Sabra I (Gebel 1988).

### Mobility Levels and Settlement Patterns

The reduced settlement mobility of populations during the Natufian is well documented (Bar-Yosef 1981; Henry 1981, 1989a, 1991; Tchernov 1992), but significant debate rotates around the permanence of Natufian occupations and overall Natufian settlement-procurement patterns (Byrd 1987, 1989b; Olszewski 1991; Edwards 1989). Various criteria have been proposed for distinguishing sites according to their different degrees of permanence. Such characteristics as site area, thickness, artifact density, toolkit diversity, tool:debitage ratio, and presence or absence of features, structures, and burials have been variously weighted in schemes designed for assessing Natufian site types and residential mobility.

The global application of such schemes is confronted with a number of problems, principal of which I think is that of sampling. Of the more than 50 Natufian sites known in the Levant, no more than 10 can be said to have been extensively excavated; and these are concentrated in a very small area consisting of the Carmel and Western Galilee of northern Israel and the upper Jordan Valley. While the presence of features, architecture, and burials are undoubtedly important indicators of sedentism, their apparent absence from sites that have not undergone extensive excavation is not terribly meaningful. Thus, given the limited excavation of Wadi Judayid (J2) and Wadi Humeima (J406a), the lengths of their occupations can best be assessed through examining site area, thickness of cultural horizon, artifact density, and toolkit diversity.

## Wadi Judayid (J2)

Although artifacts on the surface of Wadi Judayid (J2) cover an area of only ca. 400 $m^2$, the excavation confirms that an area of occupation extending over at least another 130 $m^2$ is buried under 2–4m of drift sand at the edge of the terrace. This would be, at least, a "moderate size" site (using Bar-Yosef's classification) in comparison to other Natufian sites and perhaps even a large site if the buried portions of the occupation are extensive. For an open site, the 60 cm thick cultural horizon of Wadi Judayid (J2) would also fall in the moderate range resembling that of Beidha (60 cm). This is substantially thicker than that recorded for other south Jordan Natufian sites, such as Tabaqa (30 cm) and Sabra (20 cm), but much thinner than the 1–2m thick horizons revealed at Natufian hamlets (e.g., Hayonim Terrace, Ein Mallaha, Wadi Hammeh) located in the core Mediterranean zone.

The occupation horizon of Wadi Judayid (J2) yields a remarkably high artifact density of 1,437 specimens/$m^3$ which exceeds that computed for most other occupations including hamlets such as Wadi Hammeh and Hayonim Terrace (Table 12.4). Its toolkit is clearly specialized, however, with a dominance (60%) of lunates coupled with a paucity of maintenance and processing tools (i.e., scrapers, burins, perforators, denticulates). In fact, the toolkit resembles that of a specialized, short-term hunting camp, but this is inconsistent with the size, thickness, and artifact density of the occupation, not to mention the ground stone items and rich ornamental shell assemblage. The apparent lack of diversity in the toolkit may stem from the small excavated area (4 $m^2$) from which it was recovered and the possibility that this represents a specialized activity area. Such spatially segregated activity areas, in fact, tend to char-

**Table 12.4. Comparison of chipped stone artifact densities recorded for selected Natufian deposits where comparable screen sizes were employed**

| | Provenience Units/$m^3$ | | Artifacts | Density/0.1 $m^3$ |
|---|---|---|---|---|
| Wadi Judayid[a] | 20, 21, 26, 27 | / 2.4 $m^3$ | 34,497 | 1,437 |
| Wadi Humeima | 1, 2 | / 0.5 $m^3$ | 2,042 | 408 |
| Wadi Hammeh[b] | Plot XXD | / ? $m^3$ | 71,000 | 471 |
| Hayonim Terrace[c] | Layers A-D | / 19.2 | 7,559 | 393 |
| Beidha[d] | C-01:04 | / 8.8 $m^3$ | 2,716 | 31 |
| Rosh Zin[e] | Locus 4 | / 1.2 $m^3$ | 35,633 | 2,969 |

[a] Sellars, 1989: 99,
[b] Edwards, 1992: 137,
[c] Henry, 1976: 397 - 398,
[d] Byrd, 1988: 185,
[e] Henry, 1976: 325, 330

acterize more permanent residential encampments as previously defined in the Natufian (Henry 1973b).

The high tool:debitage ratio of 1:19 of the Wadi Judayid (J2) assemblage might imply a transitory occupation as suggested by Byrd (1987, 1989a), if it were not for the fact that chert was so readily available to the site. The notion that a direct correlation exists between the duration of a prehistoric occupation and the degree to which debitage was modified into tools is probably sound as is the corollary that argues for more extensive retouch on specimens in such settings (Dibble 1988). These relationships depend, however, on the availability of raw material being constrained to the extent that groups were obliged to economize in its depletion through a broader, less selective use of debitage and through the recycling of tools.

The combined evidence from Wadi Judayid (J2) implies that while the site may not have been occupied year-round, it most likely experienced multi-season residence. For an open site, the high artifact density and 50 cm thick occupational horizon delimited to an area of 600–1,000 $m^2$ suggest a rather intensive, mostly continuous occupation. The concentration of habitation debris within an uninterrupted stratigraphic succession is inconsistent with the site having been periodically revisited, perhaps seasonally. Open sites resulting from such a palimpsest of encampments typically cover large areas, often display low–moderate artifact densities, and are likely to exhibit culturally sterile strata separating occupation horizons. In bearing all of these attributes, the Natufian occupation of Beidha furnishes a clear example of such a site (Byrd 1991).

If Wadi Judayid (J2) had not been occupied year-round, what seasons would have been most likely? Lieberman's study (Chapter 18) points to an occupation that spanned part if not all of the wet season (November through March) and dry season (April through October). The presence of ground stone implements suggests that the site was inhabited during the maturation of cereals and nuts, that is, from mid-summer through fall for elevations of 1,000–1,300 m. Emery-Barbier's pollen study (Chapter 16) indicates a dry setting for the initial settlement of the site, which might suggest an occupation limited to the winter wet-season when the nearby wadi would have been running. The nearby spring, which flowed early in this century, reduces the significance of water availability as a seasonal indicator, however. In short, the existing evidence fails to point to a specific season of occupation for Wadi Judayid (J2), but instead suggests multi-season residence, perhaps over the better part of a year.

While the seasonality of residence at Wadi Judayid (J2) may be somewhat ambiguous, the site clearly shows a break with earlier settlement-procurement patterns. The artifact density and thickness of cultural deposit indicate a more intensive occupation than that of Hamran coalesced winter camps. Rather than being located in lowland (ca. 800 masl) rockshelters, Wadi Judayid (J2) was established as an open site at an elevation of ca. 1,100 masl. This shift from the lowlands to the piedmont for long-term, coalesced encampments is likely to have been induced by the onset of warmer conditions accompanying a global rise in temperature ca. 12,500 B.P.

Locally, these warmer conditions would have allowed groups to occupy the higher elevations of the piedmont during the winter wet season with greater comfort. But, perhaps more importantly, the higher temperatures are likely to have increased evaporation rates, thus reducing the sources of water on the floor of the Hisma Valley and requiring humans and their prey species to find water (especially during the dry season) closer to its origin of runoff in the piedmont and uplands. As I have argued elsewhere (Henry 1989), this terminal Pleistocene elevation in temperature also would have enabled warmth-loving plants (e.g., cereals and nut-bearing trees) to colonize upland settings. In so doing, cereals, nuts, and perhaps legumes would have formed a rich, highly predictable food source. In light of these environmental changes and their economic consequences, the specific location of Wadi Judayid (J2) and the adaptive implications of its behavioral residue are perhaps more understandable. While the site may not evince its permanence as clearly as those inhabited somewhat later as Natufian hamlets in the core of the Mediterranean woodlands, it likely represents a nascent shift to complex foraging based upon the exploitation of cereals and nuts along the southern edge of the Mediterranean zone.

### Wadi Humeima (J406a)

The relatively small area and thin cultural deposit of Wadi Humeima (J406a) are more indicative of a seasonal camp than a long-term occupation. The artifact density of 408 specimens/0.1 $m^3$ is some three times lower than that of Wadi Judayid (J2), but quite similar to those recorded for the lowland seasonal camps of the Hamran. Clues to the seasonality of the occupation are sparse, but suggest a warm season encampment, perhaps from mid-summer to early fall. A winter occupation seems unlikely given the eastern exposure of the setting. The presence of a bedrock mortar may indicate the processing of cereals and nuts and their harvest time from mid-summer through early fall.

The apparent shift toward more mobile settlements with seasonal camps established at lower elevations in the piedmont may be an expression of the colder and drier conditions that emerged during the Late Natufian (Henry 1981, 1989a; Leroi-Gourhan and Darmon 1991; Sillen and Lee-Thorp 1991). Other Late Natufian sites, such as Hatula, have also been associated with more transient encampments (Ronen and Lechevallier 1991).

## Regional Comparisons and Natufian Origins

Although an Early Natufian presence in the highlands of southern Jordan has been known since Kirkbride's (1966) work at Beidha, it is only in the last decade that the considerable density and antiquity of such sites have been established. Wadi Judayid (J2), Beidha, Sabra I, and Tabaqa confirm a substantial Early Natufian population in the region (Figure 12.6). Somewhat surprisingly the antiquity of these sites would appear to fall as early or earlier than those in the core Mediterranean woodlands, traditionally thought to be the area of Natufian emergence. Radiocarbon dates from Wadi Judayid (J2) and Beidha cluster around 12,500 B.P.

Moreover, the Early Natufian assemblages of southern Jordan clearly show a continuity with the immediately preceding Geometric Kebaran occurrences of the area, specifically Late and Final Hamran assemblages. Beyond this continuity being expressed in the dominance of lunates among the microliths, the alternating retouch common to Late and Final Hamran lunates would appear to be a precursor to the fully bifacial, Helwan retouch that characterizes the Early Natufian. Habitual use of the microburin technique in the fabrication of lunates within the Late–Final Hamran and Early Natufian also underscores a connection. Interestingly, such high microburin indices appear more common to the Early Natufian of southern Jordan than to synchronous sites in the core of the Mediterranean woodland zone and, as previously discussed, are also unique to the Late–Final Hamran industries of the Geometric Kebaran Complex.

While these technotypologic attributes indicate an origin for the Natufian of southern Jordan within the final stages of the local Geometric Kebaran (i.e., Late–Final Hamran), this does not necessarily mean that Hamran groups gave rise to the Natufian throughout the Levant. The Natufian, representing a more tightly bound and homogeneous social entity than that of preceding Geometric Kebaran populations, was probably formed by the coalescence and integration of nascent complex foraging communities. These probably evolved more or less synchronously across the Levant, with Wadi Judayid (J2) and neighboring sites representing but one of a number of areal clusters that gave rise to complex foraging.

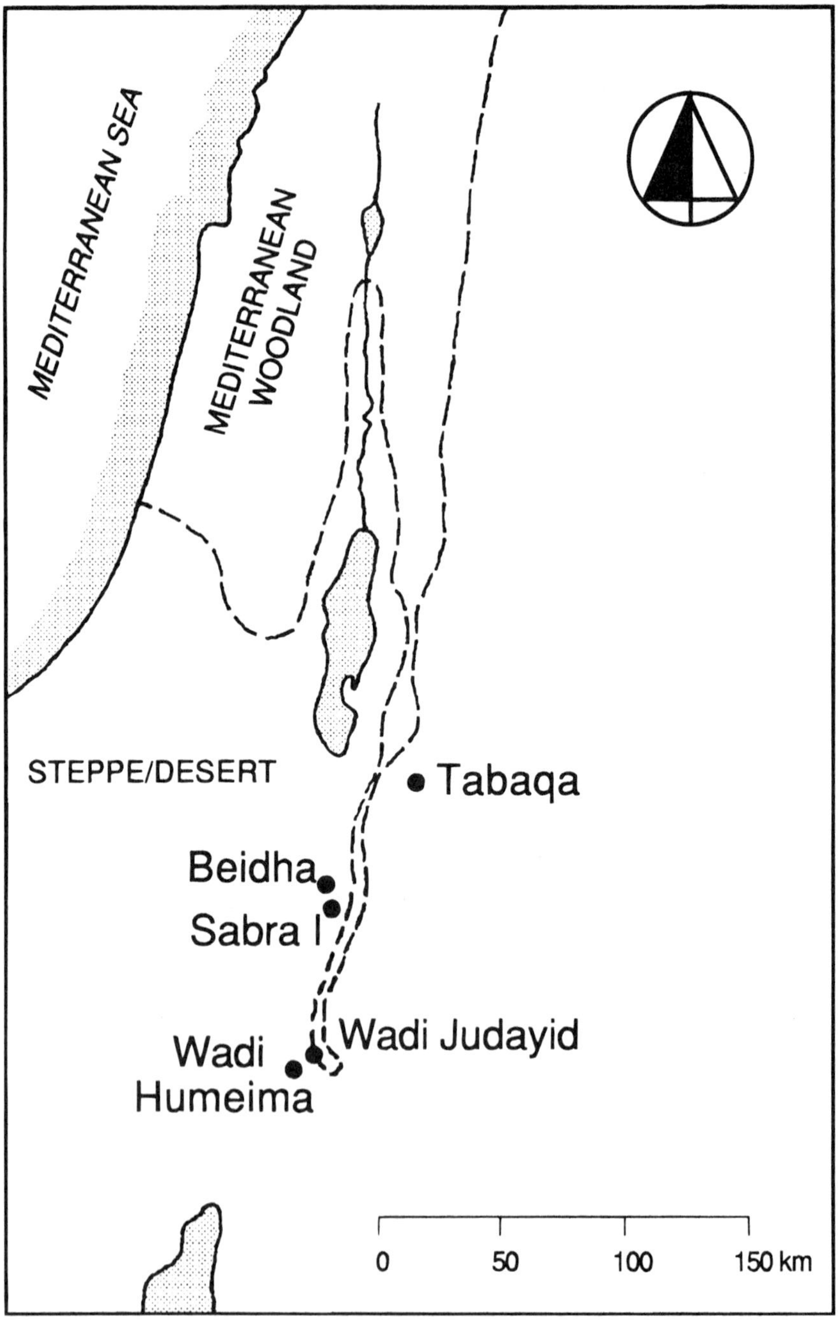


Figure 12.6. Map of Early Natufian sites (after Byrd 1992).

**Appendix 12.1. Type List for Natufian Assemblages**

| Assemblages | J2 | J406a |
|---|---|---|
| Number | 781 | 91 |
| A. Scraper Class | | |
| 1. | 0.8 | 1.0 |
| 2. | | 1.0 |
| 3.–6. | | |
| 7. | 0.1 | |
| 8.–11. | | |
| 12. | 0.1 | |
| 13. | | |
| 14. | | |
| 15. | 0.5 | |
| B. Burin Class | | |
| 16. | 0.1 | |
| 17. | 1.7 | |
| 18. | | |
| 19. | 0.3 | |
| 20. | 0.1 | |
| 21. | 0.1 | |
| 22. | 0.1 | |
| 23.–26. | | |
| C. Microlithic Geometric | | |
| 27. | 31.4 | 20.8 |
| 28. | 21.5 | |
| 29. | 5.8 | 10.9 |
| 30. | 0.8 | |
| 31. | 0.1 | |
| 32. | | |
| 33. | | |
| 34. | 0.2 | |
| 35. | | |
| 36. | 1.0 | |
| D. Nongeometric Microliths | | |
| 37. | 0.8 | 2.1 |
| 38. | | |
| 39. | | |
| 40. | 0.1 | |
| 41. | | |
| 42. | 0.2 | 3.2 |
| 43. | 0.1 | |
| 44. | 0.2 | |
| 45. | 0.1 | |
| 46. | | |
| 47. | | |
| 48. | 0.2 | |
| 49.–51. | | |

Appendix 12.1. (*Continued*)

| Assemblages | J2 | J406a |
|---|---|---|
| Number | 781 | 91 |
| 52. | 0.2 | |
| 53. | | |
| 54. | | |
| 55. | 0.6 | 6.5 |
| 56. | | |
| E. Abruptly Retouched Blades and Flakes | | |
| 57.–60. | | |
| 61. | 0.1 | |
| 62.–64. | | |
| 65. | 0.5 | |
| F. Truncations | | |
| 66. | 4.0 | 3.2 |
| 67. | 0.6 | |
| 68. | 0.2 | |
| G. Retouched Pieces | | |
| 69. | 2.4 | 6.5 |
| 70. | 2.0 | 3.2 |
| 71. | 0.7 | |
| 72. | 0.1 | 0.3 |
| 73. | | |
| 74. | 2.6 | 1.0 |
| 75. | 0.2 | |
| 76. | | |
| 77. | 0.3 | 1.0 |
| 78. | 0.2 | 0.7 |
| 79. | 0.3 | |
| 80. | 0.1 | |
| 81. | 0.1 | |
| 82. | 0.2 | |
| 83. | 0.8 | 3.2 |
| 84. | 0.3 | |
| 85.–87. | | |
| 88. | 1.5 | |
| 89. | 0.7 | 3.2 |
| H. Denticulate Class | | |
| 90. | 0.1 | 4.3 |
| 91. | | |
| I. Notch Class | | |
| 92. | 6.6 | 19.7 |
| 93. | 0.8 | |
| 94. | 1.1 | 2.1 |
| 95. | 0.1 | |
| 96. | 1.4 | 2.1 |

**Appendix 12.1.** (*Continued*)

| Assemblages | J2 | J406a |
|---|---|---|
| Number | 781 | 91 |
| J. Multiple Use Class | 0.2 | |
| K. Perforator | 1.5 | 3.3 |
| L. Massive Piece Class | | |
| 97.–100. | | |
| M. Utilized Piece Class | | |
| 101.–103. | | |
| N. Points | | |
| 104.–106. | | |
| 107. | 0.1 | |
| 108. | | |
| 109. | | |
| O. Varia | | |
| 110. | 0.1 | |

# Chapter 13

# Cultural Evolution and Interaction during the Epipaleolithic

The behavioral residue of the prehistoric populations inhabiting southern Jordan from ca. 19,000–10,000 years ago points to the existence of two distinct cultural evolutionary tracks. The one represented by the Hamran–Natufian succession is an extension of the Kebaran–Geometric Kebaran–Natufian sequence that characterizes this interval within the Mediterranean woodlands. In contrast, the other developmental sequence, represented by the Qalkhan–Madamaghan succession, appears to have been centered in the arid zone. During the early part of this succession, Qalkhan groups seemingly occupied only areas east of the Rift Valley within the arid belt along the eastern margins of the Mediterranean woodlands stretching from southern Jordan to near the Euphrates of northeastern Syria. Later, some 13,000–14,000 years ago, a population expansion west across the Rift is indicated by the strong affinities between the Madamaghan Industry and other regional facies of the Mushabian Complex concentrated in the Sinai and Negev.

## ALTERNATIVE EVOLUTIONARY TRACKS

Identification of parallel cultural evolutionary paths for Epipaleolithic populations along the southern edge of the Jordanian plateau allows for comparing adjacent and in part overlapping settlement patterns and procurement practices of groups exploiting the same general locus. The distinction of these populations and their evolutionary paths over this 6,000–9,000 year period can be seen as an expression of an adaptive dichotomy induced by their inhabiting broadly different environments at a regional scale. A more detailed comparison of their respective settlement and procurement patterns relative to their exploitation of a common region along the boundary of steppe-desert and woodland zones further highlights some of the more fundamental cultural ecologic differences of these two foraging traditions.

### Adaptation to the Arid Zone

Distribution of Qalkhan assemblages within the steppe-desert zone during the interval from ca. 19,000–15,500 B.P. confirms an arid land adaptation by late Last Glacial foragers. This also traces a different, although parallel, line of development with the Kebaran Complex of the Mediterranean woodlands. Heretofore, scholars have assumed that during this time, cold, dry conditions prevailed across the Levant and the arid zone was essentially unoccupied (Bar-Yosef and Belfer-Cohen 1989). This view drew support from the fact that little evidence of this timeframe had come from the surveys in the Sinai and Negev beyond a few late Upper Paleolithic occurrences. Lacking indicators of a distinct arid zone occupation during this interval, most researchers viewed the isolated assemblages of Jordan and Syria as variants of the better known Kebaran. This notion was maintained even though these assemblages display certain fundamental technotypological characteristics (e.g., intensive microburin technique and La Mouillah points) not in evidence within the Kebaran Complex. Given this perception of a single developmental track stemming from the foraging populations of the Mediterranean woodlands, the later Mushabian Complex with its clear ties to the arid zone was not seen as an indigenous development of Levantine foragers, but as representing an intrusive population from northeast Africa. Taken together, then, the Qalkhan and Madamaghan occurrences trace an alternative cultural evolutionary path to the one initiated by Kebaran foragers in the Mediterranean woodlands.

Recognition of this distinct evolutionary track with its arid-zone focus implies not only an indigenous Levantine development of the Mushabian Complex, but also the presence of a population of foragers well adapted to life in the steppe-desert zone of the region during the earlier part of the Epipaleolithic. From a purely evolutionary ecologic perspective, the successful settlement of the arid zone during this time should not be surprising, given the abundant evidence of earlier Upper Paleolithic sites in Sinai, the Negev, and southern and eastern Jordan.

### Ties to the Mediterranean Woodlands

Falling clearly within the developmental track of foraging populations exploiting the Mediterranean woodlands, the Early Hamran sites confirm a Kebaran presence further south into the arid region than heretofore reported (Bar-Yosef and Belfer-Cohen 1989; Hovers and Marder 1991). The high elevations of the region appear to have allowed for an extension of the Mediterranean woodlands, at least late (ca. 15,000–16,000 B.P.) within the Kebaran Complex, as is indicated by the pollen diagram of the Jebel Mishraq (J504) deposit (Chapter 16). Subsequent Middle, Late, and Final Hamran occupations point to a continued, intensive occupation of the region by foraging populations with strong ties to those inhabiting the Mediterranean woodlands as manifested in the Geometric Kebaran Complex. Apparently accompanying a moist interval from ca. 14–13,000 B.P., expansion of woodland adapted foragers deep into the steppe desert zone is documented by the great numbers of sites reported from the Sinai and the Negev (Goring-Morris 1987; Henry 1989a:176; Bar-Yosef and Belfer-Cohen 1989:486). The return of drier conditions ca. 13,000 B.P., perhaps induced by a global rise in temperature and a regional northward migration of storm paths (Henry 1983, 1989a), prompted a retreat of Geometric Kebaran groups from the arid settings of Sinai and the Negev to the better watered areas of the central Levant. While the arid south was abandoned by late Geometric Kebaran groups west of the Rift Valley, there was continued occupation, as indicated by Late and Final Hamran sites, along the southern edges of the Ma'an Plateau during this interval. Again the extremely high elevations of the region would appear to have moderated the arid interval and enabled woodland adapted foragers to continue settlement along the edges of the plateau. Perhaps the most important consequence of this continued occupation of the mountainous region of southern Jordan by woodland adapted foragers lay in the foundation it provided for the development of complex foraging in the region as reflected in the Early Natufian occupations of ca. 12,500 B.P.

## THE QALKHAN–MADAMAGHAN SUCCESSION

Qalkhan assemblages contain toolkits dominated by nongeometric microliths with narrow arched-backed and pointed bladelets, La Mouillah points, and Qalkhan points serving as diagnostic types for the industry (Figure 13.1). The microburin technique, used for the controlled

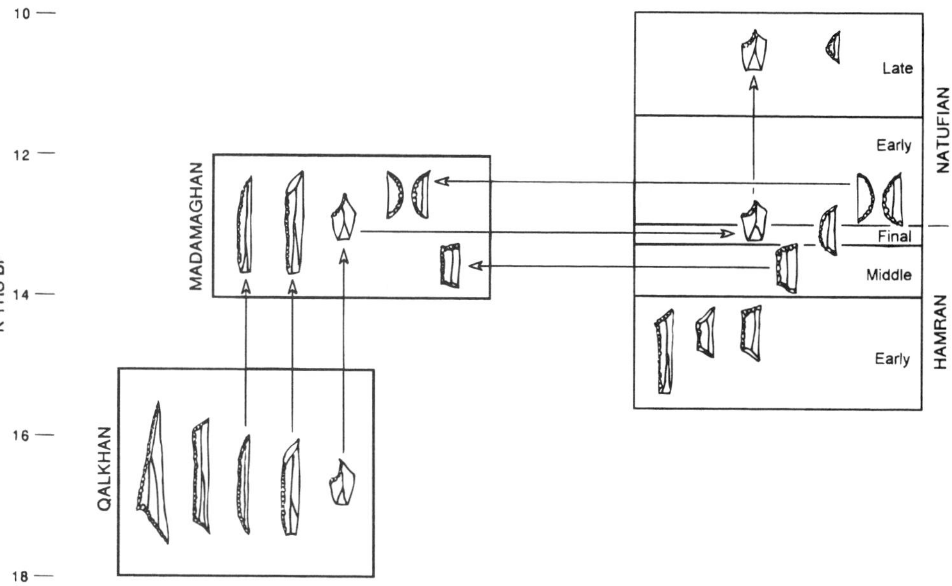


**Figure 13.1.** Schematic illustration tracing the technotypological trends and relationships of the Epipaleolithic complexes.

sectioning of bladelets, was central to the production of the points. In Madamaghan assemblages, we see technotypologic continuity with the Qalkhan reflected in the use of the microburin technique along with a dominance of arched-backed and pointed bladelets and La Mouillah points. Qalkhan points and very narrow arched-backed pointed bladelets are replaced in the Madamaghan by wider arched-backed bladelets, scalene bladelets (Ramon points), trapeze/rectangles, and ultimately Helwan lunates.

## Chronology

Assemblages of the Qalkhan Industry stratigraphically underlie an Early Hamran (Late Kebran Complex) horizon at Wadi Humeima (J406b) and are radiocarbon dated to between 20,000–15,500 B.P. in three deposits in the Azraq Basin area. Beyond those assemblages described in this volume, assemblages that should be included within the industry have been reported from the nearby Petra area in southern Jordan (Schyle and Uerpmann 1988), the Azraq Basin of northeastern Jordan (Garrard *et al.* 1985, 1986, 1987, 1988; Byrd 1990), the Yabrud rockshelter of western Syria (Rust 1950), and the El Kowm Oasis in northeastern Syria (Cauvin *et al.* 1979; Cauvin 1981; Cauvin and Coqueugniot 1988).

Madamaghan assemblages are known only from the sites of Tor Hamar (J431), Jebel Fatma (J436), and Madamagh in southern Jordan. Although resting within the broad technotypologic parameters of the Mushabian Complex, Madamaghan assemblages form a distinct regional facies of the complex characterized by high proportions of arched backed bladelets and La Mouillah points (Henry 1989:98–99; Shen 1992). Two radiocarbon dates from upper Layer C of Tor Hamar (J431) cluster ca. 12,500 B.P. and confirm a late Mushabian presence in the area. Earlier occupations (ca. 14,000–13,000 B.P.) are also indicated by those assemblages that lack Helwan lunates: the lower layers (D and E') of Tor Hamar (J431) along with Jebel Fatma (J436) and Madamagh. Additional evidence of an early Mushabian presence east of the Rift Valley comes from the site of Wadi el-Jilat 8 in the Azraq Basin (Garrard *et al.* 1988). The assemblage, radiocarbon-dated to ca. 13,300 B.P., certainly appears to fall within the Mushabian Complex, although its specific industrial affiliation remains uncertain.

## Local Settlement and Procurement Patterns

Within the study areas, Qalkhan and Madamaghan sites are confined to the extreme western end of the Wadi Hisma near the heads of tributaries that drain to the Rift Valley. The absence of evidence for a Qalkhan or Madamaghan presence further east along the edge of the plateau and into the lowlands suggests that these occupations represent only a fragment of an annual settlement cycle. The remainder of the cycle most likely encompassed portions of the Rift Valley, perhaps extending as far as the coast. The Qalkhan inhabitants of the western Hisma appear to have enjoyed a wide variety of camp types, ranging from small ephemeral occupations to longer-term, repeated encampments at some sites. The seasonal utilization of the area by Qalkhan groups also lacks clear patterning. While some sites are linked to cold season occupations, others are more likely to have served as warm season encampments. Although Qalkhan groups may have followed a transhumant cycle in moving from the Rift Valley to the western Hisma, some 1,000 m higher, the wide range of site types and seasons of residence imply that migration schedules, if existent, were not finely structured.

By Madamaghan times, settlement cycles appear to have become more patterned with groups repeatedly and intensively occupying southwestern-facing rockshelters near the heads of wadis at elevations from ca. 900–1,000 masl. Direct seasonal evidence based upon analyses of phytoliths and cementum increments of teeth from the Tor Hamar deposit confirms winter occupations. These Madamaghan encampments also apparently involved larger groups, perhaps for longer periods of residence, than the earlier Qalkhan camps as indicated by their greater sizes (Table 13.1). The much thicker cultural deposits of the Madamaghan horizons and the presence of large ground stone items also implies that the sites were visited more regularly than during Qalkhan times. But as with Qalkhan groups, Madamaghan foragers are most likely to have exploited the much lower elevations of the Rift Valley and even the Red Sea coast during the warm season. Although surface water would have been limited to springs and seeps, such oasis settings are numerous along the flanks of the southern Rift Valley due to high ground water (Raikes 1985) and are likely to have been even more common during the terminal Pleistocene. The numerous Red Sea shells in Madamaghan deposits suggest that groups had some form of

**Table 13.1. Comparison of the characteristics of sites and deposits of long-term camps for various Epipaleolithic taxa**

| | Site area ($m^2$) | | Cultural deposit (M) | | Artifact density (0.1 $m^3$) | |
|---|---|---|---|---|---|---|
| Industry | Mean | (Range) | Mean | (Range) | Mean | (Range) |
| Late Natufian | 400 | | 0.3 | | 408 | |
| Early Natufian | 530[a] | | 0.6 | | 1,437 | |
| Late-Final Hamran | 390 | | 0.6 | | 240 | |
| Middle Hamran | 485 | (320 - 650) | 0.6 | (0.3 - 0.9) | 204 | (114 - 294) |
| Early Hamran | 430[b] | (260 - 600) | 0.5 | (0.3 - 0.6) | 292 | (172 - 412) |
| Madamaghan | 257 | (214 - 300) | 1.2 | (0.7 - 1.6) | 232 | (68 - 265) |
| Qalkhan | 120 | (50 - 200) | 0.4 | (0.3 - 0.6) | 367 | (299 - 449) |

[a] Likely to be underestimated
[b] Likely to be overestimated

contact with the coast and the ground stone items recovered from Tor Hamar (J431) similarly indicate access to the granite mountains that form the western flank of the southern Rift Valley.

## THE HAMRAN–NATUFIAN SUCCESSION

The technotypologic evolution of the Hamran–Natufian succession follows a very similar path to that recognized within the Epipaleolithic of the core Mediterranean zone west of the Rift Valley. There the sequence of Kebaran, Geometric Kebaran, and Natufian complexes is reflected in nongeometric microlithic assemblages giving way to those dominated by trapeze/rectangles, which in turn are replaced by those characterized by lunates and the microburin technique.

Early Hamran assemblages, dominated by nongeometric microliths, are replaced by assemblages with trapeze/rectangles during the Middle Hamran. Lunates along with an intentional use of the microburin technique emerge in the Late Hamran and become dominant characteristics of toolkits in Final Hamran times. In the following Early Natufian, lunates continue as the principal geometric form, but most bear a unique form of bifacial (Helwan) retouch. The microburin technique continues to be intensively utilized in their production. Lunates persist as the characteristic microlith in the Late Natufian, but they are diminutive and formed by abrupt and bipolar retouch.

Although the Hamran–Natufian succession largely follows that seen in the Kebaran, Geometric Kebaran, and Natufian complexes of the core Mediterranean zone, there exist some important specific differences. The high proportions of lunates accompanied by microburins in the Late and Final Hamran assemblages is unlike any known technotypologic configuration of the Geometric Kebaran assemblages of the core Mediterranean woodlands. The intensive utilization of the microburin technique in the Early Natufian is atypical for Early Natufian assemblages of the core zone. While these local characteristics differ slightly from the broader evolutionary patterns of the late Epipaleolithic of the Levant, they confirm a smooth, mostly unbroken sequence linking the terminal phases of the Geometric Kebaran (the Late–Final Hamran) to the Early Natufian.

### Chronology

The chronology of the Hamran–Natufian succession would appear to stretch from ca. 16,000–15,000 B.P. until sometime shortly after 11,000 B.P. Early Hamran assemblages, dominated by straight-backed and truncated bladelets, show strong affinities to the those of the Late Kebaran and thus are likely to date some 16,000–15,000 B.P. With a geometric microlith component composed almost exclusively of wide trapeze/rectangles, Middle Hamran assemblages fit within the Group II Industry of the Geometric Kebaran (Henry 1989:153–55; Bar-Yosef 1981:297–298) and should date from ca. 14,500–13/12,800 B.P. (Bar-Yosef and Vogel 1987; Henry 1989a). The Late/Final Hamran assemblages, with their significant lunate and microburin representation, are unique, lacking parallels within the sequence of the core Mediterranean zone. The lunate and microburin component, both common elements of the succeeding Early Natufian assemblages of the area, do imply a very late position within the Geometric Kebaran sequence, however. Given the $C^{14}$ dates from the Early Natufian sites of Wadi Judayid (J2) and Beidha that cluster ca. 12,500 B.P., the Late/Final Hamran occupations are most likely to date ca. 12,500–13,000 B.P. The high proportions of Helwan retouched lunates in the Wadi Judayid (J2) assemblages suggest an early placement within the Early Natufian and this is supported by the radiocarbon dates of ca. 12,500 B.P. The Late Natufian occupation of Wadi Humeima (J406a), characterized by diminutive abrupt-backed lunates, is almost certain to date late in the interval stretching between 10,000–11,000 B.P.

## Local Settlement and Procurement Patterns

Unlike Qalkhan and Madamaghan site distributions, those of the Hamran trace a more complete settlement cycle and a presumably larger portion of the annual range of these Epipaleolithic foraging groups. Throughout the Hamran sequence a persistent and striking dichotomy is present between sites on the floor of the Hisma and those of the piedmont as expressed in their sizes, occupational thicknesses, exposures, features, artifact densities, tool-debitage-core production ratios, and toolkit configurations (Table 10.8). These data point to a transhumant pattern in which groups aggregated in rockshelters adjacent to seasonal lakes on the floor of the Wadi Hisma during the winter and dispersed into smaller, more mobile groups in the piedmont and uplands during the warm season. This contrasts markedly with the Qalkhan and Madamaghan settlement cycle in which groups were likely to have exploited those elevations near sea level in the Rift Valley during the warm season.

Beginning with the Early Natufian, the settlement pattern followed during the long Hamran sequence was fundamentally altered. The Early Natufian site of Wadi Judayid (J2) furnishes evidence of a long-term, if not permanent, occupation near the upper edge of the piedmont. Beyond signalling a significant elevational shift to higher elevations for long-term encampments, the site also yielded evidence from cementum increment analysis of teeth that indicates it was occupied multiseasonally. Its large size, considerable thickness of deposit (0.6m) for an open site, and extremely high artifact density are all consistent with Wadi Judayid (J2) having been occupied more permanently and by a larger group than the preceding lowland, long-term, winter camps of the Hamran. The emergence of Early Natufian hamlets around 12,500 years ago in the high elevations of southern Jordan is thought to have been induced by a colonization of warmth-loving cereals, pulses, and nut-bearing trees accompanying a global rise in temperature.

The Late Natufian occupation of Wadi Humeima (J406a), located at the lower edge of the piedmont, is smaller with a thinner cultural deposit and much lower artifact density than that of Wadi Judayid (J2). In combination these attributes indicate a less permanent residential encampment. The apparent shift toward more mobile settlements with seasonal camps established at lower elevations in the piedmont may be an expression of the colder and drier conditions that emerged during the Late Natufian (Henry 1989; Leroi-Gourhan 1992; Sillen and Lee-Thorp 1992). Other Late Natufian and Harifian sites, such as Rosh Horesha, Salibiya I, and Abu Salem, have been determined through cementum increment analysis to have been seasonally occupied (Lieberman 1993).

## Interaction between the Populations

The parallel and partially synchronous successions of Qalkhan–Madamaghan and Hamran–Natufian industries indicate the presence of ethnically distinct Epipaleolithic populations who maintained adjacent and partially overlapping ranges for some 4,000–5,000 years. While inter-fingering site distributions and synchronous radiometric dates suggest opportunities for contact and social interaction between neighboring populations, it is the artifactual clues which confirm that ideas were exchanged between living groups. Within the parallel late Epipaleolithic traditions of southern Jordan, evidence for social interaction is found principally in lithic artifacts, especially in implements that would have served as armatures for arrows. The trapeze/rectangle appears to have been adopted from Middle Hamran groups by groups of the early Madamaghan between 13,000–14,000 B.P. (Figure 13.1), whereas somewhat later the microburin technique was apparently acquired from the Madamaghan by Late and Final Hamran groups. Later during Early Natufian times, the Helwan lunate, having been developed within the Final Hamran, was adopted by late Madamaghan groups.

These examples of shared traits raise the question of how the populations were able to interact for such a long period of time without co-mingling, fusing ideas, and blurring their ethnic identities. Part of the explanation may rest in the fact that the contact took place along a fringe area where interaction was not prolonged nor intense, especially when compared to the social interaction that was maintained with the parent populations of the respective core areas. What we know of the annual settlement and procurement cycles of these adjacent populations indicates that their winter exploitation of the Wadi Hisma prompted them to establish camps within 10–20 km of one another. These camps were located on different drainage networks, however. During the warm season, the distinct populations pulled away from one another and occupied distant and quite different ranges. Whereas Qalkhan-Madmaghan groups appear to have moved south–southwest to the extremely low elevations of the Rift Valley and coast, Hamran–Natufian groups exploited the woodlands on the plateau to the north.

Another perhaps even more compelling force that acted to maintain the ethnic distinctions of the populations was embedded in their adaptive differences. This follows an old notion in which "cultural areas" were observed to be defined by natural environmental boundaries. Kroeber (1939:205), for example, proposed that culture was only intelligible in relation to the environmental factors to which it was conditioned. From a more detailed ecologic perspective, it simply can be argued that the social interaction between groups is likely to be greater if their survival rests on the exploitation of a common environment through similar means. This is also consistent with Hodder's (1979:452) view of ethnicity which he sees as a "mechanism by which interest groups symbolize their within-group organization in opposition to and in competition with other interest groups." Such interest groups, formed around sets of resources, would have varied significantly between woodland and steppe-desert settings in concert with the differences in resources. In taking this view, the spatial and temporal patterning of ethnicity might be understood best in light of the environmental background and those resources upon which populations depended for their survival.

## Chapter 14

# An Early Neolithic Hunting Camp: Jebel Queisa

Given the relatively high density of Neolithic sites in southern Jordan, it is somewhat surprising that a systematic, intensive survey of some 32 $km^2$ within the region recorded only a single occupation. And even this find, the site of Jebel Queisa (J24), represents an ephemeral occupation that most likely served as a hunting camp or vantage spot. The Neolithic horizon at the site is deeply buried under more than a meter of deposit containing Chalcolithic artifacts. The occupation is defined by a concentration of lithic artifacts within a 20 cm thick horizon that is restricted to an area of a little more than 10 $m^2$. The artifact inventory is composed largely of debitage and debris. The small toolkit is dominated by points and casually retouched specimens. The few temporally diagnostic characteristics of the assemblage (i.e., Helwan points, Netiv Hagdud truncation, and opposed platform, naviform-like blade cores) indicate an early Neolithic placement, probably on the order of 9,500 B.P.

## SITE SETTING AND RESOURCES

The site is positioned at the base of a steep cliff formed by the northeastern toe of a sandstone outlier, the Jebel Queisa (Figure 14.1). It and the Jebel el Jill to its north form the western rim of the Judayid Basin. The site, resting at 1,200 masl, has a northeastern exposure and a commanding view of the basin as well as the small embayment to its north. The embayment connects to the saddle between the two high jebels and forms a natural access route from the basin to the lowlands on the floor of the Hisma.

The nearest surface water would have been available from the nearby Wadi Judayid during its time of flow in the winter wet season and perhaps from pools into mid-spring. Sources of chert are abundant within the vicinity of the site. Cobbles from stream beds on the basin's floor as well as near continuous outcrops along its eastern flank would have provided convenient sources of fine quality cherts for the site's inhabitants.

## SITE PLAN AND STRATIGRAPHY

Jebel Queisa (J24) was discovered during survey of the Judayid Basin in 1979 and test excavated (3 $m^2$) that season. Chalcolithic artifacts, ash, and stone walls were found eroding from the surface of a steeply sloped terrace (Figure 14.1). The exposure of a thick Chalcolithic horizon underlain by Neolithic artifacts in the *sondage* prompted an expansion of the excavation the following year.

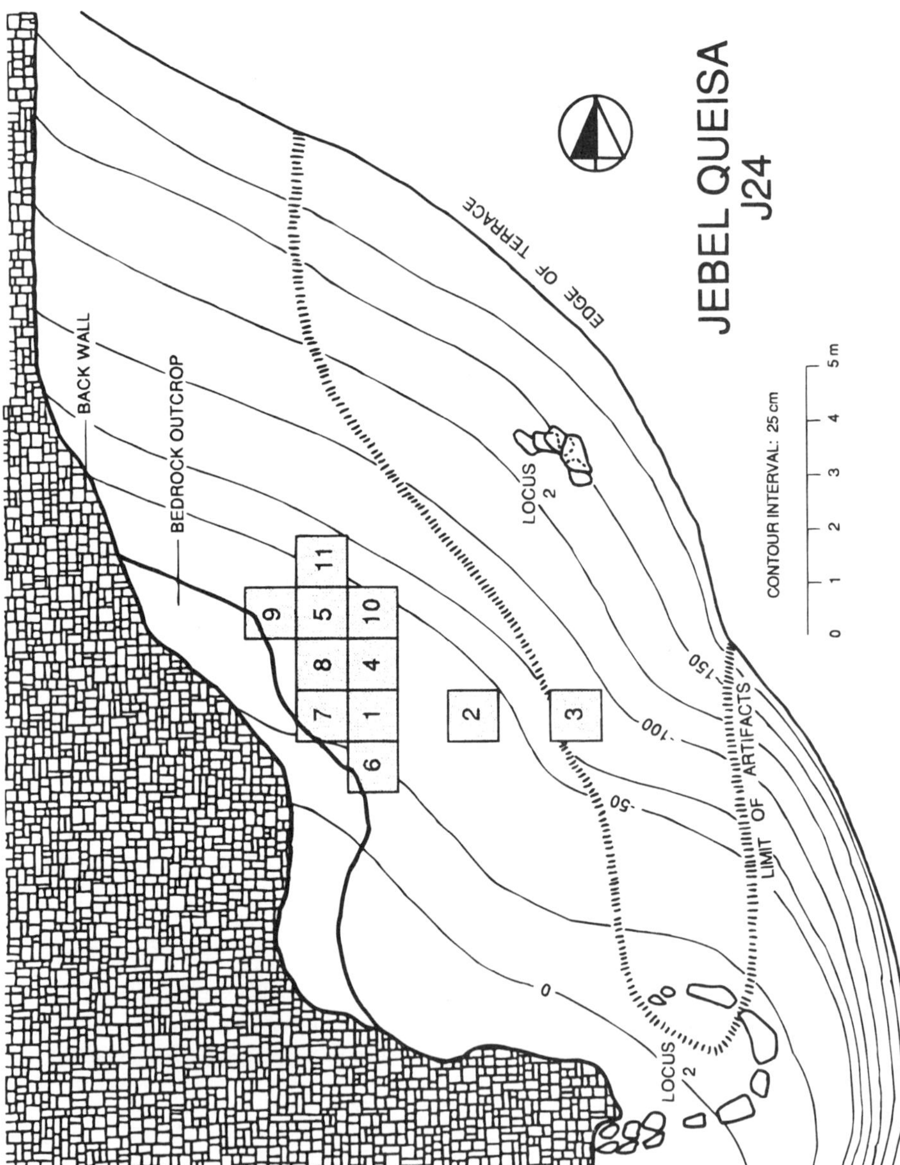


**Figure 14.1.** Topographic map showing the artifact distribution and excavation plan for the site of Jebel Queisa (J24).

## Stratigraphy

The excavation revealed three stratigraphic layers within a 120 cm thick deposit of drift sand (Figure 14.2). The upper two layers (A and B), defined by different concentrations and colors of ash, contained relatively high densities of Chalcolithic artifacts along with features mostly represented by hearths. A radiocarbon date of 5,720 ± 149 B.P. (SMU 804) was obtained from a hearth in lower Layer B, just above the layer B/C contact.

Although Layer C consists of a lithologically homogeneous stratum, the upper 30–40 cm of the layer can be distinguished from the lowest 20–30 m of the deposit on the basis of natural and cultural evidence. There appears to be an old erosional surface identified by a scree lens separating the upper from the lower part of the layer and this also corresponds to a break in the pollen diagram for the deposit. While pollen was recovered in high densities in layers A, B and upper Layer C, lower Layer C was barren of pollen (Chapter 16).

The cultural break between upper and lower Layer C is striking. The artifact densities of upper Layer C are extremely low (< 5/.1 $m^3$) and although attributed to the Chalcolithic, the artifacts most likely were introduced from the overlying occupation of layer B. Features, which are common in layers A and B, also are absent from upper Layer C. Resting just above bedrock in lower Layer C, Neolithic artifacts occur within a 20–30 cm thick band and in very high densities (225/.1 $m^3$) relative to the rest of the deposit. The Neolithic occupation was defined within only a small area, however, consisting of six units located closest to the face of the cliff on the southern portions of the terrace. While it is possible that the Neolithic horizon could extend further to the south as far as the jutting cliff face, even this distribution would encompass less than 50 $m^2$.

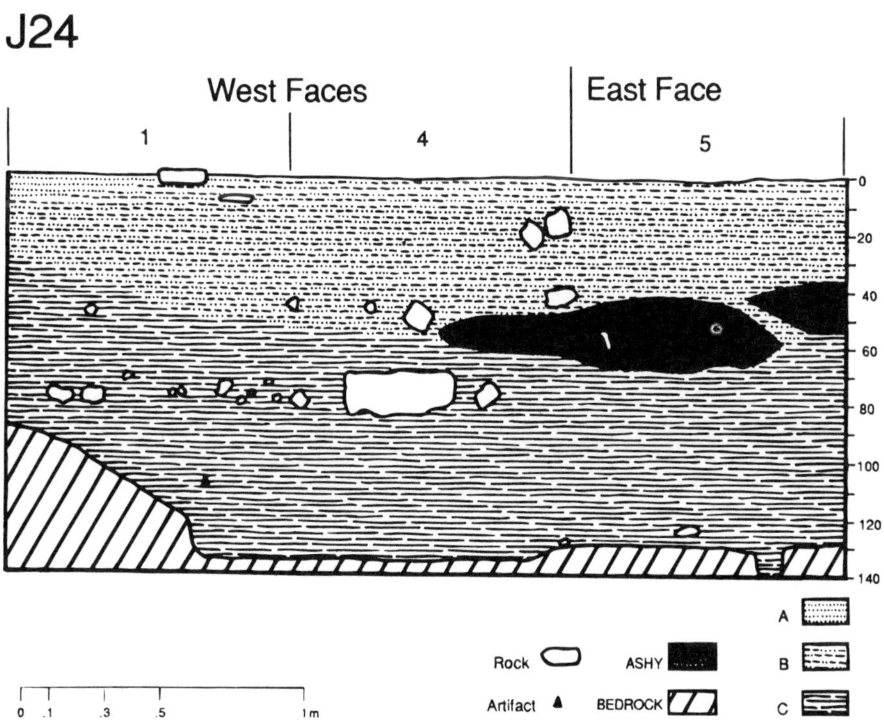


Figure 14.2. Stratigraphic profile of the deposit of Jebel Queisa (J24).

The cultural stratigraphic succession of the Jebel Queisa (J24) deposit suggests that a brief early Neolithic occupation took place ca. 9,500 B.P. as drift sand was accumulating at the base of the cliff. Sometime after this date, there was an erosional interval, creating a hiatus that accounts for as much as 3,500 years. This was followed by continued building of the drift sand as represented by upper Layer C. Sometime ca. 5,700 B.P., the site was again occupied. This time, more permanently by Chalcolithic pastoralists who appear to have regularly encamped at the site during the deposition of layers A and B. In recent times, erosion again has acted to truncate the deposit.

## ARTIFACTS

With the exception of a single shell bead, the artifact inventory is confined to 3,379 chipped stone specimens (Table 14.1).

### Typology

With a tool assemblage composed of only twenty specimens, the proportionate representation of the various classes may well be misleading. Given the small sample, it is somewhat surprising to find a relatively well balanced toolkit, however. Whereas the tool assemblage is dominated by retouched pieces and points, some six other classes of tools are represented (Figure 14.3). Most of these were formed on blades or bladelets.

Three Helwan points were recovered. One is complete, another broken, and the third is in an early stage of manufacture. Also, a single example of a bi-truncated bladelet, a Hagdud truncation, was identified. Similar truncations have been recorded at several early Neolithic (Sultanian) sites including Netiv Hagdud, Jericho, Gesher and Hatula (Bar-Yosef *et al.* 1987; Lechevallier *et al.* 1989). Although this specimen was initially classified as a broken and reworked basal section of an El Khiam point, the analysis of a much larger sample of Hagdud truncations by Bar-Yosef *et al.* (1987) indicates that the type was an unlikely by-product of broken points. Given its form, it may well have served as a transversal arrowhead, as initially suggested by Crowfoot-Payne (1983) for specimens from Jericho.

### Technology

The technology was centered on the production of blades and long bladelets from opposed platform wedge-shaped cores (Figures 14.4, 14.5, 14.6). Blade production was initiated through the *lame à crête* technique and percussion with a soft-hammer.

#### Raw Material

A fine grained, translucent flint or chalcedony was the preferred raw material. Although most artifacts display a white patination or frosting, most also exhibit very fresh edges. Stream cobbles were apparently rarely used, as few cores or primary elements show water worn cortex.

#### Reduction Strategies

Activities related to a full reduction stream appear to have been undertaken by the inhabitants of the site. The relatively high proportion of primary elements in conjunction with well formed cores and crested blades confirm initial processing activities related to core shaping and blank production. The high density of fine chips is also indicative of the shatter that would have accompanied the early stages of blank production. Tool use, manufacture, and rejuvenation were also part of the activities at the site as evidenced by the complete, broken, and reworked points.

Table 14.1. Artifact inventory for the Early Neolithic assemblage from Jebel Queisa (J24) lower Layer C

| Assemblage | J24 |
|---|---|
| *Tools* | |
| Scraper | 5.0 |
| Burin | 5.0 |
| Geometric | 0.0 |
| Nongeometric | 0.0 |
| Abrupt Ret. | 0.0 |
| Truncation | 5.0 |
| Retouched PC. | 45.0 |
| Denticulate | 0.0 |
| Notch | 10.0 |
| Multiple TL. | 0.0 |
| Perforator | 10.0 |
| Utilized PC. | 0.0 |
| Points | 15.0 |
| Varia | 0.0 |
| Total No. | 20 |
| *Debitage* | |
| Core | 0.9 |
| Blade | 57.8 |
| Flake | 31.4 |
| Core Trim Elem. | 2.0 |
| Primary Elem. | 7.9 |
| Total No. | 2,159 |
| *Debris* | |
| Chips | 88.4 |
| Chunks | 11.6 |
| Total No. | 1,200 |
| *Total Artifacts* | 3,379 |

Blade production involved fashioning one face of a core into a subpyramidal or wedge-shaped form and then creating crested ridges along parallel edges of the opposite face for blade removals (Figure 14.6). Once blade production had been initiated it was maintained through use of opposed platforms. Soft-hammer percussion represented the predominant means by which blades were struck from cores as indicated by the high proportions of their lipped bulbs. The high degree of control over the removal of blades also finds expression in their relative thinness and the high proportion of feathered terminations.

## Chronology and Regional Comparisons

Although the tool sample from Jebel Queisa (J24) is quite small, the three Helwan points and the Hagdud truncation all indicate an early Neolithic temporal placement for the occupation. In examining the chronology and distribution of Helwan points, Gopher (1989) traced their diffusion from the northern to the southern Levant over a period of some 700–900 years. Helwan points appear to have been most common in southern Levantine assemblages ca. 9,500

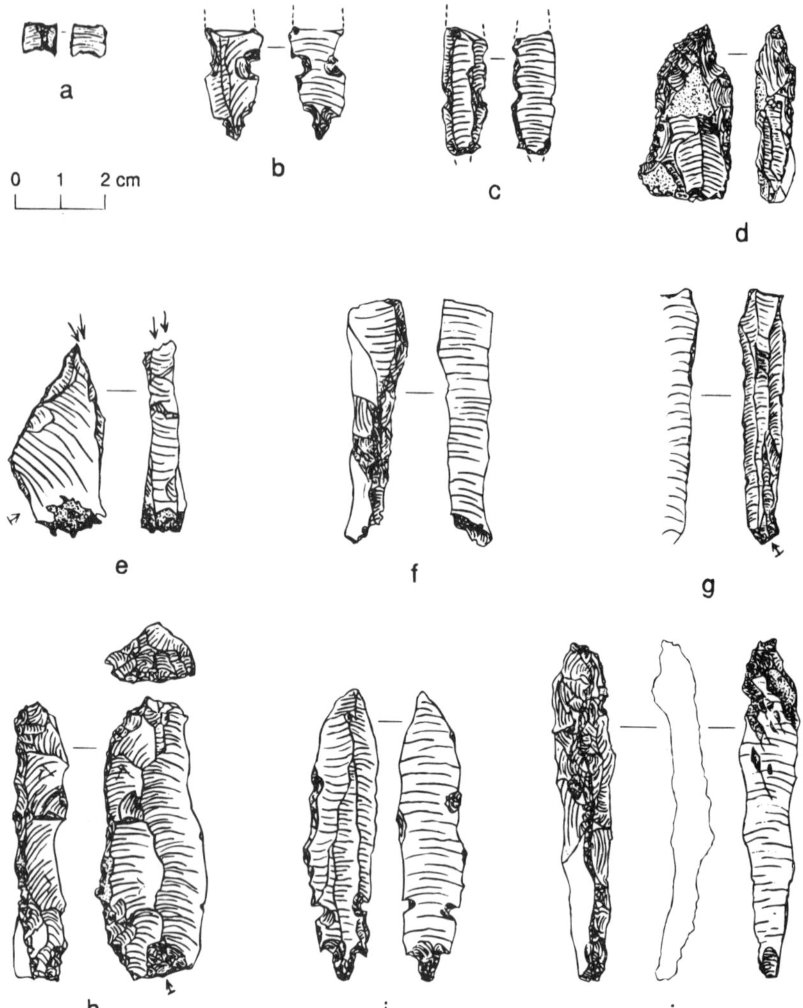


**Figure 14.3.** Illustrations of Early Neolithic artifacts recovered from Jebel Queisa (J24): (a) Hagdud truncation, (b,c,i) Helwan points, (d) perforator, (e) burin, (f,j) *lames à crête*, (g) retouched bladelet, (h) endscraper on blade.

B.P. Beyond recovery at the type site, Netiv Hagdud truncations have also been reported from the "Sultanian" horizons at Hatula and Jericho (Bar-Yosef *et al.* 1987; Lechevallier *et al.* 1989). This early Neolithic industry is dated to ca. 9,500–9,700 B.P. (Bar-Yosef and Belfer-Cohen 1989; Gopher 1989).

## SUMMARY AND CONCLUSIONS

The thin cultural deposit and small area of artifact distribution suggest that the Neolithic occupation of Jebel Queisa (J24) was quite ephemeral. Activities associated with core shaping, blank production, tool manufacture, use, and rejuvenation were all undertaken within the same

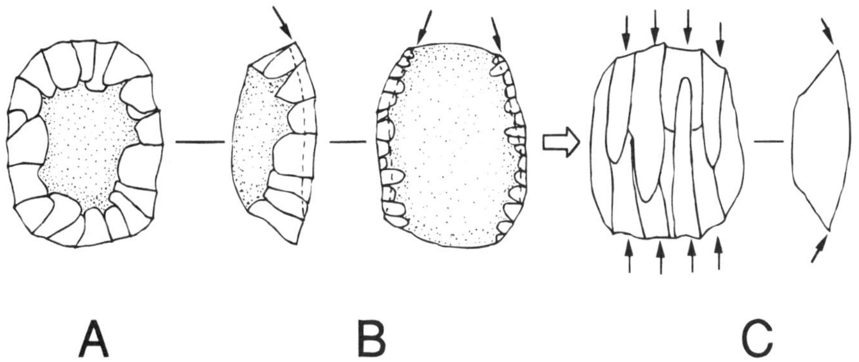


**Figure 14.4.** Schematic illustration of method of core reduction. Note that *lames à crête* are used to initiate blade production and that the core face is maintained through opposed removals.

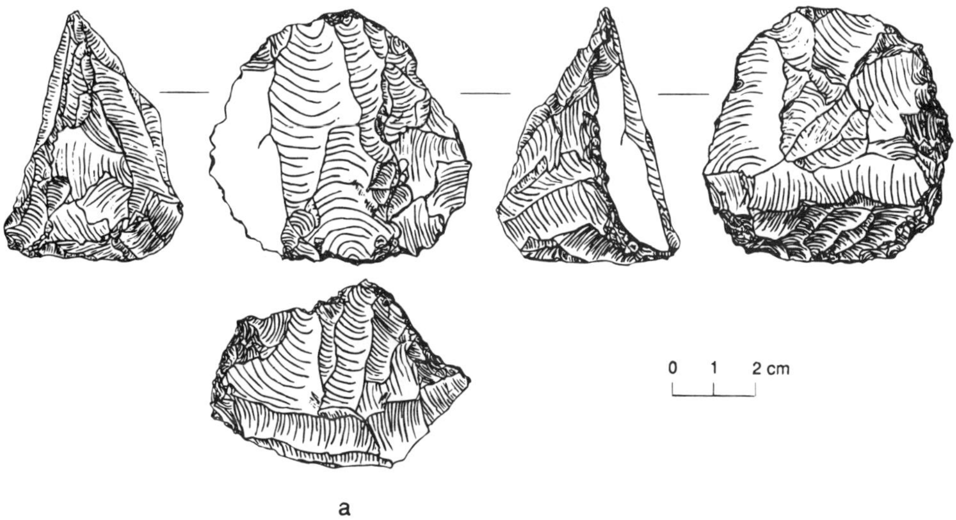


**Figure 14.5.** Illustration of Early Neolithic bladelet core recovered from Jebel Queisa (J24).

small area. The lack of spatial segregation of such activities implies that they were conducted as part of a continuous stream of related tasks and thus reflective of a rather brief interval of time.

The very small toolkit is dominated by points and casually retouched specimens. The large bifacial tools, sickle blades, and ground stone implements that are common to the large early Neolithic agricultural sites of the region are noticeably absent. The artifact configuration of Jebel Queisa (J2), coupled with the lack of a nearby permanent water source, suggests that the site served as a short-term exploitation camp, most likely a hunting station.

The paucity of Neolithic sites in the study may stem in part from the fact that few springs or permanent water sources were encompassed by the survey. In this regard, it is noteworthy that the nearby Neolithic sites of Ain Jamal (Gebel 1988) and Ain Abu Nekheileh (Kirkbride 1978) are located on springs. Although both of these sites fall well into the Pre-pottery Neolithic

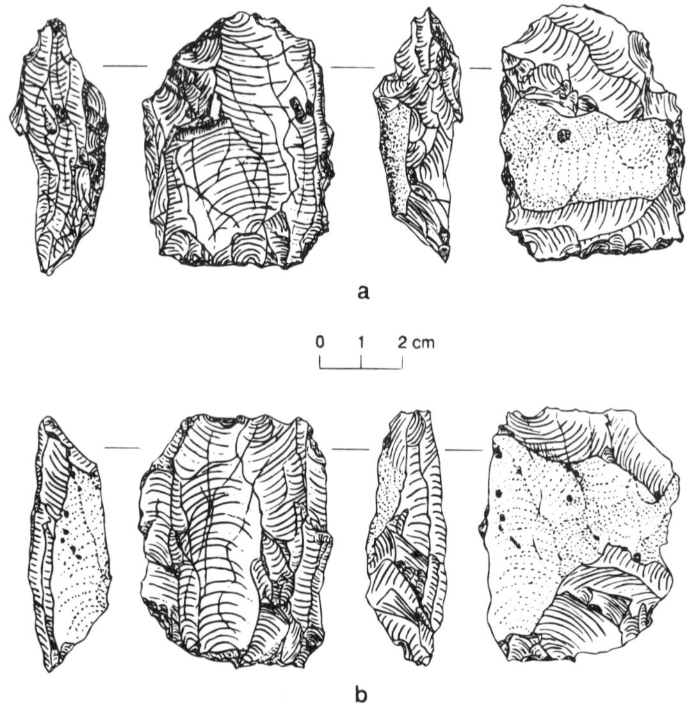


**Figure 14.6.** Illustrations of Early Neolithic bladelet cores recovered from Jebel Queisa (J24).

"B" period and thus are most likely to date somewhat later than Jebel Queisa (J24), similar constraints on Neolithic settlement may have been in place during the earlier Neolithic occupation of the area. Small, ephemeral occupations such as at Jebel Queisa (24) may have been common Neolithic site types within the region, but their residues would have been particularly vulnerable to removal or dispersion by the succeeding erosional intervals that appear to have characterized the Holocene (Chapter 2).

From a broader perspective, the occupation of Jebel Queisa (J24) provides another reference point for tracing the time–space distribution of Helwan points within the Levant. As denoted by Gopher (1989), the Helwan point type appears to have spread from northern Syria to the southern Levant between ca. 10,100–9,500 B.P. With the adoption of the point type, there also appears to have been a concomitant utilization of a new technology—one based upon blade production from opposed platform cores. Whether these technotypologic traits reflect the expansion of an interaction sphere and dissemination of ideas (Bar-Yosef 1989b) or a real expansion of human groups is indeterminable with the data at hand. In either case, these traits do confirm a clear break with the long Epipaleolithic succession of the southern Levant.

## *Chapter* 15

# *The Timnian and Pastoral Nomadism in the Chalcolithic*

Although the earliest evidence for the domestication of caprines (sheep and goats) in the Levant is associated with Neolithic occupations of the early Holocene (Clutton-Brock 1979; Ducos and Helmer 1981; Hecker 1982; Horwitz 1987; Hole 1989), it was not until very late Neolithic or even Chalcolithic times, some 5,000–7,000 years ago, that we see the first clear indications of populations that practiced pastoral nomadism (Gilead 1992; Rosen 1988; Bar-Yosef and Khazanov 1992; Koehler-Rollefson 1989, 1992). Depending upon their environmental settings, these groups displayed varying degrees of residential mobility and dependence upon the herding of caprines. Such patterns appear to have ranged from permanent settlements based largely upon a mixed economy of horticulture and animal husbandry to mobile settlements lacking horticulture and dependent upon the herding of sheep and goats. Hunting, especially of gazelles, and presumably gathering also appears to have contributed significantly to the economies of the fully nomadic pastoralists, at least as indicated by the evidence from southern Jordan. As might be expected, the more permanent settlements with their more diversified economies are found within the better watered areas of the Levant, whereas fully nomadic pastoralism is most often associated with the exploitation of the driest portions of the region.

## CULTURAL HISTORIC CONSIDERATIONS

Given its relatively late placement within the Levantine prehistoric succession, the cultural historic framework of the Chalcolithic is surprisingly poorly understood. Regionally, the southern Levant offers the most complete picture of the temporal-spatial patterns observed in material culture and economy (Gilead 1988). The Levantine Chalcolithic has been viewed as chronologically equivalent to the fourth millennium B.C., but Gilead (1988:399–405), in his recent summary of calibrated radiocarbon dates, traces the interval from as early as 4,500 B.C. to as late as 3,200 B.C.

The best defined of the southern Levantine complexes is the Ghassulian. After initial discovery at Teleilat Ghassul (Mallon *et al.* 1934), located in the Jordan Valley northeast of the Dead Sea, the complex was subsequently found to extend to sites in the western Negev (Macdonald 1932) and the Beersheva region (Perrot 1955). While early researchers argued that the Ghasullian was intrusive to the southern Levant, recent research centered on Negev sites points to indigenous development within the region (Gilead and Alon 1988; Gilead 1988). Ghasullian settlements typically consist of unplanned household clusters forming relatively small villages. These are formed mainly of rectilinear mudbrick structures, storage pits, courtyards, and retaining walls. Artifact assemblages are characterized by painted pottery and chert tools including sickle blades, fan scrapers, bladelets, and axes.

In contrast to the Ghassulian and related Chalcolithic village sites, the Timnian Complex consists of small, ephemeral hamlets and seasonal camps found in the deserts of Sinai (Kozloff 1974, 1981) and southern Jordan (Henry 1982a, 1988a, 1992b). The architecture of the hamlets is curvilinear in form and includes unhewn stone pithouses, corrals, retaining walls, and storage pits. Pottery, although uncommon, consists of a thick plain ware often tempered with chert fragments. Timnian lithic assemblages also differ substantially from those of the Ghassulian in the much greater emphasis on microliths (especially lunates) and an absence of Ghassulian blades.

## PATTERNS OF ECONOMY AND SETTLEMENT

Relative to Chalcolithic economic patterns, scholars have come to focus on two interconnected problems. One concerns defining the ties between different economic strategies, mobility levels, and environmental settings, while the other involves tracing the origin and evolution of such strategies (Bar-Yosef and Khazanov 1992). In the southern Levant (mostly from areal studies in the Negev), three economic strategies have been identified and correlated with different points along a continuum from permanent to mobile settlements. These consist of two village-based patterns, both in part dependent upon horticulture, that differ in degree of residential mobility. The third pattern is associated with high levels of mobility in which villages and horticulture are absent. The village-based systems with horticulture fall under Khazanov's (1984) category of "semi-nomadic pastoralism," whereas the most mobile system, lacking villages and horticulture, fits within his category of "nomadic pastoralism."

### Diverse Agricultural, Permanent Village Pattern

In the better watered areas of the Negev, village sites accompanied by agricultural economies based upon cereal cultivation (along with other crops) and animal husbandry (including pigs, cattle, and caprines) have been described as "village-based pastoral" (Levy 1992) and "peasant (Fellahin) village" (Gilead 1992) communities. Such communities are thought to have been occupied year-round by a relatively stable number of residents. The herding of sheep/goats from sedentary communities that were dependent upon mixed agricultural economies would have involved moving the flocks to nearby pastures on a daily or short-term basis. While such a pastoral pattern would have produced special-purpose "herding stations," such as described by Gilead (1992), longer-term residential camps would not be expected in such a settlement system.

### Agricultural, Semipermanent Village Pattern

In drier settings of the Negev, village sites that largely relied upon cereal cultivation and the herding of caprines have been interpreted as a "village-based transhumance" (Levy 1992). Levy (1992:76) argues that growth in population accompanied by the emergence of hoe-based horticulture prompted the "shift from village based pastoralism to a full transhumance system." Although in part dependent upon cereal cultivation, the pastoral component of this strategy focussed upon the herding of caprines; pigs were absent and cattle poorly represented. Levy (1992:76) sees this emphasis on sheep and goats as an indication of the greater human control over herds and a "more intensive indication of their secondary products (milk, wool, hair, and possibly blood)."

### Nomadic Pastoral Pattern

In contrast to the more sedentary patterns, nomadic pastoralism has not been well defined archaeologically within the Levant. This may stem in part from the ephemeral nature of such sites and their general lack of clarity within the archaeological record. While Chalcolithic sites located in the central Negev and Sinai are suggestive of a nomadic pastoral pattern (Gilead 1992; Rosen 1988; Kozloff 1974, 1981), these sites have not yielded detailed evidence on the economy and settlement structure of their inhabitants. The sites reported upon here, from the study areas of southern Jordan, not only furnish a reasonably clear understanding of settlement structure but also provide substantial information on economy. The evidence from these sites and the associated reconstructions may have only local significance. But the strong affinities in architecture and artifact inventories of these sites in southern Jordan with those described for the Timnian occupations of nearby northeastern Sinai imply that fully nomadic pastoralism was a region-wide phenomenon across the arid zone of the Levant during the Chalcolithic.

Data recovered from the southern Jordan Timnian sites point to a settlement structure that largely resembles the pattern of transhumance followed by historic Bedouin groups inhabiting the region (Henry 1992b). This entailed an annual cycle in which groups occupied relatively large, long-term seasonal camps during the winter in the piedmont zone followed by some 9–10 months of population dispersal into much smaller, more mobile camp groups during the dry season. The dry season segment of their transhumant cycle was initiated by movement to spring pastures at lower elevations on the floor of the broad Wadi Hisma, but as the season advanced groups reversed this movement and progressively settled at higher elevations in search of surface water and grazing areas. This eventually took them to the uplands on the edge of the plateau where perennial springs and late season pastures would have furnished critical resources. With the coming of winter rains and snow, groups again moved to the lower elevations of the piedmont where they enjoyed more comfortable temperatures while at the same time having predictable water sources from upland runoff.

Although the transhumant settlement pattern indicated by the distribution, context, and artifact content of the Timnian sites closely parallels that of the historic pastoralists of the area, the economic patterns of the two intervals differ in some important ways. The historic Bedouin economy, resembling the "agricultural, semipermanent village" pattern, relied upon the herding of sheep and goats and was complemented by cereal cultivation. Although the Timnian economy also emphasized the herding of caprines, it relied heavily upon gazelle hunting and apparently lacked horticulture. Thus in many ways the subsistence strategy of Timnian groups incorporated elements of both foraging and pastoralism.

## ORIGIN AND EVOLUTION OF PASTORALISM

When examining the problem of the origin of the domestication of caprines, many scholars argue that the cultivation of plants (especially cereals) would have necessarily preceded the tending and herding of animals (Reed 1977; Ingold 1980; Khazanov 1984; Hole 1992). This view is based upon the observation that the sedentary nature of horticultural villages would have led to the depletion of prey species and thus triggered efforts to better manage meat and perhaps even "secondary product" (milk, wool, blood) resources through the herding of sheep and goats. This emphasis upon human responses to the depletion of resources (e.g., overgrazing, competition with horticulture) is also used to explain the subsequent transition from seminomadic to nomadic pastoralism (Khazanov 1984; Bar-Yosef and Khazanov 1992; Koehler-Rollefson 1992; Levy 1992).

These arguments are certainly supported by the chronology of economic strategies, as we know them, in the Levant. Evidence for plant domestication does precede the earliest firm evidence for animal domestication by some 1,500 years (Bar-Yosef and Khazanov 1992; Helmer 1989). Although based upon more tenuous evidence, seminomadic pastoral patterns appear to precede fully pastoral patterns in both central Jordan and the Negev (Koehler-Rollefson 1992; Levy 1992). These diachronic trends and the dependence of recent nomadic pastoralists on mounted and pack animals prompt Bar-Yosef and Khazanov (1992) to question whether "pure pastoralism" ever existed in prehistory.

The evidence from the Timnian sites not only confirms that fully nomadic pastoralism did exist in prehistory, at least some 6,000 years ago, but it also defines an economic strategy that combined foraging with the herding of sheep and goats. Timnian data thus suggest that the traditionally held notion of a unilineal development of caprine domestication and the subsequent dependence upon herding by fully nomadic pastoralists needs rethinking. It may well be that the Timnian strategy which complemented foraging with the managing of caprines represents a segment of an alternative evolutionary path that characterized the arid zone of the region. This succession from forager to mobile herder may not have involved the intervening stage of settled horticulturalists.

## VARIETIES AND DISTRIBUTIONS OF SITES

Twenty-four Chalcolithic sites were recorded in the study. Sites were found in each of the three major physiographic units (i.e., lowlands, piedmont, and uplands) of the region with all but the Jebel Hamra study area furnishing occupations. The highest density of sites was situated in the Judayid Basin (Figure 15.1).

### Site Types

Two fundamentally different types of sites were identified: ephemeral camps and long-term camps (Table 15.1). The ephemeral camps display substantially smaller areas, thinner deposits, and lower artifact densities than long-term camps. A much wider range of artifacts are found in the long-term camps. Pottery sherds, portable ground stone items, and worked bone tools are common to most of the long-term camps, where they are rare or absent in the ephemeral camp inventories. Moreover, all but one of the long-term camps display some form of architecture, typically represented by oval stone-walled structures, while architecture is absent from the ephemeral camps.

### Distributional Patterns and Resources

When the distributions of these two site types are compared, they show quite different patterns relative to their natural settings. The long-term camps are found only in the piedmont within an elevational belt ranging from ca. 1,100–1,250 masl. Ephemeral camps, on the other hand, are found in lowland, piedmont, and upland settings ranging in elevation from just under 800 masl to over 1,500 masl. Most of the ephemeral sites were discovered in the lowlands on the floor of the Wadi Hisma, however.

In that our environmental reconstruction for the time-frame points to an arid or hyperarid setting similar to that of today, surface water for the lowlands would have been restricted to the winter wet-season and early spring. In being closer to the source of upland runoff, the piedmont would have enjoyed surface water until late spring, while perennial springs on the edge of the plateau would have furnished water from limited locations throughout the year.

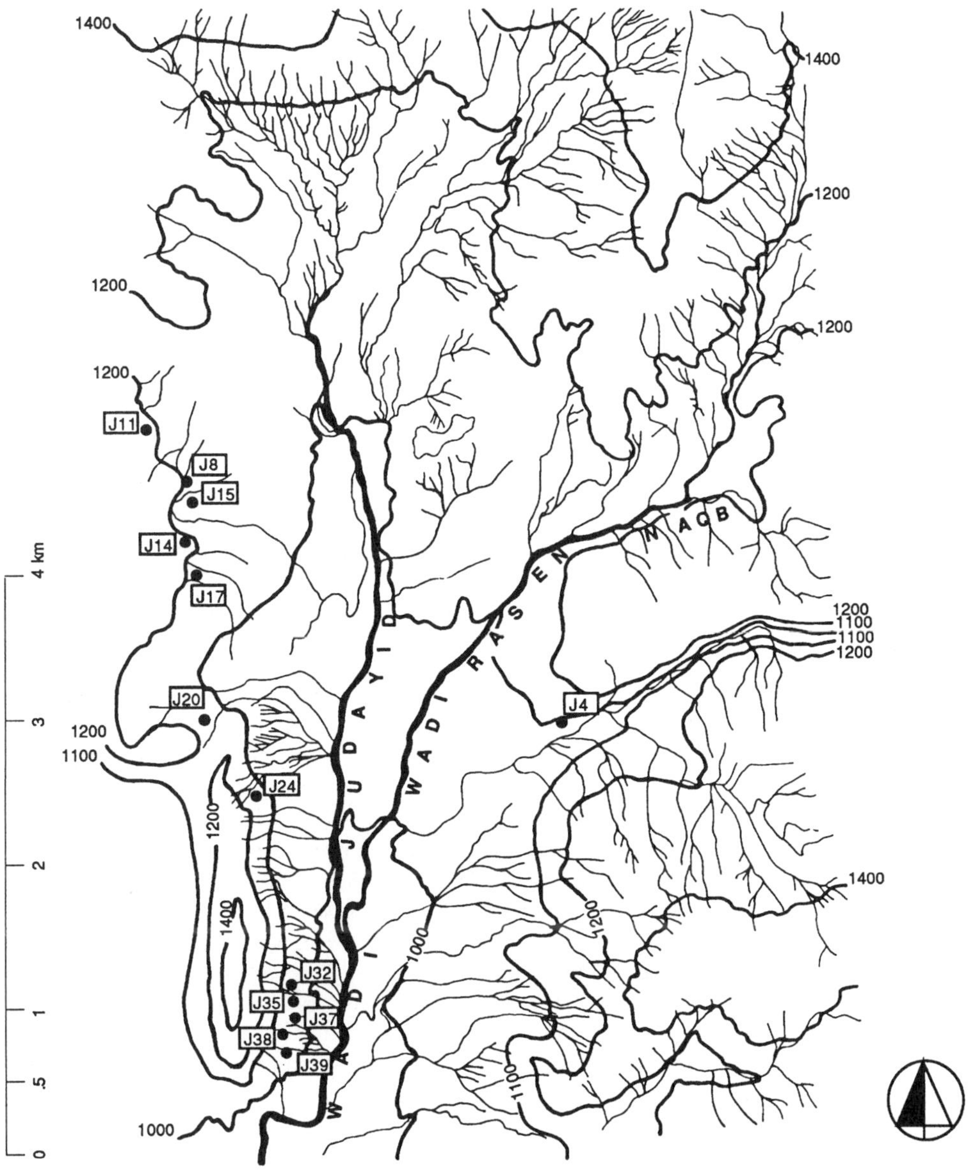


**Figure 15.1.** Map of Judayid Basin showing locations of Neolithic and Chalcolithic sites. Jebel Queisa (J24) is a stratified Chalcolithic–Early Neolithic site. The other sites are Timnian occurrences with all but J8 and J20 displaying architecture.

## PALEOENVIRONMENTAL EVIDENCE

Geomorphological (Chapter 2, Henry *et al.* 1983) and palynological (Emery-Barbier, Chapter 16 1988; Henry *et al.* 1985) studies suggest that the area experienced a warm and dry climate during Chalcolithic times. Pollen spectra from the Chalcolithic deposits at the sites of Jebel Queisa (J24), Jebel el Jill (J14), and Tor Sabiha (J8) are consistent in being dominated by

Table 15.1. Comparison of Long-term and ephemeral Timnian Camps According to Certain Site and Artifact Characteristics

| | Long-term camps | Ephemeral camps |
|---|---|---|
| Physiographic Zone & Elev. (masl) | Piedmont (1,100–1,300) | Hisma Floor (800) |
| Mean Area ($m^2$) | 3,692 | 125 |
| Depth of Cultural Deposit (cm) | 90 | 10 |
| Architecture | Pithouses, Corrals, Storage Pits | None |
| Ground Stone | Common: Querns, Slabs, and Mullers | Absent/Rare ?Bedrock Mortars |
| Pottery | Common: Large Storage and Cooking Vessels | None |
| Mean Lithic Artifact Density (per/0.1$m^3$) | 115 | 29 |

semidesert, halophilic flora, mostly represented by Chenopodiaceae. Interestingly, cereal-type pollen grains are absent or extremely scarce in these spectra. Hassan's (Chapter 2, Henry *et al.* 1983) findings in his geomorphic investigation of the Judayid Basin are consistent with the palynological interpretation. They indicate an episode of erosion of the basin's floor from mid-Holocene to Roman times. This episode was accompanied by the deposition of drift sand enclosing Chalcolithic occupations along the margins of the basin.

Regional evidence for climatic environmental reconstructions of the early and mid-Holocene lack agreement (Goldberg and Rosen 1987; Henry 1992b). These apparent inconsistencies in evidence are likely to stem from a number of factors including the poor resolution of chronologies, intraregional variability in climate and environment, lag times between climatic changes and their expressions in geologic and vegetational records, and geomorphic changes (e.g., tectonics, base level) that are unrelated to climate. In reviewing multiple lines of evidence and considering these potential biases, Goldberg and Rosen (1987:29–30) conclude that a relatively wet interval persisted from ca. 9,000–7,000 B.P. followed by relatively dry conditions which stretched from ca. 7,000–5,500 B.P. This setting was replaced by more moist conditions.

Interestingly, if the twenty $C^{14}$ dates for the Timnian are compared to the climatic reconstruction of Goldberg and Rosen (1987), the early cluster of dates displays a temporal sweep that is near identical to the 7,000–5,500 B.P. dry interval (Henry 1992b; Figure 15.2). Following this cluster there appears a hiatus of almost 1,000 years before the second cluster that stretches from ca. 4,440–4,000 B.P. Given our available evidence, it is difficult to know if the correlation between the archaeological evidence of the Timnian and dry intervals is an expression of cultural or natural factors.

## STRATIGRAPHIES, ARCHITECTURE, AND DATING

The ephemeral camp sites of the lowlands and lower piedmont display very shallow (<20 cm thick) or deflated deposits. In contrast, the occupations of the long-term camps of the higher piedmont rest in much thicker (80–110 cm) cultural deposits and are typically associated with stone-lined hearths, thick ash beds, and architecture. In the Judayid Basin, these deposits are formed of weakly weathered reworked drift sand that overlies the red Epipaleolithic age sands of the Member II alluvial fill of the basin (Chapter 2). At Jebel el Jill (J14), a Chalcolithic horizon in an 80 cm thick sand unit was found to overlie a rubified sand presumably of Epipaleolithic age. The Chalcolithic horizon of Jebel Queisa (J24) was exposed in a meter-thick unit of drift sand overlying a lithologically similar unit containing an early Neolithic horizon. The cultural

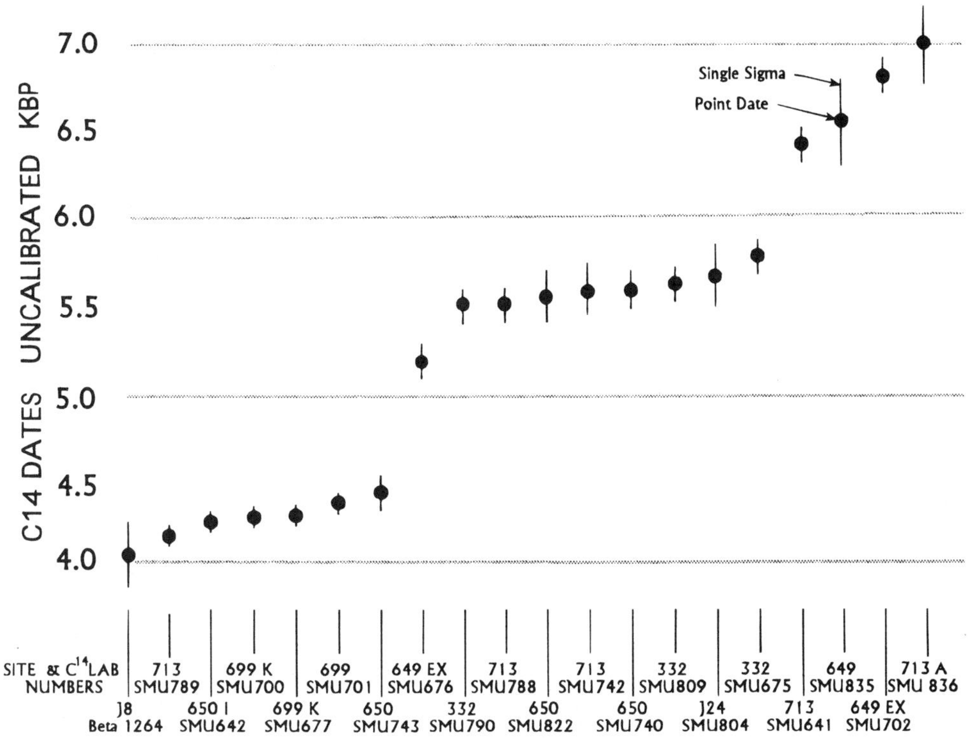


Figure 15.2. Distribution of radiocarbon dates for the Timnian Industry.

horizons of Jebel Queisa (J24) appear to be separated by an erosional hiatus of ca. 3,000–3,500 years. Probes at other long-term camp sites revealed thick cultural deposits formed of drift sand, but failed to expose earlier natural strata or cultural horizons.

## Jebel el Jill (J14)

The site is situated at 1,200 masl in a saddle between sandstone promontories at the base of Jebel el Jill. Artifacts are distributed over an erosional area on the southern slope of the saddle in association with circular stone walls (Figure 15.3). Overall the site covers an area of ca. 1,400 $m^2$.

Eight units, excavated along a transect through the site, traced the thickening of the cultural deposit from south (20–30 cm) to north (50–80 cm). A single layer consisting of a light brownish gray (10YR 6/2) sand was identified in Locus 1, while alternating ash layers and a compacted floor were exposed in Locus 7 (Figure 15.4).

The large structure of Locus 1, some 10 m in diameter, may well have served as a corral. Its relatively thin, homogeneous deposit and very low density of artifacts restricted to chert specimens is inconsistent with a storage or habitation structure. Excavation of Locus 7 revealed a semisubterranean pithouse lined with stone. An unlined firepit and several grinding stones were exposed in the southern half of the structure along with various chert, bone, and ceramic artifacts. The alternating ash lenses separated by sand suggest episodes of abandonment, whereas the compacted floor is most likely to have been formed during the winter wet season.

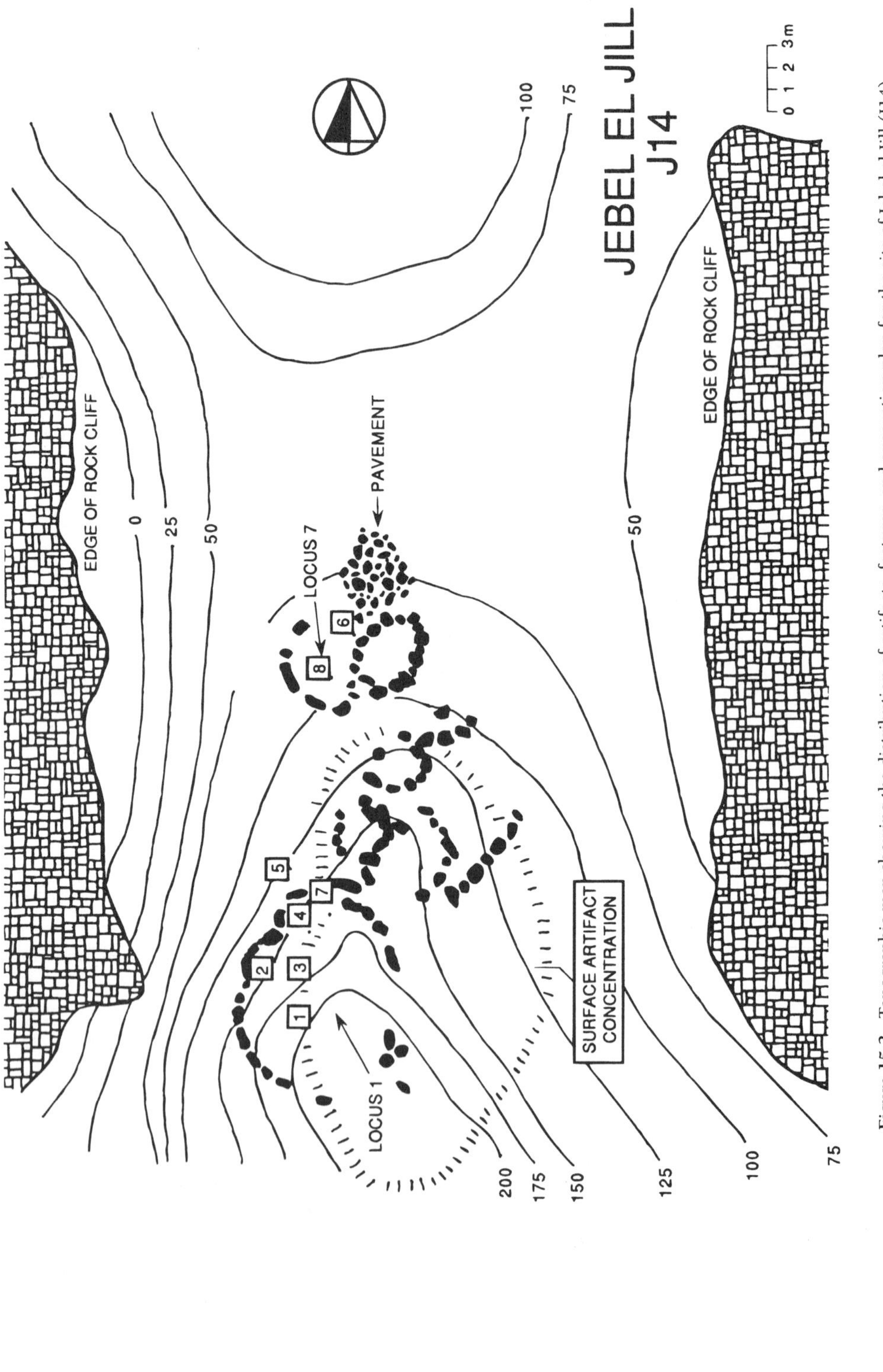


Figure 15.3. Topographic map showing the distribution of artifacts, features, and excavation plan for the site of Jebel el Jill (J14).

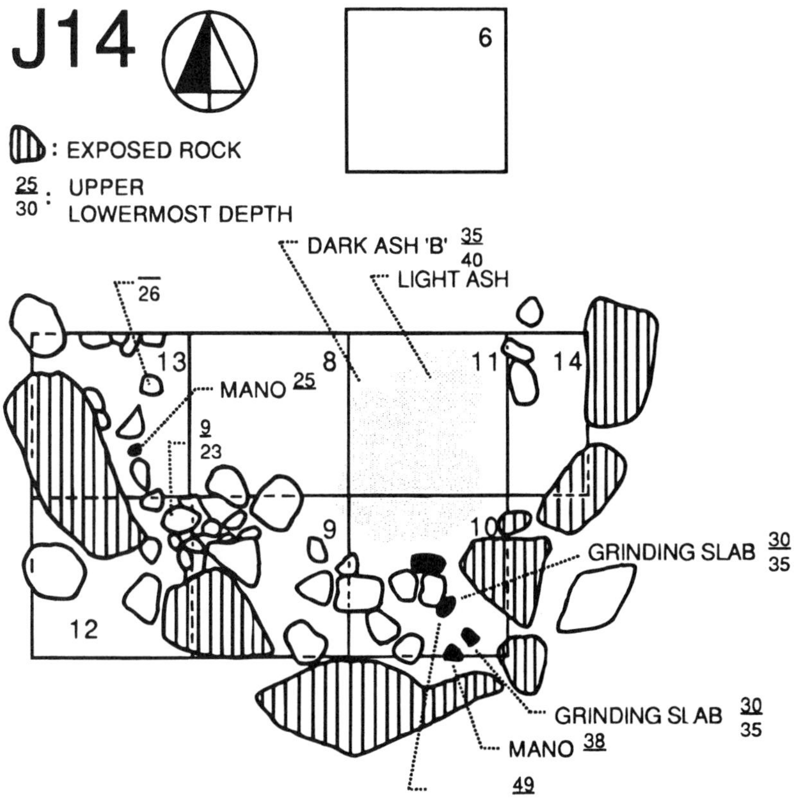


Figure 15.4. Plan and stratigraphy of the Locus 7 pithouse at Jebel el Jill (J14).

## Jebel Queisa (J24)

The site, at 1,200 masl, is located on a narrow terrace that fronts a low cliff on the northeastern edge of Jebel Queisa. Artifacts and remnants of stone walls are exposed along the steep slope that drops away from the terrace (Figure 14.1). Following excavation of three test units (1–3) in 1979, a block excavation was undertaken near the cliff face the following year.

These excavations revealed three stratigraphic layers within a 120 cm thick deposit of drift sand (Figure 14.2). The Chalcolithic horizon is confined to the upper two layers (A and B), defined by different concentrations and colors of ash, along with relatively high densities of artifacts. Several stone-lined hearths also were found within these strata. A radiocarbon date of 5,720 ± 149 B.P. (SMU 804) was obtained from one of these in lower Layer B, just above the layer B/C contact.

## Other Sites

Other long-term sites, all with thick cultural deposits and evidence of architecture, include J11, J17, J32, J35, J37, J38, J39, and J40. These are found within an elevational belt ranging from 1,100–1,250 masl along the eastern slopes of Jebel el Jill and Jebel Queisa. All display curvilinear stone-walled structures. The numbers of structures within the sites show a bimodal distribution, however, with one set of sites displaying 2–4 structures and another some 10–15. The walls of the structures in these large sites are interconnected and form a "nested" pattern, as at Jebel el Jill.

Many of the architectural sites have been probed by treasure seekers and pot-hunters. Examination of the pits and backdirt from these diggings shows a low density of artifacts in conjunction with quite thick ash layers, a pattern also found in our excavations.

## ARTIFACTS

The sites, especially the long-term camps, yielded a wide range of artifacts including chipped stone, ground stone, ceramic, worked bone, and ornamental shell specimens.

### Chipped Stone Industry

The lithic assemblages are characterized by a dichotomy in blank production that includes microlithic flakes and bladelets struck from globular *sur enclume* cores as well as large flakes produced from tabular chert. At a typological level, the microliths served as blanks for the fabrication geometrics and minute backed elements, whereas the large flakes were used mainly in the production of tabular scrapers (Table 15.2).

Aside from these technotypological characteristics, the assemblages also have a curious signature as expressed in the blockish and angular form of debris and debitage. This may be derived in part from the intensive use of a "block-on-block" technology and hard-hammer percussion. In that the cores associated with this technique often display high angles between their platforms and faces, blanks generated from *sur enclume* cores will typically be thick and, when mis-struck, angular and blockish. The lack of precision of the technique also generates a relatively high degree of shatter and fragmentation of cores. Another factor that may have contributed to the unusual morphological signature of the debitage and debris is the high incidence of thermal fracturing. Evidence for burning as denoted by fire-crazing and negative pot lids are quite common to the assemblages.

#### Typology

Given that the assemblages contain relatively small tool samples, the toolkits display a surprising diversity of classes. While they are dominated by scrapers, nongeometric microliths, and retouched pieces, several other classes (e.g., burins, geometrics, denticulates, notches, and points) are regularly present.

*Scrapers.* Although typical endscrapers on flakes and blades occur within most assemblages, tabular scrapers compose the large majority of the class and represent a typological marker for the period and industry. They appear on large, relatively thin and flat flakes which retain much of their cortex. Whereas heavy retouch typically is found along the entire perimeter of specimens (excluding their platforms), the angle of retouch and the configuration of the retouched edge show a distinctive pattern (Figures 15.5:a,b,c,e–h; 15.6:g). The retouch along one lateral edge is invasive, creates an acute angle with the inverse surface, and forms a convex profile. In contrast, retouch along the other edge is abrupt, creates a near 90° angle with the inverse surface, and forms a straight profile.

*Burins.* The class lacks patterning with dihedral and angle burins struck from truncations predominating (Figure 15.6:h).

*Nongeometric Microliths.* Significant numbers of abruptly retouched small flakes and bladelets occur within most of the assemblages, but these lack patterning in terms of the orientation of retouch or form. Most, in fact, are fragments of flakes or bladelets.

Table 15.2 Artifact Inventories for Timnian Assemblages

| Assemblage | J11 | J14 Locus 1 Corral | J14 Locus 7 Pithouse | J17 | J20 | J24 | J408 | J521 |
|---|---|---|---|---|---|---|---|---|
| *Tools* | | | | | | | | |
| Scraper | 3.0 | 0.0 | 29.1 | 40.0 | 20.0 | 20.0 | 11.9 | 11.8 |
| Burin | 0.0 | 0.0 | 1.8 | 0.0 | 0.0 | 9.0 | 2.2 | 8.9 |
| Geometric | 3.0 | 0.0 | 7.3 | 0.0 | 0.0 | 8.5 | 2.2 | 13.6 |
| Nongeom | 39.3 | 0.0 | 9.1 | 0.0 | 6.6 | 2.8 | 57.0 | 21.4 |
| ABR-RET | 0.0 | 0.0 | 0.0 | 0.0 | 0.0 | 0.0 | 0.0 | 16.1 |
| Truncation | 0.0 | 0.0 | 1.8 | 0.0 | 0.0 | 2.8 | 0.0 | 3.6 |
| Retouched Pc. | 36.3 | 0.0 | 25.5 | 60.0 | 53.3 | 28.5 | 9.7 | 14.3 |
| Denticulate | 3.0 | 0.0 | 0.0 | 0.0 | 6.6 | 8.5 | 0.0 | 0.0 |
| Notch | 3.0 | 0.0 | 7.3 | 0.0 | 6.6 | 14.2 | 5.4 | 3.6 |
| Multiple Tl. | 0.0 | 0.0 | 0.0 | 0.0 | 0.0 | 0.0 | 0.0 | 0.0 |
| Perforator | 0.0 | 0.0 | 12.7 | 0.0 | 0.0 | 0.0 | 0.0 | 0.0 |
| Utilized Pc. | 9.0 | 0.0 | 1.8 | 0.0 | 6.6 | 2.8 | 0.0 | 0.0 |
| Points | 3.0 | 0.0 | 0.0 | 0.0 | 0.0 | 11.4 | 0.0 | 0.0 |
| Varia | 0.0 | 0.0 | 3.6 | 0.0 | 0.0 | 0.0 | 11.8 | 9.8 |
| Total No. | 33 | 0 | 55 | 5 | 15 | 35 | 93 | 112 |
| *Debitage* | | | | | | | | |
| Core | 5.6 | 0.0 | 0.3 | 1.0 | 5.6 | 1.8 | 1.9 | 1.8 |
| Blade | 13.0 | 7.2 | 23.6 | 13.0 | 25.8 | 15.7 | 58.6 | 54.8 |
| Flake | 77.2 | 83.4 | 75.0 | 78.0 | 61.7 | 77.2 | 33.4 | 30.1 |
| Core Trim. El. | 4.1 | 0.0 | 1.0 | 2.0 | 0.0 | 2.2 | 2.0 | 1.5 |
| Primary El. | 0.0 | 9.4 | 0.1 | 6.0 | 6.7 | 3.0 | 8.3 | 9.6 |
| Microburin | 0.0 | 0.0 | 0.0 | 0.0 | 0.0 | 0.0 | 2.1 | 1.6 |
| Burin Spall | 0.0 | 0.0 | 0.0 | 0.0 | 0.0 | 0.0 | 0.4 | 0.6 |
| Total No. | 338 | 181 | 1226 | 100 | 89 | 496 | 991 | 947 |
| *Debris* | | | | | | | | |
| Chips | 82.6 | 59.2 | 88.0 | 75.2 | 63.3 | 83.1 | 94.9 | 95.9 |
| Chunks | 17.3 | 40.8 | 12.0 | 24.8 | 36.6 | 16.8 | 5.1 | 4.1 |
| Total No. | 1269 | 814 | 3320 | 230 | 142 | 871 | 1828 | 1638 |
| *Total Artifacts* | 1640 | 995 | 4601 | 230 | 246 | 1402 | 2912 | 2697 |

*Geometric Microliths.* Most of the toolkits contain low to moderate proportions of geometrics, composed of lunates and, more rarely, trapeze/rectangles (Figure 15.6:c–f). The lunates are small and fashioned by abrupt or bipolar retouch. Rosen (1983a) has noted that such small lunates serve as a chronological marker for Early Bronze I/II horizons falling near the end of the fourth millennium BC. Their presence in some abundance within the radiocarbon dated Timnian assemblages of south Jordan and Sinai pushes this marker back to the beginning of the fourth millennium or, when calibrated, even the mid-fifth millennium BC.

*Utilized Pieces.* The class is not well represented, but there is a real consistency in the typology of its constituent specimens. These include rather thick flakes that display steep bilateral retouch. Accompanying the retouch is evidence of bifacial battering and crushing (Figure 15.6:k).

*Points.* These consist of small partial bifacially retouched specimens with contracting stems and a few triangular transverse arrowheads (Figure 15.6:b).

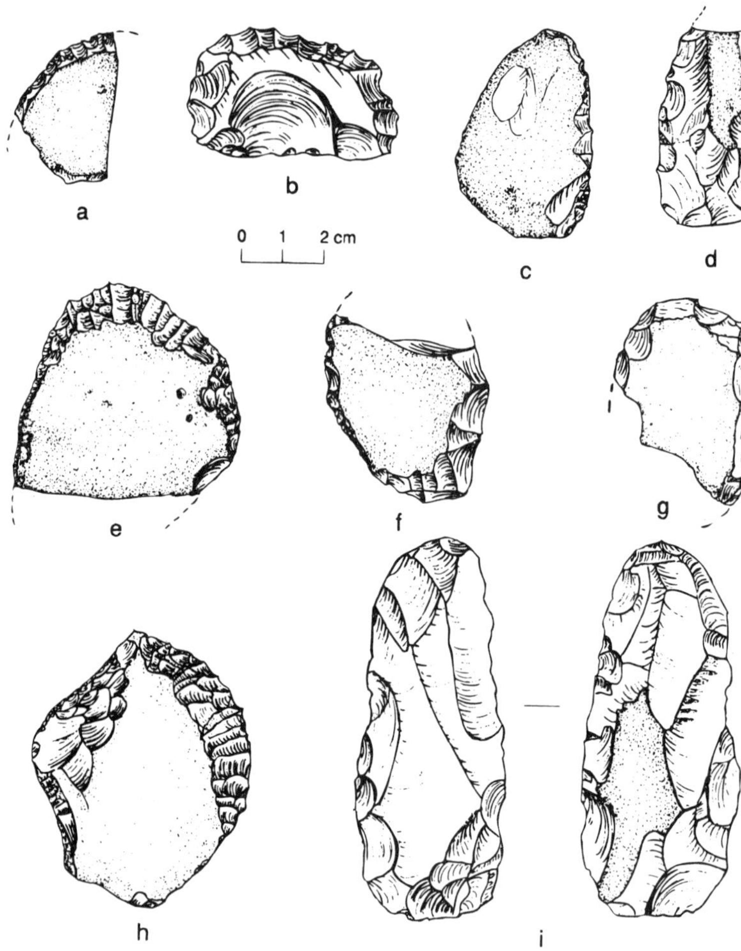


**Figure 15.5.** Illustrations of Timnian chert tools: (a–c,e–h) tabular scrapers; (d,i) elongated bifaces.

## Technology

Blank production followed two quite distinct reduction steams. One was devoted to generating microlithic flakes and bladelets from subpyramidal and globular cores, whereas the other focussed on the creation of large, flat tabular flakes from *in situ* exposures of chert. The microlithic cores are formed exclusively from a fine grained chalcedony, whereas the large flakes occur on either the chalcedony or coarser grained cherts.

The cortex displayed on the tabular flakes and other large flakes is exclusively limestone; none shows gravel or water-worn cortical cover. This indicates that either *in situ* beds of chert were exploited at the sources or chert was transported from these sources to the occupations for further reduction. The low proportion of primary elements in the assemblages, coupled with their small dimensions, points to flake production at the sources of the chert and not at the sites.

Given the ease and simplicity by which large tabular flakes can be struck from the exposed beds of chert and chalcedony along the Ras en Naqb escarpment, it would have made very little sense for the Chalcolithic occupants of the sites to have portaged large chunks back to their

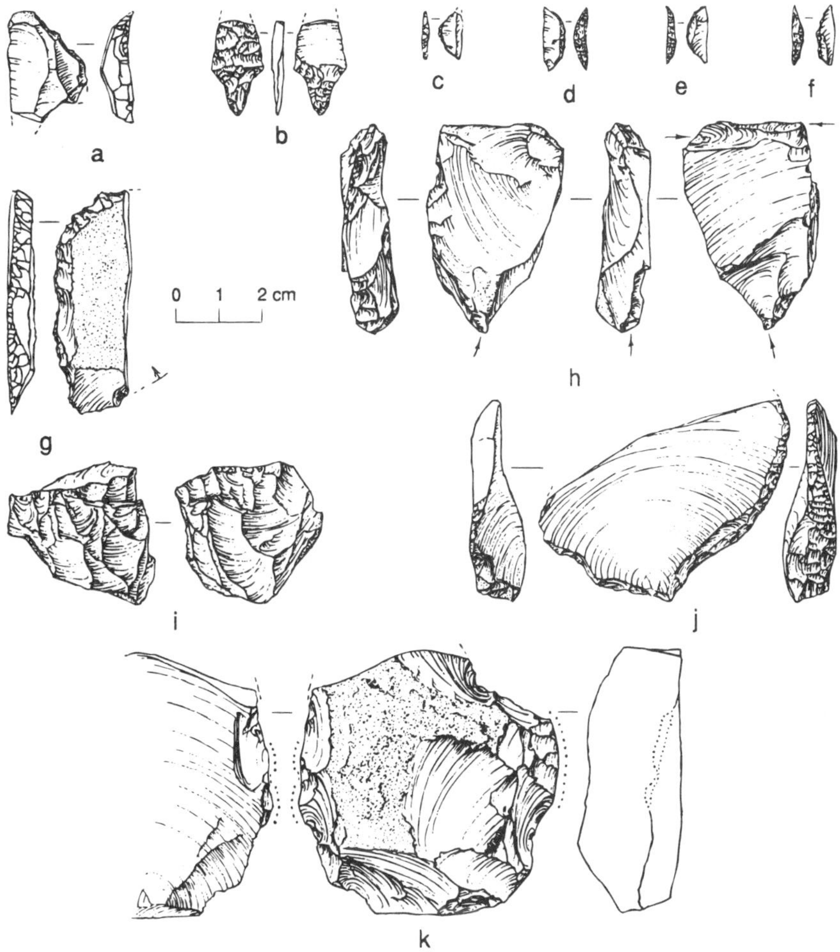


**Figure 15.6.** Illustrations of Timnian artifacts: (a) truncation, (b) point, (c–f) abrupt-backed lunates, (g) tabular scraper, (h) multiple burin, (i) bladelet core, (j) backed flake or broken scraper, (k) thick utiilized flake.

encampments. This would have been especially true for chert, as opposed to chalcedony acquisition, since the chert sources appear to have been targeted for tabular flakes and not for a full range of blanks. Thus with little to no initial processing and an absence of core shaping, the desired blanks (tabular flakes) could have been struck from the natural chert outcrops along the edge of the plateau and then more easily carried back to the sites.

## Bone Tools

Ten worked bone specimens were recovered from Jebel el Jill (J14) all coming from the pithouse structure of Locus 7 (Figures 15.7 and 15.8). All of the pieces are pointed, having been shaped from the broken shafts of small long bones. The points display two distinct forms that depend upon the degree to which the fractured margin of the bone has been modified. The weakly modified forms exhibit rather substantial but not acutely pointed tips (Figure 15.7:a–d), whereas the heavily modified specimens have near needle-like, delicate tips (Figure 15.8:d–g).

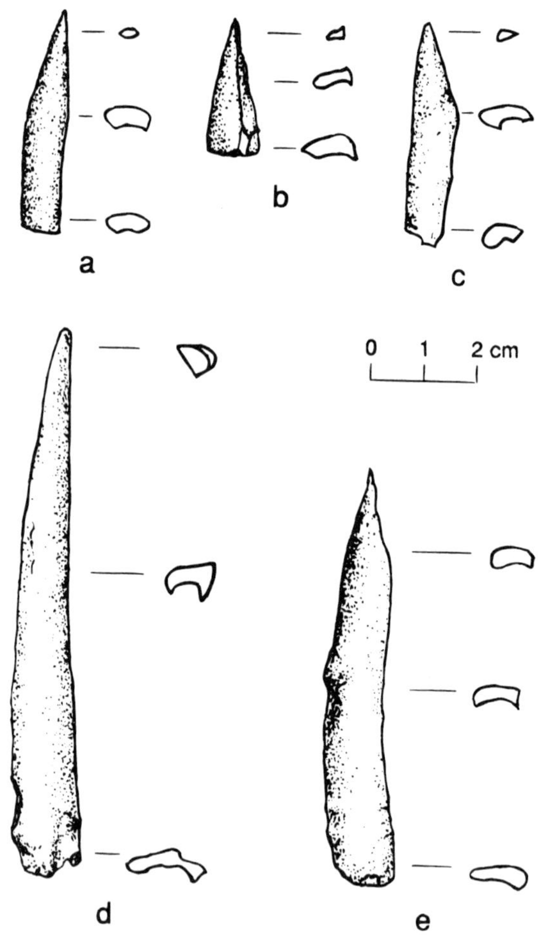


Figure 15.7. Illustrations of worked bone points from Timnian sites.

## Ground Stone Specimens

Nine ground stone specimens were also recovered from Locus 7 of Jebel el Jill (J14) (Figure 15.9). All of the specimens were fashioned from a coarse-grained sandstone that could have been obtained from outcrops immediately adjacent the site. Although the grinding surfaces of the specimens have been worn smooth and in some cases even polished through use, the overall shape of the implements was formed through pecking. Relatively large, deep peck marks are evident on all of the modified surfaces.

The specimens consist of both elements of a grinding complex: hand stones or mullers and querns or grinding slabs. Four of the elements clearly are hand stones, whereas three represent grinding slabs. Two specimens are too fragmentary for classification. It is noteworthy that the two large broken grinding slabs show signs of use following their breakage. The fractured edges have been rounded and smoothed from abrasion.

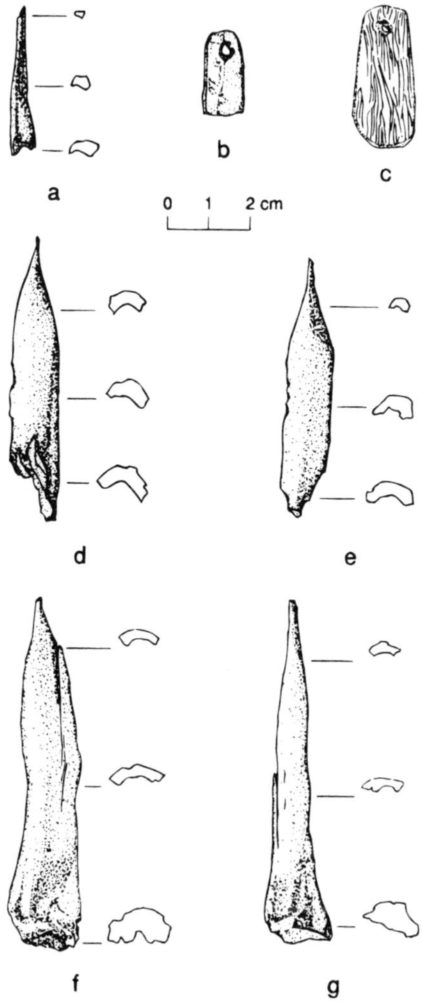


**Figure 15.8.** Illustrations of worked bone points and pendants from Timnian sites.

## Pottery

The large majority of sherds were recovered from the pithouse excavation at Jebel el Jill (J14), but small samples were found at five other long-term campsites in the piedmont. Only one of the sites in the lowlands yielded pottery. The sherds are uniform in representing thick, heavily tempered plainwares that were generally fired at low temperatures for short durations. Tempering largely consists of chert fragments. The reddish buff oxidation layers, relatively thin when compared to the black to dark gray cores, indicate that the pottery was briefly fired at low temperatures.

Only a few large sherds and those representing rims and shoulders were recovered. These indicate that the vessels were for the most part holemouth and short neck jars of middle to large sizes. The mouths of the jars are estimated to measure around 18 cm in diameter and most of

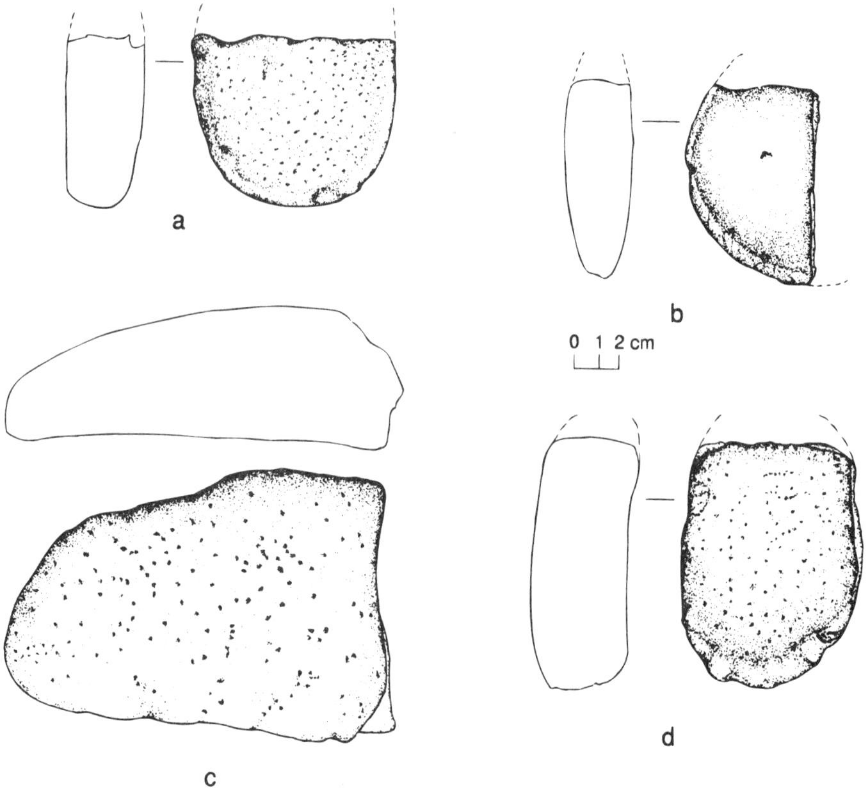


Figure 15.9. Illustrations of groundstone artifacts from Timnian sites: (a,b,d) handstones, (c) grinding slab.

their bodies would fall between 30–40 cm in maximum diameter. Very few of the sherds display soot and those that do are the thinnest of the specimens.

Functionally, most of the vessels would appear to have served as storage containers. These large thick-walled, heavy-tempered vessels would have been durable and suitable for holding a range of staple resources. A few vessels were used for cooking, however, as indicated by exterior sooting. As might be expected given the relationship between thickness and convective efficiency, they are thinner than the storage jars.

### Ornaments of Shell and Stone

Most of the shell was recovered from Jebel el Jill (J14) with the great majority having been holed or cut into beads. All of the specimens come from the Red Sea with *Dentalium*, *Conus*, and *Mitra* as the dominant genera (Chapter 17). Other ornamentation is represented by two small perforated pendants of stone and bone (Figure 15.8).

## FAUNAL EVIDENCE

As might be expected in a pastoral camp, the bones of caprines account for over 74% of the faunal remains from Jebel el Jill (J14), but surprisingly about 25% of the assemblage consists of gazelle bones (Henry *et al.* 1984). Whereas some sheep and goat remains were specifically identified within the assemblage, most of the caprine bones could not be distinguished. Moreover, the nature of the elements and the sample sizes are such that Turnbull was unable to confirm that the caprines were, in fact, domesticated. Given the prevalence of domesticated caprines in the Levant beginning in the Neolithic, it is assumed that sheep/goat bones from Jebel el Jill (J14) represent herded animals. This assumption also is consistent with the much greater representation of caprines in the Chalcolithic assemblage than in Epipaleolithic faunal assemblages from nearby sites (Table 15.3).

Although the occupants of Jebel el Jill (J14) appear to have engaged in pastoralism, the relatively high proportion of gazelle bones indicates a substantial dependence upon hunting. This differs significantly from other Chalcolithic assemblages in which gazelle remains account for only a small proportion of the assemblages (Grigson 1987).

## SUMMARY AND CONCLUSIONS

The Timnian sites of southern Jordan, dated from ca. 4,000–5,700 B.P., represent the remnants of a settlement-procurement pattern in which groups subsisted on the herding of caprines coupled with hunting and gathering. In following a strategy of transhumance, campgroups appear to have moved seasonally to different elevational belts in order to take advantage of fluctuations in the available resources and creature comfort. In addition to altering their levels of residential mobility from season to season, they also appear to have adjusted the sizes of their groups in order to accommodate seasonal differences in the densities of critical resources.

### Annual Cycle of Transhumance

Timnian groups appear to have followed a cycle of transhumance that closely parallels that practiced by historic Bedouin groups of southern Jordan (Henry 1992b,1994). This entailed a relatively long-term (8–12 week) occupation of the piedmont zone during the winter wet-season, followed by a spring–early summer movement to the lower elevations of the Wadi Hisma and then a late summer migration to the uplands on the plateau, where groups remained until the return of the wet season. This pattern of transhumance also linked fluctuations in the sizes of groups and their permanence of residence to the season and elevational belt of occupation. Relatively large, long-term camps characterized the occupation of the piedmont during the winter wet season, whereas much smaller, more ephemeral camps were inhabited during the dry season.

**Table 15.3. The Proportionate Representation of Sheep/Goat Remains Within the Faunal Assemblages of Epipaleolithic and Chalcolithic Sites**

| | Percentage of Caprines |
|---|---|
| Jebel el Jill[a] (Timnian) | 73.1 |
| Wadi Judayid[a] (Early Natufian) | 65.3 |
| Tor Hamar[b] (Mushabian) | 21.1–37.8 |
| Tor Hamar[b] (Qalkhan) | 17.4 |

[a]Henry *et al.* (1985).
[b]Klein, Chapter 20.

The seasonally and elevationally linked cycle of population coalescence and dispersal for Timnian groups finds archaeological expression in the dichotomy between the sites of the piedmont and those on the floor of the Wadi Hisma. The sites of the piedmont are substantially larger, have much thicker cultural deposits, and contain higher artifact densities than those of the lower Hisma (Table 15.1). Architecture in the form of pithouses, corrals, and storage pits is also confined to the piedmont sites. Beyond displaying greater diversity in their chipped stone toolkits, the piedmont sites also contain a much wider array of other artifacts including pottery, ground stone, worked bone, and ornaments of stone and shell.

The presence of architecture, heavy grinding slabs, and large storage jars suggest a certain degree of residential permanency for the piedmont sites. Their stratigraphies with alternating ash and sterile layers and superimposed hearths point to periodic, probably seasonal abandonment of the sites. Construction of semisubterranean structures with internal hearths are suggestive of efforts that would have been related to conserving heat in winter dwellings and the compacted floor exposed in the pithouse (Locus 7) at Jebel el Jill is also consistent with the moist conditions expected in a winter occupation. Moreover, the location of the corral (Locus 1) at Jebel el Jill implies a northerly wind direction if one considers the modern practice of positioning corrals and ash dumps downwind from the dwellings. Wind from the north is prevalent in the winter.

The small, thin occupations at lower elevations in the Hisma are indicative of ephemeral camps. Their low artifact densities and narrow range of tool types are consistent with short-term camps linked to a limited set of activities. Seasonally, the sites could have been occupied only during the spring and early summer, given that water and pasturage would have been available only during this time of year.

## Foraging, Pastoralism, and Secondary Products

Pastoralism tied to permanent and semipermanent villages has been relatively well defined in the better watered areas of the Levant during the Chalcolithic (Levy 1983, 1985; Gilead 1988, 1989, 1992). But suggestions for the presence of synchronous, fully nomadic pastoral societies have been based more on speculation than evidence. The Timnian sites of southern Jordan not only furnish a clear example of such fully nomadic pastoralism, they also confirm an unsuspected economic link between herding and foraging. The contribution of hunting to the meat diet of these pastoral groups is denoted by almost 25% of the bone from Jebel el Jill coming from gazelle. Cereal pollen is present (marginally) in only one spectrum of the three Chalcolithic site diagrams and wheat and barley phytoliths do not appear in the samples recovered from Jebel el Jill (Emery-Barbier, Chapter 16; Rosen, Chapter 19). This suggests that Timnian groups did not engage in cereal cultivation and must have depended largely upon gathering wild plant foods. An absence of sickle blades, a dominant characteristic of Chalcolithic sites in better watered regions, furnishes additional support for this argument.

In combining foraging and pastoralism, the subsistence strategy of Timnian groups fails to fit our prevailing notions on the relationships of these broad economic patterns. Foraging and herding have been viewed as mutually exclusive subsistence strategies because of the very nature of the acquisition of resources (i.e., procurement versus production) and different requirements for residential mobility and scheduling (Khazanov 1984; Ingold 1987; Cribb 1991). Cribb (1991:21), for example, suggests that the logistics of foragers and herders fundamentally differ because:

> the hunter-gatherer schedules his movements with regard to *consumption*, the pastoral nomad is tied to a schedule of pastoral *production* which is in turn dependent upon the consumption patterns of pastoral stock.

Whereas tensions between differing ways of acquiring resources may render them incompatible in certain ecological settings, I see no real reason why this should always be the case between foraging and pastoralism (cf. Meadow 1989). The principal factors that appear to have conditioned the settlement-procurement pattern of Timnian groups, in fact, are identical to those to which "pure" foragers or "pure" herders would have responded. Coupled with concerns for creature comfort, the seasonal and spatial distribution of surface water, herbaceous plants, and grasses would have been the principal factors influencing the movement and scheduling of foragers and pastoralists alike.

For the Timnians, foraging and the herding of sheep and goats appear to have been complementary subsistence modes. Hunting provided meat and other important by-products (e.g., horns and hides) to augment those acquired from pastoral stock. Gathering of wild plant foods is likely to have had an even greater economic importance, however, given the apparent lack of horticulture. Culling of sheep and goat herds also would have yielded a substantial amount of meat, but the "secondary products" of milk and wool were likely to have been critical to making herding economically viable for the Timnians.

## Milking: Origins and Evidence

Sherratt (1981:284, 1983) has noted that beyond providing fat, protein, milk-sugar, and the important amino-acid lysine, milk is a remarkably economic resource when compared to meat production. Milk yields some four to five times more protein and energy than meat from stock supported by the same amount of pasturage. In specifically comparing the energetics of "meat-only" and "meat plus milk" pastoral strategies for sheep and goats, Russell (1988:101,103) found meat plus milk strategies to have provided more than twice the annual caloric yields of meat-only strategies. The energetic efficiency of utilizing milk as a secondary product prompted Kohler-Rollefson's (1992:16) observation that

> pastoralists all over the world have one thing in common: the prime focus of their animal utilization is milk not meat.

This raises the question of when milk was first exploited from caprine herds. The emergence of the utilization of milk would have been complicated by an intolerance to the consumption of lactose by humans after infancy (Sherratt 1981). Although milk consumption may be significantly improved through fermentation, drying, or dilution with water, true tolerance for the life-long consumption of milk is linked to the production of the enzyme lactase (Sherratt 1981; Russell 1988). This production is acquired genetically, and characterizes pastoral populations stretching across the arid belt of Africa and Eurasia. The advantages of milk consumption for populations dependent upon pastoralism would have selected for those genetic characteristics which provided individuals tolerance of the consumption of lactose. In resting near the center of the modern distribution of lactose-tolerant populations, the mountain areas and the adjacent steppes of the Near East are thought to have been the area of origin for milk consumption by mid-Holocene times (Sherratt 1981:277).

Evidence from Chalcolithic sites in the Levant supports the idea that milking formed part of the pastoral economy of the region as early as the mid-Holocene (Levy 1983; Ducos 1969; Gilead 1988; Rosen 1988). Pottery vessels thought to represent churns are common to Chalcolithic assemblages (Rosen 1988; Gilead 1989). A calcium decline in caprine bones from the Early Bronze Age occupation at Arad is viewed as an indicator of intensive milking (Smith and Horvitz 1984). Morphological changes in caprines at this time also are viewed as pointing to the exploitation of secondary products (Davis 1984).

## Wool, Sheep Shears, and Tabular Scrapers

Although less important to the subsistence of Timnian groups than milk, wool and goat hair were likely to have been significant to their economy. Beyond providing materials used in fabricating clothing, bags, and tentcloth, wool and goat hair also may have been traded to more settled groups for pottery, copper items, and most likely cereals. Intensive and far-reaching trade relationships are well documented for the period (Gilead 1988, 1989; Levy 1985).

We have little direct evidence for the earliest "wool" as opposed to "hair" sheep. Hair sheep resemble their wild ancestors in having an outer coat of bristly fibers (kemps), which cover an undercoat of fine wool (Ryder 1984). What appears to be the earliest evidence for the emergence of fleeced sheep is a clay sheep figurine, dated to the sixth millennium BC from the site of Sarab (Iran), which shows a "hairy medium fleece" (Ryder 1988:70). Hair sheep are depicted in Early Dynastic Egypt (third millennium BC) with a loss of natural pigment and other morphological changes (long tails, corkscrew horns) that resemble modern domestic forms.

An indirect clue to the exploitation of wool may come from an associated artifact variety—the fan or tabular scraper. The temporal and spatial distributions of tabular scrapers in the Levant parallel that of caprine herding. Ethnographic accounts and functional studies point to their use in shearing sheep.

Tabular scrapers make their first appearance in the pottery Neolithic of the Levant and come to represent a typical component of Chalcolithic tool assemblages. From an evolutionary perspective it is surprising to see what seemingly is the reappearance of the Middle Paleolithic "sidescraper" after some 40,000 years of replacement by endscrapers. In reality Middle Paleolithic sidescrapers and Chalcolithic tabular or fan scrapers have very little in common, other than the scraper label. In contrast to sidescrapers, tabular scrapers occur on thin, relatively wide flakes with acute, invasive retouch. The presence of silica sheen on specimens from some Chalcolithic assemblages (Lee 1973) and the results of preliminary edge-wear analysis (McConaughy 1979) suggest that tabular scrapers, in fact, were not used as scrapers at all.

Although microscopic edge-wear study indicates that tabular scrapers with sheen may have been used in the cutting and processing of reeds or butchery, I think that most tabular scrapers were used in shearing sheep. Stimulated by Lawrence's (1914–15) account of Bedouin in the Negev using "flint scrapers" to shear their sheep, I replicated two tabular scrapers and participated in the shearing of a flock owned by a Bedouin friend in southern Jordan. Much to his entertainment and one unfortunate animal's displeasure, I discovered that the flint shears were about as effective in cutting the wool as the metal ones. Although they dulled rapidly from contact with the sand contained in the wool matt, resharpening was easily accomplished with a small hammerstone. Although I failed to generate a silica sheen on either of the replicated specimens, I think that through prolonged use silica sheen is likely to occur given the high density of sand in the wool. Such sheen is commonly produced on flint hoes and other implements that contact silica in sediments.

The morphology of the tabular scraper is ideal for its use as a sheep shear. The thin, uniform cross-section and acute edge enable one to lay the implement close to the skin while pulling back and tensioning the wool matt for cutting. Moreover, the convex edge of the scraper accommodates the somewhat sweeping motion employed in cutting the matt. Even the cortex on the obverse surface of the scraper facilitates its use as a shear. It provides an ideal "friction surface" for grasping the implement as it becomes covered with lanolin.

The synchronous emergence of tabular scrapers and independent evidence for the "secondary products revolution" is consistent with the scrapers having served as sheep shears. Furthermore, tabular scrapers represent a much greater proportion of Chalcolithic toolkits from sites located in the steppe-desert zone than from those of the Mediterranean woodlands. This again is consistent with the notion that the herding of caprines formed the economic core of

Chalcolithic groups inhabiting the arid zone and that their toolkits reflected this emphasis. Rosen (1983b) has intepreted the distributional pattern of tabular scrapers differently, however. He views the pattern as an expression of the progressive decline of tabular items that accompanied their increasing distance from their source, which he sees as the central Negev.

## Alternative Paths to Pastoralism

Herding of animals traditionally has been viewed as occurring initially in Neolithic horticultural villages with the taming and domestication of caprines (Bar-Yosef and Khazanov 1992:4). With subsequent natural and culturally induced environmental deterioration, pastoralism is thought to have emerged as an economic option for the exploitation of marginal settings in which agriculture or horticulture alone was impractical (Khazanov 1984; Kohler-Rollefson 1992).

While the proposed precondition of a horticultural setting for the emergence of pastoralism is consistent with the archaeological record of the better watered regions of the Levant, pastoralism in the arid zone appears to have come about along a different evolutionary path. Rosen (1988) argues that pastoralism in the arid zone of the southern Levant grew out of a foraging as opposed to a settled horticultural base. He notes that extensive investigations in the central and southern Negev and Sinai have revealed numerous Neolithic occupations, none of which contains evidence for domesticated plants or animals. Contemporary sites of the central and northern Levant are represented by agricultural villages. The Neolithic foragers of the drier settings appear to have followed a specialized adaptation involving transhumance that enabled them to maintain seasonal, long-term camps with architecture (Bar-Yosef 1984; Goring-Morris and Gopher 1983).

During the following Late Neolithic and Chalcolithic periods (ca. 6,000–4,000 B.P.), the desert settings of the Negev and Sinai appear to have experienced a population reduction, most likely related to the onset of much drier conditions (Rosen 1988). Rosen (1984) suggests that it was at this time that these desert foragers began to engage in pastoralism. In support of his argument, he points to the appearance of what are thought to represent pens or corrals in the Chalcolithic architectural sites, the dominance of caprines in the faunal assemblages, and a reduction in arrowheads within the toolkits.

Although the southern Negev and Sinai experienced a reduction in population during Chalcolithic times, the higher, somewhat better watered region of southern Jordan appears to have been intensively occupied in judging from the high density of Timnian sites. This may represent a packing of the population along the desert fringes. In many ways the Timnian strategy for inhabiting the mountainous area of southern Jordan continued the older desert Neolithic foraging pattern of transhumance, but it also incorporated the herding of sheep and goats. Whether or not the caprines were domesticated indigenously within the desert Neolithic foraging tradition or acquired from Neolithic agricultural villages is difficult to determine.

If caprines were domesticated by desert foragers, this process would have had to have taken place within the natural range of wild sheep and goats. Within the southern Levant, this would have been limited to the mountainous areas of the Negev, Sinai, Jordan, and perhaps even Arabia. We in fact know that caprines formed a large if not dominant portion of the human prey species from Upper Paleolithic times onward in southern Jordan and during late Epipaleolithic and Neolithic times in the Negev and southern Sinai. But the establishment of long-term camps within caprine habitats would have been critical to triggering the first steps to domestication. Such a degree of permanency in human settlement would have been necessary for the initial steps of "controlling" caprines to have been both practical and economic (Hecker 1982; Hole 1989). The southern Jordan settlement evidence indicates that the inhabitants of the region followed a transhumant settlement pattern from as early as Middle Paleolithic times. It was only with the elevation of temperature near the end of the Pleistocene (ca. 12,500 B.P.), however, that long-term winter camps were established at relative high elevations in the piedmont—an optimum caprine habitat. The relative permanence and location of such camps, coupled with the transhumant schedule of their occupants, may have prompted groups to progressively bring caprines under various kinds of cultural control short of domestication. This probably took place sometime in the late Neolithic, for the caprines recovered from the early Neolithic site of Beidha (dated from 9,000–8,500 B.P.) show signs of cultural manipulation or control, but had yet to be domesticated (Hecker 1982).

The appearance of caprines, well removed from their natural habitat, in Neolithic sites (8,000–7,500 B.P.) of the Azraq Basin of northeastern Jordan do indicate the presence of pastoralism among Neolithic foragers (Garrard *et al.* 1988). But in the absence of a nearby caprine habitat, it is assumed that these had been borrowed from permanent agricultural settlements located in more fertile settings to the west.

*Chapter* **16**

# *Pollen Analysis: Environmental and Climatic Implications*

ALINE EMERY-BARBIER

The results of the palynological study of samples from the archaeological deposits in southern Jordan allow for the reconstruction of specific "climatic moments" as dated by reference to temporally sensitive artifacts, stratigraphy, and chronometry. The interval under consideration stretches from the Levantine Mousterian through the Chalcolithic. While dominated by a generally arid setting, relatively humid phases were identified in early Upper Paleolithic (Early Ahmarian), early Epipaleolithic (Early Hamran), and late Epipaleolithic (Late Hamran and Early Natufian) times. Comparison with data from other settings in the Near East provides for certain region-wide correlations.

## METHODOLOGY

Pollen studies attempt to reconstruct local and regional images of the past vegetation from fossilized pollen left in ancient deposits. Samples selected for analysis, having been identified by their stratigraphic or archaeological provenience, allow changes in vegetation and climate to be traced through time. Primary results of palynological analysis are expressed statistically and used as a basis for evaluating hypotheses concerning the reconstruction of vegetation and climate. Multiple lines of analysis in conjunction with contemporary environmental and climatic analogies assist in this process.

Within the human and botanical sciences, archaeological palynology contributes significantly to our knowledge of the environmental background of prehistoric cultures and to our understanding of climatic models.

## POLLEN ANALYSIS IN SEMIARID ZONES

The successful recovery and analysis of pollen is possible in semiarid zones, but samples are sometimes marginal given their recovery from easily transported, dry sediments. In this study, extraction of pollens followed a process involving the dissolution of silica, oxidation of the organic matters and concentration by centrifuge. This typically yields samples of pollens that are sufficient for statistical examination.

Those sediment samples that were found to be sterile or very low in pollen correspond to erosional intervals that apparently marked periods of extreme aridity as noted in the geomorphic study (Chapter 2). These intervals for the most part coincide with arid periods identified elsewhere in the Near East through other methods of investigation. The samples rich in pollen are those collected in the zones where the processes of pedogenesis led the formation of grey or brown paleosols which are generally poor in organic matter. These weakly organic paleosols, deprived of humus, resemble the modern soils of the region and resulted largely from the mechanical deterioration of parent material.

The variety of the pollens encountered and their marked differences in comparison with modern types confirm that neither infiltration nor contamination has occurred. Still the phenomenon of corrosion cannot be excluded. Both the rate of sedimentation and the thickness of the deposit affect the degree of fossilization and the quality of preservation of pollens. Climatic factors at the time of deposition and those affecting the intervening time may also have influenced the differential preservation of pollens. In attempting to interpret the numeric and statistical results of the analysis, one must keep in mind the alternative environmental and climatic conditions that may have been responsible for the pollen configurations. Beyond the problem of differential preservation, the anthropogenic influence on pollen rain must be considered. In this case, samples from natural and cultural (on-site) settings are evaluated differently. Given these qualifications of palynological investigations, such studies should be conducted in concert with other lines of inquiry (e.g., geomorphology, zooarchaeology) in order to understand the relationships between vegetation, climate, and prehistoric behaviors.

## THE LIMITS OF INTERPRETATION

As determined in earlier studies, sediments from the archaeological deposits in the study area principally reflect a local flora comprising representations of entomophagous and anamophagous plants (Leroi-Gourhan 1980; Emery-Barbier 1988). Some rare grains of pollen could have been introduced in water transported to sites for consumption and others could have originated from species cultivated or selected for their alimentary properties, maybe medicinal. This anthropogenic phenomenon could be particularly informative, but it must be evaluated in the context of fluctuations in climate as indicated by the whole of the pollen curve.

The study area is characterized by its aridity accompanied by extreme temperatures. The overall vegetational community as identified from pollen allows for tracing variability in the aridity index, but it is more difficult to determine the relative fluctuations of temperature and moisture levels. This is accomplished principally through studying fluctuations in herbaceous plants, but samples containing a significant percentage of arborescent pollens (maximum 25%) are also informative as they point to a Mediterranean climate. Among the most common of these is the oak, *Quercus Ithaburensis* (variety of *Quercus aegilops*), which grows presently over much of the Mediterranean woodlands. Less abundant are the alder-tree (*Alnus*), the elm (*Ulmus*), the "chaste tree" (*Vitex agnus-castus*) and the fig-tree (*Ficus*), all of which confirm the presence of greater surface water and more humid conditions than exist in the region today. Pine (*Pinus*), another signature of the Mediterranean woodlands, also appears in the diagrams. Its representation as a component of the local flora cannot be confirmed, however, because pine pollen is widely distributed from its source by wind.

During those climatic intervals that were cooler and more humid than now, herbaceous plants occurred in great variety, but two families most often dominate the pollen spectrum: Gramineae and Compositae tubuliflorae. Samples not containing arborescent pollens and pointing to drier conditions are dominated by Chenopodiaceae and Compositae liguliflorae. Leroi-Gourhan (1980; Henry *et al.* 1981) has previously noted these contrasting representations and corroborating sedimentological data in her work in Syria and Israel. It would appear that the more xerophytic Chenopodiaceae came to replace less arid-tolerant grasses (Gramineae) with the onset of drier conditions. Certain Chenopodiaceae are also halophytes, plants adapted to wet, saline conditions, but within the region these are typically found in settings where vegetation is supported by a high water table, not a moist climate.

A review of the diagram allows for another observation of climatic import regarding the temporal succession of the different species of the Chenopodiaceae family. In fact, the rounded peripore grains of Chenopodiaceae (type *Atriplex*) alternate with the hexagonal pollens with large pores and quite visible columella of Chenopodiaceae (type *Noaea*). When Chenopodiaceae (type *Atriplex*) dominates a spectrum there is an absence of evidence for warmth-loving, thermophilous vegetation. In contrast, those spectra dominated by Chenopodiaceae (type *Noaea*) also include some clear thermophiles such as Asphodels. Therefore, it seems likely that Chenopodiaceae (type *Noaea*) provides a good indication of warmer conditions. It is not surprising then to find that Chenopodiaceae (type *Noaea*) make up 70–95% of the flora from the Chalcolithic deposits of Jebel Queisa (J24) and Jebel el Jill (J14), given that they were formed during a time of global elevation in temperature—the hypsothermal.

Interestingly enough, the sizes of the pollen grains of Chenopodiaceae also would appear to have changed in correlation with environmental and climatic conditions. The grains of the type *Atriplex* are very small (averaging some 10–12 microns although some are as small as 6 microns). Leroi-Gourhan (Henry *et al.* 1981) has observed in the terminal Pleistocene deposit of Hayonim Terrace that their sizes increased in concert with rises in temperature and humidity. The larger grains of the *Noaea* type also show a variation in their diameters between spectra and the same is true of accompanying Compositae liguliflorae grains.

While climatic fluctuations were undoubtedly responsible for changes in the composition of floral communities, these changes also may have induced direct phenotypic variation. Further research will be necessary, however, before we know whether the variability in the sizes of pollen grains can be used to trace transformations in the habitats of plants as can other phytomorphic attributes.

## THE ARCHAEOLOGICAL DEPOSITS

As in the Levant, the study area shows a much higher density of prehistoric sites for certain periods. Although this may imply that these intervals were accompanied by more favorable environments and higher biomasses than at other times, a wide variety of factors (many of which are unrelated to resource availability) are known to influence site density (Chapter 2). In any event, the following paragraphs describe the succession of plant communities and their implied climatic settings as interpreted from the diachronic variation in pollen evidence (Figure 16.1). Based upon technotypologic seriation, stratigraphic placement, and chronometric assays, the deposits and their associated pollen data have been placed in chronological order (Chapter 3, Table 3.1).

### Levantine Mousterian Site of Tor Sabiha

The three upper layers of the Tor Sabiha deposit contain archaeological evidence. The deposit, formed from the deterioration of the surrounding sandstone cliff and from windborne driftsand, rests protected in a narrow embayment, high on the western flank of the Judayid Basin (Chapter 2, Chapter 5). Layer A is recent, containing reworked sediments with Chalcolithic and Levantine Mousterian artifacts. Layer B appears to be primarily of mid-Holocene age with a $C^{14}$ assay of ca. 4,000 B.P. and a dominance of Chalcolithic materials.

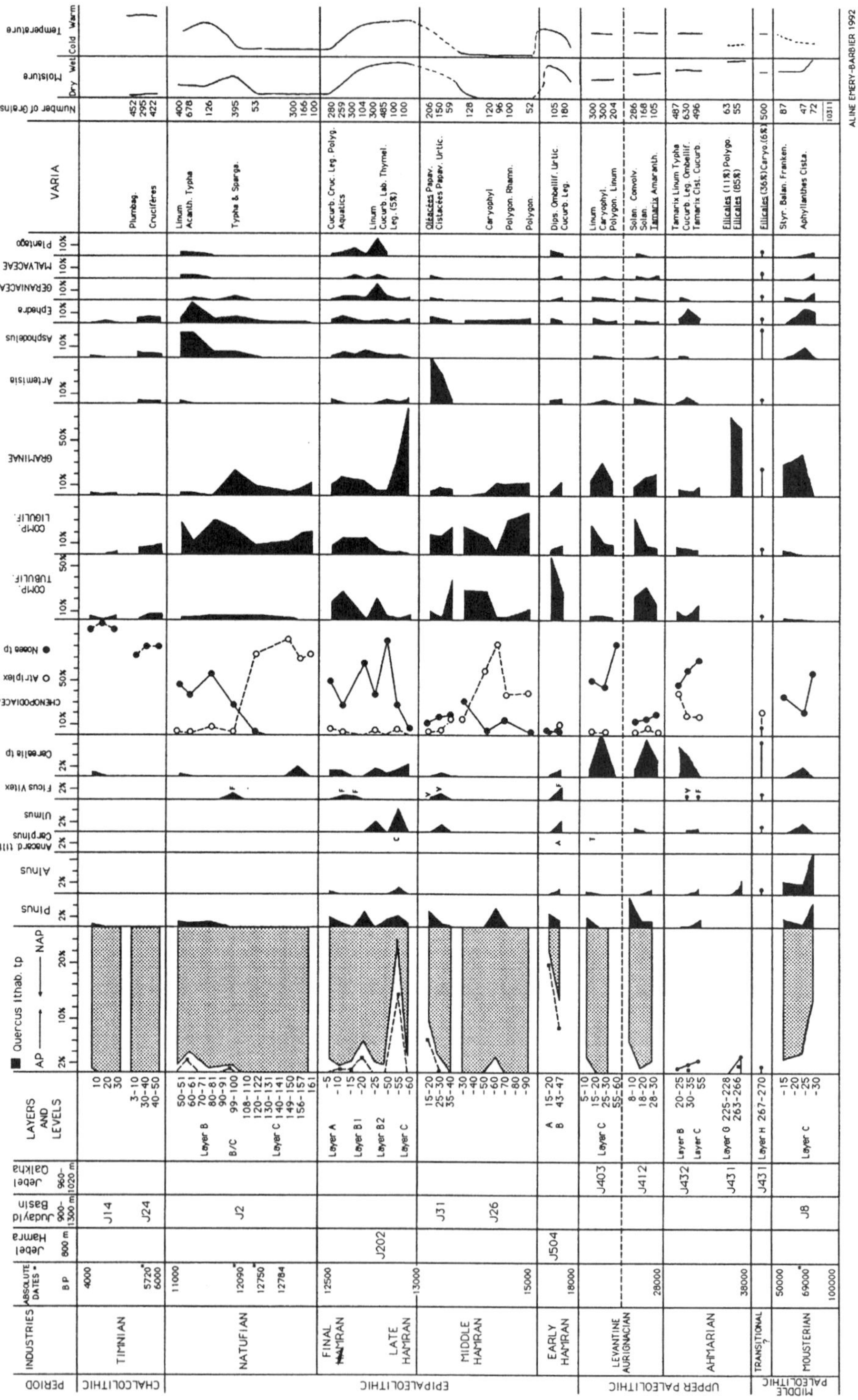


Figure 16.1. Pollen diagram of Middle Paleolithic through Chalcolithic deposits.

Sparse Levantine Mousterian artifacts occur within the layer, however, and thus indicate some mixture. This is thought to have resulted from the clearing of the cave by Chalcolithic shepherds. Layer C, separated from Layer B by a depositional hiatus, contains *in situ* Levantine Mousterian artifacts.

Of the four samples collected from Layer C, only three yielded pollen. The arboreal pollen includes alder, elm, and pine. Alder and elm trees would have been likely to have lined the streams found on the floor of the nearby basin. Regarding the herbaceous plants, pollen of Chenopodiaceae (type *Noaea*) shows frequencies that are inversely related to those of Gramineae. The remainder of the floral community is represented mainly by xerophytic families (Styracaceae, Santalaceae) and an arid adapted Liliaceae belonging to the genus *Aphyllantes*. The presence of halophytic plants, represented by Frankeniaceae, indicate at least a limited occurrence of a wet saline setting within the area.

It is difficult to characterize the climate on the basis of these three samples, given especially the relatively low counts and variable ecological indicators. The successive decline of elm pollen accompanied by a rise in grasses may imply a climatic fluctuation toward drier conditions.

Noteworthy is the presence of two grains of pollen of Gramineae classified as type *Cerealia* given their large diameters and dimensions of the annuli of their pores. Identical grains have been reported in the Mousterian layers of Nahr Ibrahim Cave (coastal Lebanon) and at Shanidar in the mountains of northern Iraq (Leroi-Gourhan 1969, 1980).

Although not presented in the diagram (Figure 16.1), the palynological results from Layer B are dominated by Chenopodiaceae (type *Noaea*) and virtually the same as those identified from other Chalcolithic deposits (e.g., Sites J14 and J24).

## Transitional Middle-Upper Paleolithic or Early Upper Paleolithic

Within the deposit of the Tor Hamar (J431) rockshelter, the artifacts from Layer H indicate a transitional or very early Upper Paleolithic placement. Arboreal species are represented by alder, elm, and wild pepper tree (*Vitex*). In combination, the Chenopodiaceae, Compositae liguliflorae, and *Artemisia* are roughly balanced with Gramineae and Compositae tubuliflorae. Gramineae (cereal type) account for 7% of the spectrum. The lower frequency of arboreal pollen may point to slightly drier conditions than those of the preceding Late Levantine Mousterian; however, this may also merely reflect the lower elevation of Tor Hamar and its inherently drier setting than that of Tor Sabiha.

## Upper Paleolithic Sites

### Early Ahmarian

The flora associated with the Early Ahmarian (38,000–30,000 B.P.) deposit of Layer G at Tor Hamar (J431) is characterized by the presence of pollens of alder and oak, and, above all, by the abundance of grasses (Gramineae). This pollen configuration is indicative of a humid period, the reconstruction that draws additional support from the high percentages of spores of ferns. At the beginning of the Upper Paleolithic, fig, tamarisk, oak, and wild pepper trees contribute to a relatively wide range of arboreal species. These are likely to have lined the wadis and canyons that drain the Jebel Qalkha. The presence of tetrades of pollen belonging to the genus *Typha*, a variety of rush, suggests that permanent water would have been found nearby.

Although the pollen recovered from these rockshelter deposits may have been derived more from human activities than from the natural pollen rain, the general indicators of a relatively humid setting are nevertheless present. The high percentages of pollen attributed to Gramineae (type *Cerealia*) may have resulted from human consumption and the same may be true for the pollen of Curcurbitaceae and flax.

## Levantine Aurignacian

Given their artifact content, the deposits of the rockshelters of Tor Fawaz (J403) and Jebel Humeima (J412) are most likely to have been formed within the interval from 32,000–28,000 B.P. The presence of alder in the deposit of Jebel Humeima suggests that the nearby wadi flowed to the extent that trees could have been supported along its banks. The diversity and relative balance in the herbaceous flora is also consistent with relatively moist conditions.

A rise in the percentage of pollen of Chenopodiaceae (type *Noaea*) and of Compositae liguliflorae within the Jebel Humeima (J412) deposit suggests a tendency toward warmer and drier conditions. The high percentages of Gramineae, especially the type *Cerealia* which accounts for 7.9% of the spectrum, reflect human activity. Such frequencies point to the concentration of straw or grains within the occupation. The type *Cerealia* pollen recovered from level 25–30 is 40 microns in size and 8/10/12 microns for the external diameter of the pores. The relatively high frequencies of Malvaceae pollen may also indicate that the plant was consumed by the shelter's inhabitants as it is today by Bedouin in the Negev (Darmon 1981).

## Epipaleolithic Sites

The deposits within the sites of Jebel Mishraq (J504), J26, J31, and Qa Salab (J202) contribute to a palynological succession that stretches from ca. 18/16,500–12,500 B.P. (Chapter 13).

Placed at the beginning of the sequence, the Early Hamran deposit of Jebel Mishraq (J504) yielded very high frequencies of arboreal pollen dominated by oak. The high percentages of oak pollen, accompanied by a considerable diversity of families among the Centaureae and Anthemideae, imply an amelioration of climate that prompted the expansion of oak woodlands throughout the piedmont. This shows evidence of a humid phase synchronous with the Kebaran, even though this period is mostly associated with a dry, cold interval.

Pollen recovered from the succeeding Middle Hamran (15,000–13,000 B.P.) deposits of sites J26 and J31 traces a cold, dry phase followed by a trend of climatic amelioration. In the deposit of Site J26, the very high percentage (peaking at 80%) of remarkably small grains of pollen of Chenopodiaceae (type *Atriplex*) indicates cold, steppic conditions. A slow amelioration in temperature and rise in moisture appears to have followed, as indicated by a decline in Chenopodiaceae (type *Atriplex*) pollen accompanied by a rise in type *Noaea*, Compositae liguliflorae, and C. tubuliflorae. This trend of climatic improvement apparently continued through the end of Middle Hamran times (ca. 13,000 B.P.), as evidenced by the pollen spectra from Site J31. Samples from the deposit show a rise in arboreal pollen (including pine and elm) and *Artemisia*. The presence of pollen of *Vitex agnus castus*, Papaveraceae, and Cistaceae also points to more humid conditions.

This trend of increasing humidity and temperature appears to have culminated in a marked humid phase during the Late–Final Hamran (13–12,500 B.P.). Layer C of the Qa Salab (J202) deposit yielded pollen showing an arboreal peak of 20%. The trees are represented largely by oak, *Quercus Ithaburensis* (14%), but alder, elm, and hornbeam pollen is also present. Chenopodiaceae and Compositae frequencies are relatively low in contrast to high percentages of grasses (Gramineae). Pollen of Leguminosae (5%), Curcurbitaceae, and *Linum* suggest human staples.

During Final Hamran times, pollen from the upper part of the Qa Salab deposit indicates the onset of drier conditions as indicated by the decline in arboreal pollen and the expansion of Chenopodiaceae (type *Noaea*). However, the moderate percentages of Compositae tubuliflorae and Gramineae along with the presence of aquatic plants point to much more humid conditions than for the area today.

In combination, the palynological evidence from the Middle and Late–Final Hamran deposits traces two quite different climatic phases: a dry, rather cold phase at the beginning of the sequence and a milder, humid one between 13,000–12,500 B.P.

## The Epipaleolithic of Tor Hamar

The Madamaghan (Mushabian Complex) layers (A–E[1]) of the Tor Hamar (J431) rockshelter were formed between ca. 14,000–12,500 B.P. and are thus partially coeval with the Middle, Late–Final Hamran, and Early Natufian deposits in the area. While Layer E[1] is dominated by pollen of Gramineae (over 90%), Artemesia pollen occurs in high percentages and characterizes the more recent layers of B (80%) and A (55%). Although the intervening layers D and C are sterile of pollen, upper Layer C yielded radiocarbon dates of 12,680 ± 320 B.P. (SMU 1399) and 12,320 ± 90 B.P. (ETH 806) on charcoal.

The palynological transition between layers E[1] and B points to a climatic shift from moist to drier, cooler conditions. Although the extraordinary representation of grass (Gramineae) pollen, which corresponds to the burnt horizon of Layer E[1], is undoubtedly a consequence of human activity, such a resource in the area of the shelter signifies a relatively moist setting. When the palynological succession identified in the Tor Hamar deposit is compared to that traced for Late–Final Hamran (J202) and Early Natufian (J2) deposits, there are clear parallels. As in Layer E[1] at Tor Hamar, the Late Hamran (J202) shows a significant Gramineae spike (ca. 80%) indicative of very moist conditions. The drier conditions identified in the upper layers (A and B) of Tor Hamar correspond to the colder, drier conditions seen in the lower part (Layer C) of the Early Natufian deposit of Wadi Judayid, which is also dated to ca. 12,500 B.P.

## The Early Natufian Site of Wadi Judayid

The deposit of the Early Natufian site of Wadi Judayid (J2) reveals two phases with quite distinct paleoflora. The earliest, found in Layer C and dominated by Chenopodiaceae (type *Atriplex*), suggests very cold and dry conditions. Associated radiocarbon dates bracket the interval between ca. 12,780–12,090 B.P. The second phase, identified in Layer B and the B/C contact, shows a sharp rise in Chenopodiaceae (type *Noaea*) accompanied by increased proportions of Compositae liguliflorae, Asphodels, and *Ephedra*. Arboreal pollen (*Pinus*, *Ficus*) also increases along with the appearance of some semi-aquatic plants (*Typha*). Although this latter phase shows signs of warmer, more humid conditions, it seems to have been less pronounced than the pluvial periods of the Early and Late Hamran. Given the radiocarbon dates, the transition to a warmer, more humid climate may well correspond to the Dryas II/Allerod climatic transition of Europe (Leroi-Gourhan and Darmon 1991).

Few cereal pollens were recovered from the sediments of Wadi Judayid. Paradoxically, it is from Layer C, associated with unfavorable climatic conditions, that the highest proportions (2%) were found. Only a single pollen grain was identified for layers B/C and B.

## The Chalcolithic

In the Chalcolithic deposits of Jebel el Jill (J14) and Jebel Queisa (J24), Chenopodiaceae (type *Noaea*) with a mean diameter of 16 microns largely dominate the pollen spectrum. Other flora are represented by Compositae liguliflorae, Asphodels, and *Ephedra*. A radiocarbon date of 5,720 ± 149 B.P. (SMU 804) was obtained on charcoal from a hearth in lower Layer B at Jebel Queisa. These palynological data are consistent with the geomorphic (Chapter 2) and faunal (Henry *et al.* 1985) evidence collected from the deposits in that they denote very warm, dry conditions. Only three grains of cereal pollen were identified, all from Jebel el Jill (J14).

## PALEOCLIMATIC RECONSTRUCTION

Palynological evidence from Mousterian sites in the Levant is rare, especially in the southern, drier reaches of the region where information is limited to Horowitz's (1976, 1979:245) analyses of deposits in the Negev and Fish's (Clark *et al.* 1988a) work on sediments from the site of Ain Difla (WH634) in west-central Jordan. Pollen samples collected from the deposits of Rosh Ein Mor (D15) and Nahal Aqev (D35) furnished spectra containing relatively high arboreal components (25%), dominated by Hermes oak (*Quercus calliprinos*) and olive, prompting Horowitz (1979:245) to suggest a relatively moist Mediterranean woodland setting for the area. Despite its much higher (600–700m) and therefore inherently moister setting than the sites of the central Negev Highlands, the arboreal component of Tor Sabiha's pollen spectrum is significantly smaller (12%) than that of the Negev sites.

These implied drier conditions appear to have supported a more steppic setting than in the Negev with only isolated stands of trees along wadi courses (elm and alder) and in better watered areas of the uplands (pine). Given that the Negev deposits are dated to ca. 85,000 B.P. (Marks 1981) in contrast to the ca. 70,000 B.P. date for the Tor Sabiha deposit (Henry and Miller 1992), a climatic shift toward drier conditions appears to have occurred sometime between 85,000–70,000 B.P. Bar-Yosef (1989) suggests that the moist interval associated with the Early Levantine Mousterian stretched from Oxygen Isotope Stage 5d/e to 5a (ca. 110,000–74,000 B.P.) followed by colder and drier conditions during Stage 4 (ca. 74,000–50,000 B.P.). Other reconstructions have placed the onset of drier conditions in Stage 4, but at later times of ca. 70,000–62,000 B.P. (Farrand 1979; Goldberg 1986). Regardless of specific chronology, Tor Sabiha would appear to fall in the early phase of the dry interval that characterized the late Levantine Mousterian. In lacking any evidence of arboreal pollen, despite its location on a spring-fed wadi, the deposit of Ain Difla was most likely formed under hyperarid conditions and therefore probably dates late (ca. 60,000–50,000 B.P.) in the Stage 4 arid interval (Clark *et al.* 1988a).

Palynological evidence from Transitional and Early Upper Paleolithic deposits in the Levant is confined to Horowitz's (1979:245–247) analysis of the Negev site of Boker Tachtit, dated to ca. 40,000 B.P. Although Horowitz (1979:247) stresses the tentative nature of the results because of the small number of grains (only 90), the moderate percentage of arboreal pollen (17%) and high proportions of Gramineae (59%) in the NAP component imply that more moist conditions had come to replace the dry interval characterizing Oxygen Isotope Stage 4. This trend of climatic amelioration appears to have continued into Early Ahmarian times (i.e., ca. 38,000–30,000 B.P.) as evidenced by moderate percentages of arboreal pollen from Upper Paleolithic sites in the Negev (Horowitz 1979:247). Within southern Jordan this slightly humid episode is defined within the Upper Paleolithic deposits of those sites in the Jebel Qalkha area by low–moderate arboreal pollen and moderate–high proportions of Gramineae along with ferns (Filicales). Within the core Mediterranean woodlands of the northern Jordan Valley this episode is reflected in subzone A-0 of the Lake Huleh pollen diagram (Tsukada diagram published in Bottema and Van Zeist 1981). The subzone is characterized by high arboreal percentages and the upper half of the zone is dated from ca. 30,000–24,000 B.P.

Following this slightly moist episode, drier conditions appear to have emerged during late Upper Paleolithic times. Pollen from the deposits of the Negev sites of Ein Aqev (dated to 18–17,000 B.P.) and Ein Aqev East point to a steppic setting with isolated stands of trees. Within southern Jordan, the rise in the percentages of Chenopodiaceae (type *Noaea*) and of Compositae liguliflorae pollen within the Jebel Humeima (J412) deposit may well represent an earlier part of this interval. Similarly, the Upper Paleolithic deposit at Site 618 in the Wadi Hasa of west-central Jordan may rest within this arid episode as evidenced by a spectrum dominated by

Chenopodiaceae and *Artemisia* and lacking trees (Clark *et al.* 1988a). In subzone A-1 of the Lake Huleh Diagram, this arid cycle is denoted by a marked decline in tree pollen percentages that reaches the lowest values recorded for the entire diagram. Associated $C^{14}$ dates place subzone A-1 from ca. 24,000 to sometime after 16,000 B.P. (Bottema and Van Zeist 1979:115). Leroi-Gourhan (1981) suggests that the interval reached its maximum aridity between 22,500 and 19,500 B.P., a time of intense cold in Europe.

With the beginning of the Epipaleolithic (characterized by the Kebaran Complex in the Mediterranean woodlands) at ca. 20,000–19,000 B.P. the climate was dry and probably cold for much of the Levant. The Mediterranean Sea fell to one of its lowest levels and interior lakes receded. Given these conditions for the Mediterranean woodland zone, there is evidence for a moist setting dated to between ca. 23,000–15,000 B.P. for sites in the Eastern Desert of Jordan (Garrard *et al.* 1988:333).

Other deposits falling within the Kebaran Complex also indicate the presence of a humid phase. In coastal Lebanon, pollen recovered from the deposits of Ksar 'Akil and Jiita point to a moist episode (Leroi-Gourhan 1980) as do the palynological (Darmon 1986) and sedimentological (Goldberg 1981) studies of the Fazael VII deposit near Jericho. The Early Hamran occupation of Jebel Mishraq (J504) probably took place during this episode as indicated by the high arboreal percentages. Although undated chronometrically, the technotypologic configurations of the assemblage suggest a placement ca. 16,000–14,500 B.P. (Chapter 13). Site 1065, located in the Wadi Hasa of west-central Jordan and radiocarbon dated between ca. 17,000–15,500 B.P., also may be related to this episode (Clark *et al.* 1988a). Although dominated by Chenopodiaceae and Compositae, the presence of oak, alder, willow, and cattail (*Typha*) pollen points to a moist setting (Clark *et al.* 1988a:272).

The interval from 14,500–12,000 B.P. (equivalent with the Geometric Kebaran and Mushabian complexes) appears to have been associated for the most part with favorable climatic conditions as indicated by an intensive occupation of the arid zone (Henry 1983; Bar-Yosef and Belfer-Cohen 1989). At the very beginning of the interval within the study area, however, cold, dry conditions are indicated for the deposit of Site J26. An episode of climatic amelioration appears to have ensued. Regionally, the palynological evidence of this episode is denoted at the base of Layer E (Geometric Kebaran) in the Hayonim Terrace diagram where a temperate, moderately humid phase is defined by a rise in arboreal pollen, especially the carob-tree and oak (Henry *et al.* 1981). Similarly, the analysis of the Fazael VIII deposit reveals a moderate percentage (13%) of arboreal pollen composed largely of olive, Mediterranean oak, and tamarisk (Darmon 1988). These results are consistent with those obtained from the Middle Hamran deposits of the study area in which *Quercus ithaburense*, Gramineae, leguminous and aquatic plants point to moist, temperate settings.

The terminal phases of the Epipaleolithic appear to have experienced relatively rapid climatic fluctuations. Layer C of the Early Natufian occupation at Wadi Judayid (J2) was associated with a short dry phase dated from ca. 12,700–12,100 B.P. A similar episode is defined in middle Layer E of the Hayonim Terrace immediately prior to ca. 12,000 B.P. (Henry *et al.* 1981). These show trees to decline or disappear, while herbaceous plants become dominated by Chenopodiaceae (type *Atriplex*) and Compositae liguliflorae.

This dry phase was succeeded by a humid period that is well defined within Early Natufian deposits dated between ca. 12,000–11,000 B.P. (Henry *et al.* 1981; Henry 1983, 1989; Leroi-Gourhan 1981; Leroi-Gourhan and Darmon 1991; Baruch and Bottema 1991). Within the Wadi Judayid (J2) deposit, this phase corresponds to Layer B and is associated with a marked increase in the pollen of trees and of aquatic plants. Although not represented in the study area by palynological evidence, the Late Natufian (ca. 11,000–10,300 B.P.) was associated with a dry phase that brought the Pleistocene to a close.

The most recent palynological evidence within the study comes from mid-Holocene, Chalcolithic deposits. Data from the Timnian deposits of Tor Sabiha (J8), Jebel el Jill (J14), and Jebel Queisa (J24) are nearly identical in showing an absolute dominance of pollen of Chenopodiaceae (type *Noaea*) coupled with an absence of trees. In defining warm, quite arid conditions, this interval most likely corresponds to the arid phase defined by Goldberg and Rosen (1987:29–30) as falling between 7,000–5,500 B.P. The $C^{14}$ date from Jebel Queisa (J24) of ca. 5,700 B.P. furnishes additional support for this placement, but the ca. 4,000 B.P. date from Tor Sabiha (J2) does not.

## CONCLUSION

Based upon the interpretations drawn from variations in proportions of pollen of herbaceous plants, humid episodes are indicated for the Upper Paleolithic between ca. 38,000–24,000 B.P. and within the Epipaleolithic between ca. 16,000–14,500 B.P. (Early Hamran), ca. 13,000–12,500 B.P. (Late Hamran), and ca. 12–11,000 B.P. (Early Natufian). These episodes appear to have regional significance in that they also have been defined palynologically in Israel, Lebanon, and Syria. The chronology of these humid cycles hints of their having been linked to periods of warmer temperatures as defined for Western Europe.

# Chapter 17

# Shells from the Wadi Hisma Sites

DAVID S. REESE

## INTRODUCTION

Shells recovered from the sites were principally used as ornaments as indicated by their relatively distant, mostly Red Sea sources and the large number that have been worked. Interest in ornamental shell appears to have been the greatest during the late Epipaleolithic. This finds expression not only in the greater density of shells in the deposits, but also in the presence of a Mediterranean species.

Twelve of the investigated sites yielded shells. Almost all of the 771 marine shells come from the Red Sea, which at its closest point (Aqaba) is some 38–67 km from the individual sites. Four sites also produced 10 *Arcularia gibbosula* which come from the Mediterranean Sea, 215–230 km away. There are also 2 fresh-water-gastropods.

There are 11 Upper Paleolithic shells (3 sites), 9 mixed Upper Paleolithic/Epipaleolithic shells (1 site), 34 Qalkhan shells (2 sites), 515 Mushabian shells (66.8%; 1 site), 78 Hamran shells (10.1%; 4 sites), 76 Natufian shells (9.9%, 2 sites), 3 Neolithic shells (1 site), and 45 Chalcolithic shells (5.8%; 2 sites).

The majority of the shells (565, 73.3%) come from Upper Paleolithic and Epipaleolithic Tor Hamar (J431), 56 from the Epipaleolithic (Late and Final Hamran) Qa Salab rockshelter (J202, 7.3%), 73 from Epipaleolithic (Early Natufian) Wadi Judayid (J2, 9.5%), and 44 from Chalcolithic Jebel el Jil (J14, 5.7%).

Additional comparanda to that noted here can be found in recent papers (Reese 1991, 1989).

## SHELL EXPLOITATION BY PERIOD

### Upper Paleolithic Shells

There are 3 shells from the two Levantine Aurignacian sites of Jebel Humeima (J412) and J403. The Early Ahmarian in Tor Hamar (J431) produced 8 shells (5 *Dentalium*, 1 small gastropod, 1 *Pecten*/cockle, 1 *Ancilla*).

The mixed Ahmarian and Early Mushabian in J431 yielded 9 shells (7 small gastropods [3 holed, 2 burnt], 1 *Dentalium*, and 1 *Pecten*).

The rarity of marine shells from the Upper Paleolithic Wadi Hisma sites is typical for most sites of this period in the Levant. The Ahmarian hunting camps of Qadesh Barnea 9 and 601 (5 shells) in the Negev produced several dentalia. The late Upper Paleolithic site of Ein Aqev (D31) in the Central Negev produced 17 or 18 shells with 2 or 4 *Dentalium*, 11 *Arcularia*, and 1 small gastropod. Terminal Upper Paleolithic Ein Aqev East (D34) produced 4 *Dentalium* and 1 bivalve.

## Qalkhan Complex Shells

There are 34 Qalkhan shells from two sites. Tor Hamar (J431) Block II produced 33 shells, with 12 *Dentalium* (36.4%; 3 beads, 2 burnt), 13 small gastropods (39.4%; 7 holed, 1 ring bead, 1 burnt), 2 *Columbella* (2 holed), 2 *Pecten* (2 fragments, 1 burnt), 1 *Ancilla* (open ventral side), 1 *Arcularia* (open body), 1 *Ostrea* (small, hole below umbo, worn), and 1 cockle (fragment). J406a yielded 1 *Cypraea* (lip fragment).

## Hamran Industry (Kebaran and Geometric Kebaran Complexes) Shells

There are 78 Hamran shells from four sites. Early Hamran J504 (Kebaran Complex) produced 6 shells with 3 small gastropods (3 holed), 2 *Dentalium* (1 bead), and 1 *Mitra* (holed).

Early Hamran J201 produced 2 shells, a *Dentalium* bead and 1 bivalve fragment. The Middle Hamran (Geometric Kebaran Complex) here produced 3 shells, with 2 *Dentalium* and 1 *Arcularia* (open dorsum).

Middle Hamran El Quweira (J203) yielded 1 shell, a holed gastropod. The Late Hamran (Geometric Kebaran Complex) here produced 10 shells with 4 *Dentalium* (3 beads), 2 small gastropods (1 holed, 1 broken), 1 *Glycymeris* (hole below umbo, water-worn), 1 *Pecten* (fragment), 1 *Cerithium/Bittium*, and 1 bivalve (fragment).

Late Hamran Qa Salab (J202) produced 15 shells with 5 *Dentalium* (33.3%; 4 beads), 2 small gastropods (1 holed), 2 *Arcularia* (2 holed), 1 *Mitra* (holed), 1 *Cypraea* (fragment), 1 *Glycymeris* (fragment), 1 *Pecten* (fragment), and 2 fragments.

There are 41 shells from the Final Hamran (Geometric Kebaran Complex) of J202 with 18 *Dentalium* (43.9%; 17 beads, 2 burnt), 8 small gastropods (1 holed, 1 burnt, 1 broken), 5 bivalves (1 holed, 4 fragments), 4 *Pecten* (4 fragments), 2 *Arcularia* (2 holed), 1 *Glycymeris* (fragment, hole at umbo), 1 *Conus* apex (holed in center, water-worn), 1 *Cypraea* (complete), 1 ?vermetid, 1 fragment (unidentifiable).

## Madamaghan Industry (Mushabian Complex) Shells

There are 485 Madamaghan shells, all from Tor Hamar (J431). The Early Mushabian here produced 261 shells, with 177 *Dentalium* (67.4%; 5 beads, 24 burnt, 1 holed on side, 2 possibly worm tubes), 54 small gastropods (20.7%; 24 holed, 9 burnt, 4 broken), 8 *Pecten* (8 fragments, 1 with slit hole below umbo, 3 burnt), 1 *Pecten*/cockle (fragment), 7 *Cypraea* (4 holed, 2 fragments, 1 burnt), 3 *Arcularia* (1 open body, 2 apical fragments, 3 burnt), 3 *Ancilla* (2 holed), 2 *Glycymeris* (small, slit hole below umbo; fragment), 2 cockles (2 fragment, 1 large individual), 1 *Gibbula* (small, holed, burnt), 1 gastropod (body fragment), 1 medium gastropod (holed), and 1 bivalve (umbonal fragment).

There are 41 shells from mixed Early/Late Mushabian levels, with 23 *Dentalium* (56.1%; 1 burnt), 12 small gastropods (29.3%; 6 holed, 4 burnt, 2 broken), 3 *Pecten* (3 fragments, 1 burnt), 2 *Ancilla*, and 1 *Cerithium* (small, holed).

There are 195 Late Mushabian shells, with 110 *Dentalium* (56.4%; 7 beads, 8 burnt), 55 small gastropods (28.2%; 24 holed, 3 burnt, 2 broken), 14 *Pecten* (14 fragments, 2 with slit holes below umbo, 2 burnt), 2 *Pecten*/cockle fragments (1 burnt), 3 *Ancilla* (2 open body, 1 holed at apex and burnt), 3 *Cerithium* (3 small, 2 holed), 2 cockles (2 fragments), 1 *Conus* (medium, worn, holed on body and on ventral side), 1 *Columbella* (holed), 1 *Gibbula* (small, holed, open apex), 1 *Cypraea* (small, holed at one end), 1 gastropod (medium, fragment) and 1 *Glycymeris* (fragment).

The Mushabian in Block II produced 18 shells: 12 *Dentalium* (5 beads), 5 small gastropods (1 holed, 1 burnt), and 1 *Pecten* (fragment).

*Dentalium* make up between 56 and 67% of the shells from the Mushabian, and small gastropods account for between 21 and 28%. One ornament type found only in the Mushabian is *Pecten* (and 1 *Glycymeris*) with slit holes, probably made by sawing or filing with a flint blade. Scallops are found in earlier periods (but not in the later Natufian and Chalcolithic here), but do not preserve these holes.

The Wadi Madamagh rockshelter near Beidha, with a Madamaghan Industry of the Mushabian Complex, produced 7 holed marine shells with 4 small gastropods and 3 *Arcularia*. The number of Mediterranean *Arcularia* here is greater than one would expect for this small collection.

### Natufian Complex Shells

There are 76 Natufian shells from two sites. Early Natufian Wadi Judayid (J2) produced 73 shells with 54 *Dentalium* (74.0%; 49 beads, 4 burnt), 13 *Nerita* (17.8%; 9 holed, 1 burnt, 4 broken), 3 gastropods (1 holed, 1 burnt), 2 bivalve fragments, and 1 coral.

Late Natufian J406a produced only 3 shells, with 2 *Dentalium* beads and 1 *Arcularia* (open dorsum).

Early Natufian Beidha produced 74 marine invertebrate fragments, including 64 *Dentalium* (12 beads, 86.5%), 3 small gastropods (2 holed, 1 burnt), 1 *Nerita* (holed), 1 *Pecten* (fragment), 1 *Cypraea*, 1 worm tube, and 1 coral (fossil) (Reese 1989).

### Neolithic Shells

Jebel Queisa (J24), an Early Neolithic hunting camp, produced 3 shells: 1 *Nerita*, 1 *Conus* apex, and 1 gastropod (holed).

### Timnian (Chalcolithic) Shells

There are 45 Chalcolithic shells from two sites. Jebel el Jill (J14) yielded 44 shells with 14 gastropods (31.8%: 8 holed, 1 broken), 10 *Dentalium* (22.7%; 9 beads), 6 *Conus* (3 holed, 1 broken), 8 *Mitra* (8 holed), 3 *Nerita* (1 holed, 1 unholed, 1 broken), 3 bivalves (?*Chama*, "*Venus*," "*Venus*"/*Tapes*), and 1 coral. J17 produced a single gastropod (unholed, burnt).

## THE SPECIES REPRESENTED

### Dentalium (Tusk or Tooth shells; 437 examples)

The majority of all shells found, 55.9%, are *Dentalium*. In Block I of J431 the Early Ahmarian produced 8 shells with 5 *Dentalium*. The mixed Early Ahmarian and Early Mushabian produced 9 shells with 1 small *Dentalium*.

The Qalkhan produced 33 shells with 12 *Dentalium* (36.4%, 3 beads, 1 burnt). The Early Mushabian produced 261 shells with 177 *Dentalium* (67.8%, 25 burnt, 5 beads, 1 holed on the side, 1 possibly a worm tube).

The Early and Late Mushabian produced 41 shells with 23 *Dentalium* (56%, 1 burnt). The Late Mushabian here yielded 195 shells with 110 *Dentalium* (56.4%, 7 beads, 8 burnt). The Mushabian levels in Block II produced 18 shells with 12 *Dentalium* (66.7%, 5 beads).

Two of the 6 Early Hamran shells from J504 are *Dentalium* (1 bead). Of the 5 Hamran shells from J201 there are 3 *Dentalium* (1 bead). Four of the 10 Late Hamran shells from J203 are *Dentalium* (4 beads). Five of the 15 shells from Late Hamran J202 (Qa Salab) are *Dentalium* (3 beads). Of the 40 Final Hamran shells here there are 18 *Dentalium* (45%, 17 beads) with 2 burnt.

Of 3 Late Natufian shells from J406a there are 2 dentalia (2 beads). There are 54 *Dentalium* (74%, 49 beads), with 3 burnt of the 73 Early Natufian shells from J2. There are 9 *Dentalium* (20.5%, 9 beads) of the 44 shells from Chalcolithic J14.

### Small Gastropods (247 shells)

At J431 the Early Ahmarian produced 1 small gastropod (holed) while the Early Ahmarian/Early Mushabian has 7 small gastropods (3 holed, 2 burnt). The Qalkhan produced 13 small gastropods (39.4%, 7 holed, 1 ring bead, 1 burnt).

The Early Mushabian here produced 54 small gastropods (20.7%, 24 holed, 9 burnt, 4 broken, 1 water-worn, 1 worn). The mixed Early and Late Mushabian here produced 12 small gastropods (29.3%, 6 holed, 4 burnt, 2 broken) and the Late Mushabian 55 small gastropods (33.7%, 25 holed, 3 burnt, 2 broken). The Mushabian in Block II produced 5 small gastropods (1 holed, 1 burnt).

### Pecten (Scallops; 39 shells with 4 *Pecten*/cockle)

The Early Ahmarian at J431 produced 1 *Pecten*/cockle fragment (burnt). The Early Ahmarian/Early Mushabian produced 1 *Pecten* (fragment). The Early Mushabian yielded 8 fragments (1 with slit hole below umbo, 3 burnt) and 1 *Pecten*/cockle fragment. The mixed Early and Late Mushabian produced 3 fragments(1 burnt).

The Late Mushabian here produced 14 fragments (2 burnt, 2 with slit holes below umbo) and 2 *Pecten*/cockle fragments. (1 possibly burnt). In Block II the Qalkhan produced 2 *Pecten* fragments (1 burnt) while the Mushabian yielded 1 fragment.

The Late Hamran of J203 produced 1 *Pecten* fragment. The Late Hamran in J202 produced 1 fragment and the Final Hamran here 4 fragments.

*Pecten* are not a significant component of most Levantine shell collections. However, they are found in considerable numbers, second to *Dentalium*, at the Harifian sites of Abu Salem (448 shells, 71% dentalia and 10.5% *Pecten*), Ramat Harif, and Shluhat Harif in the Negev (Reese 1991:622).

### Nerita (Nerites; 17 shells)

*Nerita* are only found at the latest sites in the collection. There are 13 *Nerita* (17.8%, 9 holed, 4 broken, 1 burnt) of 73 shells from Early Natufian J2. Chalcolithic J14 produced 44 shells, with 3 *Nerita* (1 ground-down and holed, 1 unholed, 1 broken). Contemporary J24 produced a total of 3 shells, with 1 example.

### Other Bivalves (13 shells)

There is 1 bivalve (umbonal fragment) from Early Mushabian J431. There are 1 fragment from Early Hamran J201 and 1 from Late Hamran J203. Final Hamran J202 produced 5 bivalves (1 holed, 4 fragments). Early Natufian J2 has 2 bivalve fragments. There are 3 bivalves from Chalcolithic J14 (1 *Venus*, 1 *Venus/Tapes* fragment, 1 ?*Chama*).

### Cypraea (Cowries; 11 shells)

The Qalkhan from J406a produced 1 *Cypraea* (lip fragment). There is 1 *Cypraea* (fragment) from the Late Hamran of J202 and 1 *Cypraea* (complete) from the Final Hamran here. Early Mushabian J431 produced 7 *Cypraea* (4 small, 4 holed at one end, 2 broken, 1 burnt) and there is 1 from the Late Mushabian here (small, holed at one end).

## Mitra (Mitra shells; 10 shells)

There is 1 *Mitra* (holed, black) from the Early Hamran of J504 and 1 holed example from the Late Hamran of J202. There are 8 *Mitra* (18.2%, 8 holed) from Chalcolithic J14.

## Arcularia (Nassa or Basket shells; 10 shells from three sites)

There are 1 *Arcularia* from the Qalkhan (3%; open body) and 2 from the Early Mushabian of J431 (1.1%; 1 with open body, 3 burnt).

The Middle Hamran of J201 yielded 1 *Arcularia* (open dorsum) of 3 shells present. There are 2 *Arcularia* of 15 shells from the Late Hamran of J202 (13.3%; 1 open dorsum, 1 ?ground-down dorsum) and 2 from the Final Hamran here (4.9%; 2 holed). The Late Natufian of J406a produced 1 *Arcularia* of 4 shells found (open dorsum).

It is noteworthy that there are no *Arcularia* in the Chalcolithic Wadi Hisma samples, suggesting there was no contact with the Mediterranean area at that time.

## Ancilla (9 shells)

All *Ancilla* come from J431. There is 1 from the Early Ahmarian, 3 (2 holed) from the Early Mushabian, 2 from mixed Early/Late Mushabian levels, 2 from the Late Mushabian (2 open body, 1 holed at the apex and burnt) and 1 from the Qalkhan (open ventral side).

## Conus (Cone shells; 8 shells)

There is 1 *Conus* apex (holed in center, water-worn) from the Final Hamran of J202. There is 1 *Conus* (medium-sized, worn, holed on body and on back) from the Late Mushabian of J431. There are 5 *Conus* (3 holed) from Chalcolithic J14 and 1 *Conus* (apex) from contemporary J24.

## Cockles (6 shells)

There is 1 cockle fragment from the Levantine Aurignacian site of J412. At J431 there are 1 fragment from the Qalkhan, 2 fragments from the Early Mushabian, and 2 fragments from the Late Mushabian.

## Glycymeris (Dog-cockles; 6 shells)

There is 1 *Glycymeris* (water-worn, hole below umbo) from the Late Hamran of J203. There is 1 *Glycymeris* (fragment) from the Late Hamran and 1 fragment (holed at umbo) from the Final Hamran of J202.

There are 2 *Glycymeris* (small, slit hole below umbo, fragment) from the Early Mushabian and 1 fragment from the Late Mushabian of J431.

## Cerithium (Horn shells; 5 shells)

There is 1 *Cerithium* sp./*Bittium* sp. from the Late Hamran of J203. There is 1 *Cerithium* (small, holed) from mixed Early/Late Mushabian levels at J431 and 3 small *Cerithium* (2 holed) from the Late Mushabian here.

## Columbella (Dove shells; 3 shells)

There are 2 *Columbella* (1 with 2 holes on body; holed) from the Qalkhan of J431 and 1 *Columbella* (holed) from the Late Mushabian here.

## Gibbula (Top shells; 2 shells)

There is 1 *Gibbula* (small, holed, burnt) from the Early Mushabian of J431 and 1 *Gibbula* (small, holed, open apex) from the Late Mushabian here.

## Coral (2 pieces)

Early Natufian J2 produced 1 coral (noted by P. Turnbull in Henry *et al.* 1985 but not seen by me). Chalcolithic J14 produced 1 *Stylophora* coral.

Late PPNB ad-Daman 1, about 90 km from the Red Sea, produced 1 *Favia favus* (Forskål 1775).

## Ostrea (Oyster; 1 shell)

There is 1 *Ostrea* (small, hole below umbo, worn) from the Qalkhan of J431.

## Freshwater Shells (2 shells)

There are only 2 freshwater shells, both *Melanopsis praemorsa*, from the Early Ahmarian of Tor Hamar (J431): one from Layer G (5A/200–210) which is holed and partly burnt, and one from Layer F (5B/180–190) which is possibly holed.

# CONCLUSIONS

The shell collection from the various Wadi Hisma sites are consistent with that from neighboring and contemporary sites. Most common are *Dentalium* shells, often cut or broken into beads. Various holed small gastropods are also frequently found.

The Red Sea shells may have been collected by the occupants of the various sites; seven of the 12 sites are less than 50 km from Aqaba. *Arcularia*, present in small numbers from the Qalkhan to the Late Natufian, indicate that these people were in contact with some individuals who occasionally visited the Mediterranean coast.

Chapter 18

# Cementum Increment Analyses of Teeth From Wadi Judayid (J2) And Tor Hamar (J431): Estimations of Site Seasonality

DANIEL E. LIEBERMAN

Cementum increment analysis has considerable potential for resolving hypotheses about ancient hunter-gatherer seasonality and mobility. This chapter reports the results of analyses of seasonally deposited cementum increments of Mountain gazelle (*Gazella gazella*) from two sites in the Wadi Hisma: Wadi Judayid and Tor Hamar. According to these data, Wadi Judayid appears to have been occupied during both the wet and dry season, and Tor Hamar appears to have been a winter season encampment.

## ESTIMATION OF SITE SEASONALITY

Cementum increment analysis is a relatively new technique that can provide accurate and plentiful data on the season of occupation of archaeological sites both in the Levant and elsewhere (Lieberman *et al.* 1990). Seasonality data has often been unobtainable from many sites in the Levant because of the poor preservation of plant materials and the relative scarcity of large faunal samples usually needed to make such estimations. Yet estimation of the season of occupation of Levantine archaeological sites is crucial for modeling the proposed changes in mobility patterns that have figured in many hypotheses of cultural evolution in the region (e.g., Bar-Yosef 1970, 1981, 1983, 1987; Vita-Finzi and Higgs 1970; Henry 1983, 1985, 1989; Marks and Friedel 1977; Coinman *et al.* 1986). Many of these archaeological models are based on patterns of transhumance observed among recent Bedouin, or on the application of archaeological correlates of mobility to either Binford's (1980) model of logistical versus residential mobility or Mortensen's (1972) dichotomy between circulating and residential mobility.

Generally applicable *a priori* data on the seasonal use of resources from Levantine archaeological sites are needed to test empirically previously proposed models of mobility and site organization. Commonly used archaeological correlates of relative nonmobility—such as substantial architecture, high lithic density, large site size, and the presence of cemeteries—are probably better indicators of intensity of site use than seasonality (Rafferty 1985; Edwards 1989). Unfortunately, botanical remains are rare or absent in all but a few Levantine sites. Archaeofaunal remains provide the best sources of information of season of occupation, but have so far been relatively scant or not widely applicable. For example, migratory animals (usually birds) are a potentially excellent source of data on seasonality, but are only present at a handful of sites such as Ain Mallaha and Hayonim (Pichon 1992). In addition, Davis (1983) has demonstrated that seasonality can be estimated by comparing the mortality profiles of gazelle from archaeological sites with different seasonal age profiles of naturally occurring populations; while this technique has much potential, it requires large samples of juveniles and assumptions about the hunting strategies employed. Furthermore, Tchernov (1984, 1991) has used the presence of human commensals at Ain Mallaha and Hayonim such as *Rattus rattus*, *Mus musculus*, and *Passer domesticus* to argue for relatively long-term or permanent human occupations at these sites. While human commensals may be good indicators of sedentism (*contra* Tangri and Wyncol 1989), they are not useful for estimating the season of occupation for most sites.

## CEMENTUM INCREMENT ANALYSIS

Analyses of cementum increments in gazelle teeth have the most potential for providing estimates of the season of site occupation. Cementum is a mineralized tissue (essentially woven bone)—comprised of approximately 60% hydroxyapatite and 40% collagen—that surrounds the roots of mammalian teeth, anchoring them to the gum (periodontal ligament) by means of collagen fiber bundles known as Sharpey's fibers (Figure 18.1). Cementum is continually laid down in bands around the tooth root from the onset of occlusion until death or the tooth falls out. Cementum can be divided into two types, cellular and acellular cementum, based on

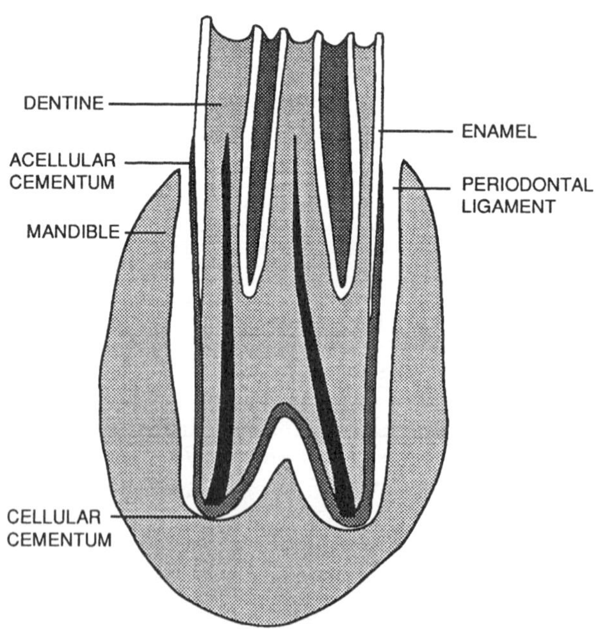


Figure 18.1. Schematic cross-section of a gazelle mandible showing the distribution of cellular cementum, acellular cementum, and other dental tissue.

function and histology. An understanding of the differences between cellular and acellular cementum is important for interpreting the results of their analysis (for a detailed review of the biology of cementum, see Lieberman and Meadow 1992; and Lieberman 1994).

## Cellular Cementum

In almost all mammals, cementum at the apical end of tooth roots is cellular (so named because of cementocytes that become entrapped in the cementum matrix) and is rapidly laid down in large quantities during periods of tooth eruption (see Figure 18.1). Bands of cellular cementum thus serve to fill voids between the periodontal ligament and the tooth root as the tooth erupts. The extremely narrow bands laid down between these cellular bands—termed Lines of Arrested Growth (LAGs) by some histologists (de Riqles *et al.* 1991)—are acellular and hypermineralized and result from periods in which no tooth eruption occurs. Given the high rate of variation of cellular cementum growth at the end of the tooth root, cellular cementum is most useful for determining the age at death of most mammalian species (Klevezal and Kleinenberg 1967; Grue and Jensen 1979), but is less useful for estimating the season of death (Lieberman and Meadow 1992).

## Acellular Cementum

Acellular cementum, which covers the rest of the tooth root up to the base of the enamel crown, is more useful for estimating the season of death in small to medium-sized mammals (large mammals tend to have predominantly cellular cementum on their tooth roots). Acellular cementum grows slowly and constantly throughout the year in bands whose optical quality varies under polarized transmitted light microscopy (Figure 18.2). In most mammals, acellular cementum bands formed during seasons of growth tend to be translucent, and bands formed during seasons of reduced growth tend to be opaque (Klevezal and Kleinenberg 1967; Lieberman *et al.* 1990). These acellular bands are most important for serving to attach the tooth root to the alveolar bone of the jaw by mineralizing around Sharpey's fibers that originate in the Periodontal Ligament. The proximate biological mechanisms of cementum growth are too complex to review here (Lieberman 1933b, 1994), but it is important to stress that the optical differences between bands result primarily from changes in the levels of strain (deformation) generated during the mastication of diets in different seasons. Seasonal changes in diet can result in alterations in the orientation of Sharpey's fibers in alternate bands, causing variations in the birefringence of bands laid down in certain seasons, and from slight differences in the mineral density of bands. Consequently, the number of bands reflects the number of seasonal growth periods since the eruption of the tooth, allowing relatively accurate estimation of age at death. The outermost acellular cementum band in a tooth indicates the season of death of the individual.

### Gazelle

The season of occupation of archaeological levels from the southern Levant can best be estimated by determining the seasons during which Mountain gazelle (*Gazella gazella*) were hunted. Gazelle are particularly suited for estimating site seasonality for several reasons. Previously published studies have demonstrated that changes in the mineralization of Sharpey's fibers in acellular cementum in Levantine gazelle produce bands that are well-correlated with their seasonal diets, which consist of primarily browsing during the winter rain season and of grazing during the spring and summer dry season (for details, see Lieberman *et al.* 1990; Lieberman and Meadow 1992). As shown in Figure 18.2, translucent acellular bands in gazelle are laid down in the spring/summer (between roughly April and October) and opaque acellular bands are deposited during the fall/winter (between roughly November and March).

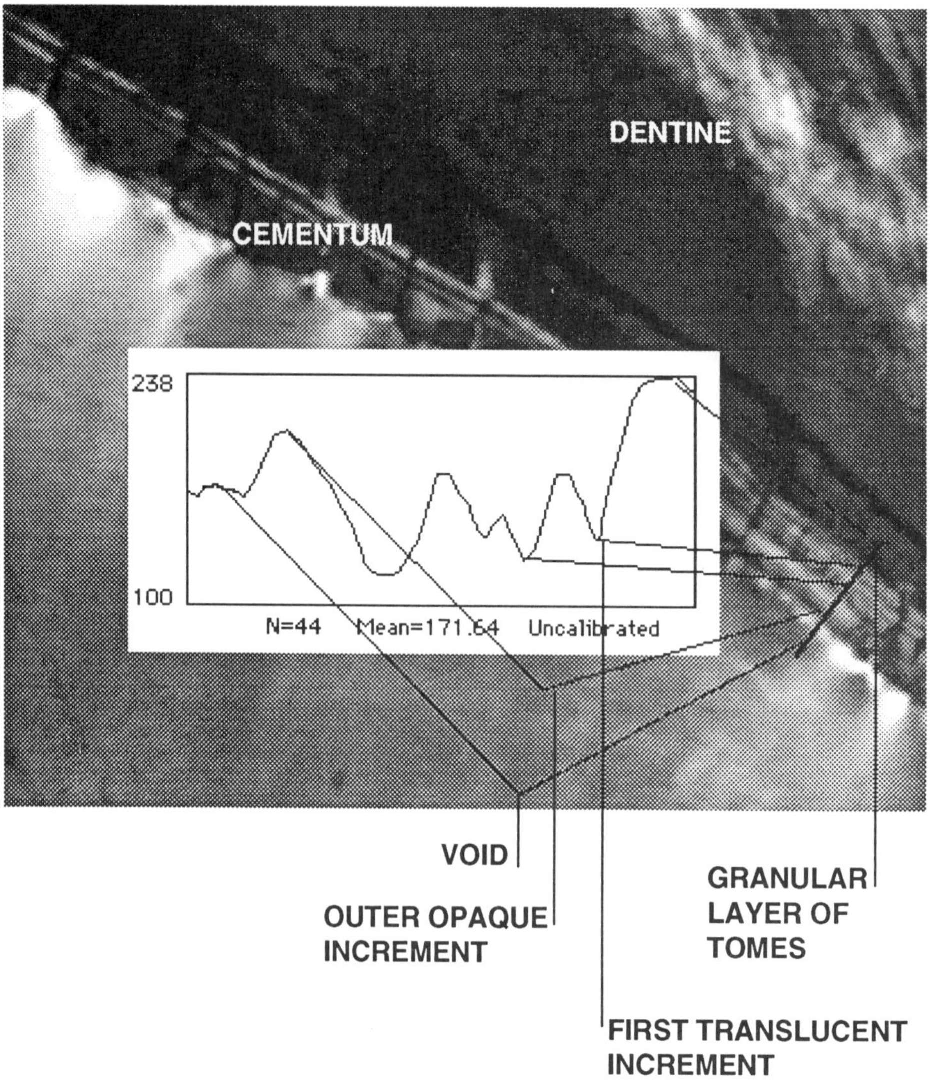


**Figure 18.2.** Transmitted light polarized micrograph (100X) of a modern Mountain Gazelle (*Gazella gazella*) molar thin-section showing acellular cementum increments, along with luminance profile which quantifies the number of increments and the nature of the outermost increment. This individual has 8 increments, of which the outermost is opaque, indicating a winter death.

Gazelle were obviously just as attractive to Levantine hunter-gatherers as they are to faunal analysts interested in estimating seasonality. Mountain gazelle are the dominant fauna in the Levantine Paleolithic and Epipaleolithic, comprising between 60–90% of the fauna in most sites (Garrard 1980; Bar-Yosef 1981). This is because gazelle are territorial, nonmigratory, extremely rapid but seasonal breeders, and plentiful in both Mediterranean and Irano-Turanian environmental zones (Baharav 1981, 1983; Mendelssohn 1974). Gazelle, moreover, are extremely easy to hunt (Spinage 1986). Therefore the absence of gazelle at a site from a particular season most likely indicates that the site was not then occupied, although this should be confirmed using other sources of data if possible to avoid the possibility of overinterpreting the absence of data. Lieberman (1993a) has demonstrated that gazelle were hunted on the coastal plain during the Upper Pleistocene throughout the year.

## ANALYSIS AND RESULTS

### Sample Studied

Sample sizes were limited by the small number of adequately preserved samples of teeth from either the mandible or maxilla from Wadi Judayid (summarized in Table 18.1) and Tor Hamar (summarized in Table 18.2). Fifteen mandibular or maxillary fragments with teeth were analyzed from Wadi Judayid (J2) and 9 samples were analyzed from Tor Hamar (J431). In the Wadi Judayid sample, 3 teeth were Ovicaprids, and 2 teeth were missing their crowns and thus could only be identified as belonging to the family Bovidae (see Henry and Turnbull 1985). In the Tor Hamar sample, 2 teeth (LTH7,8) were caprids (see Henry and Garrard 1988). These teeth were analyzed as part of a larger study (Lieberman 1993a) on the evolution of hunter-gatherer mobility strategies in the southern Levant during the Late Quaternary.

### Method of Analysis

Only teeth embedded in the alveolar bone of the mandible or maxilla were analyzed to reduce possible false estimations of season of death resulting from loss of the outermost layer of cementum in loose teeth; all mandibles were first examined carefully to ensure that there were no gnaw marks from carnivores that would imply a nonanthropogenic origin. After measurement and documentation, the teeth and surrounding bone were embedded in Epotek 301™ epoxy. The blocks were then sectioned mesio-distally through the center of the root to expose the maximum amount of cementum, polished, and affixed to petrographic slides using more epoxy. The slides were then cut and ground down to a thickness of 30 to 50 microns until the histology of the tooth was clearly visible. Some teeth with silica inclusions or mineral staining needed to be ground thinner than better preserved specimens. Finally, all sections were polished on a 16-micron diamond wheel to ensure good optical resolution.

**Table 18.1. Wadi Judayid (J2) Samples Analyzed and Results**

| Sample | Species | Tooth used | Preservation | Number of bands | Outermost band |
|---|---|---|---|---|---|
| LWJ1 | *Bov. undet.*[a] | RM1-M2 | None | No cementum | No cementum |
| LWJ2 | *Bov. undet.*[a] | LM1/M2 | None | No cementum | No cementum |
| LWJ3 | *G. gazella* | RP4-M1 | Moderate | 8 | Opaque |
| LWJ4 | *Capra sp.* | LM1/M2 | Poor | 2 | Opaque |
| LWJ5 | *G. gazella* | RM2-M3 | None | No cementum | No cementum |
| LWJ6 | *G. gazella* | LM1 | Poor | 1 | Translucent |
| LWJ7 | *G. gazella* | RM3 | Poor | 7 | Translucent |
| LWJ8 | *G. gazella* | LM1 | Moderate | 11 | Translucent |
| LWJ9 | *Capra sp.* | RP2-P3 | None | No cementum | No cementum |
| LWJ10 | *G. gazella* | Ldp4 | Poor | 3 | Translucent |
| LWJ11 | *G. gazella* | LM2 | Poor | 2 | Opaque |
| LWJ12 | *G. gazella* | RP3 | Excellent | 5 | Translucent |
| LWJ13 | *G. gazella* | RM1 | Poor | ? | Translucent |
| LWJ14 | *Capra sp.* | LP2-P4 | Moderate | 4 | Opaque |
| LWJ15 | *Bov. undet.*[a] | RM1 | Poor | 3 | Translucent |

[a] Undetermined species of *Bovidae*.

Table 18.2. Tor Hamar (J431) Samples Analyzed and Results

| Sample | Species | Tooth used | Preservation | Number of bands | Outermost band |
|---|---|---|---|---|---|
| LTH1 | *G. gazella* | RP4-M1 | Moderate | 4 | Opaque |
| LTH2 | *G. gazella* | RP3-P4 | Poor | 6 | Opaque |
| LTH3 | *G. gazella* | LP4-M1 | Excellent | 4 | Opaque |
| LTH4 | *G. gazella* | RP2-P3 | Poor | 4 | Translucent |
| LTH5 | *G. gazella* | RM1 | None | No cementum | No cementum |
| LTH6 | *G. gazella* | RM3 | Excellent | 10 | Opaque |
| LTH7 | *Capra sp.* | LM1 | Poor | 4 | Opaque |
| LTH8 | *Capra sp.* | RM1 | Moderate | 4 | Opaque |
| LTH9 | *Bov. undet.*[a] | RM1-M2 | Moderate | 8 | Opaque |

[a]Undetermined species of *Bovidae*.

The teeth were studied and digitally photographed on a Olympus™ polarized transmitted light microscope connected to a MacIntosh II™ computer. To eliminate subjectivity, the specimens were analyzed in random order along with approximately 400 teeth from roughly 20 other sites so that the author did not know from what site a particular specimen derived until after a determination of seasonality was made. In addition, subjectivity and error were also reduced by analyzing all the teeth in this study using computer-image analysis techniques (Image, version 1.23) that are described in detail in Lieberman *et al.* (1990). Luminance profiles of the gray scale values of pixels across a transect (Figures 18.2 and 18.3) are an accurate and effective technique of assessing the number of increments and the nature of the outermost increment. The width of the increments were also measured in microns.

## Results

The results are reported and summarized in Tables 18.1 and 18.2.

### Tor Hamar

The Mushabian (or Madamaghan) rockshelter site of Tor Hamar was most likely occupied during the fall/winter, as predicted by Henry and Garrard (1988) and Henry (1989). Of 9 samples analyzed, 7 had opaque outer increments and only 1 had a thin translucent outer increment, suggesting that the gazelle at the site were mostly hunted during the winter and perhaps a few during the very early spring. One sample was too damaged from post-depositional alteration of the collagen to allow estimation of season of death.

### Wadi Judayid

Wadi Judayid presents a very different pattern from Tor Hamar, but typical of many contemporary Natufian sites. Of the 11 teeth that were in good enough shape to allow estimation of the outer cementum increment with a reasonable degree of confidence, 7 (64%) had outer translucent bands and 4 (32%) had outer opaque increments. Furthermore, the relative width of the outermost increments varied widely, suggesting that gazelle at the site were killed throughout much of the dry and wet seasons. It should be mentioned that although 15 teeth were analyzed from the site, only 11 were adequately preserved to allow some estimation of season of death—indeed most of the samples were in very poor shape with considerable evidence of collagen loss through leaching of the bone.

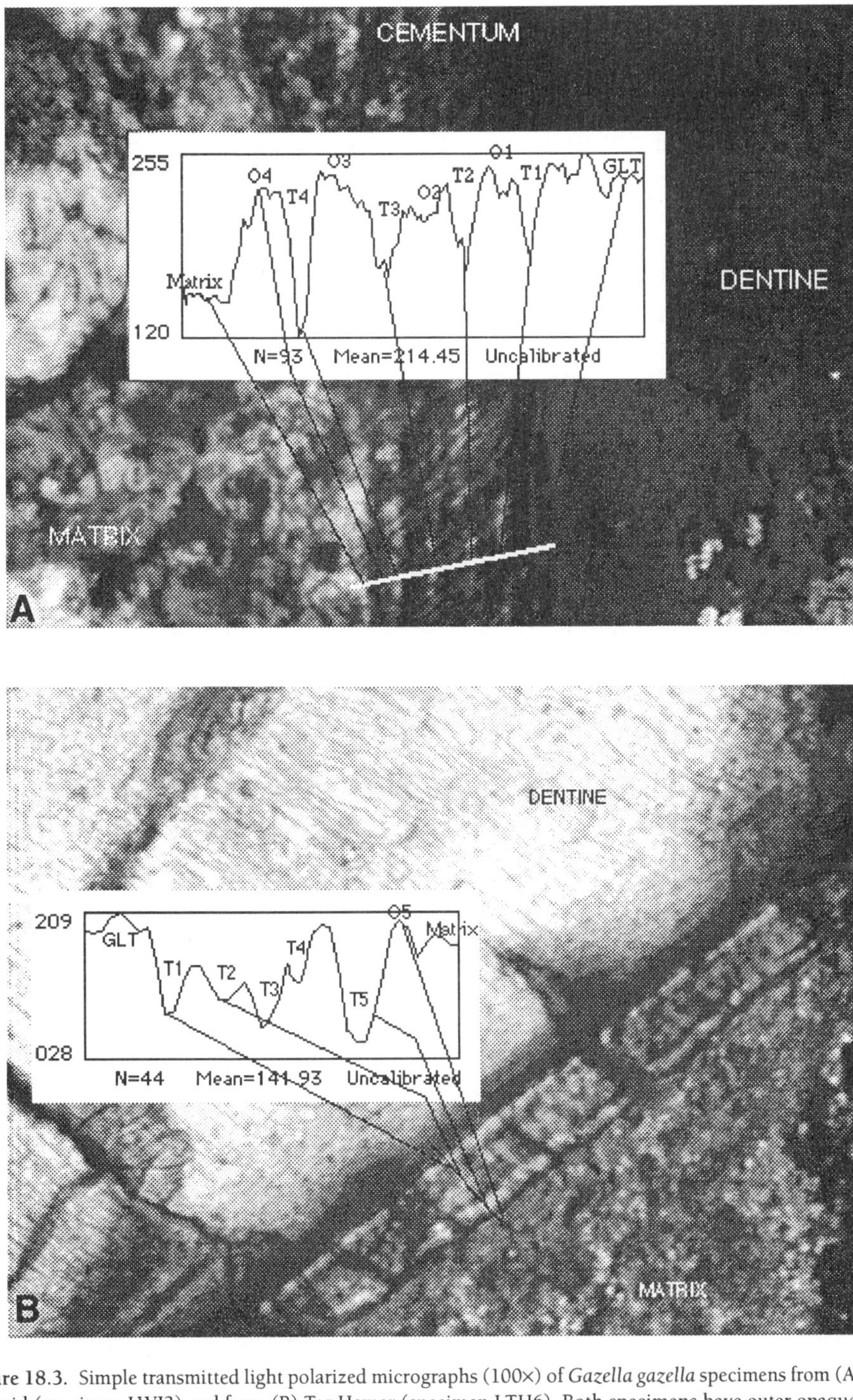


**Figure 18.3.** Simple transmitted light polarized micrographs (100×) of *Gazella gazella* specimens from (A) Wadi Judayid (specimen LWJ3) and from (B) Tor Hamar (specimen LTH6). Both specimens have outer opaque increments, indicating winter deaths.

## DISCUSSION

The estimated seasons of occupation of Tor Hamar and Wadi Judayid correspond both to previously published estimates (Henry *et al.* 1985; Henry and Garrard 1988; Henry 1989) and to the seasonality patterns observed at other Epipaleolithic Levantine sites. Wadi Judayid, like a number of other early Natufian and late Natufian sites, contains a distinct multiseasonal signature of gazelle hunting that suggests relatively permanent or at least multiseasonal occupation of the site. The proportion of winter to spring/summer gazelle at Wadi Judayid is approximately similar to Natufian sites such as el-Wad, Kebara, Hayonim Cave and Terrace, Ain Mallaha and Hatula (Lieberman 1993a). It is interesting to note that Wadi Judayid seems to have been relatively permanently occupied despite its more peripheral location from permanent water sources or the core region of the Mediterranean woodland phytogeographic zone. This would suggest that the pattern of relative sedentism in the Natufian was not entirely predicated on local environmental conditions.

Tor Hamar, in contrast, exhibits a single season pattern typical of earlier Epipaleolithic cultural complexes in which a high degree of residential mobility is suggested from both the nature of intensity at most sites as well as their seasonal signature. The winter occupation of Tor Hamar makes sense given its expected pattern of transhumance from higher elevations in the winter and spring to lower elevations associated with permanent water sources in the summer and fall (Chapters 11 and 13). Transhumance is also evident from the analysis of other Mushabian, Geometric Kebaran and Kebaran sites in different regions of the Levant.

More seasonality data are needed from other sites both within restricted regions (such as the Wadi Hisma) and throughout the southern Levant in order to investigate more fully the nature of seasonal mobility strategies in the region and how they changed over time. Cementum increment analysis of fauna from Levantine archaeological sites promises to provide important missing data necessary to constructing and testing hypotheses of hunter-gatherer mobility prior to the origins of agriculture.

*Chapter* 19

# *Preliminary Analysis of Phytoliths from Prehistoric Sites in Southern Jordan*

ARLENE MILLER ROSEN

Phytolith analysis is becoming an accepted technique of paleobotanical research at prehistoric sites in the New World (Pearsall 1989; Piperno 1988), but this method is not often exploited at sites in the Old World. Phytoliths are abundant in grasses, sedges, reeds, rushes, and palms, which are among the most useful types of plants to prehistoric peoples. They are therefore likely to have been transported and deposited at a variety of different types of sites. Since phytoliths are composed for the most part of amorphous silica, a nonorganic material, they are well-preserved in contexts where macrobotanical remains may be missing. Phytolith analyses conducted in the Near East (Helbaek 1960; Rosen 1987, 1989, 1992) have shown that phytoliths can be both abundant and identifiable in prehistoric sites of this region, including those with no other botanical remains. The following report is based on a pilot study of sediments from several different prehistoric sites in southern Jordan, to determine the kinds of paleobotanical information that can be gained from this approach in these contexts.

## METHODS

The sediments were first sieved with a 0.50 mm mesh to remove the coarse particles. They were then treated with .20% HCl to remove pedogenic carbonates which bind the particles together. Afterwards, the clays were disaggregated with a Calgon solution and poured off after allowing the suspended fine sands and silts to settle. The resulting sediment was baked in a furnace at 500 °C for two hours to remove bits of charcoal and other organic matter. The remaining particles were floated in a high density solution of sodium polytungstate set at a specific density of 2.4 which separated the opaline phytoliths from the heavier quartz particles. The phytoliths were pipetted off, washed, and mounted on microscope slides. They were examined with a petrographic microscope, and the forms were counted at 400× magnification.

The phytolith counts were divided into two major categories, the single-cell forms that represent individual silicified epidermal cells, and the multicell forms (also known as silica skeletons or spodograms) that come from sections of silicified epidermal tissue including from two to hundreds of joined cells. Each of these two categories was counted separately. Within the single-cell category there are the commonly silicified Short Cells which take on shapes primarily indicative of grass subfamilies. The short cell classification system generally follows the University of Missouri system outlined by Pearsall (1989:393). Other features included in this category are Trichomes (hairs and papillae), Bulliform Cells, Stomata Cells, and Long Cells. In addition to the above categories, I divided the Long Cells into those with markedly wavy walls, which are primarily found in the floral parts and seed husks of plants, versus those with smooth walls, found mostly in the stems and leaves of the plants. This distinction allows us to determine if plant floral parts were present at a site, and is therefore a tool for discerning the probable season of site occupation.

The multicell counts are categorized into plant taxa according to the degree of identifiability of the form. When the form is unidentifiable, it is classed as a general leaf/stem, or grass husk type.

## RESULTS

Twelve samples from seven sites of ages ranging from Middle Paleolithic to Chalcolithic were checked for phytoliths. The sites included Tor Sabiha (J8), Tor Faraj (J430), Tor Hamar (J431), Jebel Mishraq (J504), Jebel Hamra (J201), Qa Salab (J202), and Jebel el Jill (J14); see Table 19.1 for provenience information. Although almost all of the samples contained some phytoliths, only three had them in numbers considered to be statistically significant enough to count. Possible reasons for this could be that since the samples were not originally collected for phytolith analysis, they may not have come from areas with concentrations of plant material such as hearths, middens, or storage areas. Other reasons may be that the sites themselves were not occupied long enough for plant remains to accumulate, or their functions did not include the processing or use of plant material.

The three sites that did have large numbers of phytoliths were J430 (Late Levantine Mousterian), J431 (Early Madamaghan, Mushabian Complex), and J14 (Timnian, Chalcolithic). The samples were dominated by the single-cell phytolith forms, although there were a small number of multi-cell types as well. The results of the phytolith counts are displayed in Table 19.2 and Figure 19.1.

**Table 19.1. Proveniences of Phytolith Samples from Study Area**

| Sample No. | Site | Layer | Period | Phytoliths |
|---|---|---|---|---|
| JN-92-1 | J431 | E | Qalkhan | Many |
| JN-92-2 | J431 | H | Late Mousterian/ Upper Paleolithic | Few |
| JN-92-3 | J431 | A | Mushabian | Few |
| JN-92-4 | J431 | B,C,D | Mushabian | Few |
| JN-92-5 | J431 | F | Upper Paleolithic | Few |
| JN-92-6 | J431 | G | Upper Paleolithic | Few |
| JN-92-7 | J8 | C | Late Mousterian | Few |
| JN-92-8 | J14 | D, Unit 8 | Chalcolithic | Many |
| JN-92-9 | J201 | C | Few | |
| JN-92-10 | J202 | Level 20–30 | Few | |
| JN-92-11 | J430 | C | Late Mousterian | Many |
| JN-92-12 | J504 | — | Kebaran | Few |

Table 19.2. Numerical Counts and Percentages of Phytoliths from Sites in the Study Area

| Sample<br>Period | JN-92-11<br>Mid. Paleolithic | | JN-92-1<br>Mushabian | | JN-92-8<br>Chalcolithic | |
|---|---|---|---|---|---|---|
| *Single Cells* | No. | % | No. | % | No. | % |
| Long (Smooth) | 162 | 58.27 | 234 | 73.58 | 146 | 47.87 |
| Long (Wavy) | 20 | 7.19 | 8 | 2.52 | 3 | 0.98 |
| Papillae | 2 | 0.72 | 4 | 1.26 | 1 | 0.33 |
| Hairs | 4 | 1.44 | 2 | 0.63 | 5 | 1.64 |
| Stoma | 1 | 0.36 | 0 | 0 | 0 | 0 |
| Fluted Bulliform | 1 | 0.36 | 3 | 0.94 | 0 | 0 |
| Ovals (black) | 0 | 0 | 11 | 3.46 | 4 | 1.31 |
| Crenate | 7 | 2.52 | 16 | 5.03 | 5 | 1.64 |
| Bilobes | 1 | 0.36 | 15 | 4.72 | 0 | 0 |
| Polylobate | 2 | 0.72 | 1 | 0.31 | 0 | 0 |
| Total Panicoid | 3.96 | | 14.15 | | 2.95 | |
| Short Trapezoid | 1 | 0.36 | 6 | 1.89 | 3 | 0.98 |
| Hats | 28 | 10.07 | 4 | 1.26 | 71 | 23.28 |
| Rondoid | 11 | 3.96 | 8 | 2.52 | 21 | 6.89 |
| Total Pooid | 14.39 | | 5.86 | | 31.15 | |
| Chloridoid: Saddles | 7 | 2.52 | 2 | 0.63 | 6 | 1.97 |
| Cyperaceae: Cones | 25 | 8.99 | 4 | 1.26 | 21 | 6.89 |
| Palm ? | 6 | 2.16 | 0 | 0 | 0 | 0 |
| Dicot | 0 | 0 | 0 | 0 | 19 | 6.23 |
| Total Single Cells | 278 | | 318 | | 305 | |
| *Multicells* | No. | % | No. | % | No. | % |
| Leaf/Stem | 13 | 81.25 | 23 | 60.53 | 21 | 80.77 |
| Unident Husk | 0 | 0 | 0 | 0 | 1 | 3.85 |
| Cereal Straw? | 0 | 0 | 0 | 0 | 1 | 3.85 |
| Cyperus Type | 2 | 12.5 | 3 | 7.89 | 1 | 3.85 |
| Juncus Type | 0 | 0 | 2 | 5.26 | 2 | 7.69 |
| Dicot | 1 | 6.25 | 10 | 26.32 | 0 | 0 |
| Total Multicells | 16 | | 38 | | 26 | |

## Tor Faraj (J430)

The Levantine Mousterian deposit at Tor Faraj contained a greater number of phytoliths from the Pooideae (Festucoid) grass subfamily than other types. This subfamily is primarily composed of C3 grasses which are associated with cool, moist environments. There was a smaller percentage of Panicoid (primarily C4) grasses, and a moderate amount of sedges (Cyperaceae). There was also a form that closely resembles the spherical spiked phytoliths from palms (Rosen 1992), but these were so highly eroded that a conclusive identification is not possible. There was a relatively small percentage of wavy-sided long cells from the floral parts and seed husks of grasses (although the percentage is higher than in the other two samples). If the grasses in the Late Pleistocene Mousterian period flowered seasonally in the spring as they do now, then the scarcity of floral and seed parts suggests that the site was occupied in other seasons.

There are also no flowering parts represented in the multicelled forms. The multicelled phytoliths from this sample were for the most part forms from unidentifiable leaves and stems.

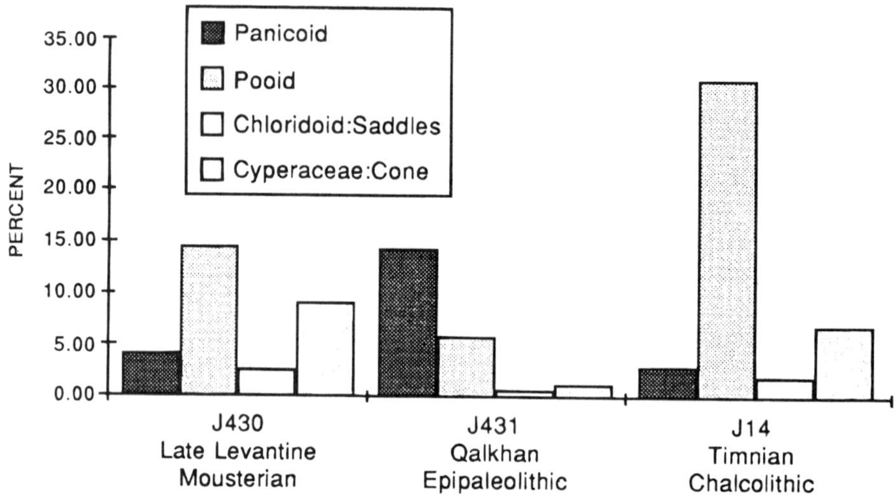


Figure 19.1. Percentages of grass subfamilies represented by single-cell phytoliths.

There were also a small number of phytoliths that are similar to those from the genus *Cyperus*, a rush that grows near standing water and produces edible tubers with medicinal properties, as well as stems that were used in antiquity for mats (Feinbrun-Dothan 1986; Greiss 1957).

## Tor Hamar (J431)

Another sample with good phytolith representation is from the Madamaghan Layer E of Tor Hamar. The single-cell phytolith assemblage of this layer is dominated by a large percentage of smooth-sided long cells derived from the stems and leaves of grasses, sedges, and/or rushes. The percentage of wavy-sided long cells is very low, suggesting that very few floral parts or seeds were brought into the site. If this sample is representative of other locations at the site it would rule out seed processing and consumption, and may even represent a late summer, fall, or winter seasonal occupation when the grasses in the vicinity were not flowering. In contrast to the earlier, Middle Paleolithic sample, the dominant grass subfamily represented by the short-cell types is the Panicoid group. These are generally C4 grasses which point to warmer and drier environmental conditions.

The majority of multicelled phytoliths are unidentifiable leaf and stem parts of grasses and sedges. There is also a moderate component of silica skeletons resembling those from the genera *Cyperus* and *Juncus*. Both of these genera consist of plants that are commonly found near desert springs, and their stems are used for mats and baskets. Finally, there is also a large component of phytoliths from dicotyledonous (woody) plants that may represent fuel for hearths.

## Jebel el Jill (J14)

The Jebel el Jill sample comes from Unit 8, Layer D, associated with the Timnian (Chalcolithic) pithouse. Again, the smooth-sided long-cell phytoliths from stems and leaves are predominant, with a negligible amount of wavy-sided long cells from seed husks. There were no wheat or barley husk phytoliths in the multicelled category, but one phytolith was typical of the type found in cereal straw. All indicators suggest that this site was not an agricultural settlement, since at agricultural sites wheat and barley phytoliths appear in varying percentages from most contexts.

In the single-celled forms, the types common to the grass subfamily Pooideae are far more prevalent than other types. These are generally C3 grasses representing cooler/moister environmental conditions. Here, as in the previous two samples there is a small component of *Juncus* and *Cyperus*-type phytoliths.

## CONCLUSIONS

Although it is unrealistic to draw broad conclusions from a single phytolith sample per site, the results of this study do show that phytoliths are present and that paleobotanical information can be obtained. This is most obviously useful in sites with no carbonized macrobotanical remains, but even in those sites with such remains, phytoliths provide additional information not attainable from seed and wood charcoal. In future studies of this kind, it will be necessary to collect phytolith samples in a systematic manner from various contexts within a given prehistoric site, as well as control samples from outside the site perimeter. More basic research is also needed in order to increase phytolith reference samples for various regions in the Near East. Finally, the phytolith information from the above three sites can form a basic framework for developing hypotheses about the paleobotany of these sites and lay the groundwork for future phytolith investigations.

# Chapter 20

# The Tor Hamar Fauna

Richard G. Klein

## INTRODUCTION

This report is based on an analysis of the 1988 faunal sample from Tor Hamar and a re-examination of the 1984 sample described by Garrard (Henry and Garrard 1988). The combined samples provided a total of 1854 pieces identified to skeletal part and taxon. Table 20.1 lists the Number of Identifiable Specimens (NISP) assigned to each taxon in each level, together with the Minimum Number of Individuals (MNI) from which these specimens must have come. The MNIs were calculated using the assumptions and computational algorithms described in Klein and Cruz-Uribe (1984, 1986). In the following sections the taxonomic identifications are explained and the implications of taxonomic abundance and skeletal part representation discussed. Potentially interesting topics like mortality (age) profiles or cut mark patterning have not been explored, because the assemblage contains too few ageable specimens (particularly measurable teeth) or visibly cut bones for meaningful analysis. A very small number of bones with macroscopic cut marks is common in stone age contexts (Lyman 1987), probably because stone age butchers tended to avoid contact with bone that might dull their tools. However, the especially small number from Tor Hamar (twenty-two) probably also reflects the relatively poor quality of bone surface preservation and the relatively high degree of fragmentation.

## IDENTIFICATIONS

Following Garrard's example (Henry and Garrard 1988), the basis for the taxonomic identifications in Table 20.1 are outlined here. The 1984 and 1988 samples provided the same taxa, with the exception of jackal and fox, which occurred only in the (larger) 1988 sample. The undated references to Garrard are to the faunal report in Henry and Garrard (1988).

### Small Rodent

The femur of a mouse-sized rodent which I could not identify more precisely for lack of adequate comparative material.

### Hare

Seven mostly fragmentary hare postcranial bones that closely match homologous elements of the Cape hare. I assigned all the pieces to Cape hare, following Garrard, who remarked that this is the only leporid species that has been conclusively identified at other regional, late Quaternary sites.

Table 20.1. The Number of Identifiable Specimens/The Minimum Number of Individuals by which Each Taxon is Represented in Each Layer at Tor Hamar Rockshelter (the totals for goat are also included in the totals for goat/sheep)

| Mammalia | A | B | C | D | E | F | G | Rock removal & below rock |
|---|---|---|---|---|---|---|---|---|
| Small rodent (*Rodentia gen.* et sp. indet.) | — | 1/1 | — | — | — | — | — | — |
| Cape hare (*Lepus capensis*) | — | 1/1 | 2/1 | 1/1 | 3/1 | — | — | — |
| Jackal (*Canus* cf. aureus) | — | — | — | 1/1 | 2/1 | 1/1 | — | — |
| Red fox (*Vulpes vulpes*) | 1/1 | 1/1 | — | 2/1 | 1/1 | — | — | — |
| Horse/ass [*Equus* sp (p. )] | 1/1 | 7/1 | 14/1 | 6/1 | 1/1 | 5/1 | — | — |
| Gazelle [*Gazella* sp(p.)] | 171/5 | 211/5 | 288/9 | 413/10 | 127/4 | 65/2 | — | 3/1 |
| Aurochs (*Bos primigenius*) | — | — | 1/1 | 4/1 | 1/1 | — | — | — |
| Goat/sheep [*Capra* sp (p. )/*Ovis* sp.] goat (*Capra aegagrus* and/or *Capra* sp.] | 106/4 | 109/5 | 116/5 | 136/7 | 25/2 | 15/1 | 2/1 | 2/1 |
| Aves | | | | | | | | |
| Partidge (*Alectoris chukar* and cf. *A. chukar*) | — | 1/1 | — | 2/1 | — | 2/1 | — | — |
| Reptilia | | | | | | | | |
| Tortoise (*Testudo* sp.) | 5/1 | 2/1 | — | 2/1 | — | 1/1 | — | — |

## Jackal

Four postcranial bones that compare closely with homologous elements of the golden jackal (*Canis aureus*), which was common in the area historically and has been identified in other regional Quaternary sites.

## Red Fox

A lower carnassial and an edentulous mandible fragment that can be confidently assigned to red fox, along with three postcranial bones that almost certainly come from the same species.

## Ass/Horse

Twenty-two mostly fragmentary postcranial bones and twelve mostly incomplete teeth from one or more species of *Eguus*. Garrard noted that wild horse (*E. przewalski*), onager (*E. hemionus*), and wild ass (*E. asinus*) have all been reported from late Quaternary sites in the Levant, and he used interspecific differences in occlusal morphology proposed by Davis (1980a, 1987:34) to assign a partial lower dentition from the 1984 sample to onager or wild ass (as opposed to wild horse). The additional teeth recovered in 1988 are too fragmentary to supplement or confirm this identification, and they probably cannot be assigned to species on strictly morphological grounds.

## Gazelle(s)

1154 mostly fragmentary postcranial bones and 116 fragmentary dentitions (mainly isolated teeth) that may be confidently assigned to one or more species of gazelle. Gazelle species are notoriously difficult to separate osteologically, and the problem is complicated for Tor Hamar and other southwest Asian sites by the possibility that more than one species could be present. Garrard noted that goitered gazelle (*Gazella subgutturosa*), dorcas gazelle (*G. dorcas*), and mountain gazelle (*G. gazella*) were all known in the Levant historically, and in theory, one or all could occur at Tor Hamar. However, he suggested that the relatively large size of the Tor Hamar bones ruled out dorcas gazelle, which tends to be smaller than the other two species. To explore this issue further, I measured relevant west Asian gazelle bones in the comparative collection of the Field Museum of Natural History. I focused on fused ("adult") distal humeri and metacarpals and on lower third molars, which together comprise the most abundant, measurable gazelle elements in the Tor Hamar sample. Through the courtesy of C. A. Wolf, I also obtained measurements on the same gazelle parts from well-known stone age cave sites in northern Israel. Bar-Yosef (1981, 1989), Garrard (1982), and Wolf (1988) provide background information on the Israeli samples. Based on adult male horn core morphology and basic sample homogeneity, Davis (1980b) has argued that the Israeli specimens derive exclusively from mountain gazelle.

Figures 20.1 and 20.2 summarize the measurements in a boxplot format described by Velleman (1989). For each sample, the vertical line near the center is the median, the vertical lines at the ends mark the range of more or less continuous data, the hachured rectangle is the 95% confidence interval for the median, and the open box encloses the middle half of the data between the 25th and 75th percentiles. Circles indicate extreme values (points that are far removed from the main body of data). When the 95% confidence limits for two samples do not overlap, their medians are significantly different in the conventional statistical sense.

Based on distal humeri and metacarpals, Figure 20.1 suggests that the Tor Hamar gazelle(s) was broadly similar in size both to recent mountain gazelle and to goitered gazelle, without allowing a choice between the two. The figure further suggests broad size similarity to the fossil

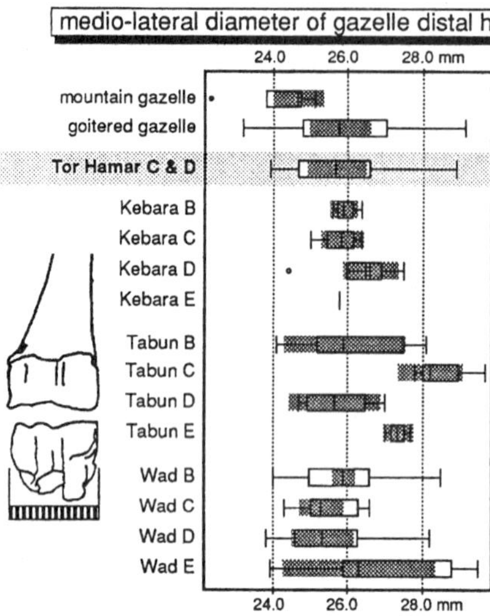


| Site | Industry (Kyrs B. P.) | mean | s.d. | N | min. | max. | median |
|---|---|---|---|---|---|---|---|
| mountain gazelle | | 24.13 | 1.17 | 5 | 22.3 | 25.1 | 24.67 |
| goitered gazelle | | 25.97 | 1.72 | 19 | 23.2 | 29.2 | 25.78 |
| **Tor Hamar C & D** | **Early Mushabian (14.5 - 13)** | **25.86** | **1.33** | **14** | **23.9** | **28.9** | **25.70** |
| Kebara B | Natufian (12 - 10.5) | 25.97 | 0.40 | 3 | 25.6 | 26.4 | 25.90 |
| Kebara C | Kebaran (19 - 14.5) | 25.78 | 0.61 | 4 | 25.0 | 26.4 | 25.85 |
| Kebara D | Aurignacian (ca. 32 - 24) | 26.28 | 1.32 | 4 | 24.4 | 27.5 | 26.60 |
| Kebara E | | 25.80 | | 1 | 25.8 | 25.8 | 25.80 |
| Tabun B | Late Mousterian (?80 - ?45) | 26.16 | 1.64 | 5 | 24.1 | 28.1 | 25.90 |
| Tabun C | Early Mousterian (?120 - ?80) | 28.57 | 1.00 | 3 | 27.8 | 29.7 | 28.20 |
| Tabun D | Early Mousterian (?160 - ?120) | 25.75 | 1.08 | 4 | 24.7 | 27.0 | 25.65 |
| Tabun E | Acheuleo-Yabrudian (?200 - ?160) | 27.35 | 0.49 | 2 | 27.0 | 27.7 | 27.35 |
| Wad B | Natufian (12 - 10.5) | 25.85 | 1.11 | 72 | 24.0 | 28.5 | 25.90 |
| Wad C | Later Aurignacian (ca. 24 - 19) | 25.49 | 0.84 | 11 | 24.3 | 26.6 | 25.30 |
| Wad D | Early Aurignacian (ca. 32 - 24) | 25.48 | 1.32 | 10 | 23.8 | 28.2 | 25.35 |
| Wad E | | 26.88 | 2.28 | 5 | 23.9 | 29.5 | 26.30 |

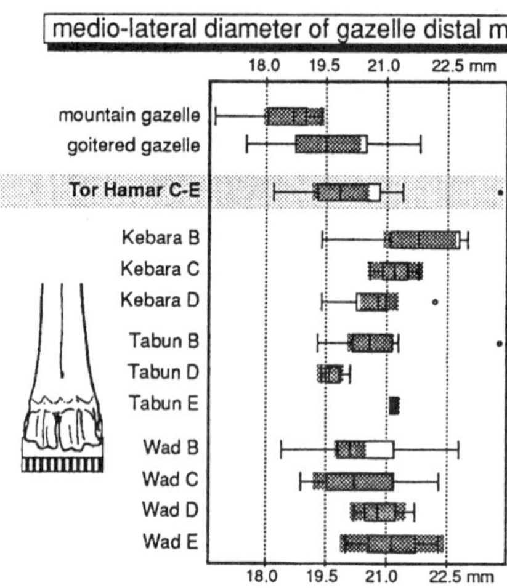


| Site | Industry (Kyrs B. P.) | mean | s.d. | N | min. | max. | median |
|---|---|---|---|---|---|---|---|
| mountain gazelle | | 18.36 | 1.57 | 4 | 16.7 | 19.4 | 18.68 |
| goitered gazelle | | 19.52 | 1.24 | 12 | 17.5 | 21.8 | 19.52 |
| **Tor Hamar C-E** | **Early Mushabian (14.5 - 13)** | **20.28** | **1.43** | **12** | **18.2** | **23.8** | **19.85** |
| Kebara B | Natufian (12 - 10.5) | 21.64 | 1.29 | 9 | 19.4 | 23.0 | 21.80 |
| Kebara C | Kebaran (19 - 14.5) | 21.20 | 0.85 | 2 | 20.6 | 21.8 | 21.20 |
| Kebara D | Aurignacian (ca. 32 - 24) | 20.72 | 0.94 | 6 | 19.4 | 22.2 | 20.80 |
| Tabun B | Late Mousterian (?80 - ?45) | 20.84 | 1.37 | 8 | 19.3 | 23.8 | 20.60 |
| Tabun D | Early Mousterian (?160 - ?120) | 19.70 | 0.36 | 3 | 19.4 | 20.1 | 19.60 |
| Tabun E | Acheuleo-Yabrudian (?200 - ?160) | 21.20 | 0.14 | 2 | 21.1 | 21.3 | 21.20 |
| Wad B | Natufian (12 - 10.5) | 20.39 | 1.05 | 33 | 18.4 | 22.8 | 20.10 |
| Wad C | Later Aurignacian (ca. 24 - 19) | 21.15 | 1.63 | 2 | 20.0 | 22.3 | 21.15 |
| Wad D | Early Aurignacian (ca. 32 - 24) | 20.41 | 1.27 | 7 | 18.9 | 22.3 | 20.20 |
| Wad E | | 20.90 | 0.76 | 3 | 20.2 | 21.7 | 20.80 |

**Figure 20.1.** Boxplots and descriptive statistics for the mediolateral diameter of distal humeri and metacarpals in recent mountain gazelle and goitered gazelle and in the fossil gazelle samples from Tor Hamar and from Kebara Cave, Tabun Cave, and Wad Cave, northern Israel.

mountain gazelles of northern Israel. Based on lower third molars, Figure 20.2 suggests that the Tor Hamar gazelle(s) may have resembled the goitered gazelle more closely than recent or fossil mountain gazelles, but this conclusion is complicated by the remarkable degree of size variability within the recent goitered gazelle sample. A geographic breakdown of this sample, which is not shown in the figure, suggests that this variability probably reflects substantial regional variation in average individual size. The mountain gazelle comparative sample is too small and geographically homogeneous to reveal similar regional size variation, but the possibility that it exists is clearly enhanced by differences in average size among the fossil Israeli humerus and metacarpal samples in Figure 20.1. Conceivably, these differences are due mainly to intersample differences in sex composition, accentuated by small sample size, but Davis (1981) has

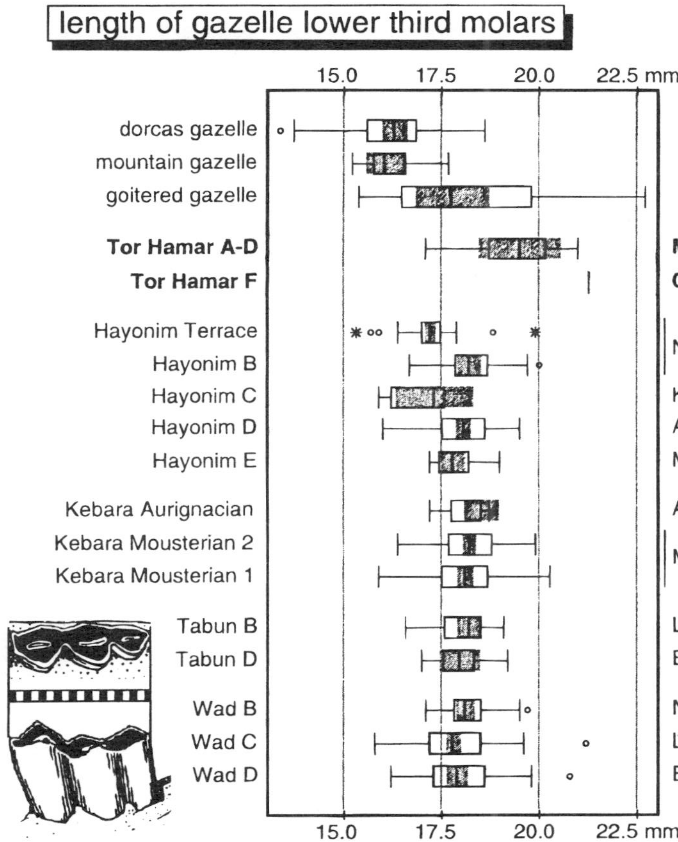


| Sample | Industry (Kyrs B. P.) | mean | s.d. | N | min. | max. | median |
|---|---|---|---|---|---|---|---|
| dorcas gazelle | | 16.10 | 1.15 | 38 | 13.4 | 18.6 | 16.30 |
| mountain gazelle | | 16.24 | 0.84 | 7 | 15.2 | 17.7 | 16.06 |
| goitered gazelle | | 18.14 | 1.95 | 31 | 15.4 | 22.7 | 17.76 |
| **Tor Hamar A-D** | **Mushabian (14.5 - 12.5)** | **19.30** | **1.50** | **5** | **17.1** | **21.0** | **19.50** |
| **Tor Hamar F** | **Qalkan (ca. 21)** | **21.30** | | **1** | | **21.3** | **21.30** |
| Hayonim Terrace | Natufian (12 - 10.5) | 17.22 | 0.90 | 30 | 15.3 | 19.9 | 17.20 |
| Hayonim B | | 18.26 | 0.90 | 16 | 16.7 | 20.0 | 18.20 |
| Hayonim C | Kebaran (19 - 14.5) | 16.92 | 0.81 | 5 | 15.9 | 17.6 | 17.30 |
| Hayonim D | Aurignacian (ca. 32 - 24) | 18.07 | 0.79 | 74 | 16.0 | 19.5 | 18.05 |
| Hayonim E | Mousterian (> 45) | 17.91 | 0.59 | 15 | 17.2 | 19.0 | 17.80 |
| Kebara Aurignacian | Aurignacian (ca. 32 - 24) | 18.13 | 0.57 | 7 | 17.2 | 18.7 | 18.50 |
| Kebara Mousterian 2 | Mousterian (?80 - ?45) | 18.23 | 0.77 | 81 | 16.4 | 19.9 | 18.20 |
| Kebara Mousterian 1 | | 18.04 | 0.94 | 74 | 15.9 | 20.3 | 18.10 |
| Tabun B | Late Mousterian (?80 - ?45) | 18.03 | 0.67 | 21 | 16.6 | 19.1 | 18.20 |
| Tabun D | Early Mousterian (?160 - ?120) | 18.00 | 0.77 | 6 | 17.0 | 19.2 | 17.95 |
| Wad B | Natufian (12 - 10.5) | 18.22 | 0.64 | 22 | 17.1 | 19.7 | 18.10 |
| Wad C | Later Aurignacian (ca. 24 - 19) | 17.84 | 0.90 | 123 | 15.8 | 21.2 | 17.80 |
| Wad D | Early Aurignacian (ca. 30 - 24) | 18.03 | 0.98 | 53 | 16.2 | 20.8 | 17.90 |

**Figure 20.2.** Boxplots and descriptive statistics for the length (anterio-posterior diameter) of lower third molars in recent dorcas gazelle, mountain gazelle, and goitered gazelle and in fossil gazelles from Tor Hamar and from Kebara Cave, Tabun Cave, Wad Cave, and Hayonim Cave, northern Israel.

argued that environmental change also affected fossil mountain gazelle size. The bottom line is that size alone may not be useful for distinguishing west Asian gazelles, and the identity of the Tor Hamar species must be left open.

### Aurochs

An upper molar and five fragmentary postcranial bones that could only come from a large bovine. In the context of Tor Hamar, aurochs is by far the most likely candidate.

### Caprine(s)

445 mostly fragmentary postcranial bones and 66 partial dentitions that clearly derive from one or more species of caprine. [I use "caprine" where many other authors use "ovicaprid." Ovicaprid is more descriptive, since it immediately suggests sheep or goat, but it also implies that sheep and goats comprise a Linnaean family. In fact, they comprise only a subfamily (Caprinae) or tribe (Caprini) within the family Bovidae. From a strictly taxonomic perspective, they are therefore more appropriately referred to as caprines.] Garrard pointed out that horn-cores of mouflon sheep (*Ovis orientalis*), wild goat (*Capra aegagrus*), and ibex (*C. ibex*) have all been found in late Quaternary sites in the Levant, and both wild goat and ibex have been reliably documented at other late Quaternary sites near Tor Hamar. Like Garrard, I was unable to separate the overwhelming majority of Tor Hamar caprine elements between sheep (*Ovis*) and goat/ibex (*Capra*), but those that I could identify, using the characters proposed by Boessneck (1969) and Payne (1985), all came from goat/ibex. As Garrard reported, the 1984 excavations provided a horn core of wild goat, though this does not rule out the presence of ibex.

### Bird, Tortoise, and Fish

Two phalanges and three limb bones of a gallinaceous bird and seven carapace/plastron fragments and three limb bones of a tortoise. Garrard assigned the single bird limb bone from the 1984 sample to the chukar partridge, and I have tentatively assigned the four additional pieces to the same species. I have also followed his lead in assigning the tortoise bones to a single species of *Testudo*. Henry (n. d.) reports that egg shell fragments of ostrich (*Struthio camelus*) and fish vertebrae were recovered from some levels, but these were not included in the present analysis.

## TAXONOMIC ABUNDANCE AND PALEOENVIRONMENT

Table 20.1 shows that only layers A through F, or more conservatively, layers A through E provided sufficient bones for paleoenvironmental assessment. Layers A–E contain Mushabian artifacts and accumulated between roughly 14,500 and 12,500 years ago. It is primarily to these layers and this time interval that the discussion here applies.

Based on the MNIs in Table 20.1, Figure 20.3 shows that gazelle and caprine bones dominate each Tor Hamar layer. In general, gazelle(s) are more numerous than caprine(s), but the degree of difference varies somewhat from layer to layer. Conceivably, this variation reflects environmental or cultural change, but it is not statistically significant, and it could be due solely to chance. Overall, the fauna suggests a hilly, rocky, semiarid landscape that need not have differed significantly from the historic one. Only the aurochs might indicate a different, somewhat more mesic setting, assuming that aurochs occurred regularly nearby. Like other bovines, aurochs almost certainly had to drink at least once a day, and they could not survive near Tor Hamar today, except perhaps as irregular, seasonal migrants. The rarity of aurochs in the deposit may in fact imply that they were infrequent nearby, but on present data, their rarity could

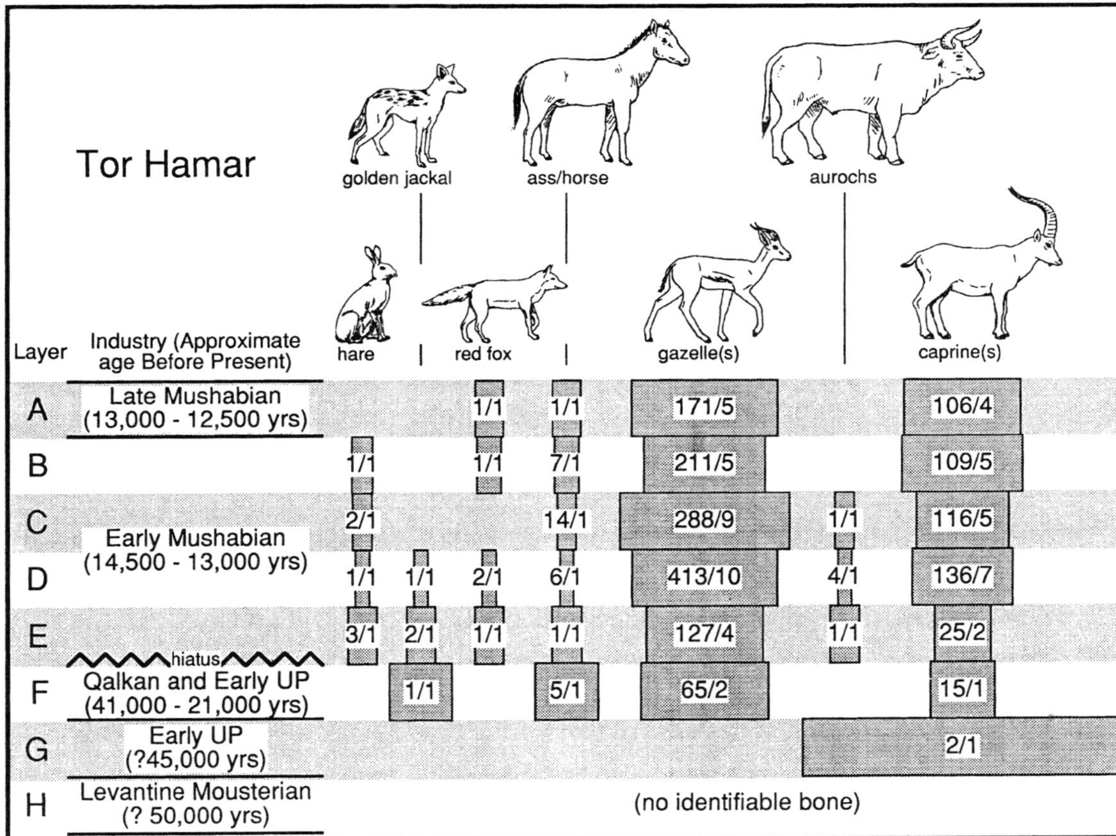


Figure 20.3. The proportional representation of large mammal taxa in each layer at Tor Hamar. For each taxon in each layer, the number before the slash is the Number of Identifiable Specimens (NISP), and the number after is the Minimum Number of Individuals (MNI) from which the specimens must have come. The bar for each taxon reflects its MNI as a percentage of the total MNI for all taxa in the same layer.

also mean that they were common, but rarely killed, or that they were frequently killed, but their large bones were rarely returned to the site. Whether or not aurochs were relatively common, however, a truly major difference between the ancient and historic Tor Hamar environments is probably ruled out by the total absence of deer. Deer were a consistent, often major element in the late Quaternary fauna of northern Israel and Lebanon (Garrard 1982; Bar-Yosef 1989), and as Garrard (Henry and Garrard 1984) pointed out, their presence at Tor Hamar would almost certainly imply a moister, probably wooded setting.

## SKELETAL PART REPRESENTATION

Only the gazelle(s) and caprine(s) are sufficiently abundant for an analysis of skeletal part representation. Table 20.2 and Figure 20.4 show that for both taxa, skeletal part representation departs significantly from anatomical expectations, that is, some parts are disproportionately common and others are disproportionately rare versus their occurrence in live animals. Such disparities are in fact the rule in archaeological faunas, but there is little agreement on how to explain them. Most authors, epitomized recently by Binford (1978, 1981, 1984), have assumed that they mostly reflect human (or sometimes carnivore) choices about what parts to remove from a carcass, while a few authors (for example, Klein and Cruz-Uribe 1984; Klein 1989; and Marean 1991) have stressed the critical role played by selective pre- and post-depositional destructive pressures. If there is any agreement, it is perhaps that selective destruction would be relatively more important than transport in determining skeletal part frequencies for small species, especially ones whose carcasses could be carried whole from kill sites to camp sites (Grayson 1989).

Except for some adult male caprine(s), the Tor Hamar gazelle(s) and caprine(s) were probably both small enough for their carcasses to be transported to the site more or less intact. The two taxa were also very similarly constructed, and the basic similarities in size and structure probably explain why their patterns of skeletal part representation appear fundamentally similar. This is suggested visually in Figure 20.4 and is supported by the insignificant values that result from the application of chi-squared (28.09, $p = 0.854$) and Kolmogorov–Smirnov (0.76, $p = 0.604$) to the paired MNI lists (in Table 20.2) on which the figure is based. Under the circumstances, the principal issue to address is the nature of the selective destructive pressures that probably produced the shared pattern.

Grayson (1988) has proposed an analytical procedure that can help distinguish between different kinds of destructive pressures. It involves plotting the abundance of various skeletal parts first against their density or hardness and second against their food (= meat, marrow, and grease) value. The abundance of various Tor Hamar gazelle and caprine parts can be estimated by the MNIs in Table 20.2. Density and food value for skeletal parts of west Asian gazelles and caprines have not been established, but figures for anatomically similar species probably provide reasonable approximations. For present purposes, I have used the bulk (or volume) density and standardized food utility estimates for deer species provided by Lyman (1984, 1985) and Metcalfe and Jones (1988), respectively.

Figure 20.5 presents the results. The plots on the left show that for both gazelle(s) and caprine(s) there is a statistically significant positive relationship between skeletal part abundance and density, while the plots on the right reveal an almost equally strong negative relationship between abundance and food value. In short, for both taxa, the most abundant parts tend to be those that may have survived because of their hardness, together with those that may have survived because they were not subjected to intense food processing. Arguably, the stronger, positive relationship between density and abundance is more important, and the somewhat weaker, negative relationship between food value and abundance may partly reflect the fact that density

Table 20.2. The Number of Identifiable Specimens (NISP)/The Minimum Number of Individuals (MNI) for Skeletal Parts of Gazelle(s) and Caprine(s) at Tor Hamar

| | Gazelle(s) | Caprine(s) |
|---|---|---|
| Frontlet | 3/2 | 6/2 |
| Occipital condyle | —/— | 1/1 |
| Auditory bulla | —/— | —/— |
| Premaxilla | —/— | —/— |
| Maxilla | 57/9 | 38/6 |
| Mandible | 59/11 | 28/5 |
| Mandibular condyle | 23/15 | 3/3 |
| Hyoid | —/— | —/— |
| Atlas | 3/2 | 1/1 |
| Axis | 7/6 | 2/2 |
| Cervical vertebrae 3–7 | 1/1 | 1/1 |
| Thoracic vertebrae | 3/1 | 2/2 |
| Lumbar vertebrae | 1/1 | 1/1 |
| Sacrum | —/— | —/— |
| Caudal vertebrae | 5/2 | —/— |
| Sternebrae | —/— | —/— |
| Ribs | 41/2 | 5/2 |
| Scapula | 7/4 | 8/2 |
| Proximal humerus | —/— | —/— |
| Distal humerus | 20/9 | 7/4 |
| Proximal radius | 42/11 | 11/3 |
| Distal radius | 9/4 | 9/6 |
| Proximal ulna | 11/1 | 5/1 |
| Distal ulna | —/— | 2/2 |
| Carpals | 51/11 | 26/6 |
| Proximal metacarpal | 70/12 | 20/4 |
| Distal metacarpal | 82/25 | 40/16 |
| First phalanges | 25/14 | 102/6 |
| Second phalanges | 115/13 | 55/6 |
| Third phalanges | 50/8 | 12/2 |
| Pelvis | 17/5 | 5/2 |
| Proximal femur | 8/5 | 4/3 |
| Distal femur | 1/1 | 3/2 |
| Patella | 12/7 | 4/2 |
| Proximal tibia | 1/1 | —/— |
| Distal tibia | 49/16 | 7/4 |
| Lateral malleolus | 1/1 | 3/2 |
| Calcaneum | 19/8 | 9/4 |
| Astragalus | 47/20 | 13/7 |
| Naviculo-cuboid | 30/16 | 10/6 |
| Cuneiform tarsals | 6/3 | 1/1 |
| Proximal metatarsal | 122/18 | 33/5 |
| Distal metatarsal | 62/16 | 29/7 |
| Proximal sesamoids | 4/1 | 4/1 |
| Distal sesamoids | —/— | 1/1 |

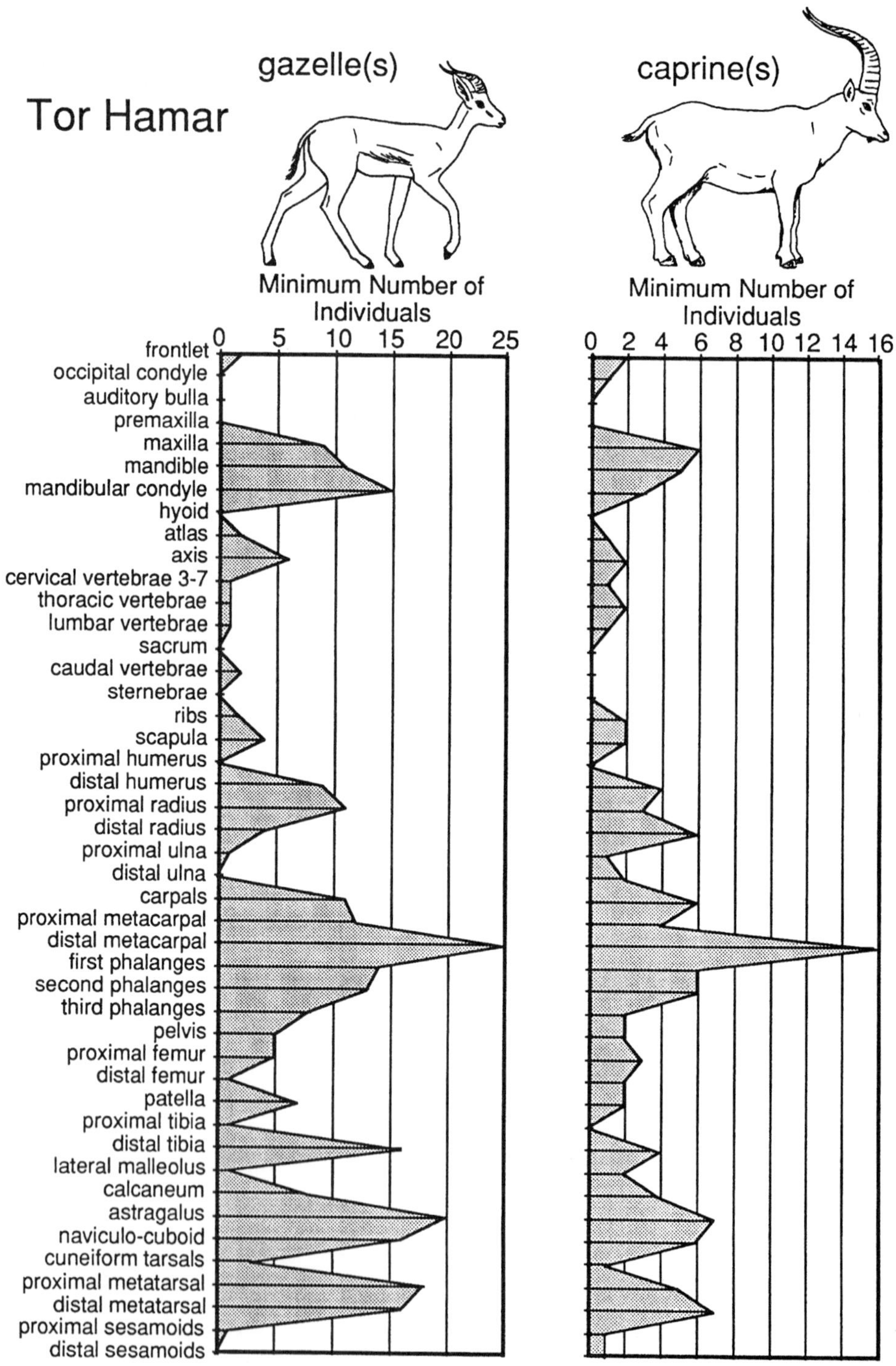


Figure 20.4. Skeletal part representation in the gazelle(s) and caprine(s) from Tor Hamar, based on the MNIs in Table 20.2.

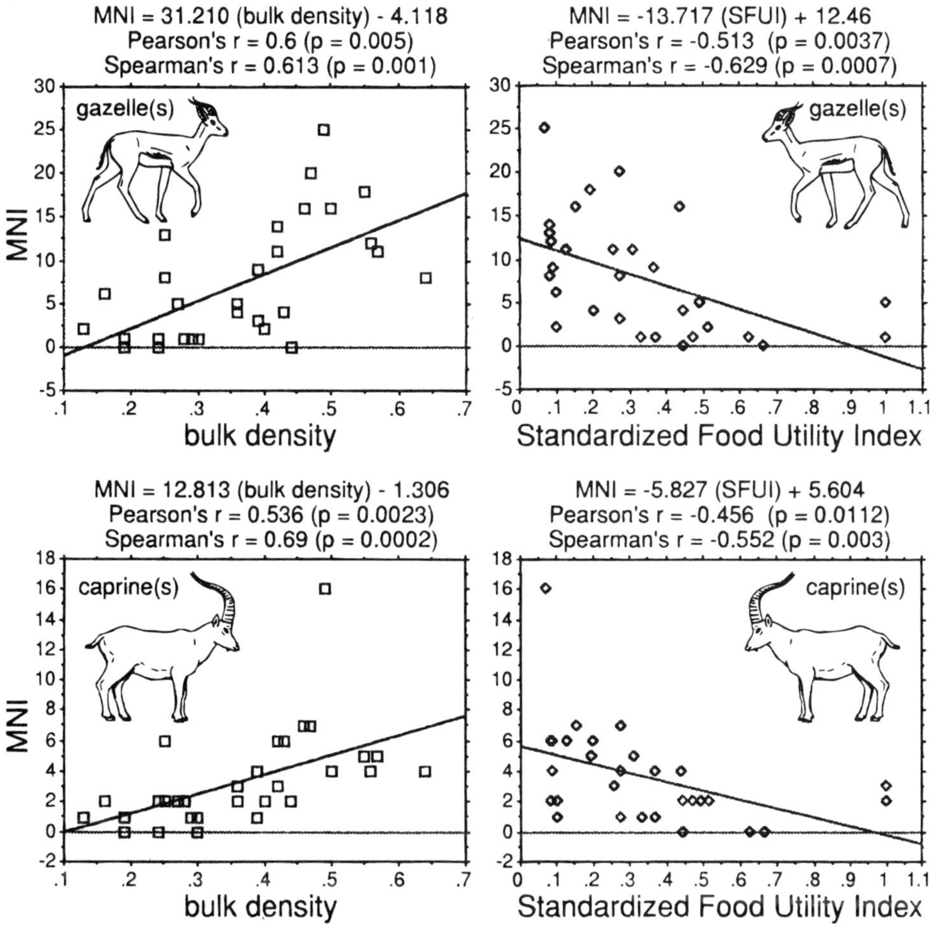


**Figure 20.5.** Left: the relationship between skeletal part abundance and bulk (or volume) density in Tor Hamar gazelle(s) and caprine(s); right: the relationship between skeletal part abundance and Standardized Food Utility in the same taxa.

and food value are themselves negatively correlated, namely, that denser parts tend to be less nutritious (Lyman 1985, 1992). The leached and highly fragmented state of the Tor Hamar bones suggests that the most important hardness- or density-mediated destructive pressures may have been post-depositional. Besides leaching and profile compaction, the post-depositional destructive forces probably included intense heating, perhaps caused by the ignition and smoldering of bedding or other flammable material in the deposit. The incidence of burnt bones and teeth is remarkably high (Table 20.3), and many specimens are deeply charred or calcined.

## SUMMARY AND CONCLUSION

The Tor Hamar fauna is heavily dominated by gazelle(s) and caprine(s) whose specific identity remains mostly uncertain. Paleoenvironmental reconstruction is possible only for Mushabian (ca. 14,500–12,500 B.P.) layers from which the majority of bones were obtained.

**Table 20.3. The Frequency of Burnt Bones and Dentitions by Taxon in the Tor Hamar Faunal Sample**

| | Bones | | | Definitions | | |
|---|---|---|---|---|---|---|
| | Number | % | Total | Number | % | Total |
| Hare | 1 | 14.3 | 7 | 0 | 0 | 0 |
| Jackal | 1 | 25.0 | 4 | 0 | 0 | 0 |
| Redfox | 0 | 0 | 3 | 0 | 0 | 2 |
| Ass/horse | 7 | 31.8 | 22 | 1 | 4.5 | 22 |
| Gazelle(s) | 408 | 35.4 | 1154 | 18 | 15.6 | 115 |
| Aurochs | 0 | 0 | 5 | 0 | 0 | 1 |
| Caprine(s) | 145 | 32.6 | 445 | 9 | 13.6 | 66 |

Overall, the fauna suggests a rocky, hilly, semiarid environment that need not have differed significantly from the historic one. The uneven pattern of skeletal part representation that the gazelle(s) and caprine(s) share probably resulted mainly from selective destruction of softer or less nutritious skeletal parts. Critical destructive pressures included post-depositional leaching, profile compaction, and intense burning.

# *Chapter* 21

# *Adaptive Behaviors, Evolution, and Ethnicity*

In many ways the interpretations presented here of the archaeological record and its environmental contexts are based upon what are often seen as distinct—some even would argue oppositional—lines of inquiry. One thrust of the study traces the cultural history of this mountainous region along the southern edge of the Jordanian Plateau over a temporal sweep of some 70,000 years. Building upon information generated by this cultural history in the context of related natural and environmental evidence, certain ecologic patterns of the region's prehistoric inhabitants are explored. Moreover, when examined diachronically, many of these patterns provide clues to understanding some of the cultural evolutionary trends of the region. This study thus incorporates by design a descriptive, cultural historic component and an explanatory, processual one.

## HISTORY AND PROCESS

Over the last thirty years, archaeology largely has adopted explanatory over descriptive goals and the principal paradigm that has come to guide searches for explanation is both anthropologic and functional (Watson *et al.* 1971; Willey and Sabloff 1974; Trigger 1978). This processual interpretation of archaeological evidence, a movement grounded in American anthropology, was especially influenced by the works of cultural ecologists and neo-evolutionists such as Julian Steward and Leslie White. Led by the writings of Lewis Binford, processual archaeology emerged in the 1960s as the prevailing way in which researchers in the Americas shaped their questions and went about seeking answers. In the Old World, "New Archaeology" also provided an alternative to the cultural historic approach as traditionally employed. This created a sharp dichotomy in the ways in which variability in the archaeological record were explained and assessed.

Traditionalists primarily regarded changes in ethnicity as the fundamental cause of artifact variability. Such changes were linked to the diffusion of ideas, to the actual migration or expansion of peoples, and to time-sensitive trends of behavior. Variability in lithic artifact assemblages was traced primarily through the definition of "guide fossils" or through specific profiles seen in the relative frequencies of artifacts composing a type list. In contrast, processualists focussed on functional explanations for variability. Temporal and spatial changes in artifact assemblages were explored as the residuals of behavioral responses to environmental pressures at various scales (e.g., biogeographic, microenvironmental, seasonal).

Within the Levant these paradigmatic differences, reflected to some degree in the nationalistic ties of scholars working in the region, have influenced the direction of recent archaeological research programs (Rosen 1992). But in my light, the theoretical orientations of archaeologists working in the Levant generally appear more eclectic than diverse and emphasize both cultural historic and processual problems. This blending of orientations is what Trigger (1978:17) predicted would evolve eventually as an "international tradition." The strength of this fusion of paradigms, I think, rests in the opportunity it provides for processualists to track prehistoric populations through time and space and thus better understand their ecologies and behavioral evolution.

It would seem that the degree to which we can trace culturally defined populations in the context of their temporal and spatial relationships determines to a great extent our ability to understand prehistoric adaptive systems and their changes. Sabloff *et al.* (1987) have pointed out the shortfalls of an inductively derived cultural history as a basis upon which to examine cultural process, but they also recognized the complementary nature of the two approaches to understanding the past. In drawing from a paleontological metaphor, it is the cultural history that produces the "taxonomy" and the processual inquiry that enables us to reconstruct the "ecology and evolution" of the population under study. From an adaptive perspective, cultures do in many ways resemble species. Thus in order to understand the functioning and the evolution of prehistoric cultures it is necessary to identify the discrete populations that formed specific adaptive systems of the past. Approaches to defining such populations and systems should not be pursued independent of examining such constructs in the context of their ecology and evolution. Again, in the realm of paleontology this recalls the potential pitfalls encountered by the taxonomist who reconstructs a phylogeny without the benefit of information on population genetics and ecology. In an archaeological context, recognition of how a cultural process influences the configurations of artifact assemblages and attributes is a large part of constructing archaeological units, that is, building taxonomies. The definition of taxa fundamentally rests upon the attributes selected for study and the meaning ascribed to the results.

From an analytic perspective, workers in the Levant have employed a somewhat eclectic and in many respects novel approach to lithic artifact analyses. Typological or morphological oriented approaches of traditional European prehistorians (e.g., modified type lists of Bordes, de Sonnville Bordes and Perrot, and Tixier) are characteristically combined with qualitative and metrical technological analyses. These analytic programs typically attempt to answer questions related to the function of an assemblage in addition to the ethnic identity (at some scale) of those who produced it. The ways in which artifact assemblages were used are inferred from variability at the "class level" of morphological typologies as well as from edge-wear data. Broader functional aspects of an assemblage, those related to raw material use and provisioning strategies, are ascertained through various technological studies that focus on the definition of reduction streams. These largely attempt to follow the steps involved in tool manufacture from raw material acquisition through core shaping and blank production. Attempts to identify prehistoric ethnicity (at scales of technocomplex, culture group, and culture) rely upon the isolation of stylistic attributes in realms of technology and secondary retouch.

## ADAPTATION, FITNESS, AND ETHNICITY

A processual interpretation of the archaeological record is based, in part, on artifact assemblages and related evidence being viewed as the material residue of the behavioral systems of past populations—that is, prehistoric cultures. Moreover, from a functional or adaptive perspective, these behaviors have come to be construed as the fundamental means by which human populations attempted to cope with their ever-changing world. Given this orientation, cultural evolution often is understood as the differential persistence of behaviors as governed by selective pressures of the environment (Clark 1991; Kirch 1980; Durham 1976). In comparison with biologic evolution along neo-Darwinian lines, cultural evolution is seen as the consequence of natural selection operating on variable behaviors, some or most of which result from deliberate efforts to solve problems of survival.

## Adaptation and Measures of Fitness

From a biologic perspective, adaptation is the process in which changes in an organism enhance its fitness within its environment. Fitness, in a technical sense, may be measured by the relative growth or decline of a genotype from one generation to another or at the level of phenotype by the growth or expansion of a population through time. Is fitness then an appropriate measure by which to evaluate cultural adaptation? Many human ecologists believe so or, at least, that such a biologic analogue provides a useful working assumption (Durham 1976; Konner 1982). Others argue, however, that beyond enhancing human survivability cultural behaviors in many ways serve to perpetuate themselves (Vayda and Rappaport 1968). In viewing culture as a process this conceptual dilemma of circularity can be avoided. As Jochim (1981:14) notes,

> humans have a valid set of solutions allowing survival and genetic perpetuation. Culture consists of the extrasomatic . . . components of these solutions, the chosen motives and the resulting strategies for survival. Human adaptation, then, might involve the perpetuation not simply of genes, but of a particular set of solutions based upon certain goals.

A human population's particular set of solutions is likely to be short-lived as the individual solutions and problems within the set are ever changing. One also would expect a population's particular set of solutions to be confined to a relatively well defined area, encompassing a niche of human exploitation, given the linkage of certain solutions to environmentally defined problems. A "culture" then might be viewed functionally as a specific set of behavioral solutions to a given constellation of problems at a certain point in time and space. If this is true and if we are interested in understanding the variability, persistence, and evolution of these solutions, then it is incumbent upon us to identify as precisely as possible the spatial, temporal, and natural contexts of these solutions. For me, this is cultural history.

The dynamism, between behavioral solutions and new problems, that gives rise to the structural fluidity of culture is masked by

> the fact that changes must build on the "evolutionary present," that is, must come about in the context of ongoing solutions (Jochim 1981:14).

This linkage to the past, along with the fact that a given population's constellation of behavioral solutions is unlikely to change in mass, may give the impression of cultural conservatism and stability. For example, behaviors related to language may remain relatively constant over an extended period of time while other behaviors tied to social organization or demography may change. From a diachronic dimension this stream of ever changing members within a set of behavioral solutions presents the greatest challenge for anthropologists and archaeologists who focus on defining "cultures." In such a succession of transformations, where does one culture end and another begin? Such definitions are arbitrary in that they rest upon the assumption that certain behaviors are more "culturally diagnostic" than others. Although arbitrary, the definition of prehistoric cultures (at various scales) is important to functionally oriented studies, for such definitions enable archaeologists to fix prehistoric populations at certain points in time and space.

## Archaeology and Adaptation

Questions of cultural evolution related to successional and systemic changes in behavioral systems have emerged as a central concern of archaeological research (Binford 1989; Clarke 1968). This, of course, is consistent with the discipline's basic functional paradigm in those situations where the archaeological record is examined diachronically. From a broader perspective this meshes with Jochim's (1981:12) observation that "the underlying theoretical basis for ecological studies is evolution, with its concepts of adaptation and fitness."

Beyond these logical paradigmatic links, the focus on evolutionary questions in prehistory is likely driven by the practical limitations and strengths of the archaeological data-base. Although the reconstructions of the behaviors of the deep past typically offer a rather "fuzzy" resolution when compared to those of recent or living populations, they furnish opportunities for comparing behavioral norms over very large intervals of time (Clarke 1968; Issac 1972). Such temporal depth also enables the evolutionary trajectories of behavioral systems to be more accurately traced, for brief anomalies or "noises," using Clarke's (1968) terminology, can be distinguished from persistent patterns or vectored changes in data. Archaeology then is uniquely positioned among the social sciences for examining cultural processes and change, and tracing the behavioral evolution of hominids.

## Archaeology and Ethnicity

When compared to examining cultural evolution from an adaptive perspective, the diachronic tracking of ethnicity within the archaeological record is often viewed as representing a different, if not mutually exclusive objective. Ethnicity here might best be understood as the ways in which membership in an interest group is advertised (Hodder 1979) or as the ways groups "distance" themselves socially (Binford 1983). Recognition or interest in prehistoric ethnicity is often pointed to as a characteristic that distinguishes historical from processual archaeology. The contrasting views produced by the now-classic Bordes–Binford debate over the interpretation of the various facies of the Mousterian epitomized the paradigmatic differences of the two traditions, but other tensions exist in the realm of methodology.

Geoffrey Clark (1989, 1991), for example, questions our ability on practical grounds to identify prehistoric ethnicity, at least in preceramic contexts, because of the difficulty in isolating lithic attributes that are encoded with social symbols. Although others argue that it is possible to identify such social or ethnic codes in Paleolithic contexts through stylistic or technical (*chaîne de opératoire*) attributes (Close 1978, 1989; Henry 1977, 1989a; Bar-Yosef *et al.* 1992; Pelegrin 1985; Pelegrin *et al.* 1988), there has been little attempt to interweave such information into questions of adaptation and cultural ecology, especially within the Levant.

The definition of prehistoric ethnicity at some scale is important to addressing questions of human adaptation and change for both theoretical and practical reasons. One of the principal forces of selection that acted on a prehistoric population was related to how well the population balanced its numbers with critical resources. Given that hominids progressively came to depend upon behavioral adaptation, mostly at the group or social level, the recognition of who belonged and who did not was critical to a group's survival. Ethnicity furnished a means for those individuals who formed an interest group and who shared information and access to common resources to recognize one another. In being able to identify their membership and determine their numbers relative to critical resources, groups were then capable of reaching solutions designed to balance their populations and resources. Ethnicity would also have enabled groups to form large and extensive alliance networks that served to reduce a population's risks through the sharing of resources and information over wide areas.

Human social interaction and organization have been proposed as critical factors allowing for the occupation of marginal, harsh, and high-risk environments. Gamble (1986) and Whallon (1989), for example, have argued that such settings as the northern latitudes of Europe and Asia and the Australian desert were not successfully occupied until the emergence of fully modern hominids along with their capacity for elaborate social organization and the development of extensive alliance networks—both of which are tied to ethnicity. Although social organization and interaction may well have been a large factor that enabled populations to inhabit settings with resources of low density and predictability, the prehistoric occupation of such settings (at least in Europe) appears to have occurred well before the emergence of modern humans (Roebroeks *et al.* 1992).

On a practical archaeological level, the definition of prehistoric ethnicity at some scale is fundamental to tracing "cultural systems" through time and space. These time-space patterns in the context of natural variables (e.g., water sources, vegetation, chert sources) form the basis of prehistoric cultural ecologic studies and, ultimately, inquiries into cultural evolution. The task of tracking prehistoric cultural systems is especially important in those settings where the behavioral residue (i.e., artifacts and sites) of such systems extends over great spans of time and where populations appear to be spatially interfingered. Both of these conditions apply in southern Jordan.

With a temporal sweep encompassing some 70,000 years, the great majority of archaeological occupations were seriated on the basis of artifactual attributes linked to ethnicity. Although stratigraphic and chronometric evidence assisted in sorting out cultural systems in time and space, the patterned variation in those attributes of lithic assemblages that cut across functional boundaries (e.g., differences in tool, activity, or site types) proved the most useful means of tracing cultural histories (Chapter 3).

For example, the identification of specific modes of retouch and techniques of blank production in the Late Levantine Mousterian occupations of Tor Faraj and Tor Sabiha suggests that their occupants belonged to a common cultural system. While this does not necessarily mean that the occupations were synchronous, it does imply that the occupations can be examined as constituent parts of a common social and adaptive system. Other artifactual evidence shows the assemblages belong to a common archaeological taxon (Late Levantine Mousterian) and chronometric dates from both sites point to their having been occupied ca. 69,000 ± 6,000 years ago, but the temporal resolution from these lines of evidence is too coarse grained to confirm that the occupations formed parts of a common system.

Even for the more recent time-frames, stratigraphic and radiometric data must be complemented with prehistoric ethnic clues in order to trace the temporal and spatial distributions of populations. The cultural history of the region during the Epipaleolithic was especially complex, given the persistence of parallel and sometimes spatially interfingered traditions stretching over some 6,000–8,000 years. Again, it is only through the definition of patterned variation in ethnically linked lithic attributes that the two traditions, represented by the Qalkhan–Madamaghan and Hamran–Natufian successions, can be distinguished. Of equal importance, these same attributes allow for assemblages from quite different site types and settings to be compared as part of a common cultural system (e.g., Madamaghan or Early Hamran).

## RECONSTRUCTION OF ADAPTIVE STRATEGIES

Many of the solutions that the inhabitants of southern Jordan found for the problems of living within the region appear to have remained relatively unchanged for some 70,000 years. This is somewhat surprising, given the marked changes that occurred in the region's environment and the basic shifts that took place in the subsistence practices of its inhabitants. In concert with fluctuations in temperature and moisture, environmental settings varied from wooded savannah to desert and subsistence practices ranged from those based on mobile foraging to pastoralism.

The most persistent, time-transgressive solutions to living in the region were tied to the general adaptive strategy of transhumance. Specifically these solutions related to: (a) settlement modes that involved episodes of coalescence and dispersal of the population, (b) seasonal scheduling of these different modes of settlement, (c) certain spatial and especially elevational positions of the different settlement modes, (d) use of natural features of the landscape, (e) shifts in provisioning strategies that co-varied with the different settlement modes, and (f) use of seasonally marginal resources.

## Modes of Coalescence and Dispersal

Foragers and pastoralists typically adjust the sizes and the permanency of their camps in order to accommodate fluctuations in the abundance of critical resources. With the exception of the Early Natufian inhabitants of the region, who appear to have occupied permanent or semi-permanent camps, this pattern persisted from Late Levantine Mousterian through Chalcolithic times. The adjustments in the sizes of residential groups and the degree of residential mobility are expressed in a dichotomy of camp types that corresponds to modes of coalescence and dispersal. During a relatively short segment of the transhumant cycle the region's inhabitants appear to have coalesced into larger groups who maintained long-term residential camps of several weeks duration. The specific locations of these encampments were regularly revisited. During the other, much longer segment of the annual cycle, the population appears to have dispersed into smaller, more mobile groups who resided in ephemeral camps. The sites of these camps were rarely revisited.

The factors thought to be responsible for the overall strategy of transhumance, as well as for the specific expressions of this strategy at certain points in time, were tied to a complex set of factors that determined the availability of resources and creature comfort. Many of the resources, along with levels of creature comfort, fluctuated both seasonally and spatially during the year. But other resources (e.g., chert) and particular settings affecting creature comfort (e.g., rockshelters) were fixed in space and available year-round.

### Defined Archaeologically

The dichotomous site types are defined archaeologically by patterned asymmetry in the occupational areas, thicknesses of cultural deposits, artifact densities, and presence of features (Table 21.1). Long-term camp sites are significantly larger, display thicker cultural deposits and yield higher artifact densities than ephemeral camp sites. Features such as hearths, structures, bedrock mortars, and petroglyphs are found only in long-term camps as are less portable ground stone artifacts.

### Site Sizes and Populations

The occupational areas of the two site types show strong asymmetry with nonoverlapping distributions. The areas of long-term camps exhibit a relatively wide range extending from 216–3,692 m$^2$, while those of ephemeral camps have a much tighter range falling between 110–175 m$^2$. When the long-term sites are partitioned into those associated with mobile foragers, sedentary foragers, and pastoralists, the Timnian pastoral sites are by far the largest with a mean area of 3,692 m$^2$. These are followed by the Early Natufian sedentary foraging site of Wadi Judayid (+530 m$^2$) and the sites of mobile foragers stretching from Levantine Mousterian through Final Hamran times (427 m$^2$).

**Table 21.1. Comparison of Various Characteristics of Long-Term and Ephemeral Camp Sites for Different Archaeological Taxa**

| | Long-term sites | | | | Ephemeral Sites | | | |
|---|---|---|---|---|---|---|---|---|
| Archaeological Taxa | Mean area ($m^2$) | Mean depth deposit (cm) | Mean artifact density (/ $1m^3$) | Features | Mean area ($m^2$) | Mean depth deposit (cm) | mean artifact density (/ $lm^3$) | Features |
| Timnian | 3,692 | 0.9 | 115 | H,S,GS | 125 | 0.11 | 29 | BM |
| Late Natufian | 400 | 0.3 | 408 | BM,GS | | not found | | |
| Early Natufian | 530 | 0.6 | 1,437 | H,GS | | not found | | |
| Late–Final Hamran | 390 | 0.6 | 240 | H,GS | | | | |
| Middle Hamran | 485 | 0.6 | 204 | — | 175 | 0.25 | 18 | H |
| Early Hamran | 430 | 0.5 | 292 | H,GS,P,BM | 110 | 0.05 | 20 | — |
| Madamaghan | 257 | 1.2 | 232 | H,GS,S | | not found | | |
| Qalkhan | 120 | 0.4 | 367 | H | | not found | | |
| Levantine Aurignacian | 807 | +0.8 | 419 | — | | | not found | |
| Ahmarian | 424 | +0.9 | 248 | H | | not found | | |
| Levantine Mousterian | 216 | +2.5 | 164 | H | 116 | 0.20 | 148 | — |

Estimates for the number of occupants who would have resided in these different site types can be made by using equations, developed from Yellen's (1977) ethnographic data, that define the correlation between the occupied areas and the number of occupants in contemporary !Kung Bushmen camps. Wiessner (1974, 1979), Casteel (1979), Read (1978), and Hassan (1981) have presented various equations that describe this relationship mathematically. Although differing in approach, both power and exponential curves yield high correlation coefficients (Hassan 1981) and similar population estimates at the range of 100–200 m²; above 200 m² the curves diverge (Figure 21.1). Population estimates indicate that the ephemeral camps had some 18–19 occupants, whereas long-term mobile foraging camps are likely to have held some 29–34 residents. With the transition to sedentary foraging, the residential population of permanent or semipermanent Early Natufian hamlets (such as Wadi Judayid) appears to have increased only slightly over that of the long-term camps of preceding mobile foraging Hamran groups. Given that the very large areas of the Timnian long-term pastoral camps greatly exceed the range of the !Kung Bushmen data, population estimates for these Timnian sites probably should be based more upon pastoral than foraging analogues.

These population estimates are consistent with a reconstructed transhumant strategy in which mobile foraging groups resided in ephemeral camps during a settlement segment of population dispersal and subsequently coalesced into larger long-term camps. More specifically, these population estimates indicate that the demographic structure of mobile foragers consisted of the aggregation of two ephemeral camp groups into a larger long-term camp. Although transhumance appears to have been briefly abandoned with the emergence of sedentary foraging

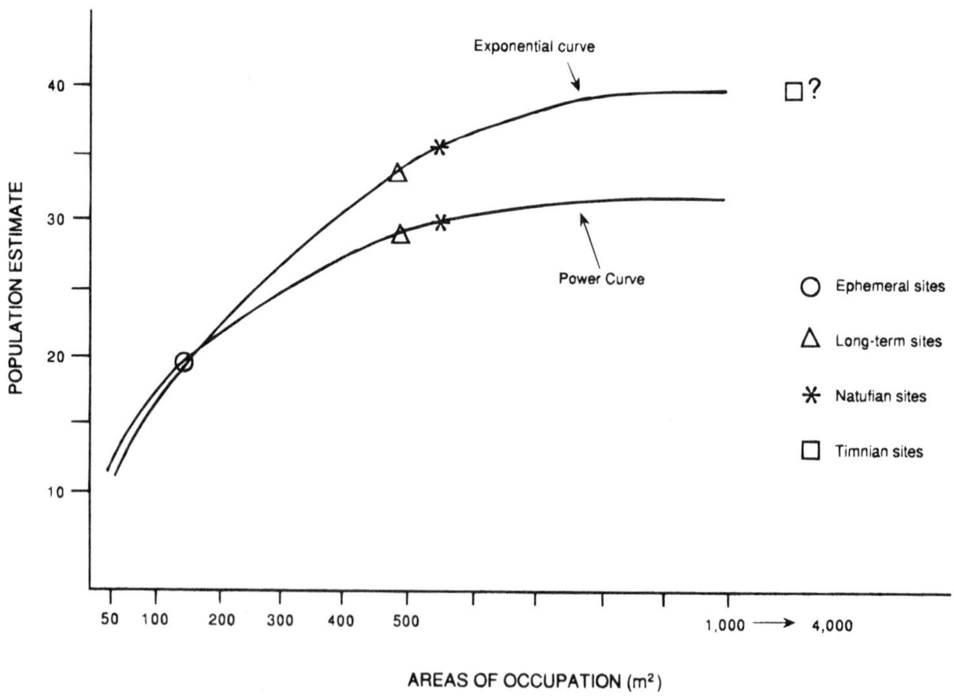


**Figure 21.1.** Population estimates for the areas of different site types based upon power (Casteel 1979) and exponential (Hassan 1981) curves. Note that the population of Natufian sedentary foraging sites only slightly exceeds that of long-term mobile foraging sites.

during the Natufian, there appears to have been only a slight increase in total residential group population over that of earlier mobile foragers. The return to a transhumant strategy during Timnian times appears to have been accompanied by a fundamental change in demographics. Whereas the long-term camps of mobile foragers contained only about twice the number residing in their ephemeral camps, long-term Timnian camps appear to have been inhabited by some 4–5 times the number of occupants of their ephemeral camps.

## Cultural Deposits and Artifact Densities

The sites identified as long-term camps show substantially thicker cultural deposits than the ephemeral camps (Table 21.1). More important than the differences in thickness, however, is persistent evidence for multiple episodes of occupation at the long-term camp sites, whereas ephemeral sites appear to have been revisited only rarely. Even though the long-term camp sites appear to been reoccupied regularly, sufficient time must have elapsed between these occupational episodes for sediments to accumulate. Stratified "living floors" can be identified within long-term camp deposits stretching from Levantine Mousterian through Chalcolithic times.

The lithic artifact densities of long-term sites are consistently higher than their ephemeral site counterparts within each of the archaeological taxa. Of the long-term sites, by far the highest density comes from the Early Natufian deposit of Wadi Judayid which is associated with a permanent or semipermanent occupation located very close to raw material. The lowest lithic artifact densities of long-term occupations are displayed by the Timnian sites. This may not be so much an expression of the lack of intensity by which these sites were occupied as a fundamental change in the importance of chipped stone tools within Chalcolithic material culture. If the Timnian and Natufian sites are excluded from the computations, artifact densities for long-term mobile foraging sites are remarkably similar, ranging from 164–419 artifacts/0.1 $m^3$.

The variation in the lithic artifact densities of the sites is consistent with the suspected differences in occupational intensities between those that served as long-term as opposed to ephemeral camp sites. This occupational intensity was likely to have been expressed in the long-term camp sites being occupied for longer durations by greater numbers than the ephemeral camps. Given that the long-term camp sites are consistently larger than those thought to represent ephemeral camp sites, their higher artifact densities not only suggest higher rates of discard (resulting from longer durations of residence), but also greater numbers of inhabitants. If the discard rates and number of inhabitants were held constant, for example, the larger long-term camp sites would be expected to show lower artifact densities than the smaller ephemeral camp sites.

## Features and Bulky Artifacts

Another indicator of the differences in mobility levels associated with the occupations of the two site types is the markedly greater presence of features and bulky artifacts in the long-term camp sites (Table 21.1). Hearths, ground stone items, bedrock mortars, structures, and petroglyphs point to prolonged occupations of the long-term camp sites. The only evidence of features to come from ephemeral camp sites consists of a hearth at the Middle Hamran site of J26 and a possible bedrock mortar at the Timnian site of J521. The paucity of such evidence in these sites seems especially telling, given that much larger areas were exposed in these sites attributed to ephemeral camps than in those classified as long-term camp sites. This is so for sites of various time-frames including Levantine Mousterian (J8), Early and Middle Hamran (J21,J22,J26,J31), and Timnian (J408,J521).

## Seasonal Schedules, Elevation, and Use of Natural Features

The natural settings of the sites and the evidence for the season(s) of their occupation indicate that the long-term camp sites were inhabited during the winter wet season and the ephemeral camp sites were formed during the warm dry season. The single exception to this pattern is found in the Natufian permanent or semipermanent camps that spanned at least part of the wet and dry seasons. In conjunction with scheduling the coalescence and dispersal of their populations seasonally, the prehistoric groups of the region selected specific elevational belts for seasonal exploitation.

During the Pleistocene, long-term, winter camps were established in the lower piedmont and lowlands of the Wadi Hisma within an elevational belt resting between 800–1,000 masl. In contrast, ephemeral, warm season camps were situated within the higher piedmont (and presumably the uplands) at elevations over 1,000–1,300 masl (Figure 21.2). Occupation of the lower elevation sites would have been restricted to the winter wet season and early spring, in that surface water would have been unavailable during the middle and latter portions of the dry season. The higher piedmont and uplands, on the other hand, would have made poor winter habitats given the cold conditions and heavy snow found at higher elevations, especially during the Pleistocene. The use of natural shelters is consistent with the proposed seasonal schedule for occupation of the different elevational belts. Without exception, the long-term lower elevation sites thought to represent cold, wet-season camps are positioned within southwest-facing natural shelters. All of these are located within 0.5 km of a major water source (e.g., stream, lake). Most of the ephemeral sites occur in the open, but when found in protected settings they have eastern or northern exposures. The direct evidence available for determining seasonality also is consistent in pointing to winter residence at the long-term sites. Phytolith data from the Levantine Mousterian deposit of Tor Faraj indicate that it was inhabited outside of the flowering season (i.e., spring–summer) and cementum increment analysis of teeth from the Madamaghan horizon at Tor Hamar signals a winter occupation.

With the onset of warmer conditions associated with the emergence of the Holocene, we see the first evidence for the establishment of long-term (even permanent) settlements in the higher elevations of the piedmont and uplands. The earliest of these in the study area is represented by the Early Natufian site of Wadi Judayid, but similar occupations at Beidha, Tabaqa, and perhaps Sabra I suggest that long-term occupations of high elevations was a regional phenomenon around 12,500 B.P. Although the following Neolithic interval is poorly represented in the study area, the nearby village sites of Beidha, Basta, and Ain Jamal confirm the continued practice of establishing long-term settlements at high elevations. The persistence of this pattern from Mid-Holocene to modern times can be traced through the Timnian sites of the Chalcolithic into the settlement patterns of Bedouin pastoralists who inhabit the region today.

The types of Timnian sites and their elevational settings suggest a schedule of transhumance that was remarkably similar to that followed by modern Bedouin herding groups. This entailed an annual cycle in which groups occupied relatively large, long-term seasonal camps during the winter in the piedmont zone followed by some 9–10 months of population dispersal into much smaller, more mobile camp groups during the dry season. The dry season segment of their transhumant cycle was initiated by movement to spring pastures at lower elevations on the floor of the broad Wadi Hisma, but as the season advanced groups reversed this movement and progressively settled at higher elevations in search of surface water and grazing areas. This eventually took them to the uplands on the edge of the plateau where perennial springs and late season pastures would have furnished critical resources. With the coming of winter rains and snow, groups again moved to the lower elevations of the piedmont where they enjoyed more comfortable temperatures while at the same time having predictable water sources from upland runoff.

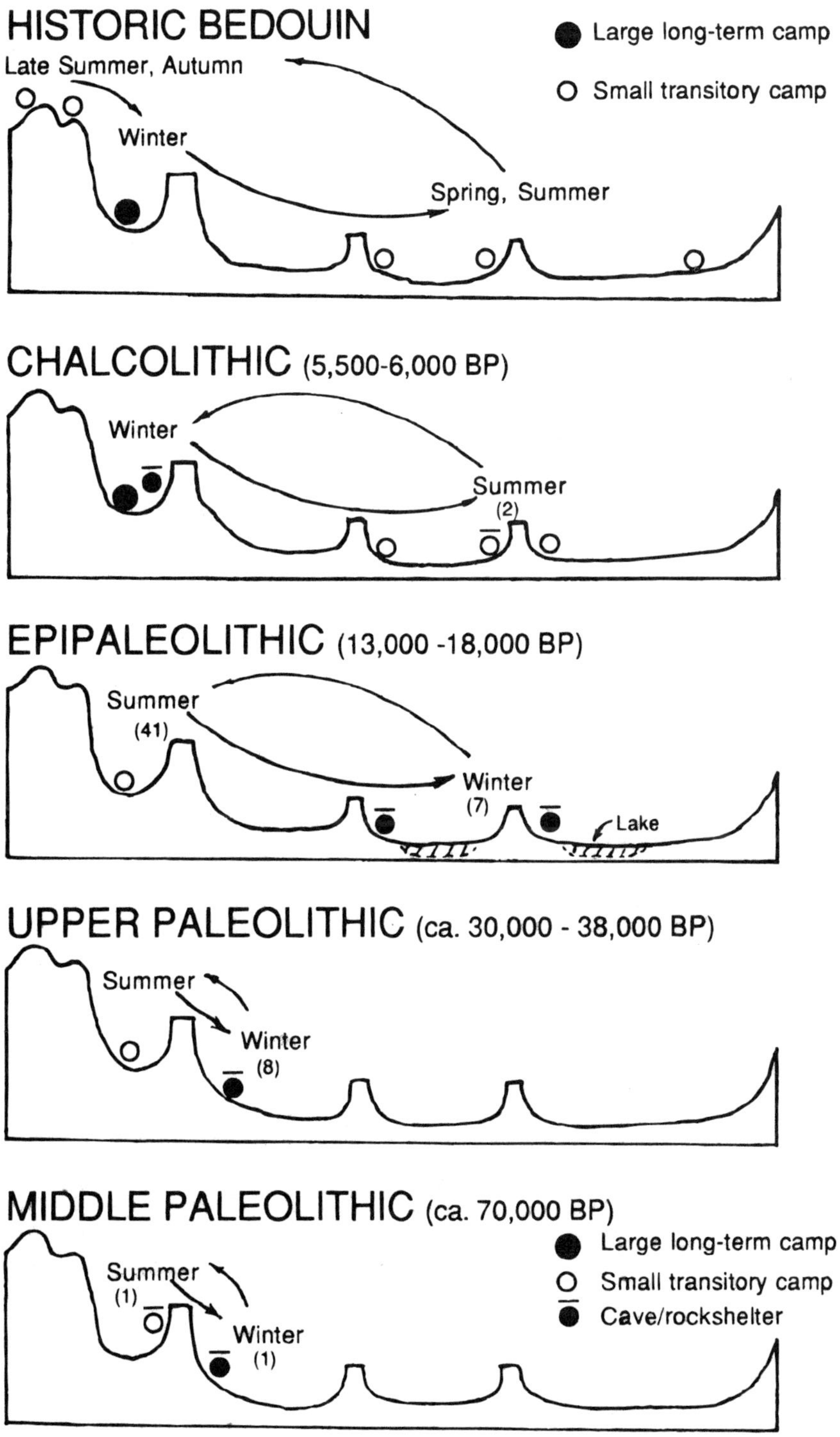


**Figure 21.2.** Idealized patterns of transhumance for the study area from Middle Paleolithic to Chalcolithic times.

A clear dichotomy between the elevational belts selected for dispersed mode as opposed to coalesced mode encampments was maintained over most of the last 70,000 years, but the specific elevations of these settlement modes fluctuated with climatic environmental shifts. The most dramatic of these appears to have taken place at the end of the Pleistocene when long-term winter camps were for the first time established at elevations above 1,000 masl in the high piedmont. This also was accompanied by a change in the elevational positioning of groups over the warm, dry season. In the Pleistocene, groups appear to have migrated from the lowlands to the higher elevations of the piedmont and uplands during the warm, dry season. Transhumance during the Holocene, in part, represented a reversal of this pattern. Over the warm, dry season groups initially moved from the higher piedmont to the lowlands of the Hisma. As water sources dried up and pasturage died off during the dry season, groups then moved from the lowlands to progressively higher elevations and perennial springs in the uplands on the edge of the plateau.

## Shifts in Provisioning Strategies

The adjustments in residential mobility levels and sizes of residential units that corresponded to the coalesced and dispersed modes of the transhumant cycle appear to have been accompanied by shifts in provisioning strategies. In following Kuhn's (1991) terminology, the residents of long-term, winter camps apparently provisioned their occupations as "places" utilizing a radiating settlement-procurement structure and the logistical acquisition of resources found beyond the limits of their site catchments. In contrast, during their occupation of ephemeral, warm season camps, groups appear largely to have provisioned "activities" in an opportunistic manner or to have provisioned "individuals" through curation of artifacts. These provisioning strategies were associated with a circulating settlement-procurement structure in which most resources were obtained from within the catchments of sites.

### Provisioning of Places

Perhaps the two most important factors that conditioned the provisioning of places rested in the anticipation that they would be inhabited for extended periods and that they were likely to be revisited in the future. Both of these conditions made it more economic to concentrate resources at a place than to depend upon (a) the opportunistic availability of resources at their sources or (b) necessarily small packages of individually portaged resources. Moreover, the occupation of a place for a prolonged duration by a large group even demanded that resources be concentrated and often even funneled from distant sources. The long-term camp sites consistently show that they were provisioned as places. Beyond the evidence of bulky and nonportable artifacts, petroglyphs, and structures, the bulk acquisition of cherts and granitics for ground stone from distant sources imply that these sites were provisioned as places in the anticipation of prolonged and even future use. This raises the question of why these specific locations were so intensively used and so regularly revisited. The answer, I think, rests in the fact that within the region only a small number of settings satisfied all of the considerations necessary for selecting a long-term winter camp. For the Pleistocene sites, the setting would have had to have fallen in a relatively narrow elevational belt (ca. 800–1,000 masl) where predictable runoff from the rains and snow of the uplands would have overlapped with higher temperatures affording reasonable creature comfort. A second set of qualifications would have demanded that the setting be naturally sheltered with a southwest exposure and in close proximity to a water source. Given that a relatively small number of settings would have met these combined conditions, it is not surprising that they were regularly and intensively occupied. Although the settings of the long-term camps of the Holocene also would have been constrained by elevation (1,000–1,250 masl), their locations within natural shelters does not seem to have been important. Protection from the elements was acquired culturally through the construction of pithouses. Proximity to water, however, appears to have continued to play a major role in site selection. Once an encampment had been established, the opportunity to reuse corrals, pithouses, storage pits, nonportable grinding slabs and ceramic storage jars is likely to have induced groups to return to these same locations during the winter segment of their transhumant cycle.

## Provisioning of Activities and Individuals

With the higher residential mobility levels and the lower probability of revisiting camp settings during the dispersed mode, dry season segment of the transhumant cycle, groups appear to have shifted their provisioning strategies from that of place to activity and individual. During the Pleistocene, ephemeral camps were positioned at high elevations in the piedmont and, most probably, the uplands. The brief duration of such occupations coupled with the infrequency with which the sites were revisited would have rendered provisioning them as places uneconomic. Most of the daily activities conducted off-site are more likely to have been opportunistically provisioned on an "as needed" basis. Relative to lithic technology, this would have represented an efficient, readily employed strategy given the ubiquitous distribution of fine quality cherts within the catchments of those sites located in the high piedmont and uplands. Although some on-site fabrication of chipped stone tools was undertaken, the prevailing, time-transgressive pattern for these ephemeral camps was one in which most tools were produced off-site. Most on-site activities were related to maintenance and rejuvenation of tools.

As might be expected given the abundance of raw material within their catchments, there is little evidence for the curation of artifacts (provisioning individuals) in the ephemeral camps of Pleistocene foragers. In the Holocene, however, Timnian groups established ephemeral camps at much lower elevations distant from the chert sources of the plateau. The assemblages from these sites do show evidence of curation. Moderate to high proportions of cores coupled with very low proportions of primary elements point to the curation of cores. Similarly, the presence of large tabular scrapers and bifaces of basalt signals the provisioning of individuals.

## Use of Marginal Resources

A persistent feature of the adaptive strategies followed by the prehistoric inhabitants of the region, one inherent to transhumance, was the partitioning of resources by season and elevational belt based upon productivity and predictability. The winter wet season and early spring corresponded to the peak in the availability of food and water. During the long dry season these resources progressively declined. During most of the Pleistocene, the high elevations of the piedmont and uplands would have been available for intensive exploitation only during the warm season, whereas the lowlands were potentially exploitable over most of the year. With the higher temperatures and generally drier conditions of the Holocene, this pattern was reversed. The lowlands were inhabitable only during and immediately following the wet season when game, pasturage, and water would have been available, whereas the high piedmont was potentially exploitable over most of the year.

Over a span of some 70,000 years, foragers as well as pastoralists incorporated resource zones that were seasonally marginal into their annual range. While the resources in these zones appear to have been less predictable than those in the zones selected for intensive exploitation during the winter wet season, their seasonal utilization would have relieved pressure on the optimum resource zones. This strategy thus allowed for a "banking" or conservation of resources that would be depended upon during the time of population coalescence. This practice demanded, however, that groups have the ability to make substantial seasonal adjustments in their settlement-procurement structure, their residential mobility levels, their residential populations, and their provisioning strategies.

## RESPONSES TO CLIMATIC-ENVIRONMENTAL CHANGES

The data on prehistoric adaptive strategies drawn from a great number of sites set in a common geographic locus and spanning some 70,000 years provide an unusual opportunity for accessing the significance of environmental changes to human cultural evolution. The residuals of human behavior that are most evident in the archaeological record and most likely to provide insights into cultural responses to environmental perturbations consist of the varieties and distributions of sites and their artifact inventories. In examining these sets of data diachronically, the evolution of human settlement-procurement systems and the evolution of technological systems both show strong correlations with paleoenvironmental fluctuations.

### Settlement-Procurement Patterns

The elevational belts selected for the locations of long-term winter camps appear to have shifted in concert with fluctuations in temperature (Figure 21.3). In following the high-resolution chronostratigraphy developed by Martinson *et al.* (1987) from oxygen-isotope data, two intervals of elevated temperature are recorded over the last 70,000 years. These were separated by a cold interval, stretching from ca. 30,000–13,000 B.P., in which temperatures were depressed some 4 °C below modern levels. This reconstruction closely parallels that developed by Emery-Barbier (Chapter 16) from pollen data with two exceptions. The pollen evidence suggests a brief cold setting ca. 12,500–13,000 B.P. and a warm episode ca. 16,500 B.P.

Long-term winter camps were positioned at their lowest elevations (800–900 masl) during this cold interval as evidenced by the Early, Middle, and Late/Final Hamran occupations situated in rockshelters overlooking dry lakes on the floor of the Hisma. Although moisture levels appear to have oscillated during this interval, there was evidently adequate runoff throughout the Hamran succession to maintain at least seasonal lakes. During the warmer interval between 30,000–38,000 B.P., Upper Paleolithic long-term camps were established at elevations from ca. 950–1,000 masl. With the marked rise in temperature that occurred during terminal Pleistocene and Holocene times, long-term winter camps of Natufian, Timnian, and even modern Bedouin groups were situated at progressively higher elevations to points reaching above 1,200 masl.

The linkage between temperature and the elevational settings of long-term winter camps may be understood in the light of groups attempting to position themselves near dependable water sources and at the same time remain within elevational belts that allowed for comfortable temperatures. During cold intervals, comfortable temperatures would obviously have been found at lower elevations. But lower temperatures would also have acted to reduce rates of evaporation and thus extend runoff from upland rains and snow to lower elevations. Historically, Bedouin groups inhabiting the region abandon the uplands in the winter to establish winter camps in piedmont below 1,400 masl. The mean minimum winter temperature of the uplands is 0–3 °C, while that of the piedmont falls between 3–6 °C. It seems then that the critical minimum temperature for determining the elevation of establishing winter camps in the region is that at or near freezing, at least for Bedouin.

When paleo-temperatures are considered in the context of the universal temperature lapse rate (0.66°:100 m), this freezing line typically would have been encountered just above the elevations of long-term winter camps (Figure 21.3). This, indeed, implies that the prehistoric inhabitants of the region situated their winter camps as high as possible to insure the availability of surface water from upland runoff, while at the same time not exceeding the elevation of the freeze line.

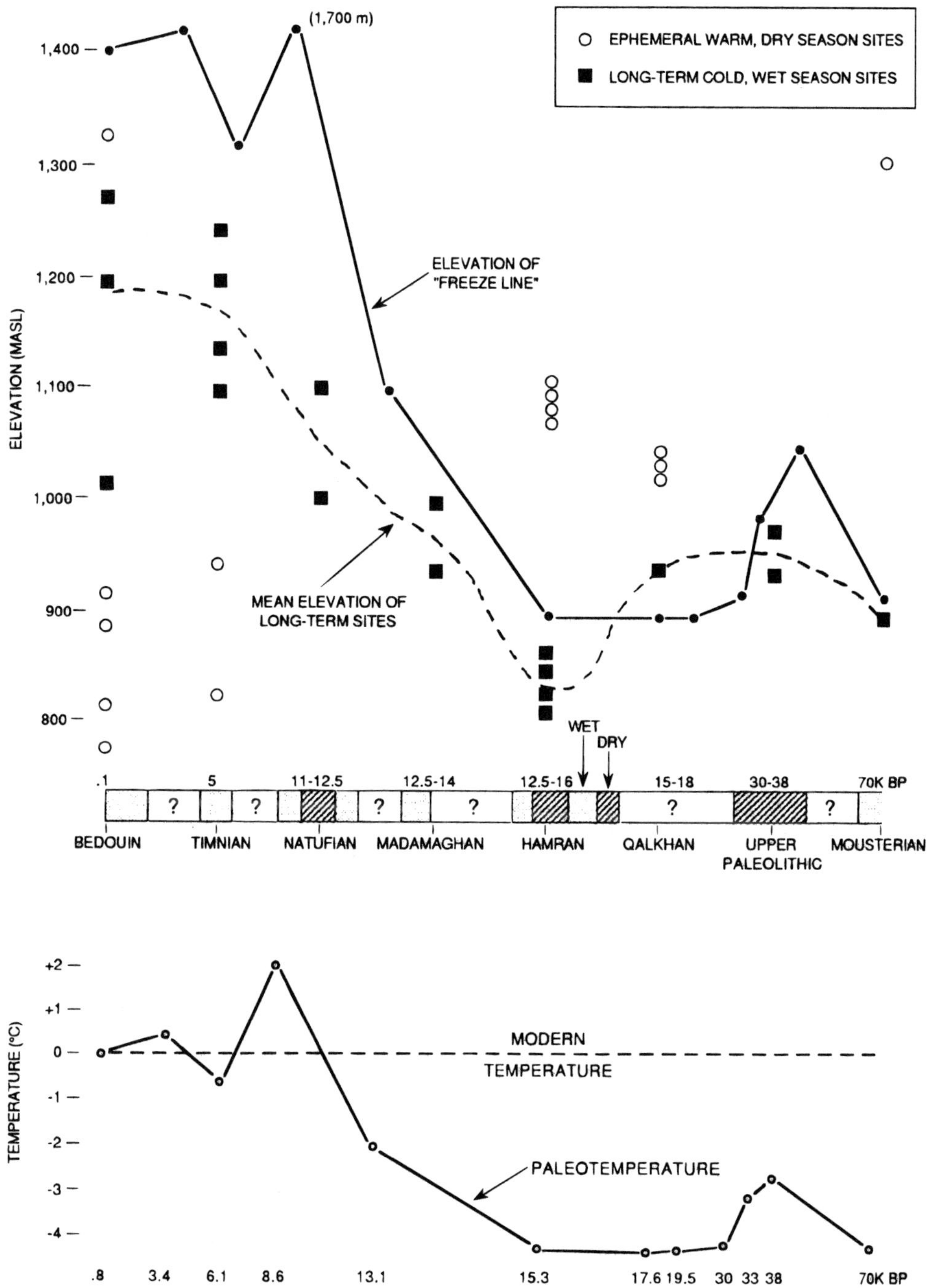


**Figure 21.3.** The elevations of winter camp sites of the major archaeological taxa identified in the study in comparison to paleotemperatures (Martinson *et al.* 1987) and the "freeze-line" as reconstructed from the paleotemperatures. Note that the sites typically rest just beneath the "freeze-line."

## Technology and Edge Production Efficiency

In that the prehistoric populations of the region appear to have adjusted the specific patterns of transhumance in response to climatic-environmental fluctuations, a question emerges as to what other adjustments were made in their adaptive strategies. Within the Levant, several economic, demographic, and social changes have been linked to climatic-environmental shifts during terminal Pleistocene and early Holocene times. These relate to the emergence of complex foraging and incipient horticulture (Henry 1989; Bar-Yosef and Belfer-Cohen 1989; McCorriston and Hole 1991) as well as to the beginnings of pastoralism (Bar-Yosef and Khazanov 1992). While these represented dramatic adaptive changes that occurred within a brief span of 5,000–6,000 years, the measures by which they have been defined (e.g., morphological evidence of domesticated plants or animals) do not allow for comparisons over much longer temporal sweeps of human evolution. Even when moving from the fields of data that are unique to these strategies to a more common, time-transgressive field such as lithic artifacts, long-term evolutionary comparisons are hindered by traditional analytic procedures which are time and/or taxon specific. Technotypologic observations tend to be tailored to the parameters of the major archaeological periods and their constituent taxa. For example, there are really no common measures for comparing Middle Paleolithic, Levantine Mousterian assemblages with those of the Epipaleolithic, Natufian.

In an effort to make such long-term evolutionary comparisons, I have focused on observations related to the efficiency by which different lithic technologies were able to generate "edge." The ultimate objective of any flaked stone technology is the production of edge in that the perimeters of blanks serve as their fundamental functional surfaces whether modified by secondary retouch or used without formal modification. Therefore, the length of edge produced for a given amount (e.g., weight) of raw material is an expression of the efficiency of the technology that produced the blank. A trend of increasing efficiency of edge production has been recognized for some time in Paleolithic studies (Butzer 1971:475), but no attempts have been made to quantify this trend within regional successions. Moreover, as far as I know, there have been no efforts to examine the tempo and mode of the trend relative to larger human ecologic considerations.

The methodology used in calculating a specimen's Edge Production Index (EPI) follows that described in an earlier study (1987) in which a specimen's perimeter (excluding its striking platform), measured in millimeters, is divided by its weight in grams. Complete, unbroken tools were selected for analysis from each assemblage.

An examination of the Edge Production Indices of the assemblages within the study area shows a general trend of increasing efficiency until ca. 12,500 B.P., when Natufian and subsequent Timnian technologies show marked declines (Figure 21.4). The tempo of this progressive rise in efficiency over almost 60,000 years followed a relatively slow pace until ca. 16,000 B.P. at which time it accelerated sharply. Within the relatively brief interval between 16,000–13,000 B.P., the EPI more than doubled and this interval is also linked to the greatest degree of variability in EPIs between assemblages.

The late Last Glacial interval of marked escalation in edge production efficiency was synchronous with the episode of depressed temperature and the attendant positioning of long-term winter camps at the lowest elevations of the 70,000-year temporal sweep. The locations of these winter camps would have placed them at a distance of some 25–30 km from the chert sources on the plateau. This distance to raw material, in fact, was substantially greater than that enjoyed by the inhabitants of earlier Middle and Upper Paleolithic long-term winter camps (i.e., 18–22 km). With these greater distances to chert sources there would have been increased selective pressures favoring more efficient techniques of generating edge.

In a large measure the need to economize the use of raw material was met by an emphasis upon the production of microlithic tools. But beyond producing smaller tools and hence greater

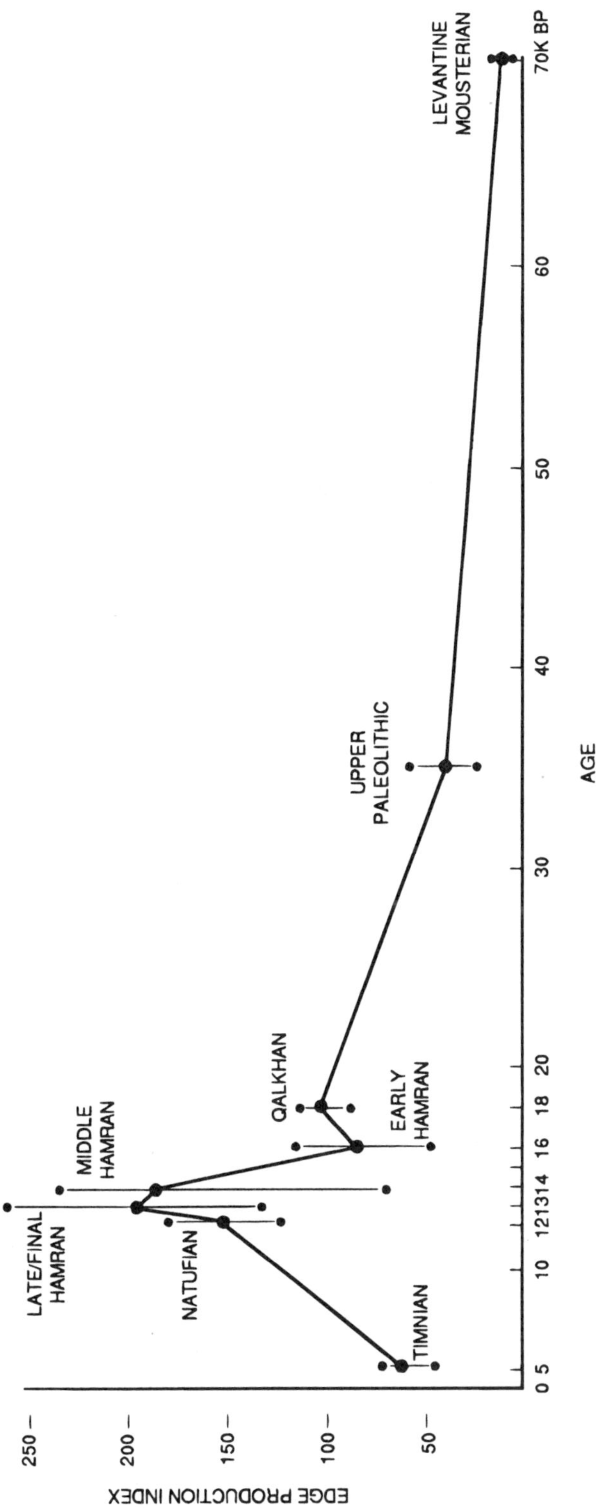


Figure 21.4. Graph plotting Edge Production Indices against time. Note the relatively slow, even rise in EPIs from ca. 70–18,000 B.P. is followed by a rapid rise and subsequent decline during the Late Pleistocene and Early Holocene.

edge, Hamran groups residing in these winter camps appear to have greatly increased the efficiency by which they provisioned the occupations, as well. Rather than importing raw material in bulk or as cores, as was the practice during Levantine Mousterian and Upper Paleolithic times, the sites were primarily provisioned through the importation of blanks in the form of bladelets (Chapter 10). The emergence of a microlithic technology then would not only have greatly increased the efficiency of generating edge, the smaller, more portable blanks also would have provided for a much more economic way of provisioning sites from distant sources of raw material. From an overarching adaptive perspective, this would have greatly reduced the degree to which groups were necessarily tethered to raw material sources. With the reduced importance of the proximity of chert sources, groups would have enjoyed much greater ranges and have been able to concentrate the positioning of their sites and catchments on other critical resources such as water, food, and fuel.

Although these technological innovations were coincident with the temperature fluctuations and adjustments of settlement-procurement patterns in southern Jordan, it is unlikely that these linkages specifically explain the regional emergence of microlithic technology. These local ecological conditions and their technological connections do, I think, provide a general model for understanding the evolution of lithic technology, however. This model essentially ties the trend and tempo of technological evolution to changes in the distributions of critical resources and hominid mobility levels (Henry 1987). Whereas in an earlier study I suggested that significant shifts in resources within the southern Levant would be most likely to have been triggered by the onset of arid cycles, this clearly was not the case along the edge of the Jordanian Plateau where temperature played a more important role. At lower elevations in other regions of the Levant, climatic changes inducing fluctuations in moisture levels may well have had a similar effect in stretching the distances between raw material sources and other critical resources. It is under these conditions that we might expect to see major changes in technology. Perhaps it is not by coincidence then that the rapid oscillations in climate of late Last Glacial times were accompanied by the widespread adoption of a microlithic technology within the region.

## ETHNICITY, ADAPTATION, AND CONCEPTUAL LOCKS

In analyzing the assemblages that were recovered from deposits stretching over some 70,000 years, specific technotypologic configurations were identified that allowed for placing the assemblages within various archaeological taxa and reconstructing a cultural history of the region (see Chapter 3; Figure 3.1). The patterned variability in artifacts upon which these taxa are defined is thought to reflect the behaviors that formed discrete socio-adaptive systems of prehistoric humans. This is largely consistent with a functional notion of how variability in the archaeological record is to be interpreted, but it can also be argued from a cultural historic perspective that these taxa at their finer scales (i.e., industry, phases/facies) have real meaning in terms of prehistoric ethnicity. In primarily relying upon behavioral adaptation at the social level, prehistoric populations would have required a mechanism for identifying those with access to limited critical resources such that these interest groups could balance their numbers with their available resources. Although this is admittedly a "functionalist" view of ethnicity, it also is consistent with what is perhaps the central cultural historic bias when it comes to the interpretation of patterned variability in the archaeological record. This relates to the assumption that real peoples or cultures can be identified on the basis of patterned variability in artifacts, even in the Paleolithic. Here it is important to make some distinctions relative to the degree to which we can define prehistoric ethnicity, however.

Our ability to identify and trace ethnicity in prehistory is complicated by the fact that, behaviorally, ethnicity exists at different scales and that, archaeologically, we rarely have sufficient evidence to address the finest of these. David Clarke (1968) underscored this in his hierarchical classification of archaeological cultures into scales of technocomplex, culture group, and culture. He also observed that from the material residuals of Paleolithic behaviors, archaeologists rarely are able to reconstruct behavioral systems to a finer degree than that of culture group. Of the taxa defined within this study (see Chapter 3) only that of the Natufian can be related to a culture, the rest classified either as to culture group (Hamran, Madamaghan, Timnian) or technocomplex (Levantine Mousterian, Ahmarian, Levantine Aurignacian).

## Culture Areas, Adaptation, and Ethnicity

An examination of the regional biogeographic distributions of the archaeological taxa identified in the study shows a near continuous representation of those tied to the steppe-desert zone, but taxa representative of the Mediterranean woodlands also occur, especially during late Last Glacial times (Table 21.2). As might be expected, it was during the driest intervals of the succession in early Last Glacial (Late Levantine Mousterian) and in mid-Holocene (Timnian) times that the study area was occupied exclusively by those taxa linked to steppe-desert zones. During moister and colder intervals, we see evidence for an expansion into the study area of archaeological taxa associated with the Mediterranean woodlands. Apparently during these intervals populations that had come to exploit quite different environmental zones came into close contact at least for portions of the year. What is so surprising is that despite their maintenance of abutting ranges for several thousands of years, coupled with clear evidence of interaction, these populations failed to fuse or melt into a common behavioral system. While the contrasts in the biogeographic distributions of the Ahmarian and Levantine Aurignacian hint of adaptively different populations with abutting ranges during the Upper Paleolithic, such a situation is most clearly traced during the Epipaleolithic. Within this interval in southern Jordan, evidence for the presence of parallel traditions represented by the Qalkhan–Madamaghan and the Hamran–Natufian sequences extends over some 6,000 years.

On a regional scale the Qalkhan–Madamaghan tradition was tied to the steppe-desert zone. Qalkhan sites are found distributed from the Syria Desert on the eastern periphery of the Mediterranean woodlands southward into southern Jordan. In the latter part of the succession, during Madamaghan (Mushabian Complex) times (ca. 14,000 B.P.), populations employing this arid zone adaptation appear to have extended their distribution westward across the Rift Valley into Sinai and the Negev. The Kebaran–Geometric Kebaran–Natufian tradition (represented locally by the Hamran–Natufian succession) traced a near synchronous development centered within the Mediterranean woodlands. During the latter phases of this succession (especially during Geometric Kebaran and Late Natufian times), groups appear to have been able to extend their ranges into the arid zone, at least along the woodland margins.

**Table 21.2. Biogeographic Affiliations of the Archaeological Taxa Found in the Study Area**

| Steppe desert | Mediterranean woodlands |
|---|---|
| Timnian | |
| | Natufian |
| Mushabian (Madamaghan Industry) | Geometric Kebaran (Middle, Late/Final Hamran) |
| | Kebaran (Early Hamran) |
| Qalkhan | |
| | Levantine Aurignacian (?) |
| Early Ahmarian | |
| Levantine Mousterian | |
| "D" Type | |

The study area in southern Jordan rests along the boundary separating the Mediterranean woodlands from the steppe-desert. Because of the area's marked elevational diversity, these zones are tightly packed. Mediterranean woodlands occupy the uplands and the higher piedmont, but give way to steppe below 1,300 masl. The steppe, in turn, is replaced by desert below 1,000 masl. Although the elevational boundaries of the zones shifted in the past in concert with fluctuations in temperature and moisture parameters, because of its steep gradient and elevational amplitude of ca. 900 m (800–1,700 masl) the area appears to have contained portions of woodland, steppe, and desert zones over most of the last 70,000 years.

Given this environmental setting, the presence of overlapping ranges of prehistoric populations with contrasting adaptive strategies should be expected. This appears to have been most pronounced, or at least captured with the greatest clarity, during the Epipaleolithic. Not only do site distributions and chronology imply an interfingering of the ranges of woodland and arid-zone adapted populations, but the time-trends of certain artifact attributes indicate social interaction between these parallel traditions (Chapter 13). The microburin technique and the lunate furnish good examples of such exchanges in material culture. Although certain attributes such as these were adopted from neighboring groups, a central core of technotypologic attributes remained unchanged and therefore definitive of specific taxa and their diachronic linkages (i.e., traditions).

In combination, archaeological data for the Epipaleolithic point to the persistence of ethnically distinct populations that occupied interfingering (if not overlapping) ranges for a temporal sweep of some 6,000 years. An important question centers on how it was that these populations did not fuse or at least become blurred with respect to their ethnicity. The differing cultural ecologies of the groups may provide an answer. Given their distinctive biogeographic ranges (i.e., woodland versus steppe-desert) or culture areas, the groups are likely to have exploited the fringe area along the edge of the Ma'an Plateau in quite different ways. The distinctive settlement patterns of the Hamran–Natufian and Qalkhan–Madamaghan traditions hint of differing exploitive strategies. Hamran groups apparently wintered around lakes on the floor of the broad Hisma valley at elevations of ca. 800 masl and then migrated to the higher elevations of the piedmont and uplands during the warm season. Madamaghan groups also appear to have wintered in the area near the heads of canyons that fall away steeply to the Rift Valley. But instead of occupying the piedmont and uplands over the warm dry season, they appear to have abandoned the area presumably to reside at lower elevations near springs in the Rift and along the Red Sea coast. The settlement structures of groups of the two traditions differ then in a very fundamental way. Although residing in close proximity during the winter, they pulled apart during the warm, dry season, most likely in the search for water. The woodland-adapted groups migrated in the direction of the core of the Mediterranean woodlands, whereas those groups geared to exploiting the steppe-desert moved to permanent springs deeper within the arid zone.

These different settlement structures imply that the groups of the two traditions retained their fundamentally different adaptive strategies, although they appear to have resided for a portion of the year in close proximity. In exploiting such different resources over the warm, dry season the groups would have necessarily maintained different interest groups, information-sharing networks, and solutions for their survival. Differences in ethnicity would have been reinforced by these distinctive behaviors, while at the same time contributing to their distinctiveness.

What is so interesting here from a cultural ecologic perspective is the strong correlation that persisted between ethnicity and environment. In the study area, groups representing two distinct traditions existed in close proximity for a very long time. At any time in this long 6,000-year succession, the material residues left by groups of either tradition show stronger affinities to those left by similarly adapted groups from distant points than to the residues from nearby sites of the other tradition. For example, Early Hamran assemblages show much greater technotypologic similarity to late Kebaran assemblages of the woodlands of northern Israel and Lebanon than to those recovered from nearby Qalkhan sites. Similarly, Madamaghan assemblages from Tor Hamar and Tor Fatma show strong technotypologic affinities to Mushabian Complex assemblages of Sinai and the Negev, but display striking differences with the synchronous Middle and Late–Final Hamran assemblage found at nearby sites within the study area. Thus it would appear that ethnic ties among these groups were based more upon their degree of similarity in adaptation than proximity.

At a more specific scale than that of culture area, the ecological concepts of "habitat" and "niche" may provide additional insights as to how ethnic differences persisted in such a limited area for so long a time. The concept seems particularly relevant for the study area, given its environmental diversity and tightly packed zones. A habitat is seen as a biogeographic area, whereas a niche represents the specific manner in which an organism draws its livelihood from the area (Odum 1959; Pianka 1978). In the context of cultural ecology, Barth's (1956) classic study of three distinct ethnic groups in Pakistan, with interfingering and partially overlapping ranges, showed that their exploitation of different niches maintained their differences in ethnicity. It is suggested here that these same linkages between adaptive and ethnic differences largely account for variability seen at the taxonomic level in the Epipaleolithic assemblages of the study area.

The correlation between ethnic boundaries and environmental zones formed the basis of the cultural area concept as initially recognized in the New World (Mason 1894; Wissler 1917; Kroeber 1931), but similar relationships appear less well defined in the Old World, especially in Asia (Barth 1956). These differences may well be attributed to the greater range of techno-economies that have cumulatively become available to Old World populations coupled with Harris's (1968:375) observation that "it is the techno-environmental interaction which is decisive, not merely the environment." For the Pleistocene Levant, one might expect a stronger correspondence between ethnicity and environment given that the region's populations at a general level followed a common techno-economy, i.e., foraging supported by stone age implements. In fact, rather than having the capacity for significantly altering their environments, the technologies employed by these prehistoric foragers appear largely to have been shaped by environmental forces. When terminal Pleistocene and Holocene times are considered, Levantine populations had access to a much wider range of technologies and economies for their survival. Horticulture and pastoralism emerged as significant techno-economic alternatives, but even in this context distinct cultural areas showing a strong correspondence to woodland and steppe-desert environments persisted. This regional dichotomy in culture areas remains in evidence today as shown by the distributions of Bedouin pastoral nomads and Fellahin farmers. The Bedouin and Fellahin populations display ethnicity at a large scale as represented by striking differences in economy, technology, settlement pattern, social organization, language, and dress. Therefore within the Levant over the last 70,000 years or so, variability in cultural ecology appears to offer the best explanation for the region's marked ethnic diversity.

# *References*


Ahler, S. (1975). *Pattern and variety in extended coalescent lithic technology*. Ph.D. dissertation. Columbia: University of Missouri-Columbia.

Ahler, S. (1989). Mass analysis of flaking debris: Studying the forest rather than the tree. In D. O. Henry and G. H. Odell (Eds.), *Alternative Approaches to Lithic Analysis* (pp. 85–118). Archeological Papers of the American Anthropological Association Number 1. Washington D.C.: American Anthropological Association.

Al-Eisawi, D. M. (1985). Vegetation in Jordan. In A. Hadidi (Ed.), *Studies in the History of Jordan*, Vol. II (pp. 45–58). Avon: Department of Antiquities.

Alexander, R. O. (1974). The evolution of social behavior. *Annual Review of Ecology and Systematics, 5*, 325–383.

Alexander, R. O., and Borgia, G. (1978). Group selection, altruism and the levels of organization of life. *Annual Review of Ecology and Systematics, 9*, 449–474.

Anderson-Gerfaud, P. (1983). A consideration of the uses of certain backed and "lustred" stone tools from late Mesolithic and Natufian levels of Abu Hureyra and Mureybet (Syria). In M.-C. Cauvin (Ed.), *Traces d'utilisation sur les outils Néolithiques du proche orient* (pp. 77–106). Lyon: Maison de l'Orient.

Anderson-Gerfaud, P. (1990). Aspects of behavior in the Middle Palaeolithic: Functional analysis of stone tools from southwest France. In P. Mellars (Ed.), *The Emergence of Modern Humans: An Archaeological Perspective* (pp. 389–418). Edinburgh: Edinburgh University Press.

Arieh, A. B. (1980) A Chalcolithic site near Serabit El Khadim. *Tel Aviv, 7*, 45–64.

Azoury, I. (editor). (1986). *Ksar Akil, Lebanon. A technological and typological analysis of the Transitional and Early Upper Palaeolithic Levels of Ksar Akil and Abu Halka*. Volume I, Levels XXXV–XII. BAR international Series 289. Oxford: British Archaeological Reports.

Baharav, D. (1981). Food habits of the mountain gazelle in semi-arid habitats of eastern Lower Galilee, Israel. *Journal of Arid Environments, 4*, 63–69.

Baharav, D. (1983). Reproductive strategies in female mountain and Dorcas gazelle. *Journal of the Zoological Society of London, 200*, 445–453.

Bamforth, D. (1988) Investigating microwear polishes with blind tests: The institute results in context. *Journal of Archaeological Science, 15*, 38–50.

Barth, F. (1956). Ecologic relationships of ethnic groups in Swat, North Pakistan. *American Anthropologist, 58*, 1079–1089.

Baruch, U., and Bottema, N. (1991). Palynological evidence for climatic changes in the Levant ca. 17,000–9,000 B.P. In O. Bar-Yosef and F. R. Valla (Eds.), *The Natufian Culture in the Levant* (pp. 11–20). Ann Arbor: International Monographs in Prehistory.

Bar-Yosef, O. (1970). *The Epipalaeolithic cultures of Palestine*. Ph.D. Dissertation, Hebrew University of Jerusalem.

Bar-Yosef, O. (1979). Excavations in Hayonim Cave. *Mitekufat Haeven, 16*, 88–100. (In Hebrew).

Bar-Yosef, O. (1981). Prehistory of the Levant. *Annual Review of Anthropology, 9*, 101–133.

Bar-Yosef, O. (1981). The Epi-Palaeolithic complexes in the southern Levant. In J. Cauvan and P. Sanlaville (Eds.), *Préhistoire du Levant* (pp. 389–408). Paris: Édition du C. N. R. S.

Bar-Yosef, O. (1983). The Natufian in Southern Levant. In T. C. Young, P. E. L. Smith and P. Mortensen (Eds.), *The Hilly Flanks and Beyond* (pp. 11–42). Chicago: Oriental Institute Studies in Ancient Oriental Civilization.

Bar-Yosef, O. (1987). Late Pleistocene adaptations in the Levant. In O. Soffer (Ed.), *The Pleistocene Old World: Regional Perspectives* (pp. 219–236). New York: Plenum.

Bar-Yosef, O. (1989a). Geochronology of the Levantine Middle Paleolithic. In P. Mellars and C. B. Stringer (Eds.), *The Human Revolution* (pp. 589–625). Princeton, NJ: Princeton University Press.

Bar-Yosef, O. (1989b). Upper Pleistocene cultural stratigraphy in southwest Asia. In E. Trinkaus (Ed.), *The Emergence of Modern Humans* (pp. 154–180). Cambridge: Cambridge University Press.

Bar-Yosef, O. (1991). The search for lithic variability among Levantine Epi-Paleolithic industries. In *25 Ans d'Etudes Technologiques en Préhistoire* (pp. 319–336). Juan-les-Pins: Editions APDCA.

Bar-Yosef, O., and Belfer, A. (1977). The Lagaman industry. In O. Bar-Yosef and J. Phillips (Eds.), *Prehistoric Investigations in Gebel Maghara, Northern Sinai, Qedem* 7 (pp. 42–84). Jerusalem: Institute of Archaeology of the Hebrew University.

Bar-Yosef, O., and Belfer-Cohen, A. (1988). The early Upper Paleolithic in Levantine caves. In J. F. Hoffecker and C. A. Wolf (Eds.), *The Early Upper Paleolithic* (pp. 23–37). Oxford: BAR International Series 437.

Bar-Yosef, O. (1989). The origin of sedentism and farming communities in the Levant. *Journal of World Prehistory, 3, 4*, 447–498.

Bar-Yosef, O., Belfer-Cohen, A., Goren A., Hershkovitz, I., Ilan, O., Mienis, H. K., and Sass, B. (1986). Nawamis and habitation sites near Gebel Gunna, southern Sinai. *Israel Exploration Journal, 36*, 121–167.

Bar-Yosef, O., Gopher, A., and Nadel, D. (1987). The "Hagdud Truncation"—a new tool type from the Sultanian Industry at Netiv Hagdud, the Jordan Valley. *Mitekufat Haeven, 20*, 151–157.

Bar-Yosef, O., and Goring-Morris, A. N. (1977). Geometric Kebaran A occurrences. In Y. Yadin, N. Avigad, J. Aviram, and A. Ben-Tor (Eds.), *Qedem* (pp. 115–148). Jerusalem: The Institute of Archaeology, The Hebrew University of Jerusalem.

Bar-Yosef, O., and Khazanov, A. (1992). Introduction. In O. Bar-Yosef and A. Khazanov (Eds.), *Pastoralism in the Levant* (pp. 1–10). Wisconsin: Prehistory Press.

Bar-Yosef, O., and Meignen, L. (1989). Levantine Mousterian variability in the light of new dates from Qafzeh and Kebara caves, Israel. Paper presented at the 54th Annual Meeting of the Society for American Archaeology, Atlanta, April 1990.

Bar-Yosef, O. (1992). Insights into Levantine Middle Paleolithic cultural variability. In H. Dibble and P. Mellars (Eds.), *The Middle Paleolithic: Adaptation, Behavior, and Variability* (pp. 163–182). Philadelphia: University of Pennsylvania, University Museum Monograph 72.

Bar-Yosef, O., and Phillips, J. L. (1977). *Prehistoric investigations in Gebel Maghara, Northern Sinai. Qedem 7*. Jerusalem: Monographs of the Institute of Archaeology, Hebrew University.

Bar-Yosef, O., and Valla, F. (1979). L'Evolution du Natoufien. Nouvelles perspectives. *Paléorient, 5*, 145–152.

Bar-Yosef, O. (1991). *The Natufian culture in the Levant*. Ann Arbor International Monographs in Prehistory.

Bar Yosef, O., and Vandermeersch, B. (1981). Notes concerning the possible age of the Mousterian layers in Qafzeh Cave. In J. Cauvin and P. Sanlaville (Eds.), *Préhistoire du Levant*. Paris: CNRS.

Bar-Yosef, O. (1993). Modern humans in the Levant. *Scientific American, 4*, 94–100.

Bar-Yosef, O., Vandermeersch, B., Arensburg, B., Goldberg, P., Laville, H., Meignen, L., Rak, Y., Tchernov, E., and Tillier, A. M. (1986). New data on the origin of modern man in the Levant. *Current Anthropology, 27*, 63–64.

Bar-Yosef, O., and Vogel, J. C. (1987). Relative and absolute chronology of the Epi-Palaeolithic in southern Levant. In O. Aurenche, J. Evin, and F. Hours (Eds.), *Chronology of the Near East* (pp. 219–246). Oxford: British Archaeological Research 379.

Beheiri, S. E. (1973). *Geography of Jordan*. Amman, Jordan: El-Shark Publishers.

Belfer-Cohen, A. (1991). Art items from Layer B, Hayonim Cave: A case study of art in a Natufian context. In O. Bar-Yosef and F. R. Valla (Eds.), *The Natufian Culture in the Levant* (pp. 569–588). Ann Arbor: International Monographs in Prehistory.

Bender, F. (1974). *Geology of Jordan*. Berlin: Gerbruden Borntaeger.

Bergman, C. (1981). Point types in the Upper Palaeolithic sequence at K'sar Akil, Lebanon. In J. Cauvin and P. Sanlaville (Eds.), *Préhistoire du Levant* (pp. 319–330). Paris: Centre National de la Récherche Scientifique.

Beyries, S. (1988). Functional variability of lithic sets in the Middle Paleolithic. In H. Dibble and A. Montet-White (Eds.), *The Prehistory of Western Eurasia* (Vol. 1) (pp. 213–223). Philadelphia: University of Pennsylvania, University Museum Monograph 54.

Binford, L. (1973). Interassemblages variability—the Mousterian and the "functional" argument. In C. Renfrew (Ed.), *The Explanation of Cultural Change: Models in Prehistory* (pp. 227–254). London: Duckworth.

Binford, L. (1977). Forty-seven trips: A case study in the character of archaeological formation processes. In R. Wright (Ed.), *Stone Tools as Cultural Markers* (pp. 24–36). Canberra: Australian Institute of Aboriginal Studies.

Binford, L. (1978). *Nunamiut Ethnoarchaeology*. New York: Academic Press.

Binford, L. (1979). Organization and formation processes: Looking at curated technologies. *Journal of Anthropological Research, 35*, 255–273.

Binford, L. (1980). Willow smoke and dogs' tails: Hunter-gatherer settlement systems and archaeological site formation processes. *American Antiquity, 45*, 4–20.

Binford, L. (1981). *Bones: Ancient Men and Modern Myths*. New York: Academic Press.

Binford, L. (1983) *In Pursuit of the Past*. New York: Thames and Hudson.

Binford, L. (1984). *Faunal Remains from Klasies River Mouth*. New York: Academic Press.

Binford, L. (1985). Human ancestors: Changing views of their behavior. *Journal of Anthropological Archaeology, 4*, 292–327.

Binford, L. (1987). Searching for camps and missing the evidence? Another look at the Lower Paleolithic. In O. Solfer (Ed.), *The Pleistocene Old World* (pp. 17–32). New York: Plenum.

Binford, L. (1989). Isolating the transition to cultural adaptations: An organizational approach. In E. Trinkaus, (Ed.), *The Emergence of Modern Humans* (pp. 18–41). Cambridge: Cambridge University Press.

Binford, S. (1968). Early Upper Pleistocene adaptations in the Levant. *American Anthropologist, 70*, 707–717.

Bleed, P. (1986). The optimal design of hunting weapons: Maintainability or reliability? *American Antiquity, 51*, 737–747.

Boëda, E., Geneste, J.-M., and Meignen, L. (1990). Identification de chaînes opératoires lithiques du Paléolithique Ancien et Moyen. *Paléo, 2*, 43–80.

Boessneck, J. (1969). Osteological differences between sheep (*Ovis aries* Linne) and goat (*Capra hircus* Linne). In E. S. Higgs and D. Brothwell (Eds.), *Science in Archaeology* (pp. 331–358). London: Thames and Hudson.

Bordaz, J. (1970). *Tools of the Old and New Stone Age*. Garden City, New York: Natural History Press.

Bordes, F. (1961). *Typologie du Paléolithique ancien et moyen* (2 Vols.). Bordeaux: Delmas.

Bordes, F. (1969). Reflections on typologie in the Paléolithique. *Artic Anthropology, 1.1*, 1–23.

Brace, C. (1988). 2nd. ed., revised. *The Stages of Human Evolution*. Englewood Cliffs, NJ: Prentice Hall.

Brace, C. (1991). Comments: New models and metaphors for the Neanderthal debate. *Current Anthropology, 32(5)*, 526–528.

Bradley, B. (1975). Lithic reduction sequences: A glossary and discussion. In E. Swanson (Ed.), *Lithic Technology*. The Hague: Mouton.

Brezillon, M. (1968). La denomination des objets de pierre taillee. Paris, IVth Supplement to *Gallia Préhistoire*, Édition de Centre National de la Récherche Scientifique.

Brooks, A. S., Hare, P. E., Kokis, J., Miller, G. H., Ernst, R. D., and Wendorf, F. (1990). Dating Pleistocene archaeological sites by protein diagenesis in ostrich eggshell. *Science, 248*, 60–64.

Brose, D. (1978). A model of changing subsistence technology in the Late Woodland of northeastern Ohio. In D. Davis (Ed.), *Lithics and Subsistence: An Analysis of Stone Tool Use in Prehistoric Economies*. Nashville: Vanderbilt University Publications in Anthropology.

Brown, L. (1982). Struthioniformes. In L. Brown, E. Urban, and K. Newman (Eds.). *The Birds of Africa, Vol. 1.*, London: Academic Press.

Burdon, D. J. (1959). *Handbook of the Geology of Jordan*.

Butler, B. H., Tchernov, E., Hietala, H., and Davis, S. (1977). Faunal exploitation during the late Epipaleolithic in the Har Harif. In A. E. Marks (Ed.), *Prehistory and Paleoenvironments in the Central Negev, Israel* (pp. 327–346). Dallas: Department of Anthropology Institute for the Study of Earth and Man, Southern Methodist University.

Butzer, K. (1971). *Environment and archaeology: An ecological approach to prehistory*. Chicago: Aldine.

Byrd, B. (1987). *Beidha and Natufian: Variability in Levantine settlement and subsistence*. Unpublished Ph.D. thesis, University of Arizona, Tucson.

Byrd, B. (1989). The Natufian: Settlement variability and economic adaption in the Levant at the end of the Pleistocene. *Journal of World Prehistory, III*, 159–198.

Byrd, B. (1990). Late Pleistocene settlement diversity in the Azraq Basin. *Paléorient, 14/2*, 257–264.

Byrd, B. (1991). Beidha: An early Natufian encampment in southern Jordan. In O. Bar-Yosef and F. R. Valla (Eds.), *The Natufian Culture in the Levant* (pp. 245–264). Ann Arbor: International Monographs in Prehistory.

Byrd B., and Colledge, S. M. (1991). Early Natufian occupation along the edge of the southern Jordanian steppe. In O. Bar-Yosef and F. R. Valla (Eds.), *The Natufian Culture in the Levant* (pp. 265–276). Ann Arbor: International Monographs in Prehistory.

Camps, G. (1975). The prehistoric cultures of North Africa: Radio-carbon chronology. In F. Wendorf and A. E. Marks (Eds.), *Problems in Prehistory: North Africa and the Levant* (pp. 181–192). Dallas: Southern Methodist University Press.

Carey, S. (1985). *Conceptual changes in childhood*. Cambridge: MIT Press.

Case, P. (1985). *Intelletual development: Birth to adulthood*. New York: Academic Press.

Casteel, R. (1979). Relationships between surface area and population size: A cautionary note. *American Antiquity, 44*, 803–807.

Cauvin, J., Cauvin, M.-C., and Stordeur, D. (1979). Récherches préhistoriques à El Kowm (Syrie). Première campagne 1978. *Cahiers de l'Euphrate*, 2, 79–117. Paris: Éditions du C. N. R. S.

Cauvin, M.-C. (1981). L'Épipaléolithique de Syrie d'après les prèmieres récherches dans la cuvette d'El Kowm (1978–1979). In J. Cauvin and P. Sanlaville (Eds.), *Prehistore du Levant* (pp. 375–388). Lyon: Colloque International du C. N. R. S.

Cauvin, M.-C., and Coqueniot, E. (1990). L'oasis d'El Kowm et le Kebarien géométrique. In *Paléorient, Préhistoire du Levant* (pp. 270–281). Lyon: C. N. R. S. Maison de l'Orient.

Chase, P., and Dibble, H. (1987). Middle Pleistocene symbolism: A review of current evidence and interpretations. *Journal of Anthropological Archaeology, 6*, 263–293.

Clarke, D. L. (1968). *Analytical archaeology*. London: Methuen and Co. Ltd.

Clarke, D. L. (1979). *Analytical archaeology: Collected papers of David L. Clarke*. London: Academic Press.

Clark, G. A. (1989). Romancing the stones: Biases, style, and lithics at La Riera. In D. O. Henry and G. H. Odell (Eds.), *Alternative Approaches to Lithic Analysis* (pp. 27–50). Archaeological Papers of the American Anthropological Association, No. 1.

Clark, G. A. (1991). Paradigms, realism, adaptation, and evolution. In G. A. Clark (Ed.), *Perspectives on the Past: Theoretical Biases in Mediterranean Hunter-Gatherer Research* (pp. 411–439). Philadelphia: University of Pennsylvania Press.

Clark, G. A., Lindly, J., Donaldson, M., Garrard, A., Coinman, N., Schuldenrein, J., Fish, S., and Olszewski, D. (1988a). Excavations at Middle, Upper and Epipalaeolithic sites in the Wadi Hasa, west-central Jordan. In A. N. Garrard and H. C. Gebel (Eds.), *The Prehistory of Jordan* (pp. 209–285). Oxford: British Archaeological Reports.

Clark, G. A., Majchrowicz, D., and Coinman, N. R. (1988b). A typological and technological study of Upper Paleolithic collections from the WHS with observations on adjacent time-stratigraphic units. In B. MacDonald (Ed.), *The Wadi el Hasa Archaeological Survey 1979–1983, West-Central Jordan* (pp. 87–127). Waterloo: Wilfrid Laurier University Press.

Clark, G. A., and Lindly, J. (1989a). The case for continuity: Observations on the biocultural transition in Europe and western Asia. In P. Mellars and C. Stringer (Eds.), *The Human Revolution: Behavioral and Biological Perspectives on the Origins of Modern Humans* (pp. 627–676). Edinburgh: University of Edinburgh Press.

Clark, G. A. (1989b). Modern human origins in the Levant and western Asia. *American Anthropologist, 91*, 962–965.

Clark, J. D. (1970). *The Prehistory of Africa*. New York: Praeger.

Clark, J. D., Phillips, J., and Staley, P. S. (1974). Interpretations of prehistoric technology from ancient Egyptian and other sources. *Paléorient, 2, 2*, 323–388.

Clark, J. E. (1986). Another look at small debitage and microdebitage. *Lithic Technology, 15, 1*, 21–33.

Close, A. E. (1978). The identification of style in lithic artefacts. *World Archaeology, 10*, 223–237.

Close, A. E. (1989). Identifying style in stone artifacts: A case study from the Nile Valley. In D. O. Henry and G. H. Odell (Eds.), *Alternative Approaches to Lithic Analysis*. Archaeological Papers of the American Anthropological Association, No. 1.

Clutton-Brock, J. (1979). The mammalian remains from the Jericho Tel. *Proceedings of the Prehistoric Society, 45*, 135–157.

Clutton-Brock, J. (1989). Introduction to pastoralism. In J. Clutton-Brock (Ed.), *The Walking Larder* (pp. 113–118). London: Unwin Hyman.

Coinman, N. R. (1990). *Rethinking the Levantine Upper Paleolithic*. Unpublished Ph.D. dissertation, Department of Anthropology, Arizona State University, Tempe.

Coinman, N. R., Clark, G. A., and Lindly, J. (1986). Prehistoric hunter-gatherer settlement in the Wadi el'Hasa, west-central Jordan. In A. Garrard and H. G. Gebel (Eds.), *The Prehistory of Jordan* (pp. 209–285). Oxford: British Archaeological Reports (IS), No. 284.

Coinman, N. R., Clark, G. A., and Lindly, J. (1988). A diachronic study of Paleolithic and early Neolithic site placement patterns in the southern tributaries of the Wadi el Hasa. In B. MacDonald (Ed.), *The Wadi el Hasa Archaeological Survey 1979–1983, West-Central Jordan* (pp. 48–86). Waterloo: Wilfred University Press.

Collins, M. (1975). Lithic technology as a means of processional inference. In E. Swanson (Ed.), *Lithic Technology*. The Hague: Mouton.

Copeland, L. (1975). The Middle and Upper Paleolithic of Lebanon and Syria in the light of recent research. In F. Wendorf and A. Marks (Eds.), *Problems in Prehistory: North Africa and the Levant* (pp. 317–350). Dallas: Southern Methodist University Press.

Copeland, L. (1981). Chronology and distribution of the Middle Paleolithic, as known in 1980, in Lebanon and Syria. In J. Cauvin and P. Sanlaville (Eds.), *Préhistoire du Levant* (pp. 299–302). Paris: CNRS.

Copeland, L., and Hours, F. (1971). A microlithic flint site in the Wadi Rum, Jordan, and a review of the Epipaleolithic of northern Arabia. *Proceedings of the Seminar for Arabian Studies, 1970*, 7–21.

Copeland, L., and Hours, F. (1981). La fin de l'Acheuleen et l'avenement du Paléolithique moyen en Syrie. In J. Cauvin and P. Sanlaville (Eds.), *Préhistoire du Levant* (pp. 225–238). Paris: CNRS.

Cresswell, R. (1976). Avant propos. *Techniques et Culture, 1*, 5–6.

Crew, H. L. (1976). The Mousterian site of Rosh Ein More. In A. Marks (Ed.), *Prehistory and Paleoenvironments in the Central Negev, Israel*, Vol. 1 (pp. 113–140). Dallas: Southern Methodist University Press.

Cribb, R. (1991). *Nomads in archaeology*. Cambridge: Cambridge University Press.

Darmon, F. (1981). *L'Homme du Kebarien et son environment au Sud du Levant*. Mémoire de Maîtrese d'Archeologie. Paris: Université de Paris I.

Darmon, F. (1986). *Le cadre écologique des sites Épipaléolithique et du Néolithique Ancien dans le Basse Vallée du Jordain*. Thèse, Université de Paris I, Panthéon-Sorbonne.

Davis, S. J. M. (1980a). Late Pleistocene and Holocene equid remains from Israel. *Zoological Journal of the Linnaean Society, 70*, 289–312.

Davis, S. J. M. (1980b). Late Pleistocene and Holocene gazelles of northern Israel. *Israel Journal of Zoology, 29*, 135–140.

Davis, S. J. M. (1981). The effects of temperature change and domestication on the body size of Late Pleistocene to Holocene mammals of Israel. *Paleobiology, 7*, 101–114.

Davis, S. J. M. (1983). The age profiles of gazelle predated by ancient man in Israel: Possible evidence for a shift from seasonality to sedentism in the Natufian. *Paléorient, 9*, 55–62.

Davis, S. J. M. (1984). The advent of milk and wool production in western Iran: Some speculations. In J. Clutton-Brock and C. Grigson (Eds.), *Animals and Archaeology: 3, Early Herders and Their Flocks* (pp. 265–278). Oxford: BAR International Series 356.

Davis, S. J. M. (1987). *The Archaeology of Animals*. London: B. T. Batsford.

Davis, S., Goring-Morris, N., and Gopher, A. (1982). Sheep bones from the Negev Epipaleolithic. *Paléorient, 8, 1*, 87–93.

Dibble, H. L. (1984). Interpreting typological variation of Middle Paleolithic scrapers: Function, style, or sequence of production. *Journal of Field Archaeology, 11*, 431–436.

Dibble, H. L. (1988). Typological aspects of reduction and intensity of utilization of lithic resources in the French Mousterian. In H. L. Dibble and A. Montet-White (Eds.), *Upper Pleistocene Prehistory of Western Eurasia* (pp. 181–198). University of Pennsylvania: The University Museum.

Dibble, H. L. (1989). The implications of stone tool types for the presence of language during the Lower and Middle Palaeolithic. In P. Mellars and C. Stringer (Eds.), *The Human Revolution* (pp. 415–432). Princeton, New Jersey: Princeton University Press.

Dobzhanksky, T. (1972). On the evolutionary uniqueness of man. *Evolutionary Biology* 6:415–430.

Ducos, P. (1968). *L'origine des animaux domestiques en Palestine*. Bordeaux: Travaux de L'Institut de Préhistoire de L'Université de Bordeaux.

Ducos, P., and Helmer, D. (1981). Le point actuel sur l'apparition de la domestication dans le Levant. In J. Cauvin and P. Sanlaville (Eds.), *Préhistoire du Levant* (pp. 523–528). Paris: CNRS.

Dunnell, R. C. (1978). Style and function: a fundamental dichotomy. *American Antiquity, 43(2)*, 192–202.

Dunnell, R. C. (1980). Evolutionary theory and archaeology. In M. B. Schiffer (Ed.), *Advances in Archaeological Method and Theory*, Vol. 3 (pp. 35–99). New York: Academic Press.

Durham, W. H. (1976). The adaptive significance of cultural behavior. *Human Ecology, 4(2)*, 89–121.

Edwards, P. C. (1989). Problems of recognizing the earliest sedentism: The Natufian example. *Journal of Mediterranean Archaeology, 2*, 5–48.

El-Eisawi, D. M. (1985). Vegetation in Jordan. In A. Hadidi (Ed.), *Studies in the History and Archaeology of Jordan*, Vol. 2 (pp 67–78). Amman: Department of Antiquities of Jordan. II.

Ember, C. R. (1977). Cross-cultural cognitive studies. *Annual Review of Anthropology, 6*, 33–56.

Emery-Barbier, A. (1988). Analyses polliniques du Quaternaire Supérieur en Jordanie Méridionale. *Paléorient, 14*, 111–117.

Ewing, J. F. (1947). Preliminary note on the excavations at the Paleolithic site of Ksar Akil, Republic of Lebanon. *Antiquity, 84*, 186–195.

Farrand, W. R. (1979). Chronology and paleoenvironments of Levantine prehistoric sites as seen from sediment studies. *Journal of Archaeological Science, 6*, 369–392.

Feinbrun-Dothan, N. (1986). *Flora Palestina* IV. Jerusalem: The Israel Academy of Sciences and Humanities.

Fenenga, F. (1953). The weights of chipped stone projectile points: A clue to their functions. *Southwestern Journal of Anthropology, 9*, 309–323.

Ferring, C. R. (1976). Sde Divshon: An Upper Paleolithic site on the Divshon Plain. In A. Marks (Ed.), *Prehistory and Paleoenvironments in the Central Negev, Israel*, Vol. 1 (pp. 199–226). Dallas: Southern Methodist University Press.

Ferring, C. R. (1977). The late Upper Paleolithic site of Ein Aqev East. In A. Marks (Ed.), *Prehistory and Paleoenvironments in the Central Negev, Israel*, Vol. 1 (pp. 131–158). Dallas: Department of Anthropology, Southern Methodist University.

Field, H. (1960). North Arabian Desert archaeological survey, 1925–1950. *Papers of the Peabody Museum, 45(2)*.

Fisher, K. W. (1980). A theory of cognitive development: The control and construction of hierarchies of skills. *Psychological Review, 87*, 477–531.

Fladmark, K. (1982). Microdebitage analysis: Initial considerations. *Journal of Archaeological Science, 9*, 205–220.

Flavell, J. H. (1978). Developmental stage: Explanans or explanandum? *The Behavioral and the Brain Sciences, 2*, 187.

Foley, R. (1984a). Putting people into perspective: An introduction to community evolution and ecology. In R. Foley (Ed.), *Hominid Evolution and Community Ecology* (pp. 1–24). London: Academic Press.

Foley, R. (1984b). Early man and the Red Queen: Tropical African community evolution and hominid adaptation. In R. Foley (Ed.), *Hominid Evolution and Community Ecology* (pp. 85–110). London: Academic Press.

Foley, R. (1989). The ecological conditions of speciation: A comparative approach to the origins of anatomically-modern humans. In P. Mellars and C. Stringer (Eds.), *The Human Revolution*. Princeton, N. J.: Princeton University Press.

Fujimoto, T. (1979). The Epi-Paleolithic assemblages of Douara Cave. In K. Hanihara and T. Akazawa (Eds.), *Paleolithic Site of Douara Cave and Paleogeography of the Palmyra Basin in Syria* (pp. 47–75). Tokyo: University of Tokyo Press.

Fullagar, R. (1991). The role of silica in polish formation. *Journal of Archaeological Science, 18*, 1–24.

Gamble, C. S. (1986). *The Paleolithic settlement of Europe*. Cambridge: Cambridge University Press.

Garrard, A. (1980). *Man-Animal-Plant Relationships During the Upper Pleistocene and Early Holocene of the Levant*. Unpublished Ph.D. Dissertation, University of Cambridge.

Garrard, A. (1982). The environmental implications of a reanalysis of the large mammal fauna from the Wadi el-Mughara caves, Palestine. In J. Bintliff and W. Van Zeist (Eds.), *Palaeoclimates, Paleoenvironments and Human Communities in the Eastern Mediterranean Region in Later Prehistory* (pp. 185–187). Oxford: British Archaeological Reports International Series 133.

Garrard, A. (1984). Community ecology and Pleistocene extinctions in Levant. In R. Foley (Ed.), *Hominid Evolution and Community Ecology* (pp. 261–278). London: Academic Press.

Garrard, A., Betts, A., Byrd, B., Colledge, S., and Hunt, C. (1988). Summary of paleoenvironmental and prehistoric investigations in the Azraq Basin. In A. N. Garrard and H. G. Gebel (Eds.), *The Prehistory of Jordan* (pp. 311–337). Oxford: BAR International Series, 396.

Garrard, A., Betts, A., Byrd, B., and Hunt, C. (1987). Prehistoric environment and settlement in the Azraq Basin: An interim report on the 1985 excavation season. *Levant, 19*, 5–25.

Garrard, A., and Byrd, B. (1992). New dimensions to the Epipalaeolithic of the Wadi El-Jilat in central Jordan. *Paléorient, 18*, 47–62.

Garrard, A., Byrd, B., and Betts, A. (1986). Prehistoric environment and settlement in the Azraq Basin: An interim report on the 1984 excavation season. *Levant, 18*, 5–24.

Garrard, A., Byrd, B., Harvey, P., and Hivernel, F. (1985). Prehistoric environment and settlement in the Azraq Basin: A report on the 1982 survey season. *Levant, 17*, 1–28.

Garrod, D. A. E. (1951). A transitional industry from the base of the Upper Paleolithic in Palestine and Syria. *Journal of the Royal Anthropological Institute, 81*, 121–129.

Garrod, D. A. E. (1954). Excavations at the Mugharet Kebara, Mount Carmel, 1932: The Aurignacian industries. *Proceedings of the Prehistoric Society, 20*, 2, 155–192.

Garrod, D. A. E., and Bate, D. M. A. (1937). *The Stone Age of Mount Carmel* (Vol. 1). Oxford: Clarendon Press.

Garrod, D. A. E., and Kirkbride, D. (1961). Excavation of Abri Zumoffen, a Palaeolithic rock shelter near Adlun, in south Lebanon, 1958. *Bulletin du Musée de Beyrouth, 16*, 7–46.

Gebel, H. G. (1983–84). Sabra 1 und die Wadi-Systeme um Petra/Wadi Musa. *Archiv fur Orientforschung* (pp. 282–284). 29–30.

Gebel, H. G. (1986a). Die Jungsteinzeit im Petra-Gebiet. In M. Lindner (Ed.), *Petra. Neue Ausgrabungen und Entdeckungen* (pp. 273–308). Muenchen: Delp.

Gebel, H. G. (1986b). Petra region 1983–85. Untersuchungen zur siedlungs- und umweltgeschichte des Fruhneolithikums. *Archiv fur Orientforschung,*

Gebel, H. G., Muheisen, M. S., and Nissen, H. J. (1988). Preliminary report on the first season of excavations at Basta. In A. N. Garraed and H. G. Gebel (Eds.), *The Prehistory of Jordan* (pp. 101–133). Oxford: BAR International Series, 396.

Gebel, H. G., and Stark, J. M. (1985). Investigations into the stone age of the Petra area (early Holocene research). A preliminary report on the 1984 campaigns. *Annual of the Department of Antiquities of Jordan, 29,* 89–114.

Geneste, J.-M. (1985). *Analyse lithique d'industries Mousterien-nes du Perigord: Une approche technologique du comportement des Groupes Humains au Paléolithique Moyen.* Thèse de Docteur. Bordeaux: Université de Bordeaux.

Geneste, J.-M. (1988). Ѐconomie dus resources dans le Mousterien du Sud-Ouest de la France. In M. Otte (Ed.), *L'homme de Neandertal.* Vol. 6: *La subsistance.* ERAUL 33. Liége: Université de Liége.

Gilead, D. (1970). Handaxe industries in Israel and the Near East. *World Archaeology, 2, 1,* 1–11.

Gilead, D. (1975). Lower and Middle Paleolithic settlement pattern in the Levant. In F. Wendorf and A. Marks (Eds.), *Problems in Prehistory: North Africa and the Levant* (pp. 273–282). Dallas: Southern Methodist University Press.

Gilead, I. (1981). Upper Paleolithic tool assemblages from the Negev and Sinai. In J. Cauvin and P. Sanlaville (Eds.), *Préhistoire du Levant* (pp. 331–342). Paris: CNRS.

Gilead, I. (1983). Upper Paleolithic occurrences in Sinai and the transition of the Epi-Paleolithic in the southern Levant. *Paléorient, 9, 1,* 39–53.

Gilead, I. (1988). The Chalcolithic period in the Levant. *Journal of World Prehistory, 2(4),* 397–443.

Gilead, I. (1989). Grar: A Chalcolithic site in the Northern Negev, Israel. *Journal of Field Archaeology, 16,* 377–394.

Gilead, I. (1991). The Upper Paleolithic period in the Levant. *Journal of World Prehistory, 5(2),* 105–154.

Gilead, I. (1992). Farmers and herders in southern Israel during the Chalcolithic period. In O. Bar-Yosef and A. Khazanov (Eds.), *Pastoralism in the Levant* (pp. 29–42). Wisconsin: Prehistory Press.

Gilead, I., and D. Alon. (1988). Excavations of protohistoric sites in the Nahal Besor and the Late Neolithic of the Northern Negev and Sinai. *Mitekufat Haeven 21*

Gilead, I., and Bar-Yosef, O. (1987). Prehistoric sites in the Kadesh Barnea area. In G. Gvirtzman, A. Shmueli, Y. Grados, I. Beit-Arieh, and M. Har-El (Eds.), *Sinai, Volume 2* (pp. 579–588). Tel Aviv: Ministry of Defense. (Hebrew)

Gilead, I., and Bar-Yosef, O. (1993). Early Upper Paleolithic sites in the Qadesh Barnea area, NE Sinai. *Journal of Field Archaeology, 20, 3,* 265–280.

Gilead, I., and Grigson, C. (1984). Far'ah II, a Middle Palaeolithic open-air site in the Northern Negev, Israel. *Proceedings of the Prehistoric Society, 50,* 71–97.

Gladilin, V. N., and Demidenko, Y. (1989). Upper Paleolithic stone tool complexes from Korolevo. *L' Anthropologie, 27,* 143–178.

Goldberg, P. (1981). Late Quaternary stratigraphy of Israel: An eclectic view. In J. Cauvin and P. Sanlaville (Eds.), *Préhistoire du Levant* (pp. 55–66).

Goldberg, P. (1986). Late Quaternary environmental history of the Southern Levant. *Geoarchaeology, 1,* 225–244.

Goldberg, P., and Rosen, A. (1987). Early holocene palaeoenviroments of Israel. In T. E. Levy (Ed.), *Shiqmim I* (pp 23–33). Oxford: BAR International Series 356.

Gopher, A. (1989). Neolithic arrowheads of the Levant: Results and implications of a seriation analysis. In O. Aurenche, M.-C. Cauvin and P. Sanlaville (Eds.), *Préhistoire du Levant: Procèssus des Changements Culturels* (pp. 47–60). Paris: Editions du CNRS.

Goren, N. (1981). The Lower Paleolithic in Israel and adjacent countries. In J. Cauvin and P. Sanlaville (Eds.), *Préhistoire du Levant* (pp. 193–206). Paris: CNRS.

Goren-Inbar, N. (1988). Too small to be true? Reevaluation of cores on flakes in Levantine Mousterian assemblages. *Lithic Technology, 17, 1,* 37–44.

Goren-Inbar, N. (1990a). The lithic assemblages. In N. Goren-Inbar (Ed.), *Quneitra: A Mousterian Site on the Golan Heights, Qedem 31.* Jerusalem: Monographs of the Institute of Archaeology, Hebrew University.

Goren-Inbar, N. (1990b). Notes on decision making by Lower and Middle Paleolithic hominids. In O. Aurenche, M.-C. Cauvin and P. Sanlaville (Eds.), *Préhistoire du Levant: Procèssus des Changements Culturels* (pp. 99–108). Paris: Editions du CNRS.

Goren-Inbar, N. (1990c). The Lower Paleolithic—Synthesis. In O. Aurenche, M.-C. Cauvin and P. Sanlaville (Eds.), *Préhistoire du Levant: Procèssus des Changements Culturels* (pp. 109–111). Paris: Editions du CNRS.

Goring-Morris, N. (1987). *At the edge: Terminal Pleistocene hunter-gatherers in the Negev and Sinai*. Oxford: British Archaeological Reports.

Gowlett, J. A. F. (1984). Mental abilities of early man: A look at some hard evidence. In R. Foley (Ed.), *Hominid Evolution and Community Ecology* (pp. 167–192). London: Academic Press.

Grant, V. (1985). *The evolutionary process*. New York: Columbia University Press.

Grayson, D. K. (1988). Danger Cave, Last Supper Cave, and Hanging Rock Shelter: The faunas. *Anthropological Papers of the American Museum of Natural History, 66*, 1–130.

Grayson, D. K. (1989). Bone transport, bone destruction, and reverse utility curves. *Journal of Archaeological Science, 16*, 643–661.

Greiss, E. (1957). Anatomical identification of some ancient Egyptian plant materials. *Mémoires de l'Institut d'Égypte* 55. Cairo: Costa Tsoumas.

Grigson, C. (1987). Shiqmim: Pastoralism and other aspects of animal management in the Chalcolithic of the Northern Negev. In T. E. Levy (Ed.), *Shiqmim I: Studies Concerning Chalcolithic Societies in the Northern Negev Desert, Israel (1982–1984)* (pp. 219–241). Oxford: BAR International Series 356.

Grue, H., and Jensen, B. (1979). Review of the formation of incremental lines in tooth cementum of terrestrial mammals. *Danish Review of Game Biology, 11*, 1–48.

Grun, R., and Stringer, C. B. (1991). Electron spin resonanace dating and the evolution of modern humans. *Archaeometry, 33(2)*, 1–47.

Harding, G. (1959) *The antiquities of Jordan*. London: Lutterworth Press.

Harlan, J. R. (1988). Natural resources. In B. MacDonald (Ed.), *The Wadi el Hasa Archaeological Survey 1979–1983, West-Central Jordan* (pp. 40–47). Waterloo: Wilfrid Laurier University Press.

Hassan, F. (1976). The study of lithic artifacts: An analytical model and two case studies. In J. S. Raymond et al. (Eds.), *Primitive Art and Technology*. Calgary: University of Calgary.

Hassan, F. (1979). Geoarchaeology: The geologist and archaeology. *American Antiquity, 44*, 267–270.

Hassan, F. (1981). *Demographic archaeology*. New York: Academic Press.

Harris, M. (1968). *The rise of anthropological theory*. New York: Thomas Y. Crowell Company.

Harris, M. (1979). *Cultural materialism*. New York: Random House.

Hecker, H. (1982). Domestication revisited: Its implications for faunal analysis. *Journal of Field Archaeology, 9*, 217–236.

Helbaek, H. (1960). Cereals and weed grasses in Phase A. In R. J. Braidwood and L. S. Braidwood (Eds.), *Excavations in the Plain of Antioch I* (pp. 540–543). Chicago: University of Chicago.

Helmer, D. (1989). Le developement de la domestication au Pouche-Orient de 9500 a 7500 B.P.: Les nouvelles donnees d'El Kown et de Ras Shamra. *Paléorient, 15, 1*, 111–121.

Henry, D. O. (1973). *The Natufian of Palestine: Its material culture and ecology*. Department of Anthropology, Southern Methodist University, Dallas.

Henry, D. O. (1974). The utilization of the microburin technique in the Levant. *Paléorient, 2*, 389–398.

Henry, D. O. (1977). An examination of the artifact variability in the Natufian of Palestine. In B. Arensburg and O. Bar-Yosef (Eds.), *Eretz-Israel*. Jerusalem: Hebrew University.

Henry, D. O. (1979). Paleolithic sites within the Ras en Naqb Basin, southern Jordan. *Palestine Exploration Quarterly, 111*, 79–85.

Henry, D. O. (1981). An analysis of settlement patterns and adaptive strategies of the Natufian. In J. Cauvin and P. Sanlaville (Eds.), *Préhistoire du Levant* (pp. 420–432). Paris: CNRS.

Henry, D. O. (1982a). The prehistory of southern Jordan and relationships with the Levant. *Journal of Field Archaeology, 9(4)*, 417–444.

Henry, D. O. (1982b). Paleolithic adaptive strategies in southern Jordan: Results of the 1979 field season. In A. Hadidi (Ed.), *Studies in the History and Archaeology of Jordan* (pp. 38–41). Amman: Department of Antiquities of Jordan.

Henry, D. O. (1983). Adaptive evolution within the Epipaleolithic of the Near East. In F. Wendorf and A. E. Close (Eds.), *Advances in World Archaeology* (pp. 99–160). London: Academic Press.

Henry, D. O. (1985a). Preagricultural sedentism: The Natufian example. In T. D. Price and J. A. Brown (Eds.), *Prehistoric Hunter-Gatherers* (pp. 365–384). Orlando, Fl.: Academic Press.

Henry, D. O. (1985b). Late Pleistocene environment and Paleolithic adaptations in southern Jordan. In A. Hadidi (Ed.), *Studies in the History and Archaeology of Jordan*, Vol. 2 (pp. 67–78). Amman: Department of Antiquities of Jordan. II.

Henry, D. O. (1986). Prehistory and paleoenvironments of Jordan: An overview. *Paléorient, 12*(2), 5–26.

Henry, D. O. (1987a). Topographic influences on Epipaleolithic land-use patterns in southern Jordan. In A. Hadidi (Ed.), *Studies in the History and Archaeology of Jordan* (pp. 1–21). London: Routledge and Kegan Paul. III.

Henry, D. O. (1987b). Climatic change, settlement mobility, and technological evolution during the Paleolithic. *Israel Journal of Earth Sciences, 36*, 121–129.

Henry, D. O. (1988a). Summary of prehistoric and paleoenvironmental research in the northern Hisma. In A. N. Garrard and H. G. Gebel (Eds.), *The Prehistory of Jordan in 1986*. Oxford: British Archaeologic Research International Series 396(i).

Henry, D. O. (1988b). The Epipaleolithic sequence within the Ras en Naqb-El Quweira area, southern Jordan. *Paléorient, 14*(2), 245–256.

Henry, D. O. (1989a). *From foraging to agriculture*. Philadelphia: University of Pennsylvania Press.

Henry, D. O. (1989b). Correlations between reduction strategies and settlement patterns. In D. Henry and G. Odell (Eds.), *Alternative Approaches to Lithic Analysis* (pp. 106–113). Washington, D. C.: American Anthropological Association. Archeological Research Series, Number 1.

Henry, D. O. (1990). The Epipaleolithic sequence within the Ras en Naqb-Quweira area, southern Jordan. In O. Aurenche, M.-C. Cauvin, and P. Sanlaville (Eds.), *Préhistoire du Levant: Procèssus des Changements Culturels* (pp. 245–256). Paris: Editions du CNRS.

Henry, D. O. (1991). Foraging, sedentism, and adaptive vigor in the Natufian: Rethinking the linkages. In G. A. Clark (Ed.), *Perspectives on the Past: Theoretical Biases in Mediterranean Hunter-Gatherer Research* (pp. 353–370). Philadelphia: University of Pennsylvania Press.

Henry, D. O. (1992a). Transhumance during the late Levantine Mousterian. In H. Dibble and P. Mellars (Eds.), *The Middle Paleolithic: Adaptation, Behavior, and Variability* (pp. 143–162). Philadelphia: The University Museum, University of Pennsylvania, Univerisity Museum Monograph 72.

Henry, D. O. (1992b). Seasonal Movements of Fourth Millennium Pastoral Nomads in the Wadi Hisma, Southern Jordan. In M. Zaghloul and K. 'Amr (Eds.), *History and Archaeology of Jordan*, Vol. IV, (pp. 137–143). Amman: Department of Antiquities.

Henry, D. O. (1994) Prehistoric cultural ecology in southern Jordan. *Science, 265*, 336–341.

Henry, D. O. (N. d.). Field report of the 1988 excavation at Tor Hamar (J431). Unpublished manuscript.

Henry, D. O., and Garrard, A. N. (1988). Tor Hamar: An Epipaleolithic rockshelter in southern Jordan. *Palestine Exploration Quarterly, 120*, 1–25.

Henry, D. O., Hassan, F. A., Henry, K. C., and Jones, M. (1983). An investigation of the prehistory of southern Jordan. *Palestine Exploration Quarterly, 115*, 1–24.

Henry, D. O., Haynes, C. V., and Bradley, B. (1976). Quantitative variations in flaked stone debitage. *Plains Anthropologist, 21*, 57–61.

Henry, D. O., and Miller, G. H. (1992). The implications of amino acid racemization dates of Levantine Mousterian deposits in southern Jordan. *Paléorient, 18*, 2, 45–52.

Henry, D. O., and Odell, G. H. (editors). (1989). *Alternative approaches to lithic analysis*. Archeological Papers of the American Anthropological Association, No. 1.

Henry, D. O., Leroi-Gourhan, A., and Davis, S. (1981). Hayonim Terrace: Late Pleistocene changes in environment and economy. *Journal of Archaeological Science, 8*, 33–58.

Henry, D. O., Turnbull, P., Emery-Barbier, A., and Leroi-Gourhan, A. (1985). Archaeological and faunal remains from Natufian and Timnian sites in southern Jordan. *Bulletin of the American Schools of Oriental Research, 257*, 45–64.

Hietala, H., and Marks, A. (1981). Changes in spatial organization at the Middle to Upper Paleolithic transitional site of Boker Tachtit, Central Negev, Israel. In J. Cauvin and P. Sanlaville (Eds.), *Préhistoire du Levant* (pp. 305–318). Paris: CNRS.

Hodder, I. (1979). Economic and social stress and material culture patterning. *American Antiquity, 44*(3), 446–454.

Hole, F. (1989). A two-part, two-stage model of domestication. In J. Clutton-Brock (Ed.), *The Walking Larder* (pp. 97–104). London: Unwin Hyman.

Horowitz, A. (1976). Late quaternary paleoenvironments of prehistoric settlements. In A. E. Marks (Ed.), *Prehistory and Paleoenvironments in the Central Negev, Israel* (vol. 1) (pp. 57–68). Dallas, Tex.: Southern Methodist University Press.

Horowitz, A. (1979). *The quaternary of Israel*. New York: Academic Press.

Horwitz, L. K. (1987). Animal remains from the Pottery Neolithic levels at Tel Dan. *Mitekufat Haeven, 20*, 114–118.

Hovers, E., and Marder, O. (1991). Typo-chronology and absolute dating of the Kebaran complex: Implications from the second season of excavation at Urkan E-Rub IIA. *Mitekufat Haeven, 24*, 34–58.

Huckriede, R., and Wiesemann, G. (1968). Der Jungpleistozane pluvial-see von El Jafr und weitere daten zum Quatar Jordaniens. *Geologica et Palaeontologica, 2*, 73–95.

Ingold, T. (1980). *Hunters, pastoralists, and ranchers*. Cambridge: Cambridge University Press.

Ingold, T. (1987). *The appropriation of nature: Essays in human ecology and social relations*. Manchester: Manchester University Press.

Isaac, G. L. (1972). Chronology and tempo of cultural change during the Pleistocene. In W. W. Bishop and J. A. Miller (Eds.), *Calibration of Hominoid Evolution* (pp. 381–430). Toronto: University of Toronto Press.

Jelinek, A. (1976). Form, function, and style in lithic analysis. In C. Cleland (Ed.), *Cultural Change and Continuity* (pp. 19–34). New York: Academic Press.

Jelinek, A. (1977). The Lower Paleolithic: Current evidence and interpretation. *Annual Review of Anthropology, 6*, 11–32.

Jelinek, A. (1981). The Middle Paleolithic of the Levant: Synthesis. In J. Cauvin and P. Sanlaville (Eds.), *Préhistoire du Levant* (pp. 299–302). Paris: CNRS.

Jelinek, A. (1982a). The Middle Paleolithic in the southern Levant with comments on the appearance of modern Homo Sapiens. In A. Ronen (Ed.), *The Transition from the Lower to the Middle Paleolithic and the Origin of Modern Man* (pp. 57–101). Oxford: BAR International Series 151.

Jelinek, A. (1982b). The Tabun cave and Paleolithic man in the Levant. *Science, 216*, 1369–1375.

Jelinek, A. (1988). Technology, typology, and culture in the Middle Paleolithic. In H. L. Dibble and A. Montet-White (Eds.), *Upper Pleistocene Prehistory of Western Eurasia* (pp. 199–212). Philadelphia: University Museum, University of Pennsylvania.

Jelinek, A. (1991). Observations on reduction patterns and raw materials in some Middle Paleolithic industries of the Perigord. In A. Montet-White and S. Holen (Eds.), *Raw Material Economies Among Prehsitoric Hunter-Gatherers* (pp. 7–31). Lawrence: University of Kansas Publications in Anthropology.

Jelinek, A. (1992). Problems in the chronology of the Middle Paleolithic and the first appearance of early modern Homo sapiens in southwest Asia. In T. Akazawa (Ed.), *The Evolution and Dispersal of Modern Humans in Asia*. Tokyo: Hokisen-sha.

Jelinek, A., Farrand, W., Haas, G., Horowitz, A., and Goldberg, P. (1973). New excavations at the Tabun Cave, Mt. Carmel, Israel 1967–1972: A preliminary report. *Paléorient, 1*, 151–183.

Jochim, M. A. (1981). *Strategies for Survival*. New York: Academic Press.

Johnson, J. K. (1989). The utility of production trajectory modeling as a framework for regional analysis. In D. Henry and G. Odell (Eds.), *Alternative Approaches to Lithic Analysis* (pp. 119–138). Archaeological Papers of the American Anthropological Association 1. Washington, D.C.

Jones, M. (1983). *The Qalkhan and Hamran*. Unpublished Ph.D. thesis, Department of Anthropology, Southern Methodist University, Dallas.

Jones, M., A. Marks, and D. Kaufman. (1983). Boker: The artifacts. In A. Marks (Ed.), *Prehistory and Paleoenvironments in the Central Negev, Israel, Vol. III* (pp. 283–332). Dallas: Southern Methodist University Press.

Kaufman, D. (1992). Hunter-gatherers of the Levantine Epipaleolithic. *Journal of Mediterranean Archaeology, 5*, 2, 165–201.

Keeley, L. (1980). *The experimental determination of stone tool uses: A microwear analysis*. Chicago: University of Chicago Press.

Kelly, R. L. (1983). Hunter-gatherer mobility strategies. *Journal of Anthropological Research, 39*, 277–306.

Khazanov, A. (1984). *Nomads and the outside world*. Cambridge: Cambridge University Press.

Kirch, P. V. (1980). The archaeological study of adaptation: Theoretical and methodological issues. In M. B. Schiffer (Ed.), *Advances in Archaeological Method and Theory*, Vol. 3 (pp. 101–156). New York: Academic Press.

Kirkbride, D. (1958). A Kebaran rockshelter in Wadi Madamagh, near Petra, Jordan. *Man, 63*, 55–58.

Kirkbride, D. (1959). Short note on some hitherto unrecorded prehistoric sites in Transjordan. *Palestine Exploration Quarterly, 91*, 52–55.

Kirkbride, D. (1960). Chronique archéologique. *Revue Biblique* (pp. 67).

Kirkbride, D. (1966). Five seasons at the Pre-Pottery Neolithic village of Beidha in Jordan. *Palestine Exploration Quarterly, 98*, 8–72.

Kirkbride, D. (1978). The Neolithic in Wadi Rumm: 'Ain Abu Nekheileh. In R. Moorey and P. Parr (Eds.), *Archaeology in the Levant* (pp. 1–10). Warminster: Aris and Phillips LTD.

Kirkbride, A. S., and Harding, L. (1947). Hasma. *Palestine Exploration Quarterly, 79*, 7–26.

Klein, R. G. (1989). Why does skeletal part representation differ between smaller and larger bovids at Klasies River Mouth and other archeological sites? *Journal of Archaeological Science, 6*, 363–381.

Klein, R. G., and Cruz-Uribe, K. (1984). *The analysis of animal bones from archeological sites*. Chicago: University of Chicago Press.

Klein, R. G., and Cruz-Uribe, K. (1986). Pascal programs for computing taxonomic abundance in samples of fossil mammals. *Journal of Archaeological Science, 13*, 171–187.

Klevezal, G. A., and Kleinenberg, S. E. (1967). *Age determination of mammals by layered structure in teeth and bone*. Academy of Sciences, U.S.S.R. Translation, 1969 Foreign Language Division, Department Secretary State Canada. Quebec: Fisheries Research Board.

Koehler-Rollefson, I. (1989). Changes in goat exploitation at 'Ain Ghazal between the early and late Neolithic: A metrical analysis. *Paléorient, 15, 1*, 141–146.

Koehler-Rollefson, I. (1992). A model for the development of nomadic pastoralism on the Transjordanian Plateau. In O. Bar-Yosef and A. Khazanov (Eds.), *Pastoralism in the Levant* (pp. 11–18). Madison, Wisconsin: Prehistory Press.

Konner, M. (1983). *The tangled wing*. New York: Harper Colophon Books.

Koucky, F. L., and Smith, R. H. (1986). Lake Beisan and the prehistoric settlement of the northern Jordan Valley. *Paléorient, 12, 2*, 27–36.

Kozloff, B. (1974). A brief note on the lithic industries of Sinai. *Museum Haaretz Yearbook, 15/16*, 47–48.

Kozloff, B. (1981). Pastoral nomadism in the Sinai: An ethno-archaeological study. *Bulletin de L'Équipe Écologie et Anthropologie des Sociétés Pastorales, 8*, 19–24.

Kroeber, A. L. (1931). The culture-area and age-area concepts of Clark Wissler. In S. Rice (Ed.), *Methods in Social Science* (pp. 248–265). Chicago: University of Chicago Press.

Kroeber, A. L. (1939). *Cultural and natural areas of native North America*. Berkeley: University of California Publications in American Archaeology and Ethnology, Volume 38.

Kuhn, S. (1990). *Diversity within uniformity: Tool manufacture and use in the "Pontinian" Mousterian of Latium (Italy)*. Unpublished PhD dissertation, Albuquerque: Department of Anthropology, University of New Mexico.

Kuhn, S. (1991). On planning and curated technologies in the Middle Paleolithic. *Journal of Anthropological Research, 48, 3*, 185–214.

Lechevallier, M., Philibert, M. D., Ronen, A., and Samsun, A. (1989). Une occupation Khiamienne et Sultanenne a Hatoula (Israel)? *Paléorient, 15, 1*, 1–18.

Lee, C. M. (1987). *A functional analysis of Levallois points from two Middle Paleolithic sites in South Jordan: Results and interpretations*. Unpublished M. A. thesis, Department of Anthropology, University of Tulsa, Oklahoma.

Lee, R. (1973). *Chalcolithic Ghassul: New aspects and master typology*. Unpublished Ph.D. dissertation. Jerusalem: Hebrew University of Jerusalem.

Lernau, H. (1985). Fish remains from Hatoula. In M. Lechevallier and A. Ronen (Eds.), *Le Site Natoufien-Khiamien de Hatoula près de Latroun, Israel; Fouilles 1980–1982 Rapport Préliminaire* (p. 102). Jerusalem: Centre de Récherches Français de Jerusalem.

Leroi-Gourhan, Andre (1964). *Le geste et la Parole I, technique et lang-age*. Paris: Albin Michel.

Leroi-Gourhan, Arl. (1969). Pollen grains of gramineae and cerealia from Shanidar and Zawi Chemi. In P. Ucko and G. Dimbleby (Eds.), *The Domestication and Exploitation of Plants and Animals* (pp. 143–148). London: Duckworth.

Leroi-Gourhan, Andre. (1980). Les analyses polliniques au Moyen Orient. *Paléorient, 6*, 79–91.

Leroi-Gourhan, Andre. (1981). Le Levant a la fin du Pleistocene et a l'Holocene d'apres la palynologie. In J. Cauvin and P. Sanlaville (Eds.), *Préhistoire du Levant* (pp. 107–109). Lyon: Maison de l'Orient.

Leroi-Gourhan, A., and Darmon, F. (1991). Analyses polliniques de station natoufiennes au Proche-Orient. In O. Bar-Yosef and F. R. Valla (Eds.), *The Natufian Culture in the Levant* (pp. 21–26). Ann Arbor: International Monographs in Prehistory.

Levy, T. (1983). The emergence of specialized pastoralism in the southern Levant. *World Archaeology, 15*, 15–36.

Levy, T. (1985). The Chalcolithic period. *Biblical Archaeologist, 49*, 82–108.

Levy, T. (1992). Transhumance, subsistence, and social evolution. In O. Bar-Yosef and A. Khazanov (Eds.), *Pastoralism in the Levant* (pp. 65–82). Wisconsin: Prehistory Press.

Liben, L. S. (1981). Copying and reproducing pictures in relation to subjects' operative levels. *Developmental Psychology, 17*, 357–365.

Lieberman, D. E. (1991). Seasonality and gazelle hunting at Hayonim Cave: New evidence for "sedentism" during the Natufian. *Paléorient, 17*, 47–57.

Lieberman, D. E. (1993a). The rise and fall of hunter-gatherer seasonal mobility: The case of the southern Levant. *Current Anthropology, 34*, 599–631.

Lieberman, D. E. (1993b). Life history variables preserved in dental cementum microstructure. *Science, 261*, 1162–1164.

Lieberman, D. E. (1994). The biological basis for seasonal increments in dental cementum and their application to archaeological research. *Journal of Archaeological Science 21, 4*, 525–540.

Lieberman, D. E. (in press). In G. Peterkin, H. Bricker and P. Mellars (Eds.), *Variability in hunter-gatherer seasonal mobility in the southern Levant: From the Mousterian to the Natufian: Hunting and animal exploitation in the Late Pleistocene Old World*. Archaeological Papers of the American Anthropological Association.

Lieberman, D. E., Deacon, T. W., and Meadow, R. H. (1990). Computer image enhancement and analysis of cementum increments as applied to teeth of *Gazella gazella*. *Journal of Archaeological Science, 17*, 519–533.

Lieberman, D. E., and Meadow, R. H. (1992). The biology of cementum increments (an archaeological perspective). *Mammal Review, 22*, 1–21.

Lindly, J., and Clark, G. (1987). A preliminary lithic analysis of the Mousterian site of Ain Difla (WHS Site 364) in the Wadi Ali, west-central Jordan. *Proceedings of the Prehistoric Society, 53*, 279–292.

Lyman, R. L. (1984). Bone density and differential survival in fossil classes. *Journal of Anthropological Archaeology, 3*, 259–299.

Lyman, R. L. (1985). Bone frequencies: Differential transport, in situ destruction, and the MGUI. *Journal of Archaeological Science, 12*, 221–236.

Lyman, R. L. (1987). Archaeofaunas and butchery studies: A taphonomic perspective. In M. B. Schiffer (Ed.), *Advances in Archaeological Method and Theory 10* (pp. 249–337). San Diego: Academic Press.

Lyman, R. L. (1992). Anatomical considerations of utility curves in zooarchaeology. *Journal of Archaeological Science, 19*, 7–22.

MacDonald, B. (1988). *The Wadi el Hasa Archaeological Survey 1979–1983, west-central Jordan*. Waterloo: Wilfrid Laurier University Press.

MacDonald, B., Rollefson, G., and Roller, D. (1982). The Wadi el Hasa Survey 1981: A preliminary report. *Annual of the Department of Antiquities of Jordan, 26*, 117–131.

MacDonald, B., Rollefson, G., Banning, E., Byrd, B., and d'Annibale, C. (1983). The Wadi el Hasa Archeological Survey, 1982: A preliminary report. *Annual of the Department of Antiquities of Jordan, 27*, 311–324.

Marean, C. W. (1991). Measuring the post-depositional destruction of bone in archaeological assemblages. *Journal of Archaeological Science, 18*, 677–694.

Marks, A. E. (editor). (1977a). *Prehistory and paleoenvironments in the central Negev, Israel*, Vol. I. Dallas: Department of Anthropology, Southern Methodist University.

Marks, A. E. (editor) (1983a). *Prehistory and paleoenvironments in the central Negev, Israel*, Vol. III. Dallas: Department of Anthropology; Southern Methodist University.

Marks, A. E. (1976a). Site D5, a Geometric Kebaran "A" occupation in the Nahal Zin. In A. E. Marks (Ed.), *Prehistory and Paleoenvironments in the Central Negev, Israel*, Vol. 1 (pp. 227–293). Dallas: Southern Methodist University Press.

Marks, A. E. (1976b). Ein Aqev: A late Levantine Upper Paleolithic site in the Nahal Aqev. In A. Marks (Ed.), *Prehistory and Paleoenvironments in the Central Negev, Israel*, Vol. I (pp. 227–292). Dallas: Southern Methodist University Press.

Marks, A. E. (1976c). Glossary. In A. Marks (Ed.), *Prehistory and Paleoenvironments in the Central Negev, Israel*, Vol. I (pp. 371–383). Dallas: Southern Methodist University Press.

Marks, A. E. (1977b). Introduction: A preliminary overview of central Negev prehistory. In A. E. Marks (Ed.), *Prehistory and Paleoenvironments in the Central Negev, Israel*, Vol. I (pp. 3–34). Dallas: Department of Anthropology, Southern Methodist University.

Marks, A. E. (1981). The Middle Paleolithic of the Negev, Israel. In J. Cauvin and P. Sanlaville (Eds.), *Préhistoire du Levant* (pp. 287–298). Lyon: Maison de l'Orient.

Marks, A. E. (1983b). The sites of Boker Tachtit and Boker: A brief introduction. In A. E. Marks (Ed.), *Prehistory and Paleoenvironments in the Central Negev, Israel*, Vol. III (pp. 15–38). Dallas: Department of Anthropology, Southern Methodist University.

Marks, A. E. (1983c). The Middle to Upper Paleolithic Transition in the Levant. In *Advances in World Archaeology*, Vol. 2 (pp. 51–98). New York: Academic Press.

Marks, A. E. (1988). The curation of stone tools during the upper Pleistocene. In H. Dibble and A. Montet-White (Eds.), *Upper Pleistocene Prehistory of Western Eurasia* (pp. 275–285). Philadelphia: The University Museum.

Marks, A. E. (1989). Early Mousterian settlement patterns in the Avdat/Aqev area. In M. Otte (Ed.), *L'homme de Neandertal*. Vol. 6: *La subsistance* (pp. 115–126). ERAUL 33. Liége: Université de Liége.

Marks, A. E. (1992). Upper Pleistocene archaeology and the origin of modern man: A view from the Levant and adjacent areas. In T. Akazawa (Ed.), *The Evolution and Dispersal of Modern Humans in Asia* (pp. 229–252). Tokyo: University Museum.

Marks, A. E., and Ferring, C. R. (1976). Upper Paleolithic sites near Ein Avdat. In A. E. Marks (Ed.), *Prehistory and Paleoenvironments in the Central Negev, Israel*, Vol. 1 (pp. 141–198). Dallas: Southern Methodist University Press.

Marks, A. E., and Ferring, C. R. (1988). The early Upper Paleolithic in the Levant. In J. F. Hoffecker and C. A. Wolf (Eds.), *The Early Upper Paleolithic: Evidence From Europe and the Near East* (pp. 43–72). British Archaeological Reports International Series 437. Oxford: British Archaeological Reports.

Marks, A., and Friedel, D. (1977). Prehistoric settlement patterns in the Avdat/Aqev area. In A. E. Marks (Ed.), *Prehistory and Paleoenvironments in the Central Negev, Israel*, Vol. II (pp. 131–158). Dallas: Department of Anthropology, Southern Methodist University.

Marks, A. E., and Kaufman, D. (1983). Boker Tachtit: The artifacts. In A. E. Marks (Ed.), *Prehistory and Paleoenvironments in the Central Negev, Israel*, Vol. III (pp. 69–126). Dallas: Southern Methodist University.

Marks, A. E., and Larson, P. A. (1977). Test excavations at the Natufian site of Rosh Horesha. In A. E. Marks (Ed.), *Prehistory and Paleoenvironments in the Central Negev, Israel*, Vol. II (pp. 191–232). Dallas: Department of Anthropology, Southern Methodist University.

Marks A. E., and Simmons, A. H. (1977). The Negev Kebaran of the Har Harif. In A. E. Marks (Ed.), *Prehistory and Paleoenvironments in the Central Negev, Israel*, Vol. II (pp. 233–270). Dallas: Department of Anthropology, Southern Methodist University.

Marks, A. E., and Volkman, P. (1983). Changing core reduction strategies: A technological shift from Middle to the Upper Paleolithic in the Southern Levant. In E. Trinkaus (Ed.), *The Mousterian Legacy. Human Biocultural Change in the Upper Pleistocene*. Oxford: British Archaeological Reports.

Marshack, A. (1988). The Neanderthals and the human capacity for symbolic thought: Cognitive and problem solving aspects of Mousterian symbol. In M. Otte (Ed.), *L'homme de Neandertal*. Vol. 5: *La pensée* (pp. 57–91). ERAUL 33. Liége: Université de Liége.

Martinson, D. G., Pisias, N. G., Hays, J. D., Imbrie, J., Moore, J. T. C., and N. J. Shackleton. (1987). Age dating and the Orbital Theory of the Ice Ages: Development of a high-resolution 0 to 300,000-year chronostratigraphy. *Quaternary Research*, 27, 1–29.

Mason, O. (1894). Technogeography, or the relation of earth to the industries of mankind. *American Anthropologist*, 7, 137–161.

Masters, P. M. (1982). An amino acid racemization chronology for Tabun. In A. Ronen (Ed.), *The Transition from the Lower to the Middle Paleolithic and the Origin of Modern Man* (pp. 433–456). British Archaeological Reports International Series 151. Oxford: British Archaeological Reports.

McConaughy, M. (1979). *Formal and functional analyses of the chipped stone tools From Bab edh-Dhra, Jordan*. Unpublished PhD Dissertation, Pittsburgh: University of Pittsburgh.

McCorriston, J., and Hole, F. (1991). The ecology of seasonal stress and the origins of agriculture in the Near East. *American Anthropologist, 93,1*, 46–69.

Meadow, R. H. (1989). Osteological evidence for the process of animal domestication. In J. Clutton-Brock (Ed.), *The Walking Larder* (pp. 80–87). London: Unwin Hyman.

Meadow, R. H. (1992). Inconclusive remarks on pastoralism, nomadism, and other animal-related matters. In O. Bar-Yosef and A. Khazanov (Eds.), *Pastoralism in the Levant* (pp. 261–268). Wisconsin: Prehistory Press.

Meignen, L. (1988). Un exemple de comportement technologique différentiel selon les matières premières: Marillac, couches 9 et 10. In M. Otte (Ed.), *L'homme de Neandertal*. Vol. 4: In L. Binford and J.-P. Rigaud (Eds.), *La technique*. ERAUL 31. Liége: Université de Liége.

Meignen, L., and Bar-Yosef, O. (1989). Kebara et le Paléolithique Moyen du Mont Carmel (Israel). In O. Aurenche, M.-C. Cauvin and P. Sanlaville (Eds.), *Préhistoire du Levant: Procèssus des changements culturels* (pp. 123–129). Paris: Editions du CNRS.

Mellars, P. (1973). The character of the Middle-Upper Paleolithic transition in Southwest France. In A. C. Renfrew (Ed.), *The Explanation of Culture Change*. London: Duckworth.

Mellars, P. (1989a). Major issues in the emergence of modern humans. *Current Anthropology, 30*, 349–384.

Mellars, P. (1989b). Technological changes at the Middle-Upper Palaeolithic transition: Economic, social, and cognitive perspectives. In P. Mellars and C. Stringer (Eds.), *The Human Revolution* (pp. 338–366). Princeton, New Jersey: Princeton University Press.

Mendelssohn, H. (1974). The development of the populations of gazelles in Israel and their behavioral adaptations.

In V. Geist and F. R. Walther (Eds.), *The Behavior of Ungulates and its Relationship to Management* (pp. 722–742). Morges Switzerland: IUCN.

Metcalfe, D., and Jones, K. T. (1988). A reconsideration of animal body-part utility indices. *American Antiquity, 53*, 486–504.

Miller, G. H. (1985). Aminostratigraphy of Baffin Island shell-bearing deposits. In J. T. Andrews (Ed.), *Quaternary Environments, Eastern Canadian Arctic, Baffin Bay and Western Greenland* (pp. 394–427). Boston: Allen and Unwin.

Miller, G. H., Wendorf, F., Ernst, R., Schild, R., Close, A. E., Friedman, I., and Schwarcz, H. P. (1991). Dating lacustrine episodes in the Eastern Sahara by the epimarization of isoleucine on ostrich eggshells. *Paleography, Paleoecology, and Paleoclimatology, 84*, 175–189.

Mortenson, P. (1970). A preliminary study of the chipped stone industry from Beidha. *Acta Archaeolocia, 41*, 1–54.

Mortenson, P. (1972). Seasonal camps and early villages in the Zagros. In P. J. Ucko, R. Tringham, and G. W. Dimbleby (Eds.), *Man, Settlement and Urbanism* (pp. 293–297). London: Duckworth.

Muheisen, M. (1985). L'épipaléolithique dans le gisement de Kharaneh IV. *Paléorient, 14,2*, 149–160.

Muhesen, S. (1981). The Upper Acheulian in Syria. In J. Cauvin and P. Sanlaville (Eds.), *Préhistoire du Levant* (pp. 185–192). Paris: CNRS.

Munday, F. (1976). Intersite variability in the Mousterian occupation of the Avdat/Aqev area. In A. Marks (Ed.), *Prehistory and Paleoenvironments in the Central Negev, Israel*, Vol. II. Dallas: Department of Anthropology, Southern Methodist University.

Munday, F. (1977). Nahal Aqev (D35): A stratified, open-air Mousterian occupation in the Avdat/Aqev area. In A. Marks (Ed.), *Prehistory and Paleoenvironments in the Central Negev, Israel*, Vol. 2. Dallas: Department of Anthropology, Southern Methodist University.

Munday, F. (1979). Levantine Mousterian technological variability: A perspective from the Negev. *Paléorient, 5*, 87–104.

Neuville, R. (1934). Le prehistorique de Palestine. *Revue Biblique, 43*, 237.

Neuville, R. (1951). *Le Paléolithique et le Mesolithique du desert de Judée*. Archives de l'Institut de Paléontologie Humaine 24. Paris: Masson.

Noy, T. (1991). Art and decoraation in the Natufian of Nahal Oren. In O. Bar-Yosef and F. Valla (Eds.), *The Natufian Culture in the Levant* (pp. 557–568). Ann Arbor: International Monographs in Prehistory.

Odell, G. (1981). The mechanics of use-breakage of stone tools: Some testable hypotheses. *Journal of Field Archaeology, 8*, 197–210.

Odell, G., and Cowan, F. (1986). Experiments with spears and arrows on animal targets. *Journal of Field Archaeology, 13*, 194–212.

Odum, E. P. (1959). *Fundamentals of ecology*. 2nd edition. Philadelphia: Saunders.

Ohnuma, K., and Bergman, C. (1990). A technological analysis of the Upper Paleolithic levels (XXV–VI) of Ksar Akil, Lebanon. In P. Mellars and C. Stringer (Eds.), *The Emergence of Modern Humans: An Archaeological Perspective* (pp. 91–138). Edinburgh: Edinburgh University Press.

Olszewski, D. I. (1986). A reassessment of average lunate length as a chronological indicator. *Paléorient, 12*, 39–44.

Olszewski, D. I. (1991). Social complexity in the Natufian? Assessing the relationship of ideas and data. In G. A. Clark (Ed.), *Perspectives on the Past: Theoretical Biases in Mediterranean Hunter-Gatherer Research* (pp. 322–340). Philadelphia: University of Pennsylvania Press.

Olszewski, D. I., Clark, G. A., and Fish, S. (1990). WHS 784X (Yutil al-Hasa): A late Ahmarian activity site in the Wadi Hasa, west-central Jordan. *Proceedings of the Prehistoric Society 56* (pp. 33–49).

Oren, E. D., and Gilead, I. (1981). Chalcolithic sites in northeastern Sinai. *Tel Aviv, 8*, 25–44.

Osborn, G. (1985). Evolution of the late Cenozoic inselberg landscape of southwestern Jordan. *Paleogeography, Palaeoclimatology, Palaeoecology, 49*, 1–23.

Osborn, G., and Duford, J. M. (1981). Geomorphical processes in the inselberg region of south-western Jordan. *Palestine Exploration Quarterly*: 1–17.

Oswalt, W. H. (1973). *Habitat and technology, the evolution of hunting*. New York: Holt, Rhinehart, and Winston.

Patterson, L. (1982). Replication and classification of large size lithic debitage. *Lithic Technology, 11*, 3, 50–58.

Patterson, L. (1985). Distinguishing between arrow and spear points on the Upper Texas coast. *Lithic Technology, 14*, 2, 81–89.

Payne, S. (1985). Morphological distinctions between the mandibular teeth of young sheep, *Ovis*, and goats, *Capra*. *Journal of Archaeological Science, 12*, 139–147.

Pearsall, D. M. (1989). *Paleoethnobotany: A handbook of procedures*. New York: Academic.

Pelegrin, J., Karlin, C., and Bodu, P. (1988). Chaînes opératoires: Un outil pour le prehistorien. In J. Tixier (Ed.), *Technologie Prehistorique*. Paris: Editions du CNRS.

Perrot, J. (1968). *La préhistoire Palestiniènne*. In *Supplement au Dictionnaire de la Bible* (pp. 286–446). Paris: Letouzey et Ane.

Perrot, J. (1984). Structures d'habitat, mode de vie et environnement. Les villages souterrains des pasteurs de Beersheva, dans le sud d'Israel, au IV Millénaire avant l'ère chrétienne. *Paléorient, 10, 1*, 75–96.

Phillips, J. L. (1973). *Two final Paleolithic sites in the Nile Vally and their external relations*. Cairo: Geological Survey of Egypt Paper No. 57.

Phillips, J. L. (1987a). Upper Paleolithic hunter-gatherers in the Wadi Feiran, southern Sinai. In O. Soffer (Ed.), *The Pleistocene Old World* (pp. 169–182). New York: Plenum.

Phillips, J. L. (1987b). Sinai during the Paleolithic: The early periods. In A. Close (Ed.), *Prehistory of Arid North Africa: Essays in Honor of Fred Wendorf* (pp. 105–121). Dallas: Southern Methodist University Press.

Phillips J. L. (1988). The Upper Paleolithic of the Wadi Feiran, southern Sinai. *Paléorient, 14*, 2, 183–200.

Phillips, J. L., and Mintz, E. (1977). The Mushabian. In O. Bar-Yosef and J. L. Phillips (Eds.), *Prehistoric Investigations in Gebel Maghara, Northern Sinai* (pp. 149–163). Qedem 7, Jerusalem: Monographs of the Institute of Archeology, Hebrew University.

Piaget, J., and Inhelder, B. (1973). *Memory and intelligence*. New York: Basic Books.

Pianka, E. R. (1978). *Evolutionary ecology*. New York: Harper and Row, Publishers.

Picard, L. (1943). The structure and evolution of Palestine. *Bulletin of the Geology Department, Hebrew University* 2–4. Jerusalem.

Pichon, J. (1987). L'avifaune. In J. Bouchud (Ed.), *La Faune du Gisement de Mallaha (Eynan), Israel* (pp. 115–150). Mémoires et Travaux du Centre de Récherche de Jerusalem. Paris: Paléorient.

Pichon, J. (1992). Les oiseaux au Natoufien, avifaune et sedentarite. In O. Bar-Yosef and F. Valla (Eds.), *The Natufian Culture in the Levant* (pp. 371–380). Ann Arbor: International Monographs in Prehistory.

Piperno, D. R. (1988). *Phytolith analysis: An archaeological and geological perspective*. New York: Academic.

Plisson, H. (1988). Technologie et tracéologie des outils lithiques moustériens en Union Soviétique: Les travaux de V. E. Shchelinskii. In M. Otte (Ed.), *L'Homme de Néandertal*, Vol. 4: *La Technique* (pp. 121–168). Liége: Études et Récherches Archéologique de l'Université de Liége.

Price, N., and Garrard, A. (1975). A prehistoric site in the Rum area of the Hisma. *Annual of the Department of Antiquities of Jordan, 20*, 19–23.

Price-Williams, D. (1961). A study concerning concepts of conservation of quantities among primitive children. *Acta Psychologica, 18*, 297–305.

Rafferty, J. E. (1985). The archaeological record on sedentariness: Recognition, development, and implications. *Advances in Archaeological Method and Theory, 8*, 113–156.

Read, D. W. (1978). Towards a formal theory of population size and area of habitation. *Current Anthropology, 19(2)*, 312–317.

Reed, C. A. (1977). A model for the origin of agriculture in the Near East. In C. A. Reed (Ed.), *Origins of Agriculture* (pp. 543–567). The Hague: Mouton.

Reese, D. (1989). The Natufian shells from Beidha. Appendix D. In B. Byrd, *The Natufian Encampment at Beidha: Late Pleistocene Adaptation in the Southern Levant* (pp. 102–104). Moesgard, Arhus: Jutland Archaeological Society Publications.

Reese, D. (1991). Marine shells in the Levant: Upper Paleolithic, Epipaleolithic, and Neolithic. In O. Bar-Yosef and F. R. Valla (Eds.), *The Natufian Culture in the Levant* (pp. 613–628). Ann Arbor: International Monographs in Prehistory.

Rhotert, H. (1938). *Transjordanien: Vorgeschichtliche Forschungen*. Stuttgart: Strecker und Schroeder.

Rigaud, J. (1989). From the Middle to the Upper Paleolithic: Transition or convergence? In E. Trinkaus (Ed.), *The Emergence of Modern Humans* (pp. 142–153). Cambridge: Cambridge University Press.

de Riqlés, A., Meunier, F. J., Castanet, J., and Francillon-Vieillot, H. (1991). Comparative microstructure of bone. In B. K. Hall (Ed.), *Bone, Volume 3: Bone Matrix and Bone Specific Products* (pp. 1–78). Boca Raton: C.R.C. Press.

Roebroeks, W., Kolen, J., and Rensink, E. (1988). Planning depth, anticipation and the organization of Middle Paleolithic technology: The "archaic natives" meet Eve's descendants. *Helinium, 28(1)*, 17–34.

Roebroeks, W., Nicholas, C. J., and van Kolfschoten, T. (1992). Dense forests, cold steppes, and the Palaeolithic settlement of northern Europe. *Current Anthropology, 33(5)*, 551–585.

Rolland, N. (1981). The interpretation of Middle Paleolithic variability. *Man, 16*, 15–42.

Rolland, N., and Dibble, H. (1990a). New synthesis of Middle Paleolithic variability. *American Antiquity, 55*, 480–499.

Rollefson, G. (1981). The Late Acheulean site at Fjaje, Wadi el-Bustan, southern Jordan. *Paléorient, 7, 1,* 5–21.

Rollefson, G. (1982). Late Acheulean artifacts from Ain el-Assad ("Lion's Spring") near Azraq, eastern Jordan. *Bulletin of the American School of Oriental Research, 240,* 1–20.

Rollefson, G. (1983). Two seasons of excavations at Ain el-Assad near Azraq. *Bulletin of the American School of Oriental Research, 252,* 25–34.

Ronen, A. (1976). The Upper Paleolithic in northern Israel: Mt. Carmel and Galilee. In *Deuxième Colloque sur la Terminologie de la Préhistoire du Proche-Orient* (pp. 153–186). Nice: IXe Congres Union Internationale des Science Prehistorique et Protohistorique.

Ronen, A. (1979). The Paleolithic industries. In A. Horowitz (Ed.), *The Quaternary of Israel* (pp. 296–307). New York: Academic Press.

Ronen, A., and Lechavallier, M. (1985). *Le Site Natoufien-Khiamien de Hatoula pres de Latroun, Israel; Fouilles 1980–1982 rapport preliminaire.* Jerusalem: Centre de Récherches Français de Jerusalem.

Ronen, A., and Lechavallier, M. (1991). The Natufian of Hatula. In O. Bar-Yosef and F. R. Valla (Eds.), *The Natufian Culture in the Levant* (pp. 149–160). Ann Arbor: International Monographs in Prehistory.

Rosen, A. M. (1987). Phytolith studies at Shiqmim. In T. E. Levy (Ed.), *Shiqmim I: Studies Concerning Chalcolithic Societies in the Northern Negev Desert, Israel (1982–1984)* (pp. 243–249). Oxford: BAR International Series 356.

Rosen, A. M. (1989). Microbotanical evidence for cereals in neolithic levels at Tel Teo and Yiftahel in the Galilee, Israel. *Mitekufat Haeven, 22,* 68–77. (Journal of the Israel Prehistoric Society)

Rosen, A. M. (1992). Preliminary identification of silica skeletons from Near Eastern archaeological sites: An anatomical approach. In G. Rapp, Jr. and S. Mulholland (Eds.), *Advances in Phytolith Systematics* (pp. 129–147). New York: Plenum.

Rosen, S. A. (1983a). The microlithic lunate: An old-new tool type from the Negev, Israel. *Paléorient, 9, 2,* 81–83.

Rosen, S. A. (1983b). Tabular scraper trade: A model of material cultural dispersion. *Bulletin American School of Oriental Research, 249,* 79–86.

Rosen, S. A. (1984). Kvish Harif: Preliminary investigation at a late Neolitic site in the central Negev. *Paléorient, 10, 2,* 111–121.

Rosen, S. A. (1988). Notes on the origins of pastoral nomadism: A case from the Negev and Sinai. *Current Anthropology, 29(3),* 498–506.

Rosen, S. A. (1991). Paradigms and politics in the Terminal Pleistocene archaeology of the Levant. In G. A. Clark (Ed.), *Perspectives on the Past: Theoretical Biases in Mediterranean Hunter-Gatherer Research* (pp. 307–321). Philadelphia: University of Pennsylvania Press.

Rosenberg, M. (1994). Pattern, process, and hierarchy in the evolution of culture. *Journal of Anthropological Archaeology, 13,* 307–340.

Russell, K. W. (1988). *After Eden.* Oxford: British Archaeological Reports International Series, 202.

Rust, A. (1950). *Die hohlenfunde von Jabrud (Syrien).* Neumunster: Karl Wachholtz Verlag.

Ryder, M. L. (1988). Sheep representations, written records and wool measurements. In J. Clutton-Brock and C. Grigson (Eds.), *Animals and Archaeology: 3, Early Herders and Their Flocks* (pp. 69–82). Oxford: British Archaeological Research International Series 356.

Sabloff, J. A., Binford, L. R., and McAnany, P. A. (1987). Understanding the archaeological record. *Antiquity, 61,* 203–209.

Sackett, J. (1977). The meaning of style in archaeology: A general model. *American Antiquity, 42,* 369–380.

Sackett, J. (1985). Style, ethnicity, and stone tools. In M. Thompson et al. (Eds.), *Status, Structure, and Stratification: Current Archaeological Reconstructions* (pp. 277–282). Calgary: University of Calgary.

Sackett, J. (1988). The Mousterian and its aftermath: A view from the Upper Paleolithic. In H. L. Dibble and A. Montet-White (Eds.), *Upper Pleistocene Prehistory of Western Eurasia* (pp. 62–83). Philadelphia: University Museum, University of Pennsylvania.

Said, R. (1962). *The geology of Egypt.* Amsterdam: Elsvier.

Sahlins, M. and E. Service. (1960). *Evolution of culture.* Ann Arbor: University of Michigan Press.

Saragusti, I., and Goren-Inbar, N. (1990). Conjoinable pieces from the lithic assemblage of Quneitra, Area B. In N. Goren-Inbar (Ed.), *Quneitra: A Mousterian Site on the Golan Heights, Qedem 31.* Jerusalem: Monographs of the Institute of Archaeology, Hebrew University.

Schiffer, M. (1987). *The formation processes of the archaeological record.* Albuquerque: University of New Mexico Press.

Schroeder, B. (1969). *The lithic industries from Jerf Ajla and their bearing on the problem of a Middle to Upper Paleolithic transition.* Unpublished Ph.D. Dissertation. New York: Department of Anthropology, Columbia University.

Schuldenrein, J., and Goldberg, P. (1981). Late Quaternary paleoenvironments and prehistoric site distributions in the lower Jordan Valley: A preliminary report. *Paléorient, 7, 1*, 57–71.

Schwarcz, H. (1980). Absolute age determination of archaeological sites by uranium series dating of travertines. *Archaeometry, 22(1)*, 3–24.

Schwarcz, H. P., Blackwell, B., Goldberg, P., and Marks, A. E. (1979). Uranium series dating of travertine from archaeological sites, Nahal Zin, Israel. *Nature, 277*, 558–560.

Schwarcz, H., Goldberg, P., and Blackwell, B. (1980). Uranium series dating of archaeological sites in Israel. *Israel Journal of Earth Sciences, 29*, 157–165.

Schwarcz, H. P., Grun, R., Vandermeersch, B., Bar-Yosef, O., Valladas, H., and Tchernov, E. (1988). ESR dates for the hominid burial site of Qafzeh in Israel. *Journal of Human Evolution, 17*, 733–737.

Schyle, D., and Uerpmann, H.-P. (1988). Paleolithic sites in the Petra area. In A. N. Garrard and H.-G. Gebel (Eds.), *The Prehistory of Jordan in 1986*. Oxford: British Archaeologic Research International Series 396(i).

Scott, T. R. (1977). The Harifian of the central Negev, Israel. In A. E. Marks (Ed.), *Prehistory and Paleoenvironments in the Central Negev, Israel*, Vol. 2 (pp. 271–322). Dallas: Southern Methodist University Press.

Sellars, J. (1989). *A study of the Wadi Judayid site: A Natufian occupation in southern Jordan*. M.A. Thesis, Tulsa: Department of Anthropology, University of Tulsa.

Sellars, J. (1991). An examination of lithics from the Wadi Judayid site. In O. Bar-Yosef and F. R. Valla (Eds.), *The Natufian Culture in the Levant* (pp. 445–458). Ann Arbor: International Monographs in Prehistory.

Selley, R. C. (1972). Diagnosis of marine and non-marine environments from the Cambro-Ordovician sandstones of Jordan. *Journal of the Geological Society of London, 128*, 135–150.

Sharer, R. J., and Ashmore, W. (1979). *Fundamentals of archaeology*. Menlo Park: Benjamin Cummings.

Shea, J. (1988). Spear points from the Middle Paleolithic of the Levant. *Journal of Field Archaeology, 15*, 441–450.

Shea, J. (1989a). A new evolutionary perspective on Neanderthals from the Levantine Mousterian. *A report to the L. S. B. Leakey Foundation* 1–26.

Shea, J. (1989b). A functional study of the lithic industries associated with hominid fossils in the Kebara and Qafzh Caves, Israel. In P. Mellars and C. Stringer (Eds.), *The Human Revolution* (pp. 611–625). Princeton, New Jersey: Princeton University Press.

Shea, J. (1991). *The behavioral significance of Levantine Mousterian industrial variability*. Ph.D. Dissertation, Harvard University. Ann Arbor: University Microfilms.

Shea, J., and Klenck, J. (1993). An experimental investigation of the effects of trampling on the results of lithic use-wear analysis. *Journal of Archaeological Science, 20(2)*, 175–194.

Shehadeh, N. (1985). The climate of Jordan in the past and present. In A. Hadidi (Ed.), *Studies in the History and Archaeology of Jordan II* (pp. 25–38). Amman: Department of Antiquities.

Shen, C. (1992). *Tor Hamar (J431): An assemblage of the Mushabian Complex from southern Jordan*. M.A. Thesis. Tulsa: Department of Anthropology, University of Tulsa.

Sherratt, A. (1981). Plough and pastoralism: Aspects of the secondary products revolution. In I. Hodder, G. Isaac and N. Hammond (Eds.), *Pattern of the Past* (pp. 261–306). Cambridge: Cambridge University Press.

Sherratt, A. (1983). The secondary exploitation of animals in the Old World. *World Archaeology, 15, 1*, 90–104.

Shott, M. (1986). Technological organization and settlement mobility: An ethnographic examination. *Journal of Anthropological Research, 42, 1*, 15–50.

Sillen, A., and Lee-Thorp, J. A. (1991). Dietary change in the Late Natufian. In O. Bar-Yosef and F. R. Valla (Eds.), *The Natufian Culture in the Levant* (pp. 399–410). Ann Arbor: International Monographs in Prehistory.

Simmons, T., and Smith, F. H. (1991). Human population relationships in the Late Pleistocene. *Current Anthropology, 32(5)*, 623–627.

Smith, P., and Horvitz, L. (1984). Radiographic evidence for changing patterns of animal exploitation in the southern Levant. *Journal of Archaeological Science, 11*, 467–475.

Smith F. H., and Paquette, S. P. (1989). The adaptive basis of Neandertal facial form, with some thoughts on the nature of modern human origins. In E. Trinkaus (Ed.), *The Emergence of Modern Humans* (pp. 181–210). Cambridge: Cambridge University Press.

Solecki, R. L., and Solecki, R. S. (1970). A new secondary flaking technique at Nahr Ibrahim cave site. *Bulletin du Musée de Beyrouth*, 23, 137–142.

Spinage, C. A. (1986). *The Natural History of Antelopes*. London: Croom Helm.

Speth, J. (1974). Experimental investigations of hard-hammer percussion flaking. *Tebiwa, 17*, 7–36.

Stahle, D. W., and Dunn, J. E. (1984). An analysis and application of size distribution of waste flakes from biface reduction. *World Archaeology, 14*, 84–97.

Steward, J. H. (1955). *Theory of culture change*. Urbana: University of Illinois Press.

Stockton, E. D. (1969). A bibliography of the flint industries of Transjordan. *Levant, 1*, 100–103.

Stringer, C. (1984). Human evolution and biological adaptation in the Pleistocene. In R. Foley (Ed.), *Hominid Evolution and Community Ecology* (pp. 55–84). London: Academic Press.

Stringer, C. (1989). Documenting the origins of modern humans. In E. Trinkaus (Ed.), *The Emergence of Modern Humans* (pp. 67–96). Cambridge: Cambridge University Press.

Stringer, C., and Andrews, P. (1988). Genetic and fossil evidence for the origin of modern humans. *Science, 239*, 1263–1268.

Stringer, C. B., Grun, R., Schwarcz, H. P., and Goldberg, P. (1989). ESR dates for the hominid burial site of Es Skhul in Israel. *Nature, 338*, 756–758.

Tangri, D., and Wyncoll, G. (1990). Of mice and men: Is the presence of commensal animals in archaeological sites a positive correlate of sedentism? *Paléorient, 15*, 85–94.

Tchernov, E. (1979). The fauna. In A. Horowitz (Ed.), *The Quaternary of Israel* (pp. 259–290). New York: Academic Press.

Tchernov, E. (1981). The biostratigraphy of the Middle Wast. In J. Cauvin and P. Sanlaville (Eds.), *Préhistoire du Levant* (pp. 67–97). Paris: CNRS.

Tchernov, E. (1984). Commensal animals and human sedentism in the Middle East. In J. Clutton-Brock and C. Grigson (Eds.), *Animals and Archaeology 3: Early Herders and Their Flocks* (pp. 91–115). Oxford: British Archaeological Reports International Series, 202.

Tchernov, E. (1989). The Middle Paleolithic mammalian sequence and its bearing on the origin of *Homo sapiens* in the southern Levant. In O. Bar-Yosef and B. Vandermeersch (Eds.), *Préhistoire du Sud-Levant* (pp. 25–42). Oxford: British Archaeological Reports International Series, 497.

Tchernov, E. (1991). Biological evidence for human sedentism in southwest Asia during the Natufian. In O. Bar-Yosef and F. R. Valla (Eds.), *The Natufian Culture in the Levant* (pp. 315–340). Ann Arbor: International Monographs in Prehistory.

Thomas, D. H. (1978). Arrowheads or atlatl darts: How the stones got the shaft. *American Antiquity, 43*, 461–472.

Tixier, J. (1963). *Typologie du l'Épipaléolithique du Maghreb.* Mémoires du Centre de Récherches Anthropologiques, Préhistoriques et Ethnologiques, Alger. Paris: Arts et Métiers Graphiques.

Trigger, B. (1978). *Time and tradition.* New York: Columbia University Press.

Trinkaus, E. (1989). The Upper Pleistocene transition. In E. Trinkaus (Ed.), *The Emergence of Modern Humans* (pp. 42–66). Cambridge: Cambridge University Press.

Uerpman, H.-P. (1981). The major faunal areas of the Middle East during the Late Pleistocene and Early Holocene. In J. Cauvin and P. Sanlaville (Eds.), *Préhistoire du Levant* (pp. 99–106). Paris: CNRS.

Usik, V. I. (1989). Korolevo-transition from Lower to Upper Paleolithic according to reconstruction data. *L' Anthropologie, 27*, 179–212.

Valla, F. R. (1987). *Les industries de Silex de Mallaha (Eynan).* Jerusalem: Mémoires et Travaux du Centre de Récherches Français de Jerusalem.

Valladas, H., Joron, J. L., Valladas, G., Arensburg, B., Bar-Yosef, O., Belfer-Cohen, A., Goldberg, P., Laville, H., Meignen, L., Rak, Y., Tchernov, E., Tillier, A. M., and Vandermeersch, B. (1978). Thermoluminescence dates for the Neanderthal burial site at Kebara in Israel. *Nature, 330*, 159–160.

Valladas, H., Reyss, J. L., Joron, J. L., Valladas, G., Bar-Yosef, O., and Vandermeersch, B. (1988). Thermoluminescence dating of Mousterian "Proto-Cro-Magnon" remains from Israel and the origin of modern man. *Nature, 331*, 614–616.

Vandermeersch, B. (1981). *Les hommes fossils de Qafzeh.* Paris: CNRS.

Vayda, A., and Rappaport, R. (1968). Ecology, cultural and noncultural. In J. Clifton (Ed.), *Introduction to Cultural Anthropology* (pp. 477–497). Boston: Houghton Mifflin.

Velleman, P. F. (1989). *Data desk handbook.* Volume 1. Northbrook, IL: Odesta Corporation.

Vita-Finzi, C., and Higgs, E. S. (1970). Prehistoric economy in the Mount Carmel area of Palestine: Site catchment analysis. *Proceedings of the Prehistoric Society, 36*, 1–37.

Volkman, P. (1983). Boker Tachtit: Core reconstructions. In A. E. Marks (Ed.), *Prehistory and Paleoenvironments in the Central Negev, Israel*, Vol. III (pp. 69–126). Dallas: Southern Methodist University.

Waechter, J., Seton-Williams, V., Bate, D. M., and Picard, L. (1938). The excavations at Wadi Dhobai 1937–1938 and the Dhobaian industry. *Journal of the Palestine Oriental Society, 18*, 172–186, 292–298.

Watson, P. J., LeBlanc, S. A., and Redman, C. L. (1971). *Explanations in archaeology.* New York: Columbia University Press.

Weinstein-Evron, M. (1988a). Middle Paleolithic sequence from the Hula Valley, Israel. In M. Otte (Ed.), *L'Homme de Neandertal*, Vol. 2 (pp. 207–222). Liége: Service de Préhistoire Université de Liége.

Whallon, R. (1989). Elements of cultural change in the later Paleolithic. In P. Mellars and C. Stringer (Eds.), *The Human Revolution: Behavioural and Biological Perspectives On the Origin of Modern Humans* (pp. 433–454). Vol. 1. Edinburgh: Edinburgh University Press.

White, L. (1959). *The evolution of culture*. New York: McGraw-Hill.

White, R. (1982). Rethinking the Middle/Upper Paleolithic transition. *Current Anthropology, 23*, 169–192.

Wiessner, P. (1974). A functional estimater of population from floor area. *American Antiquity, 39(2)*, 343–350.

Wiessner, P. (1979). Reply to Casteel. *American Antiquity, 44(4)*, 808–810.

Willey, G., and Sabloff, J. (1974). *A history of American archaeology*. San Francisco: Freeman.

Wissler, C. (1917). *The American Indian: An introduction to the anthropology of the New World*. New York: D. C. McMurtrie.

Wolf, C. A. (1988). Analysis of faunal remains from early Upper Paleolithic sites in the Levant. In J. F. Hoffecker and C. A. Wolf (Eds.), *The Early Upper Paleolithic: Evidence from Europe and the Near East* (pp. 73–95). Oxford: British Archaeological Reports International Series 437.

Wolpoff, W. (1989). Multiregional evolution: The fossil alternative to Eden. In P. Mellars and C. B. Stringer (Eds.), *The Human Revolution: Behavioral and Biological Perspectives on the Orgings of Modern Humans* (pp. 62–108). Edinburgh: Edinburgh University Press.

Wynn, T. (1979). The intelligence of later Acheulean hominids. *Man, 14*, 371–379.

Wynn, T. (1985). Piaget, stone tools and the evolution of human intelligence. *World Archaeology, 17, 1*, 32–43.

Yellen, J. (1977). *Archaeological approaches to the present*. New York: Academic Press.

Yizraeli, T. (1967). Mesolithic hunters' industries at Ramat Matred (The Wilderness of Zin), first report. *Palestine Exploration Quarterly, 99*, 78–85.

Zeuner, F. E., Kirkbride, D., and Park, B. (1957). Stone age exploration in Jordan, 1. *Palestine Exploration Quarterly, 89*, 17–54.

Zohary, D. (1962). *Plant life of Palestine*. New York: Ronald Press Company.

Zohary, D. (1973). *Geobotanical foundation of the Middle East*. Amsterdam: Swets and Zeitlinger.

# Index

## *INTERDISCIPLINARY CONTRIBUTIONS TO ARCHAEOLOGY*

*Chronological Listing of Volumes*

THE PLEISTOCENE OLD WORLD
Regional Perspectives
Edited by Olga Soffer

HOLOCENE HUMAN ECOLOGY IN NORTHEASTERN NORTH AMERICA
Edited by George P. Nicholas

ECOLOGY AND HUMAN ORGANIZATION ON THE GREAT PLAINS
Douglas B. Bamforth

THE INTERPRETATION OF ARCHAEOLOGICAL SPATIAL PATTERNING
Edited by Ellen M. Kroll and T. Douglas Price

HUNTER–GATHERERS
Archaeological and Evolutionary Theory
Robert L. Bettinger

RESOURCES, POWER, AND INTERREGIONAL INTERACTION
Edited by Edward M. Schortman and Patricia A. Urban

POTTERY FUNCTION
A Use-Alteration Perspective
James M. Skibo

SPACE, TIME, AND ARCHAEOLOGICAL LANDSCAPES
Edited by Jacqueline Rossignol and LuAnn Wandsnider

ETHNOHISTORY AND ARCHAEOLOGY
Approaches to Postcontact Change in the Americas
Edited by J. Daniel Rogers and Samuel M. Wilson

THE AMERICAN SOUTHWEST AND MESOAMERICA
Systems of Prehistoric Exchange
Edited by Jonathon E. Ericson and Timothy G. Baugh

FROM KOSTENKI TO CLOVIS
Upper Paleolithic–Paleo-Indian Adaptations
Edited by Olga Soffer and N. D. Praslov

EARLY HUNTER–GATHERERS OF THE CALIFORNIA COAST
Jon M. Erlandson

HOUSES AND HOUSEHOLDS
A Comparative Study
Richard E. Blanton

THE ARCHAEOLOGY OF GENDER
Separating the Spheres in Urban America
Diana diZerega Wall

ORIGINS OF ANATOMICALLY MODERN HUMANS
Edited by Matthew H. Nitecki and Doris V. Nitecki

PREHISTORIC EXCHANGE SYSTEMS IN NORTH AMERICA
Edited by Timothy G. Baugh and Jonathon E. Ericson

STYLE, SOCIETY, AND PERSON
Archaeological and Ethnological Perspectives
Edited by Christopher Carr and Jill E. Neitzel

REGIONAL APPROACHES TO MORTUARY ANALYSIS
Edited by Lane Anderson Beck

DIVERSITY AND COMPLEXITY IN PREHISTORIC MARITIME SOCIETIES
A Gulf of Maine Perspective
Bruce J. Bourque

CHESAPEAKE PREHISTORY
Old Traditions, New Directions
Richard J. Dent, Jr.

PREHISTORIC CULTURAL ECOLOGY AND EVOLUTION
Insights from Southern Jordan
Donald O. Henry